Chemistry and Pharmacology of Natural Products

Lignans

Chemical, biological and clinical properties

CHEMISTRY AND PHARMACOLOGY OF NATURAL PRODUCTS

Series Editors: Professor J.D. Phillipson, *The School of Pharmacy, University of London*; Dr D.C. Ayres, *Department of Chemistry, Queen Mary College, University of London*; H. Baxter, *formerly at the Laboratory of the Government Chemist, London.*

Also in this series
Edwin Haslam *Plant polyphenols: vegetable tannins revisited*

Lignans

Chemical, biological and clinical properties

D.C. AYRES
Department of Chemistry
Queen Mary College, London

and

J.D. LOIKE
Department of Physiology and Cellular Biophysics
College of Physicians and Surgeons of Columbia University, New York

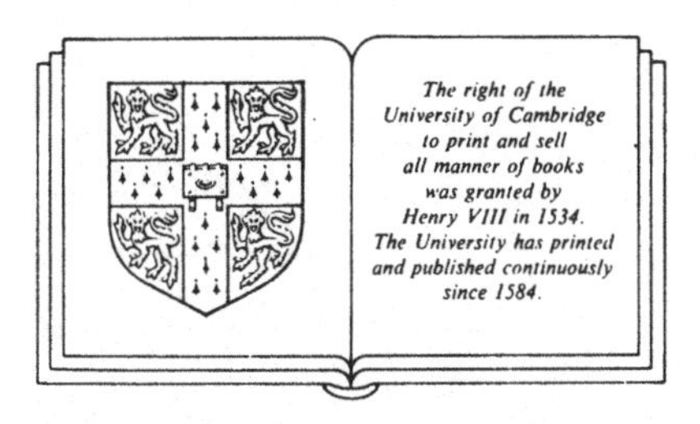

CAMBRIDGE UNIVERSITY PRESS
Cambridge
New York Port Chester
Melbourne Sydney

CAMBRIDGE UNIVERSITY PRESS
Cambridge, New York, Melbourne, Madrid, Cape Town, Singapore, São Paulo

Cambridge University Press
The Edinburgh Building, Cambridge CB2 8RU, UK

Published in the United States of America by Cambridge University Press, New York

www.cambridge.org
Information on this title: www.cambridge.org/9780521304214

First published 1990
This digitally printed version 2008

A catalogue record for this publication is available from the British Library

Library of Congress Cataloguing in Publication data
Ayres, D.C.
Lignans: Chemical, biological and clinical properties/D.C. Ayres and J.D. Loike.
p. cm. – (Chemistry and pharmacology of natural products)
Includes bibliographical references.
ISBN 0 521 30421 0
1. Lignans. I. Loike, J.D. II. Title. III. Series.
[DNLM: 1. Plant Extracts. QV 766 A985L]
QP801.L53A94 1990 615′.7–dc20 89-22249 CIP

ISBN 978-0-521-30421-4 hardback
ISBN 978-0-521-06543-6 paperback

Contents

To students and colleagues in Westfield College,
1965–1984.

Preface

The first systematic review of the naturally occurring lignans was presented by Professor R.D. Haworth in his Tilden lecture of 1942. There have been a number of subsequent review articles, notably that by W.M. Hearon and W.S. MacGregor in 1955. Their chemistry was covered in a collection of learned reviews published in honour of Professor L.R. Row by Andhra University Press in 1978. The present work is the first to cover the whole field of lignan chemistry including the application and promise of lignans as pharmaceutical agents. It is anticipated that expansion will continue through the application of modern methods of chromatography including HPLC, combined with the use of 2D-NMR and NOE for structure evaluation. These techniques are of especial relevance to the study of oligomeric lignans which are touched upon in the text.

The principal classes are defined in Chapter 1 with an explanation of the system of nomenclature that has been adopted. The contribution of Dr G.P. Moss who took on the considerable task of rationalising the often conflicting systems is gratefully acknowledged. It is hoped that readers who find that we have diverged from their own preference will accept that changes had to be made in order to be self-consistent. The system used throughout the book evolved with the help of some twenty active researchers who kindly responded to our requests for criticism of draft proposals.

Chapter 2 is a registry of lignans described up to April 1988 and includes at least one leading literature reference and plant source for each entry. A comprehensive review of the sources of lignans and neolignans has recently been published (p. 84) and Professor Richard Gottlieb is thanked for his help in making the manuscript available before publication. The senior author (DCA) would be particularly grateful if any sins of commission or omission in the registry are brought to his attention.

Chapter 3 (DCA and JDL) describes the general aspects of the pharmacology of lignans. The fourth chapter (JDL) describes the development of the clinically effective podophyllotoxin derivatives, Etoposide and Teniposide. Moreover, this chapter describes the current understanding of the mechanism of action of these drugs. Subsequent chapters (DCA) deal with the isolation, characterisation and synthesis of lignans; here the volume of material necessitated a selective approach and this has been based on examples of general application largely chosen from the more recent literature. Dr Paul Dewick offered valuable criticism of the section on biosynthesis and the lignin scheme which appears in it was published with the approval of the American Chemical Society.

The infrared spectrum of wuweizisu C was provided by Professor H. Taguchi and material for the SFORD Scheme 6.14 by Professor D.N. Kirk, who also read part of the manuscript. Acknowledgement is also made to the Royal Society of Chemistry for the data in Scheme 6.32, to the Editor of the *Journal of Natural Products* for Scheme 6.33 and to the Editor of *Tetrahedron* for Scheme 6.34. Permission to publish Scheme 6.39 was given by the Pharmaceutical Society of Japan and material for figures 6.36 and 6.37 was kindly provided by Dr Peter B. Hulbert. We also wish to thank those authors who approved of the use of their material and to numerous others who responded to chemical queries which arose during the preparation of the manuscript.

The support of the Chemistry Department in Queen Mary College is gratefully acknowledged and Professor B. J. Aylett and Mr G. Coumbarides are thanked for help with computing.

The production of the text owes much to the services of the library of the Royal Society of Chemistry, to the staff of the Cambridge University Press and to Editorial colleagues Professor J. David Phillipson and Mr Herbert Baxter.

London D.C. Ayres January 1990
New York J.D. Loike

Glossary for lignans

ACTINOMYCIN D (structure as given)

Actinomycin D

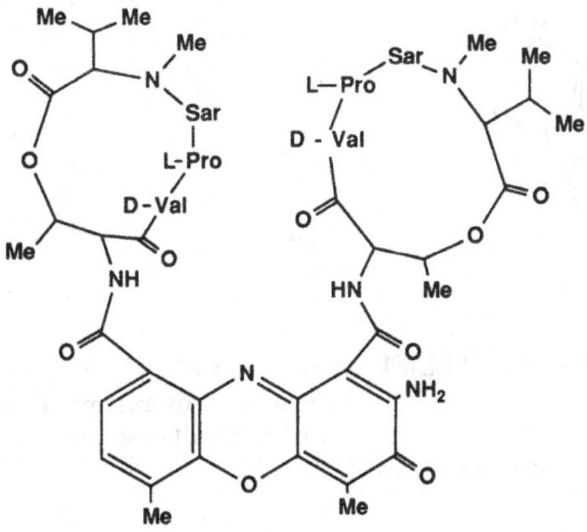

ADENOCARCINOMA Malignant cancer of epithelial cell origin, derived from any of the three germ layers with a glandular growth pattern.

ADRIAMYCIN (structure as given)

Adriamycin

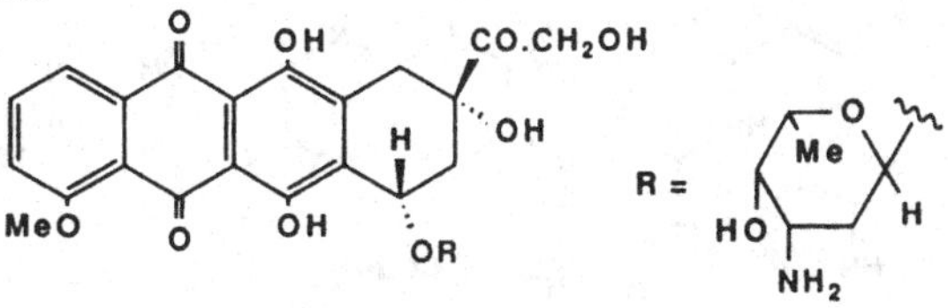

A1BN Azo *bis*-isobutylnitrile

ALOPAECIA Loss of hair

AMPHIPATHIC molecules that contain both a hydrophilic and a hydrophobic moiety.

AMSACRINE (m-AMSA; structure as given)

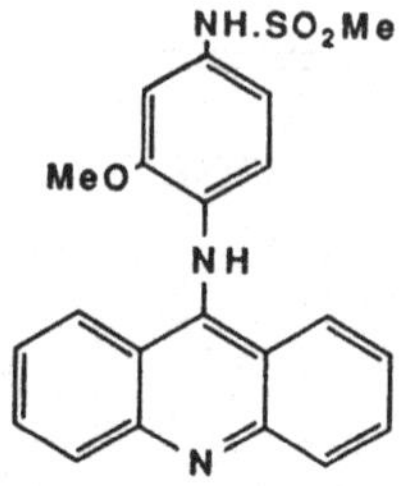

meta- isomer

ANTIPYRETIC An agent used to control body temperature.
ARA-C CYTOSINE ARABINOSIDE (structure as given)

Ara - C or cytosine arabinoside

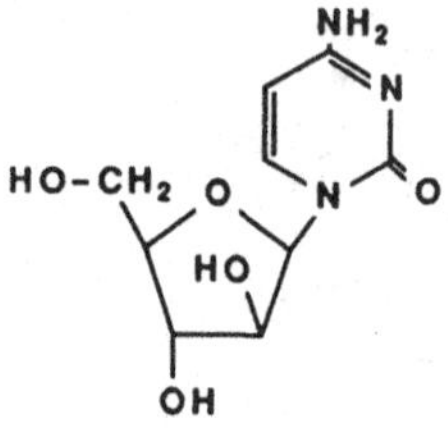

BIOGENETIC EQUIVALENT A substance available to the plant which may be converted by enzymic action to a metabolite needed for biogenesis.
BLEOMYCIN (structure as given)

Bleomycin A_2

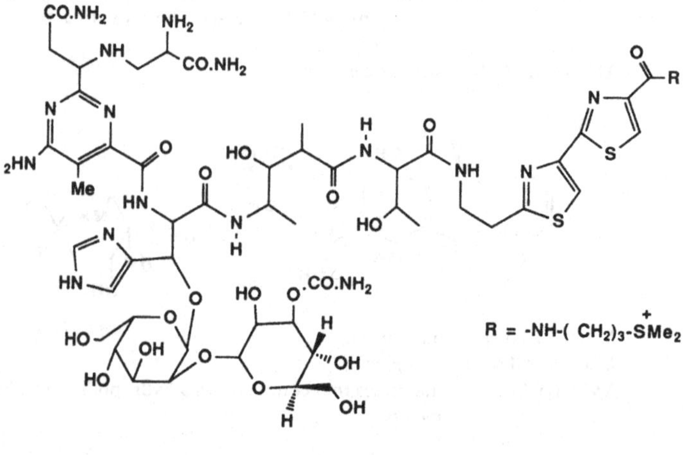

$R = -NH-(CH_2)_3-\overset{+}{S}Me_2$

CEREBELLAR ATAXIA Motor abnormality associated with lesions in the cerebellum

CHLOROTIC Pertaining to a kind of anaemia sometimes affecting girls at puberty, characterised by a pale or greenish hue of the skin.

CHROMATIN Chromosal material composed of DNA and proteins.

CISPLATIN (structure as given)

Cisplatin or *cis*- diamminedichloroplatinum

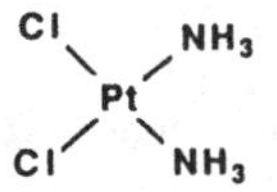

COLLIN'S REAGENT for the selective oxidation of alcohols using chromium trioxide/pyridine.

CONCANAVALLIN A Globular protein of molecular weight 26,000 with two identical subunits, each containing 237 aminoacids.

COREY'S METHOD for the stepwise oxidation of allylic alcohols and aldehydes to carboxylic acids using manganese dioxide/hexane followed by manganese dioxide/CN^-/methanol/acetic acid.

COSY A two dimensional NMR technique used to identify coupled protons

CYCLOPHOSPHAMIDE (structure as given)

Cyclophosphamide

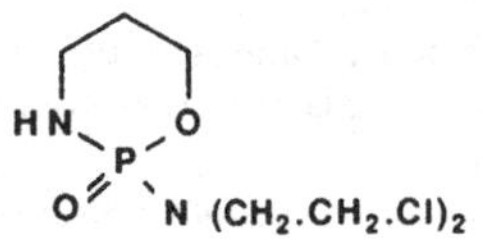

R- form

CYTOTOXIC NUCLEOSIDES Nucleosides which can damage cells.

DABCO 1,4-Diazabicyclo-[2,2,2]-octane.

DDQ 2,3-Dichloro-5,6-dicyano-1,4-benzoquinone

DIBAL Di-isobutylaluminium hydride

DIPYRIDAMOLE (structure as given)

Dipyridamole

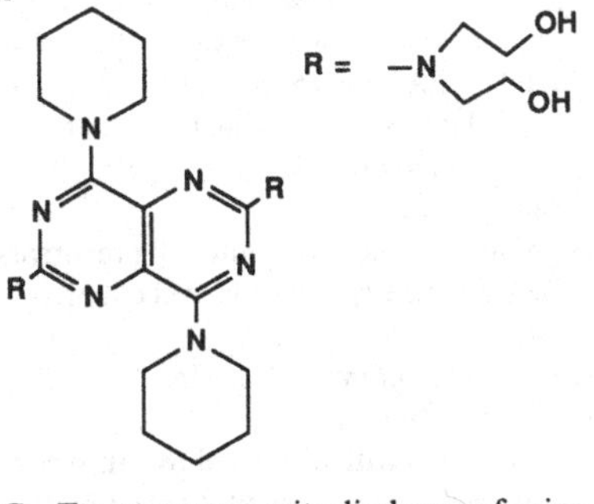

DIURETIC Treatment to excite discharge of urine.

DMAD Dimethyl acetylenedicarboxylate.

DMSA Dimethylsilylamide.
DOXORUBICIN See adriamycin.
ENTEROHEPATIC CIRCULATION The cycling of compounds through the liver and intestine.
ERYTHROPHAGOCYTIC LYMPHOSTIOCYTOSIS A malignancy of white cells that are highly phagocytic.
EUKARYOTIC (eucaryotic) pertaining to an organism whose cells contain a limiting membrane around the nuclear material.
FETIZON'S METHOD Oxidation with silver carbonate freshly precipitated onto Celite.
FIBROBLAST CELL LINE derived from elongated cells present in connective tissue and capable of forming collagen fibre.
FREMY'S SALT Potassium nitrosodisulphonate $K_4[(SO_3)_2NO]_2$.
HARDWOOD includes *both* coniferyl and sinapyl residues in its lignin in contrast to softwood lignin, which is largely derived from coniferyl alcohol.
HELA CELLS from a patient, Helen Lane, with carcinoma of the uterine cervix.
HEPATOTOXIC relates to an agent capable of damaging the liver.
HMPA Hexamethylphosphoramide.
HMDS Hexamethyldisilylamide $HN(SiMe_3)_2$ commonly used as the lithio derivative.
IFOSFAMIDE An isomer of cyclophosphamide where both amide groups carry one chloroethyl substituent.
IMMUNOBLOTTING Transferring proteins to special nitrocellulose filters where they can be tested for their capacity to react with specific antibodies.
IMMUNOMODULATOR A bioreactive substance that effects the physiological response of cells derived from the immune system.
INDOR Internuclear double resonance used mainly in proton spectra to detect coupling by monitoring the amplitude of one transition, while sweeping a low power excitation through the frequency range of the other.
KARPLUS RELATION relates the magnitude of the coupling between vicinal protons to the dihedral angle between the linking C—H bonds.
L 1210 CELL LINE A lymphocytic mouse leukemic cell line which has been used extensively for routine screening programs of chemical agents and natural products for cytotoxic activity.
LAMELLAE (middle) The inner of two membranes which enclose the chloroplasts, which are the sites of photosynthesis.
LANTHANIDE SHIFT REAGENTS Paramagnetic lanthanide β-diketoenolate complexes which associate with basic organic functional groups.
LDA Lithium di-isopropylamide.
LEUKEMIA Malignant neoplasms of white cell precursors.
LEUKOPAENIA An abnormally low white cell count.
LHDS Lithium hexamethyldisilylamide.
LTBA Lithium tri-*t*-butoxyaluminium hydride.
LTBH Lithium triethyl borohydride.
LYMPHOCYTIC Associated with or related to lymphocytes.
LYMPHOMA Malignancies that are characterised by the proliferation of cells native to the lymphoid tissues.

LYMPHOCYTIC S49 CELL LINE A transformed cell line established from a lymphoma induced in a BALB/c mouse. These cells retain many of the properties of thymocytes.

LYMPHOCYTIC LEUKAEMIA Leukaemic cells that arise from white blood cell precursors of the lymphocyte.

MACROPHAGE any large mononuclear phagocyte.

MAYTANSINE (structure as given).

Maytansine

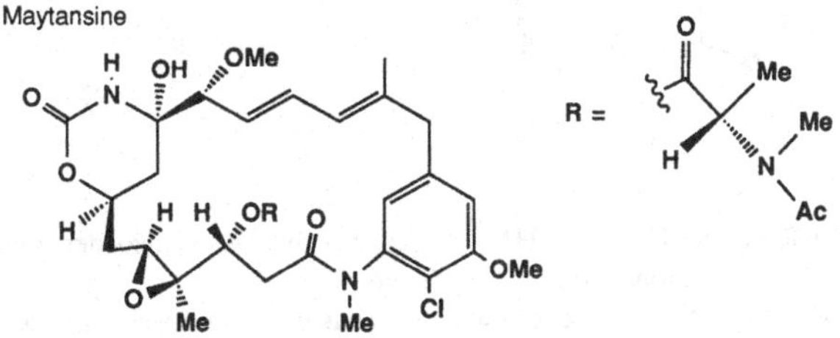

METHOTREXATE (structure as given).

Methotrexate

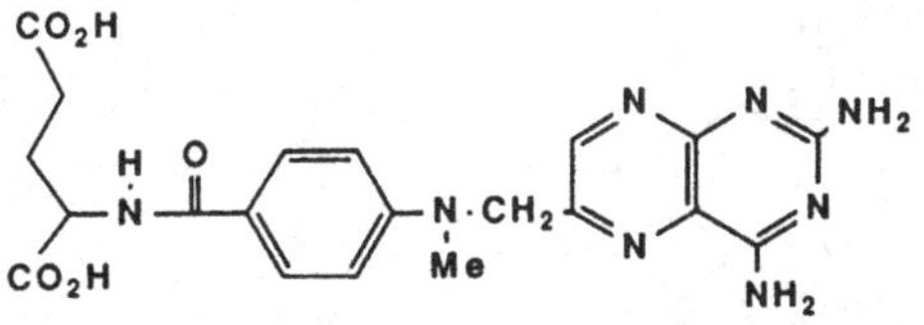

MITOGEN a substance which stimulates mitosis and the transformation of lymphocytes including those associated with lectin.

MURINE relating to mice.

NADH The reduced form of nicotinamide-adenine dinucleotide.

NASAL EMPYEMA A collection of pus within the nasal passage.

NASOPHARYNX Part of the pharynx lying directly behind the nasal passages and above the soft palate.

NBMPR 4-nitrobenzylthioinosine (structure as given).

NBMPR or 4-nitrobenzylthioinosine

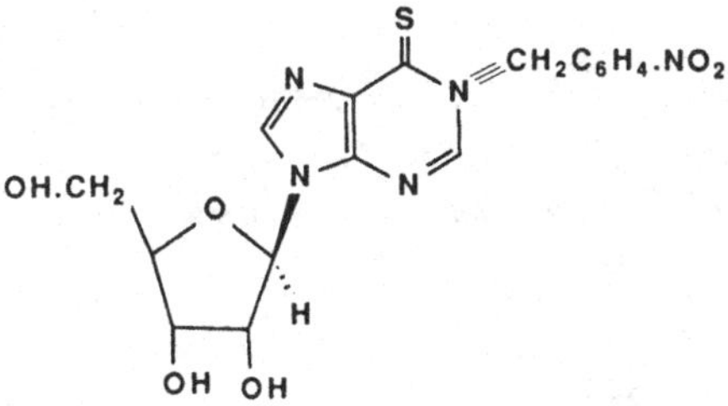

NBS N-bromosuccinimide.

NEUTROPHILS White blood cells that contain horseshoe-shaped nuclei and neutrophilic granules.

OCULOCUTANEOUS TELANGIECTASIA A group of abnormal prominent capillaries, venules and arterioles that create small focal red lesions in the eye.

OXOLINIC ACID (structure as given).

Oxolinic acid

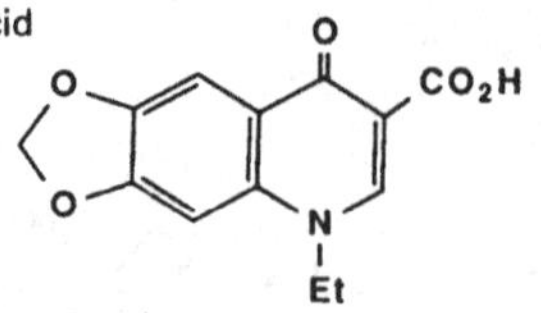

PERIPHERAL NEUROPATHY A nervous disease or disorder involving the periphereal nervous system.

PHAGOCYTE A scavenger cell which ingests bacteria, foreign particles etc.

SPLENOCYTE A phagocytic mononuclear leukocyte of the spleen.

TAL Tyrosine ammonia lyase

TAXOL (structure as given).

Taxol

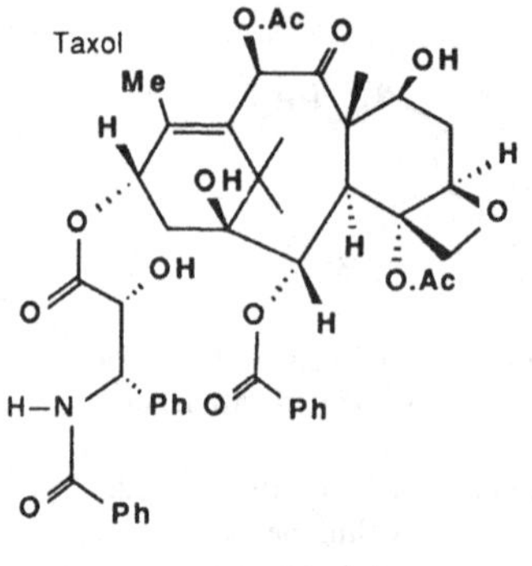

TFA Trifluoroacetic acid

THROMBOCYTOPENIA A reduction in platelet count.

VERAPAMIL (structure as given)

Verapamil

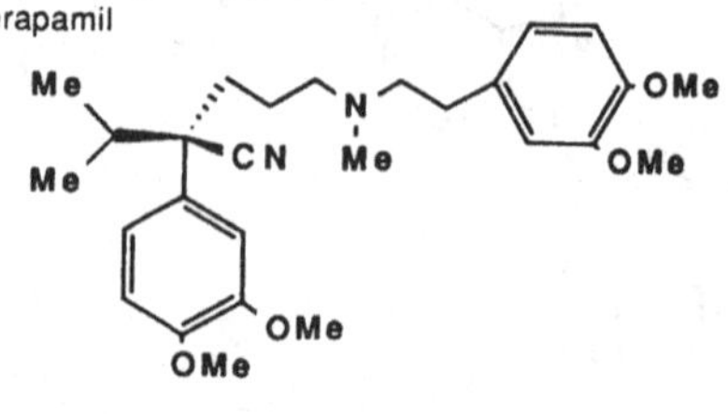

VESICANT Any agent used in chemical warfare to blister and burn body tissues by contact with the skin or by inhalation.

VINCRISTINE (structure as given).

Vincristine

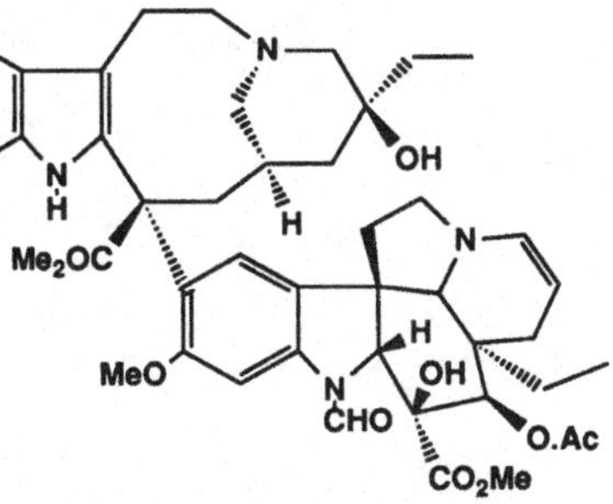

1

Introduction

The lignans form a group of plant phenols whose structure is determined by the union of two cinnamic acid residues or their biogenetic equivalents.

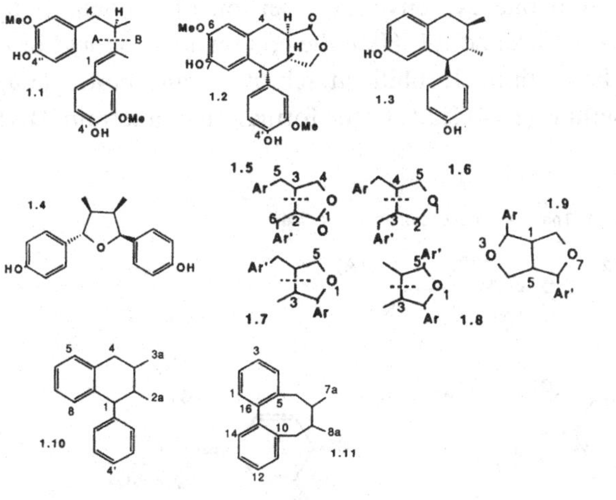

This unifying definition was first made by R.D. Howarth (1936) and is illustrated by the structures (**1.1**) and (**1.2**), in which the bar line separates the two β,β-linked cinnamic residues. Thus guaiaretic 'acid' (**1.1**), with the terminal groups fully reduced, is structurally one of the simplest members of the group; whilst conidendrin (**1.2**; Lindsey, 1892) is more complex in that the terminal groups differ and are also at higher oxidation levels. Other common variations occur naturally in which the aromatic substituents are modified in whole or in part to methylenedioxy and also where additional oxysubstitution takes place with or without O-methylation. The orientation of these substituents is also subject to variation. No lignan has been isolated with an unsubstituted phenyl ring: attenuol (**2.290**; Joshi *et al.*, 1978), the compound (**1.3**; Vieira *et al.*, 1983) and a group of furans (e.g. **1.4**) isolated from *Krameria cystisoides* (Achenbach *et al.*, 1987) are the only monosubstituted examples. Removal of phenolic hydroxyl groups *in vitro* is known to be a difficult energy-demanding process. One is not therefore surprised to find that although reduction with deoxygenation proceeds readily *in vivo* in the side chain, it does not do so in the aryl groups.

The ubiquity of aromatic oxysubstituents especially those *para*-related to the side chain gave the lead to Erdtmann (1933) who postulated a mode of lignan biogenesis. He proposed that ionisation of the phenolic hydroxyl group followed by a single electron transfer produces a mesomeric radical (**1.12**; Scheme 1.1).

This may dimerise to form structures such as (**1.1**). The point of insertion of substituents (X, Scheme 1.2) and of modification of the terminal groups are still largely matters of conjecture; they will be discussed more fully in Chapter 7.

The principal lignan structures that may be formed as linked dimers are shown (**1.5**–**1.11**). They are all known with a variety of oxysubstitution patterns in the aromatic rings and hydroxyl and glycosyl substituents may also be inserted in the side chain.

One possible biogenetic route to furans (**1.8**) is the reductive hydroxylation of a quinone methide (**1.13**) formed during initial dimerisation, followed by intramolecular attack upon the inserted –OH group (in **1.13A**) by the residual radical at the γ-position. Lignan alcohols hydroxylated at benzylic positions in this way are known.

There is a limited amount of evidence (Chapter 7) that oxygenation of the aromatic rings is complete before coupling takes place *in vivo*, but by way of illustration one can envisage insertion of a phenolic hydroxyl group as shown in Scheme 1.3. Here the combination of acid or enzymic catalysis together with nucleophilic attack by a water molecule upon the *bis*-quinone methide (**1.14**) leads to the formation of a tetralin (**1.15**). This

Scheme 1.1 The mesomeric cinnamate radical

1.12 (A) (B) (C)

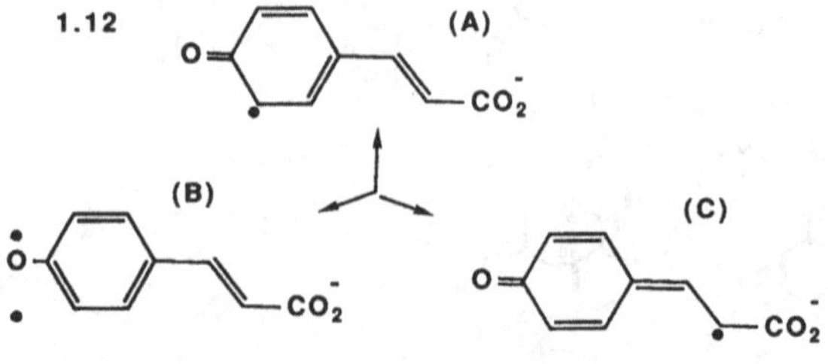

Scheme 1.2 Hypothetical mechanism for the elaboration of a coupled dimer

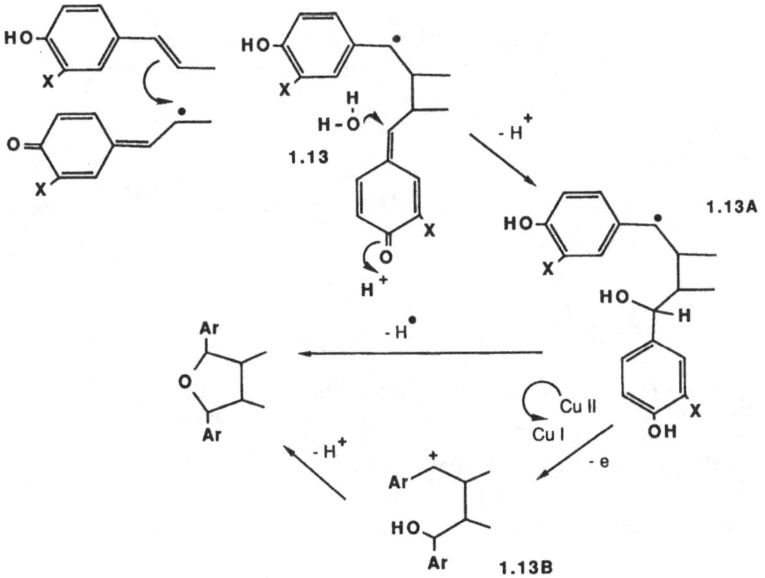

Scheme 1.3 A possible cyclisation mode for a bis-quinone methide

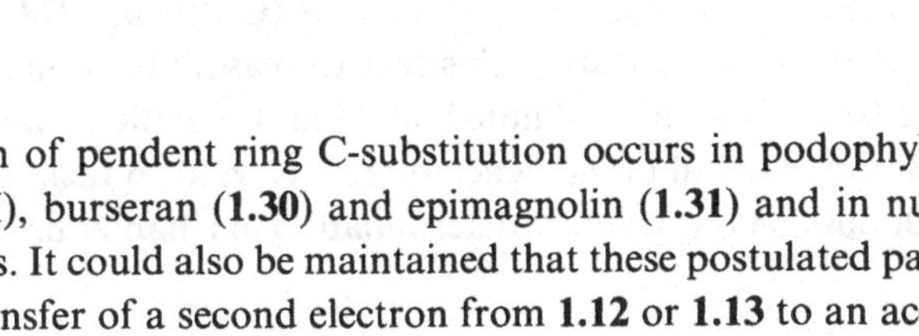

pattern of pendent ring C-substitution occurs in podophyllotoxin (**1.36**; X = H), burseran (**1.30**) and epimagnolin (**1.31**) and in numerous other lignans. It could also be maintained that these postulated pathways follow the transfer of a second electron from **1.12** or **1.13** to an acceptor such as Cu(II), with subsequent changes proceeding through a mesomerically stabilised carbocation. It is also evident that either intermediate type may dimerise in other ways and so give rise to structures such as **1.16–1.18.**

These three dimeric structures do occur naturally, together with some dozen others none of which are β,β-linked and which are known as neolignans (Gottlieb, 1978; Whiting, 1987). The structure **1.18** is found in nature as magnalol; it was isolated in 1983 from *Sassafras taiwanensis* (Watanabe *et al.*, 1983) and shows depressant activity on the central nervous system

Scheme 1.4 Alternative linking modes for the cinnamate radical

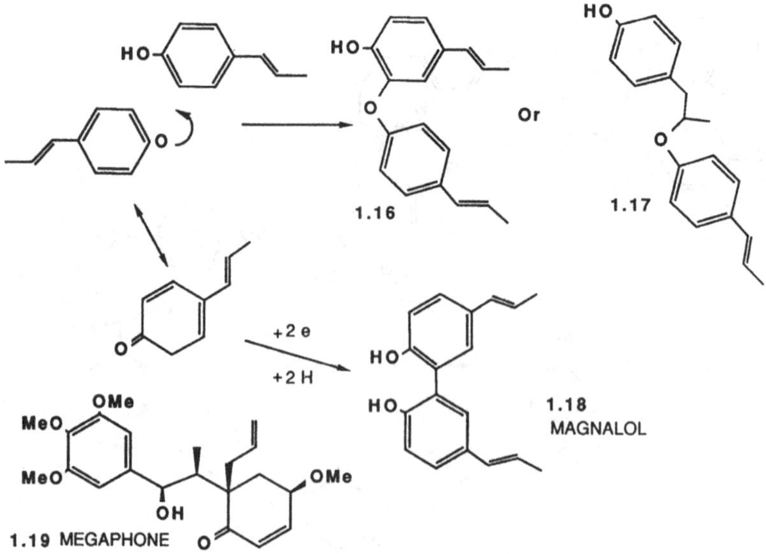

1.16 Or 1.17

+2 e +2 H

1.18 MAGNALOL

1.19 MEGAPHONE

(CNS). Another product of the attack of the cinnamyl precursor by the mesomeric radical is the structure **1.19** which corresponds to megaphone (Kupchan *et al.*, 1978), which is active against human carcinoma of the nasopharynx.

The neolignans, although more diverse in structural type, are less numerous than the classical β,β-linked lignans and to date they have been isolated only from the *Magnoliales* and *Piperales*. The monomeric allyl- and propenylphenols are also found in plants of these orders (Gottlieb, 1974). In view of the existence of thorough reviews (Gottlieb, 1978; Whiting, 1987) and of the need to restrict this text to reasonable bounds the neolignans have been given only a limited mention. It should however be pointed out that the material on nomenclature is relevant to them as also is much of the discussion on structure determination in Chapter 6.

Nomenclature

The problem may be defined by reference to structures **1.1**, **1.2** and **1.5–1.11** in which the carbon atoms are numbered in a manner which many authors have accepted. Unfortunately even within the distinct groups and subgroups of lignans a diversity of other systems has been used. Thus the numeration of the C2–C3 linked butanes (as in **1.1**) is often associated with chemically related cyclised compounds such as **1.6–1.8** in which the heteroatom has priority and hence the β-carbons are now designated C3–C4. This lack of correspondence extends to the butyrolactones (**1.5**) where the writer has found four different systems in use, including those which accord priority to the endocyclic oxygen atom and others where this goes to the carbonyl carbon. It should also be noted that when the aryl groups are differently situated their substituent locants will generate two sets of primed numbers (cf. **1.1**).

In tetralins (**1.10**) the common practice is shown but in many substituted compounds priority is allocated at the point of substitution rather than at the point of attachment of the pendent ring. Furofurans are commonly described as 2,6-diaryl-3,7-dioxabicyclo-[3,3,0]-octanes with priority given to the bridgehead carbon atom, yet this is at variance with the furans (cf. **1.6–1.8**) despite the close chemical and biochemical relationship. Several conflicting methods have been used for dibenzocyclo-octadienes (**1.11**) and the inescapable conclusion is that there is need for a system which relates to the β,β-linkage common to all lignans and which will stand without modification during interconversions between compounds in different lignan groups.

A suitable scheme was proposed in 1961 by Freudenberg and Weinges where the two cinnamyl residues are numbered as in **1.21** and where the 8,8′-linkage is assigned to all lignan classes as a matter of definition. Although this system has been commended by other authors it has not won general acceptance possibly due to the proposed retention of trivial names such as pipero – (for methylenedioxyphenyl) and guaia – (for 4-hydroxy-3-methoxyphenyl) in conjunction with systematic usage. There is now urgent need for revision owing to the welcome subsequent increase in the number and structural diversity of known compounds (see Figure 2.1). At the time of writing a modification of the Freudenberg and Weinges system is being considered by a IUPAC nomenclature committee. This has attracted favourable comment from numerous workers in the field and has been adopted in the writing of this book. As a result it was necessary to change the numeration chosen by many authors in order to interrelate their findings and the writer in apologising for this hopes that they will see the end as justifying this means.

The proposals (Moss, 1987) now being considered set out rules for assigning a priority for unprimed locants; these are to be taken in the following sequence until a distinction can be made:

(a) A phenyl group which is fused to another carbocyclic ring has unprimed locants (cf. **1.22**).
(b) A phenyl group directly substituted on a heterocyclic ring has unprimed locants (cf. **1.23**).

(c) Where both phenyl groups are similarly placed the priority is to be decided using the sequence rules (IUPAC, 1974).

The distinction is clear for guaiaretic acid (**1.1**) since the priority goes to the aryl group associated with the functionalised side chain and guaiaretic acid is described as 3,3′-dimethoxylign-7-ene-4,4′-diol, whilst α-conidendrin (**1.2**) becomes 4,4′-dihydroxy-3,3′-dimethoxy-2,7′-cyclolignan-9,9′-olide. Note that the carbonyl group of the lactone is placed at C-9, the locant 9 preceding 9′. The preferred lactone group is written as a suffix here and it supplants the hydroxyl group which had pride of place in guaiaretic acid. In the open chain lactone **1.24** the priority follows the principal function at C-9.

In structure **1.23** the priority is clear but for burseran (**1.30**) the sequence rule must be invoked. The aryl group with the substitution pattern **1.26** is preferred to the 3,4,5-trimethoxylated residue (**1.27**) and so burseran is described as (8α,8β)-3,4-methylenedioxy-3′,4′,5′-trimethoxy-9,9′-epoxylignan. This is established by taking the atoms in sequence from C-1 when a difference first appears beyond the first ether oxygen. The priority then goes to the methylene ether group at C-3 with a substituent set of (O, H, H) rather than to the methyl group at C-3′ whose set is (H, H, H). The same analysis defines a lower priority to the dihydroxyphenyl group in **1.20** and so it carries primed locants.

If the furofuran lignan (+)-epimagnolin (**1.31**) is written with α-hydrogen atoms at the ring junctions it is then described as −(7α,7′β)-3,3′,4,4′,5-pentamethoxy-7,9′:7′,9-diepoxylignan. Owing to the symmetry of this model its rotation leads to ambiguity, which can be avoided by always

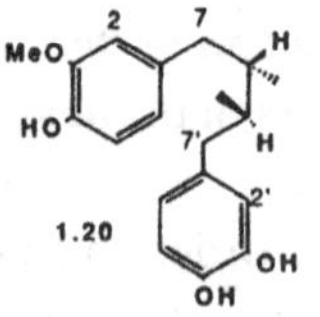

1.20 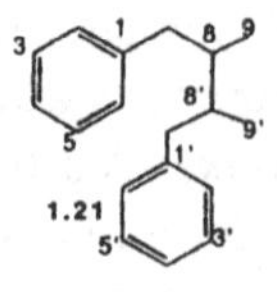 1.21 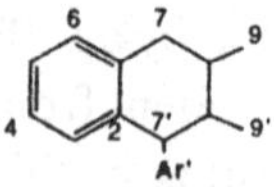 1.22

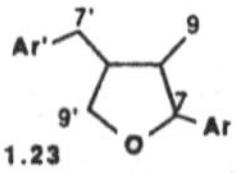

1.23 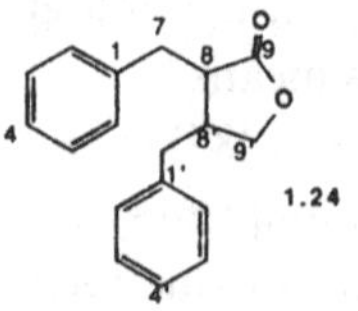 1.24

Order of priority for aryl groups

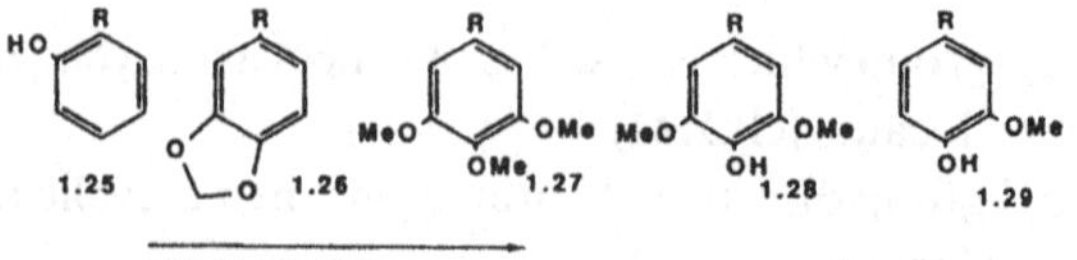

1.25 1.26 1.27 1.28 1.29

order of preference

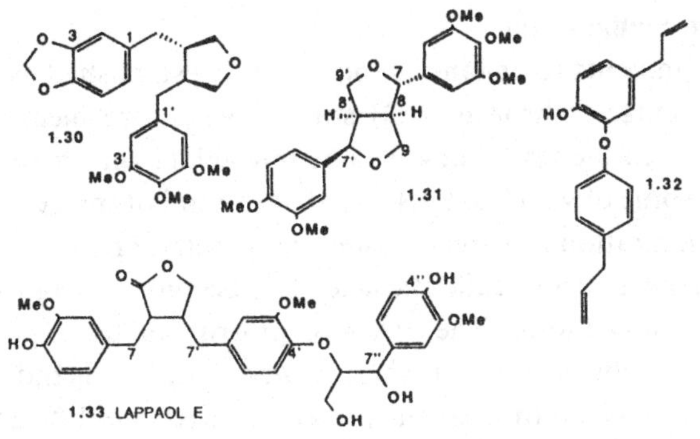

writing the aryl substituents at two and eight o'clock and by allocating the former position to the principal group where a difference exists. The convention of designating α-substituents as below the plane and β-substituents as above the plane can now be adopted (see also Chapter 2, p. 15).

In the present context where lignans are distinguished from neolignans it is not necessary to specify their 8,8′-coupling. Should the distinction between the two be set aside and the term lignan be broadened so as to cover all classes this simplification would not be admissible.

Although the neolignans are not covered in detail in the text, it should be pointed out that they are equally well described using this system. Thus isomagnolol (**1.32**) becomes 3,4′-epoxylign-8,8′-diene-4-ol, whilst megaphone (**1.19**) is 3,3′4,5-tetramethoxy-8,1′-lign-8′-ene-7-ol-6′-one.

It is also possible to extend the system so as to describe sesqui- and dineolignans. Lappaol E (**1.33**) becomes 4,4″,7″,9″-tetrahydroxy-3,3′3″-trimethoxy-4′,8″-epoxy-8,8′-sesquineolignan-9,9′-olide.

An ambiguity of nomenclature has been introduced by Gottlieb (1978) whose group has made a substantial contribution to the chemistry of neolignans. They have chosen to define neolignans as compounds formed by the oxidative dimerisation of allyl or propenylphenols. This was based on a speculation that this coupling mode followed a biogenetic path distinct from the formation of lignans from cinnamyl alcohols (pp. 3, 278). One especially difficult consequence of regarding neolignans as dimers of allylphenols is that all tetralin lignans of the type (**1.15**, Scheme 1.3) would require redefinition. There is therefore a preference for an earlier proposal by Gottlieb (1972) which is consistent with Howarth's original definition and which maintains the unique 8,8′-linkage of lignans.

Additional illustrations of the application of the modified Freudenberg system are given in the registry of naturally occurring lignans in Chapter 2.

Absolute configuration

The assignments to lignans are now largely established as a result of X-ray and circular dichroism (CD) studies which are mentioned at appropriate places in the text. These results interrelate with many chemical correlations some of which are discussed at the end of this section.

Absolute configuration is assigned using the sequence rules of Cahn *et al.* (1966) and in these terms guaiaretic acid (**1.1**) becomes (7*E*)-(8′*R*)-3,3′-dimethoxylign-7-ene-4,4′-diol. The order of priority of the C-8′ substituents is shown in Scheme 1.5 and when viewed with the ligand of least atomic number (d = H) to the rear the preference path a > b > c follows in a clockwise (*R*) sense. Note that in accordance with the rules the methyl group is of lower priority than the substituted carbon atoms; in guaiaretic acid the alkene carbon is preferred to the alkane. In the important example of (−)-dihydroguaiaretic acid (**1.34**), the priority order is the same (8*R*,8′*R*), since that of (a) with its substituent group (C, C, H) is greater than (b; C, H, H).

The lactone (−)-dimethylmatairesinol (**1.39**, Scheme 1.6) also has the (8*R*,8′*R*) configuration as the highest priorities at C-9 and C-9′ are determined by the greater atomic weight of the oxygen atoms. However, contradictions which are more apparent than real can arise from the operation of the rules. Thus, although the carbon skeleton of (+)-trachelogenin (**1.35**) is superposable with that of dimethylmatairesinol, the insertion of the hydroxy group at C-8 requires that it be designated as (8*S*,8′*S*) because the sequence of substituents a > b > c is now counterclockwise.

Attention must be drawn to the important aryltetralin class which includes podophyllotoxin (**1.36**; X = H): (7*R*,7′*R*,8*R*,8′*R*)-7-hydroxy-

Scheme 1.5 Assignment of priority by the sequence rules

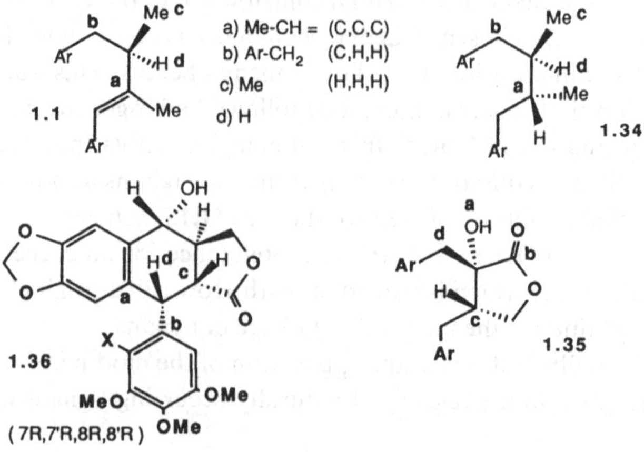

a) Me-CH = (C,C,C)
b) Ar-CH$_2$ (C,H,H)
c) Me (H,H,H)
d) H

(7R,7′R,8R,8′R)

3′,4′,5′-trimethoxy-4,5-methylenedioxy-2,7′-cyclolignan-9′,9-olide. In determining the order of substituent priority for C-7′ each benzene ring carbon is taken as having three substituent carbon atoms – (C, C, C), where that of the fused ring ranks first because it is *ortho*-carbon substituted. The last priority goes to C-8′, as it has the lowest ranked substituent group (C, C, H). The other assignments are straightforward although note that with regard to C-8 the attached carbons have the sets:

C-7 (O, C, H)
C-9 (O, **H**, H)
C-8′(C, C, H)

and therefore the priority for C-7 is determined by the atoms of second preference **C** > **H**.

In applying the sequence rules to lignans one must always be wary of the effect of the insertion of additional substituents. As we saw above, the side chain insertion in (−)-trachelogenin led to an apparent anomaly and a similar situation can arise if the status of an aryl group is changed. Should podophyllotoxin be substituted at the 2′-position as in the chloro-derivative (**1.36**; X = Cl), then the priority now goes to the pendent ring and so this compound is properly described as 7′-S.

Correlation of absolute configurations

A whole matrix of chemical transformation exists which interrelates the lignan classes and which links them to guaiaretic acid (**1.1**) and many of these have been recorded (Hearon and MacGregor, 1955). Some examples of the methods used are shown in Scheme 1.6. *R*(−)-Dimethylguaiaretic acid (**1.37**) on hydrogenation (Schroeter *et al.*, 1918) gave a dihydrocompound (**1.38**) together with the isomeric *meso*-addition product. (−)-Dimethylmatairesinol (**1.39**) was then shown (Schrecker and Hartwell, 1955) to have the same absolute configuration at C-8 by reduction with lithium aluminium hydride to give the *trans*-diol (**1.40**). More vigorous treatment of the ditosylate of this compound with the hydride afforded dimethyldihydroguaiaretic acid (**1.38**) as a common product.

The diol (**1.40**) was also a key substance in the correlation of the absolute configuration of furofurans with other lignans. (+)-Dimethylpinoresinol (**1.41**) gave a furan, (+)-lariciresinol (**1.42**), on initial hydrogenation with a palladium catalyst (Howarth and Woodcock, 1939). Correlation of both these classes is possible, since continued hydrogenation afforded the reference compound (**1.40**; Freudenberg and Dietrich, 1953). Fully alkylated lignans are preferred in work of this kind for they are less susceptible to oxidation than the parent phenols and there is no risk of the

Scheme 1.6 Chemical correlation of absolute configuration

DIMETHYLPINORESINOL **1.41** rotation$_D$ + 64°

DIMETHYL DIHYDROGUAIARETIC ACID **1.38** rotation$_D$ – 31°

DIMETHYLGUAIARETIC ACID **1.37** rotation$_D$ -55°

H_2 / Pd

LAH on ditosylate

LAH

LARICIRESINOL **1.42** rotation$_D$ +22°

1.40 rotation$_D$ -26

DIMETHYLMATAIRESINOL **1.39** rotation$_D$ -30°

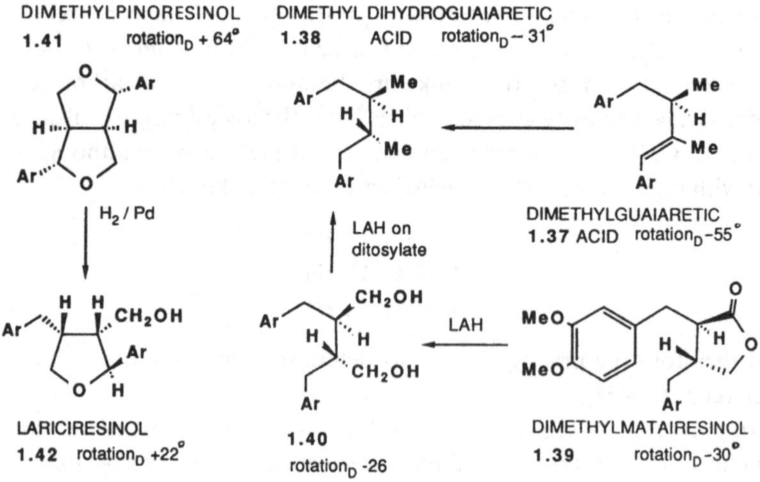

separation of phenolate salts during hydride reductions. Other examples of chemical methods of correlation of configuration will be given in further chapters (pp. 213, 228–30).

More modern techniques for assigning absolute configuration include those of optical rotatory dispersion and circular dichroism; numerous examples are included in the series published by Klyne and his collaborators (Hulbert *et al.*, 1981). The earlier work has also been substantiated recently by the application of X-ray crystallography to lignans (pp. 37, 49, 199, 230).

The techniques for establishing relative configurations depend very largely on proton magnetic resonance and the nuclear Overhauser effect and they are discussed in detail in Chapter 6.

References

Achenbach, H., Grob, J., Dominguez, X.A., Cano, G., Star, J.V., Brussolo L. del C., Munoz, G., Salgado, F. and Lopez, L. (1987). Lignans, neolignans and norlignans from *Krameria cystisoides*. *Phytochemistry*, **26**, 1159–63.

Cahn, R.S., Ingold, C.K. and Prelog, V. (1966). Specification of molecular chirality. *Angew. Chem. Intern. Edit.* **5**, 385–415.

Erdtman, H. (1933). Dehydrogenation of conifer substances. II. Dehydrodi-isoeugenol. *Liebigs Ann. Chem.* **503**, 283–94.

Freudenberg, K. and Dietrich, H. (1953). On syringaresinol, an oxidation product of sinapyl alcohol. *Chem. Ber.* **86**, 4–10.

Freudenberg, K. and Weinges, K. (1961). A system of nomenclature for lignans. *Tetrahedron* **15**, 115–28.

Gottlieb, O. R. (1972). Chemosystematics of the *Lauraceae*. *Phytochemistry* **11**, 1537–70.

Gottlieb, O.R. (1974). Lignans & neolignans. *Rev. Latinoamer. Quim.* **5**, 1–11.

Gottlieb, O.R. (1978). *Neolignans. Fortschr. Org. Chem. Naturst.* **35**, 1–72.

Hearon, W.M. and MacGregor, W.S. (1955). The naturally occurring lignans. *Chem. Rev.* **55**, 957–1068.

Howarth, R.D. (1936). Natural resins. *Ann. Rep. Progr. Chem.* **33**, 266–79.

Howarth, R.D. and Woodcock, D. (1939). The constituents of natural phenolic resins. Part 15. The stereochemical relationship of lariciresinol and pinoresinol. *J. Chem. Soc.* 1054–7.

Howarth, R.D. and Wilson, L.J. (1950). The constituents of natural phenolic resins. Part 22. Reduction of some lactonic lignans with lithium aluminium hydride. *J. Chem. Soc.* 71–2.

Hulbert, P.B., Klyne, W. and Scopes, P.M. (1981). Chiroptical studies. Part 100. Lignans. *J. Chem. Research(S)* **27**; *J. Chem. Research(M)*, 0401–49.

International Union of Pure & Applied Chemistry (1974). *Rules for the Nomenclature of Organic Chemistry*, pp. 13–30.

Joshi, B.S., Ravindranath, K.R. and Viswanathan, N. (1978). Structure & stereochemistry of attenuol, a new lignan from *Knema attenuata* Wall. *Experientia* **34**, 422–3.

Kupchan, S.M., Stevens, K.L., Rohlfing, E.A., Sickles, B.R., Sneden, A.T., Miller, R.W. and Bryan, R.F. (1978). New cytotoxic neolignans from *Aniba megaphylla* Mez. *J. Org. Chem.* **43**, 586–90.

Lindsey, J.B. and Tollens, B. (1892). On the wood-sulphite liquor & lignin. *Liebigs Ann. Chem.* **267**, 341–66.

Maiden, J.H. and Smith, H.G. (1896). On aromadendrin or aromadendric acid from the turbid group of eucalyptus kinos, *Am. J. Pharm.* 679–87.

Moss, G.P. (1987). Paper to the Biochemistry Committee of the International Union of Pure & Applied Chemistry.

Schrecker, A.W. and Hartwell, J.L. (1955). Application of tosylate reductions and molecular rotations to the stereochemistry of lignans. *J. Amer. Chem. Soc.* **77**, 432–7.

Schroeter, G., Lichtenstadt, L. and Irineu, D. (1918). On the constitution of the Guaia heart substances. *Chem. Ber.* **51**, 1587–613.

Vanzetti, B.L. and Dreyfuss, P. (1934). On the constitution of olivil. Products of oxidation of isoolivil. *Gazzetta* **64**, 381–99.

Vieira, P.C., Gottlieb, O.R. and Gottlieb, H.E. (1983). Tocotrienols from *Iryanthera grandis*, *Phytochemistry* **22**, 2281–6.

Watanabe, K., Watanabe, H., Goto, Y., Yamaguchi, M., Yamamoto, N. and Hagino, K. (1983). Pharmacological properties of magnolol & hinokiol extracted from *Magnolia officinalis*: central depressant effects. *Planta Med.* **49**, 103–8.

Whiting, D.A. (1987). Lignans, neolignans & related compounds. *Nat. Prod. Repts.* 191–211.

2

A registry of the natural lignans

The sustained and growing interest in these compounds is evident from study of the Figure (2.1) which records the numbers of lignans listed in the given reviews through the fifty years from 1936.

This review is also concerned primarily with the structures of the lignans themselves and the glycosides are usually treated as derivatives. They are listed and numbered as separate entities when they have been given a trivial name which differs from that of the aglycone. It should however be born in mind that the pharmacological properties of glycosides may not necessarily be simply equated with those of the associated lignan. This is well illustrated by the different modes of action of the podophyllotoxins, which are spindle poisons, and the DNA breaking activity of their epiglucosides.

FIGURE 2.1. Growth of lignan chemistry 1936–1987.

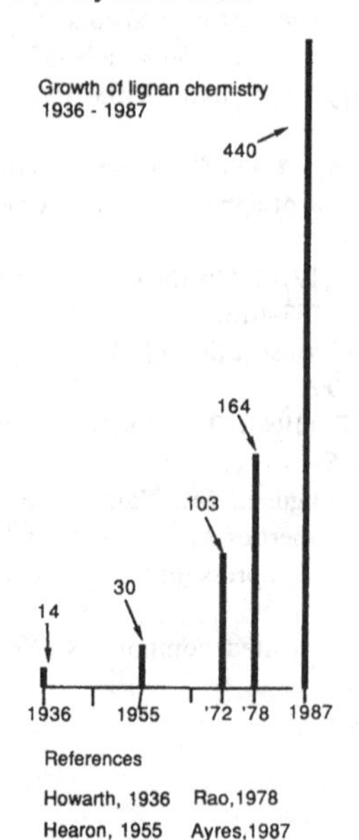

The order of the material

Entries are made in alphabetical order of the trivial names within each of the following sections or subsections:

Dibenzylbutanes

(a) with unsubstituted side chains

(b) oxygenated in the side chain

Dibenzylbutyrolactones

(a) includes all types except
(b) unsaturated lactones

Minor variants are treated as derivatives. Thus entry **2.063** is – **MATAIRESINOL** DI-*O*-METHYL (MACULATIN) – not DI-*O*-**METHYLMATAIRESINOL** (MACULATIN).

Tetrahydrofurans

(a) 9,9′-epoxylignans
(b) 7,9′-epoxylignans
(c) 7,7′-epoxylignans

Furofurans

(a) with unsubstituted side chains
(b) oxygenated in the side chain
(c) glycosides

A separate subsection was included for the glycosides of the furofurans because of their significance and because these names do not always reveal their nature. For example, this is clear for the **ACANTHOSIDES** and for **LANTIBESIDE** but it is not evident for **EUCOMMIN A** or for **MAGNOLENIN C.**

In this section the prefixes **DIA-**, **EPI-** and **ISO-** are retained to ensure that like can readily be compared with like.

Aryltetrahydronaphthalenes (Tetralin lignans)

(a) with unsubstituted side chain
(b) oxygenated in the side chain
(c) lactones

The prefix **CYCLO-** is preferred where a clear relationship exists with a lignan of another class for example, **LARICIRESINOL (2.119)** – **CYCLOLARAICIRESINOL (2.313)**. At the same time **ISO-** must continue in use for those lignans of one class which are so related; long term usage has necessitated its retention as a prefix in some instances.

A few compounds have been included which do not occur naturally but which are significant for the proof of structure of the whole class. Examples of this are **ISO-GALBULIN** and **PODOPHYLLOTOXIN – EPI**.

Apolignans and naphthalenes These two sections include lignans related to the tetralin class but with the B ring partially or wholly dehydrogenated.

Dibenzocyclo-octadienes This is the smallest group and it has not been subdivided. A possible division into the (*R*)- and (*S*)-series of biphenyls was not adopted because both types commonly occur in the same source plant and this would have led to needless repetition.

Unnamed lignans These are included at the end of each relevant section and the systematic nomenclature has been used to describe each of them. Systematic names have not been included for the great majority with trivial names, partly to avoid unnecessary repetition and also to emphasise the structures by limiting the volume of text. Compounds of interest which lack the full lignan structure are also located here.

Absolute configuration

The D line rotation is given for many of the entries although optical rotatory dispersion (ORD) or circular dichroism (CD) curves are better criteria. The configurations shown in this registry are based on the evidence of these spectra (Chapter 6) and the numerous chemical interconversions between lignans of different classes (Chapters 1 and 8). X-ray measurements have substantiated assigned relative configurations in a number of instances and others give data on the absolute configuration of bromolignans. The absolute configuration of lignan glycosides may be obtained by reference to that established for the sugar residue, although it has been suggested (Hursthouse, 1987) that such a concentration of oxygen substituents could be equivalent to a 'heavy' bromine atom.

Some confusion has arisen in the literature because the printed page limits authors to one choice of configuration and it is not always made clear when this is of necessity an arbitrary one. Another source of error can arise in lignans such as the furofurans with symmetrical skeletons (cf.

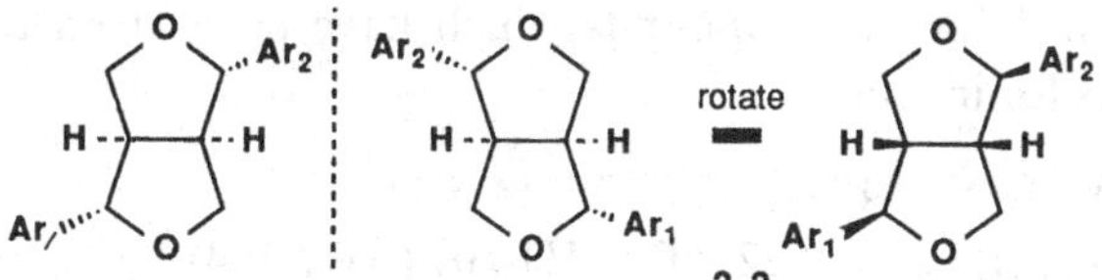

2.1 reflect rotate 2.2

p. 7) and the aryl substituents must be placed at two and eight o'clock (**2.1**) rather than at four and ten o'clock (**2.2**).

Inspection of models will confirm that the latter is the *ent*-form of the former and that failure to observe this convention will result in an 'inversion' of configuration. This relationship is seen **(+)-PIPERITOL (2.219)** and **(−)-PLUVIATILOL (2.222)**.

Arrangement of individual entries

A typical example is shown below:

(+)-**AUSTROBAILIGNAN 7** 55890-25-0
2.139 C20 H22 05

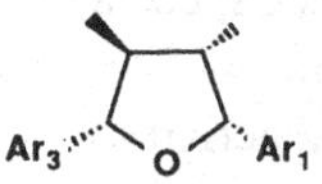

Canad. J. Chem. 1981, **59**, 1680(CD)
Aust. J. Chem. 1975, **28**,81(isol)

Schizandra sp.
Austrobaileya scandens

(i) *The Chemical Abstracts Registry Number* (top RH)

This provides an entry for a computerised search of the literature. These were taken from the C.A. service publication *Registry Handbook – Common Names*, August, 1988, which is updated at six-monthly intervals. Registry numbers are revised from time to time as new information becomes available. They provide a useful check on substances for which more than one trivial name is in use and the handbook also provides a registry number – name index; this can be used to find all the existing synonyms.

Enantiomers will have different registry numbers and racemic material will differ again. Thus four entries are given for the eudesmins:

EUDESMIN 526-06-7 C22 H26 06
(−)-EUDESMIN 526-06-7
rac-EUDESMIN 38759-91-0
(+)-EUDESMIN 29106-36-3

The molecular formula provides a useful check against the structure given, an entry into the formula index of the abstracts and also serves to distinguish between different compounds which have been given the same name. One notes for instance:

DIPHYLLIN 479-18-5 C10 H14 N4 04
DIPHYLLIN 22055-22-7 C21 H16 07 (see **2.394**).

One also finds that there exist *five* EXCELSINS and *three* EXCELSINES.

(ii) *Literature references*

Where only one appears it follows that the plant source (lower LH) is established here with adequate supporting evidence for the structure given. Additional references appear where they significantly extend knowledge of the compound in its class. Thus for AUSTROBAILIGNAN 7 support for the absolute configuration comes from circular dichroism (CD). Other specialist interests are abbreviated as follows:

(chromatog) – detail on separation
(CMR) – ^{13}C magnetic resonance with the ^{1}H spectrum
(MS) – mass spectral fragmentation details
(NOE) – use of nuclear Overhauser difference spectra
(ORD) – optical rotatory dispersion spectra
(pharm) – data of relevance to physiological activity
(PMR) – ^{1}H magnetic resonance spectra
(synth) – general synthetic procedures
(X-ray) – data on relative or absolute configuration.

(iii) *Structural diagrams*

The usual conventions have been followed but the four most common aromatic structures may be abbreviated to:

Ar_1 = 4-hydroxy-3-methoxyphenyl (vanillyl)
Ar_2 = 3,4-dimethoxyphenyl (veratryl)
Ar_3 = 3,4-methylenedioxyphenyl (piperonyl)
Ar_4 = 3,4,5-trimethoxyphenyl.

When the aromatic residues differ from the above they are set out in full. When both aryl substituents are the same the abbreviation – R may stand for one of them.

References for the preamble

Devon, T.K. and Scott, A.I. (1972). *Handbook of Naturally Occurring Compounds*, pp. 71–84. New York and London: Academic Press.
Howarth, R.D. (1936). Natural resins, *Ann. Rep. Progr. Chem.* **33**, 266–79.

Hearon, W.M. and MacGregor, W.S. (1955). The naturally occurring lignans. *Chem. Rev.* **55**, 957–1068.

Hursthouse, M.B. (1987). Personal communication.

Rao, C.B.S. (ed. 1978). *Chemistry of Lignans.* Andhra University Press: Waltair.

Structures and references in the registry

ANWULIGNAN 90365-59-6
2.001 C20 H24 O4

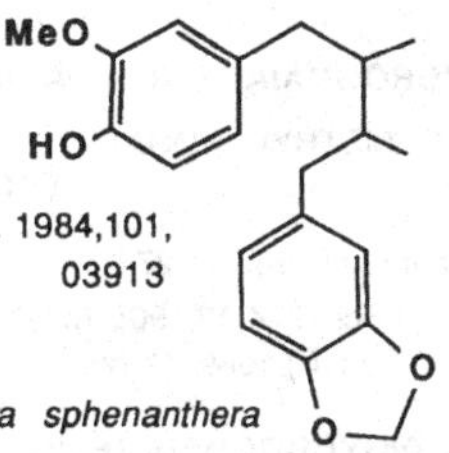

Chem.Abs. 1984,101, 03913

Schizandra sphenanthera

(-) - AUSTROBAILIGNAN 6 55890-24-9
2.003 C20 H24 O4

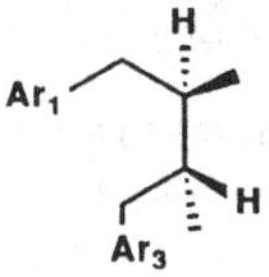

Aust.J.Chem.1975,28,81

A. scandens

(-) - DIHYDROGUAIARETIC ACID
2.006 36469-60-0
C20 H26 O4

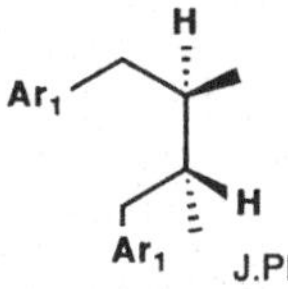

J.Pharm.Sci.1974,63,1905

Larrea divaricata

(-) - AUSTROBAILIGNAN 5 55890-23-8
2.002 C20 H22 O4

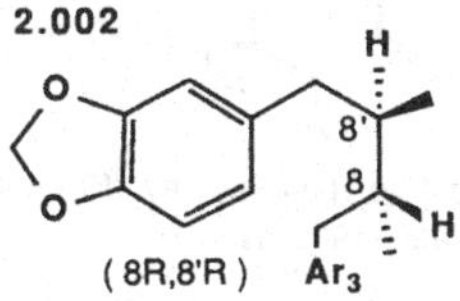

(8R,8'R)

Aust.J.Chem.1975,28,81

Austrobaileya scandens

CALOPHYN 114127-24-1
2.004 C20 H24 O4

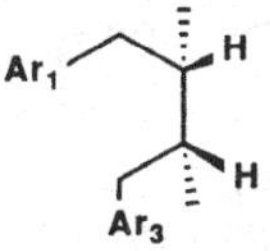

Chem.Abs.1988,108,183594

Virola calophylla

CONOCARPOL 56319-00-7
2.005 C18 H22 O2

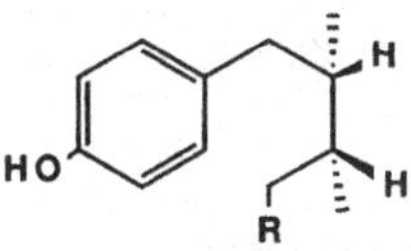

Phytochem.1975,14,1085

Conocarpus erectus

CONOCARPOL 2 - METHOXY
2.007

See 2.005

C. erectus

DIHYDROGUAIRETIC ACID DEMETHYL 71113-15-0
2.008 C19 H24 O4

HO, HO, H, H, Ar_1

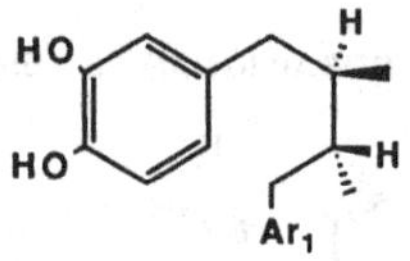

Biochim.Biophys.Acts.1984,788,167 (Physiol)
J.Pharm.Sci.1974,63,1905 (Isol)

(-) - **GUAIRETIC ACID** 500-40-3
2.010 C20 H24 O4

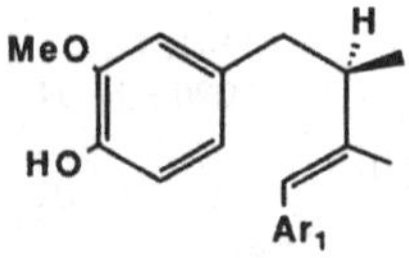

MeO, HO, H, 8'-R, Ar_1

JACS.1957,79,3823(Struct)
Tetrahedron Lett.1980,21,4017 (synth)
Guaiacum officinale

MACELIGNAN 107534-93-0
2.013 C20 H24 04

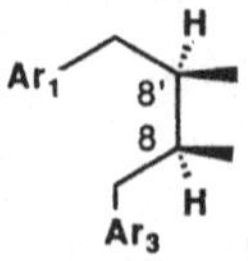

Ar_1, 8', 8, H, H, Ar_3 8S,8'R

Phytochemistry 1987,26,1542 (X-ray)
Myristica fragrans

meso - **DIHYDROGUAIARETIC ACID**
2.009 cf 2.010 66322-34-7
8S,8'R C20 H26 O4

JACS.1957,79,3823 (struct & isol)

DIHYDROGUAIARETIC ACID
bis- DEMETHYL (NDGA) 500-38-9
2.011 C18 H22 O4

J.Chem.Ecol.1985,11,27
J.Pharm.Sci.1974,63,1905 (physiol)
J.Chem.Soc.(B),1969, (X-ray)

(+) - **DIHYDROGUAIARETIC ACID**
2.012 ENT 2.006

Yaoxue Xuebao,1986,21,68
Sargentodoxa cuneata (Oliv)

MACHILIN A 110269-50-6
2.014 C20 H22 O4

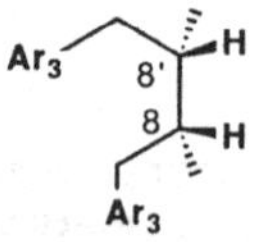

Ar_3, 8', 8, H, H, Ar_3 8R,8'S

Machilus thunbergii

DIBENZYLBUTANE

2.015 4'-Hydroxy-3'-methoxy-3,4-methylenedioxylignan

104569-15-5

C20 H 24 O4

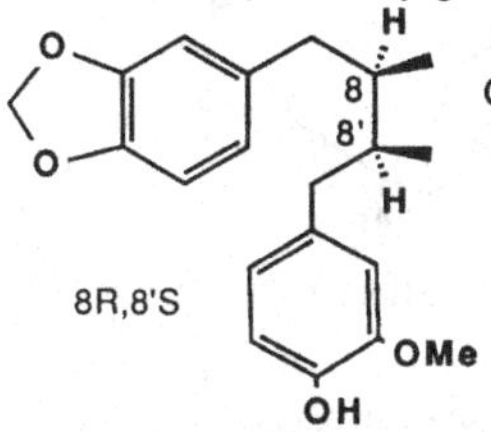

8R,8'S

Saengyak Hakkoechi,1986,17,91 (pharm)

Chem.Abs.1986,105,146184

Myristica fragrans

OXY- and OXODIARYLBUTANES

ARIENSIN 81410-43-7

2.016 X = Ac C24 H26 08

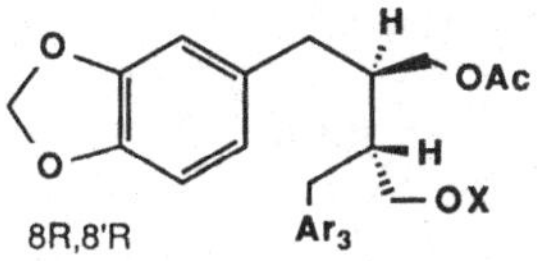

8R,8'R

Planta Med.1983,47,215

Bursera ariensis

HEMIARIENSIN

2.017 As 2.016 but X = H

Phytochem.1987,26,2033

Piper cubeba

(-) - CARINOL 58139-12-1

2.018 C20 H26 O7

tetrahydro-(+)-gmelinol

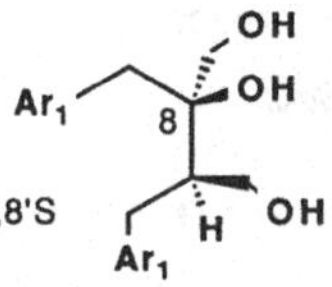

8S,8'S

Phytochem.1983,22,749

Carissa carandas

CARISSANOL 87402-76-4

2.019 C20 H24 O7

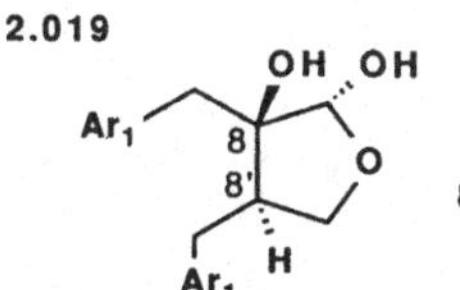

8R,8'R

Phytochem.1983,22,749

Carissa edulis

(-)- CLUSIN 86992-94-1
2.020 C22 H26 O7

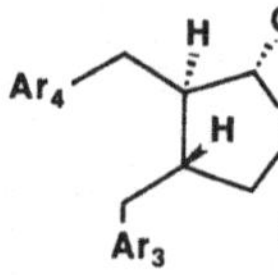

8R,8'R

Phytochem.1984,23,2099 (CD,PMR)
Piper clusii

(-)-DIHYDROCLUSIN 73149-51-6
2.022 X = H C22 H28 07

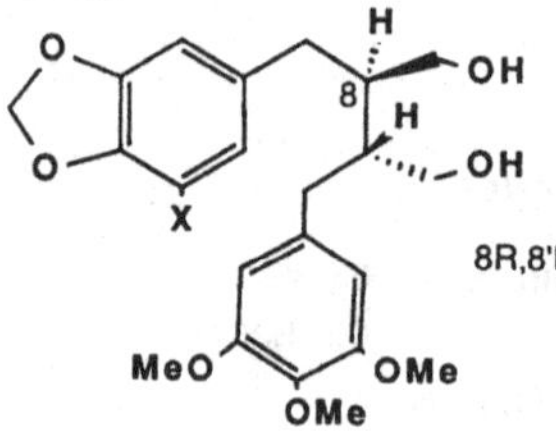

8R,8'R

Phytochem.1985,24,329 (CD)
Piper cubeba; P. clusii

DIHYDROCLUSIN - 3-METHOXY
2.024 93395-18-7
C23 H30 O8

As 2.022 but with X = OMe
Phytochem.1984,23,2099 (CD; PMR)

(-)- CUBEBIN 18423-69-3
2.021 C20 H20 O6

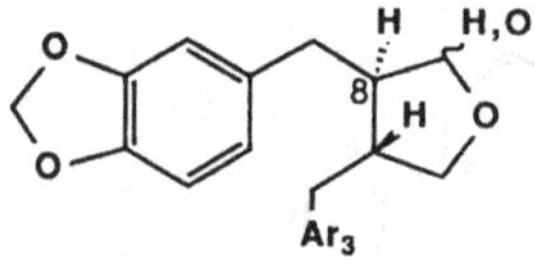

8R,8'R

J.Chem.Soc.(C),1969,2470
Piper cubeba

CUBEBIN 3',4'-DIMETHOXY-
3',4'-DEMETHYLENEDIOXY
2.023
C21 H24 O6

As 2.021, R = Ar_2

Tetrahedron Lett.1978,457(ORD)
Aristolochia triangularis

CUBEBIN 3,4-DIMETHOXY-
3,4-DEMETHYLENEDIOXY
2.025

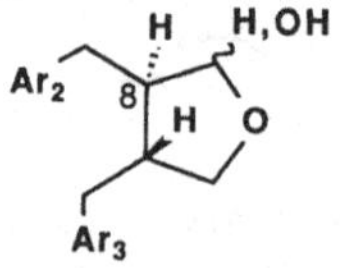

(-)-CUBEBININ 96238-92-5
2.026 C24 H32 O8

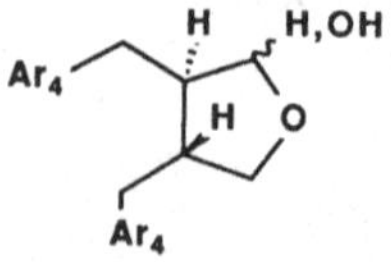

Phytochem.1985,24,329 [CD cf (-) hinokinin]
Piper cubeba

(-)-DIHYDROCUBEBIN 24563-18-6
2.027 $C_{20} H_{20} O_5$

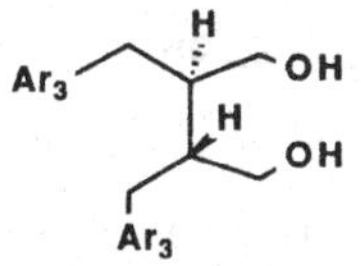

Phytochem.1985,24,329
P.cubeba

ENTERODIOL 80226-00-2
2.028 $C_{18} H_{20} O_4$

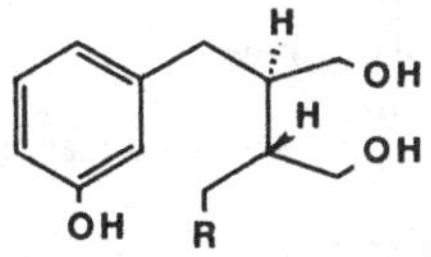

J.Chem.Soc.Perk.Tr.1984,489
Human metabolite

FUROGUAIAOXIDIN 58096-91-6
2.029 $C_{20} H_{20} O_6$

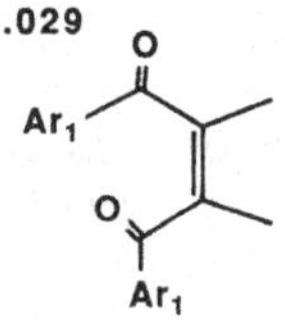

cis- or Z

J.Chem.Soc.1964,4011

Guaiacum officinale

α -INTERMEDIANOL 103019-34-7
2.030 $C_{22} H_{28} O_7$

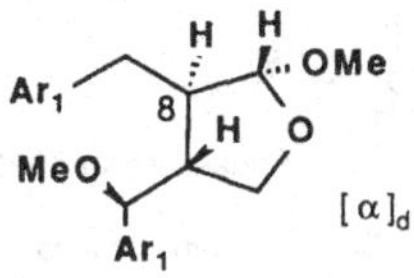

$[\alpha]_d$ -96°

β - INTERMEDIANOL 103019-33-6
2.031 As above but 9-β-OMe

Aust.J.Chem.1985,38,1631 $[\alpha]_d$ +24°
Dacrydium intermedium

(+)-NIRANTHIN 50656-77-4
2.032 $C_{24} H_{32} O_7$

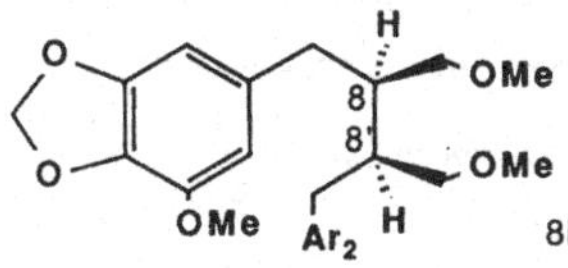

8R,8'S

Tetrahedron,1973,29,1291
Phyllanthus niruri

NIRANTHIN 7-HYDROXY
2.033

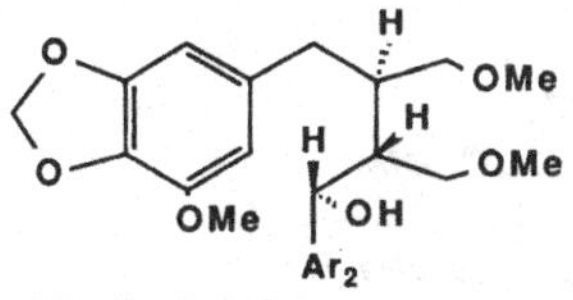

J.Nat.Prod.1988,51,44

P.niruri

(+)- PHYLLANTHIN 10351-88-9
2.034 C24 H34 O6

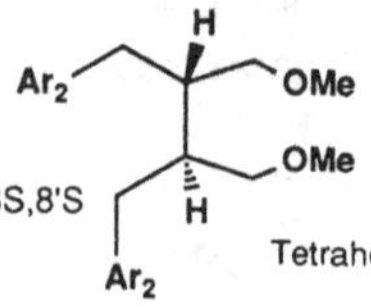

8S,8'S

Tetrahedron,1966,22,2899

Phyllanthus niruri

(-)- SECOISOLARICIRESINOL
2.035 29388-59-8
C20 H26 O6

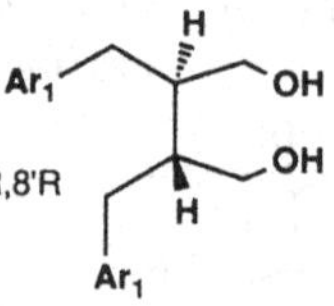

8R,8'R

Tetrahedron,1959,7,262
Larix desidua

(-)- SECOISOLARICIRESINOL
2.036 0- DIMETHYL 58311-18-5
C22 H30 O6

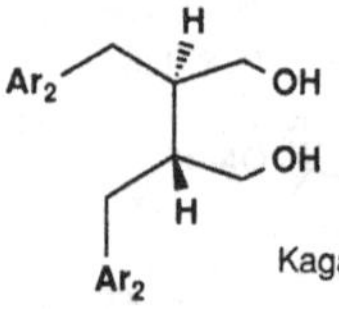

Kagaku Kaishi,1975,12,2192

Cinnamomum camphora (Sieb)

meso- **SECOISOLARICIRESINOL**
2.037 8S,8'R 57759-55-4
C20 H26 O6

Phytochem.1982,21,1459

Cedrus deodara

(-) - SECOISOLARICIRESINOL
2.038 O - TRIMETHYL

J.Nat.Prod.1988,51,44

Phyllanthus niruri

SECOISOLARICIRESINOL 73354-08-2
2.039 O - METHYL C21 H28 O6

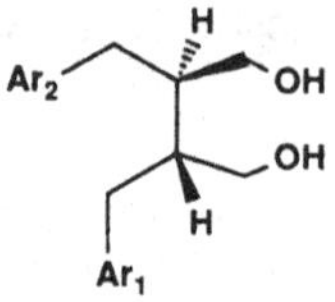

Phytochem.1979,18,1703 (CMR)

Araucaria augustifolia

SECOLIGNAN
2.040

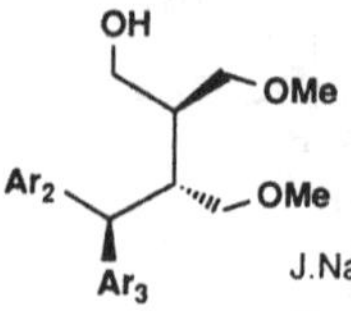

J.Nat.Prod.1988,51,44
P. niruri

ZUIHONIN D 84607-65-8
2.042 $C_{20}H_{16}O_6$

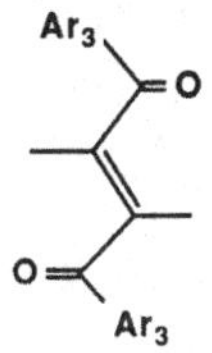

Chem.Abs.1983,98,74043

Machilus zuihoensis

DIARYLBUTYROLACTONES

(-)-ARCTIGENIN 7770-78-7
2.045 $C_{21}H_{24}O_6$

cf 2.067, 2.075

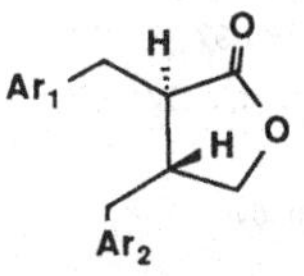

Tetrahedron Lett.1983,24,3237 (synth)
J.Pharm.Soc.Japan,1929,49,565 (isol)

Steganataenia araliacea
Arctium lappa

4'-GENTIOBIOSIDE 41682-24-0

4-GLUCOSIDE 20362-31-6
(ARCTIIN)

Chem.Pharm.Bull.,1981,29,3586 (pharm)

(+)-ARCTIGENIN 84413-77-4
2.046 $C_{21}H_{24}O_6$

Phytochem.1982,31,1824

Wikstroemia indica

(-)-ARCTIGENIN
also as 4'-O-METHYL MATAIRESINOL

(-) - BURSEHERNIN 40456-51-7
2.047 C21 H22 O6

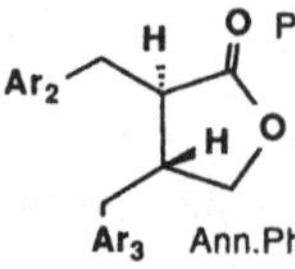

H O Planta Med.1984,50,20 (MS)
Ar_2 H O
Ar_3 Ann.Pharm.Fr.1984,42,317 (isol)

Hernandia guianensis

(-) - HAPLOMYRFOLIN
2.047A C20 H20 O6

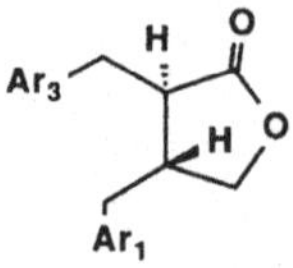

H O Phytochem.1986,25,1949
Ar_3 H O
Ar_1

Haplophyllum myrtfolium

(rac)-ENTEROLACTONE 78473-71-9
2.049 C21 H18 O4

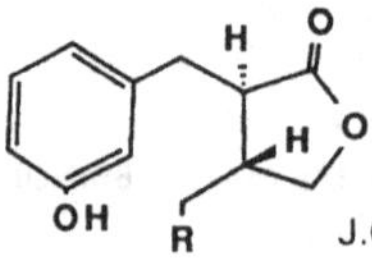

H O
H O
OH R J.Chem.Soc.Perkin Tr.2 1984,489

Mammalian metabolite

CORDIGERIN 93395-16-5
2.048 C24 H30 O8

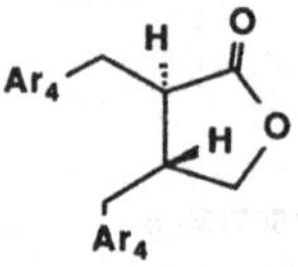

H O
Ar_4 H O
Ar_4 J.Nat.Prod.1984,47,879

Hernandia cordigera

GNIDIFOLIN 58268-84-1
2.050 C20 H22 O7

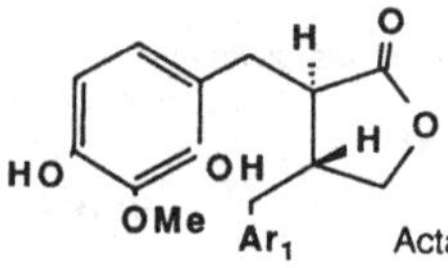

H O
H O
HO OH
OMe Ar_1 Acta Cryst.Sect.B 1978,B34,327 (X-ray)

HELIOANTHOIDIN 20585-97-1
2.051 C26 H28 O7

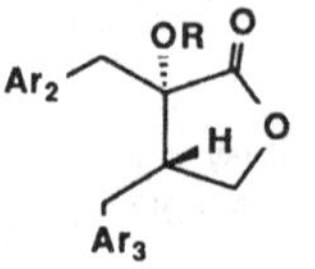

OR O
Ar_2 H O R = angeloyl
Me O
H Me
Ar_3

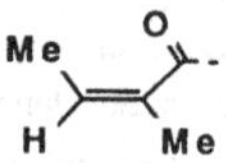

J.Chem.Soc.(C),1969,693

Heliopsis scabra Dunal

HELIOBUPHTHALMIN LACTONE
2.052 (Ent- HINOKININ) 103063-07-6
C20 H18 O6

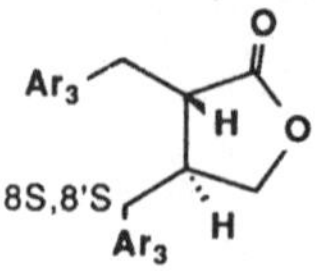

O
Ar_3 H O
8S,8'S H Planta Med.1986,52,18
Ar_3

Heliopsis buphthalmoides

HELIOBUPHTHALMIN 66547-91-9
2.053 (R = H) C22 H22 O8

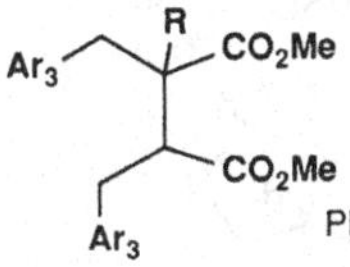

Ar3 R CO2Me CO2Me Ar3

Phytochem.1978,17,330

HELIOBUPHTHALMIN 8 - HYDROXY
2.055 (R = OH) C22 H22 O9

Planta Med.1986,52,18

Heliopsis buphthalmoides

HELIOBUPHTHALMIN C23 H26 O8
2.054 3,4 - DIMETHOXY - 3,4 - DEMETHYLENEDIOXY

Ar2 H CO2Me H CO2Me Ar3

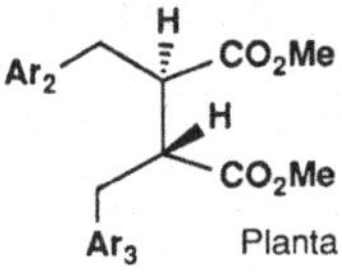

Planta Med.1986,52,18

HERNOLACTONE 112667-54-6
C24 H30 O8
2.056

MeO HO OMe H O H O Ar4

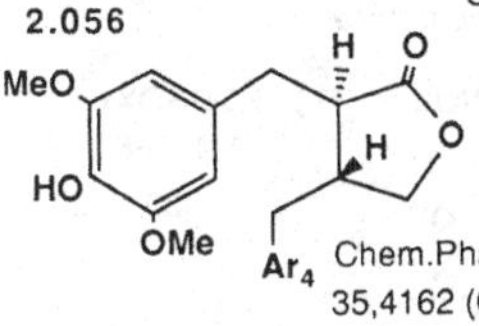

Hernandia ovigera

Chem.Pharm.Bull.1987, 35,4162 (CD,PMR,DEPT)

HINOKININ 3',4'- DIMETHOXY- 3',4' - DEMETHYLENEDIOXY
2.059 58311-20-9
C21 H22 O6

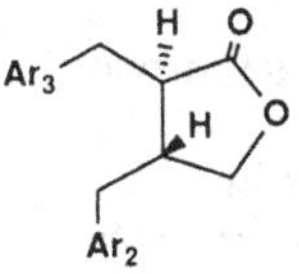

Ar3 H O H O Ar2

Phytochem.1983,22,1516

Virola sebifera

HINOKININ 3,3',4,4'- TETRAMETHOXY-bis - DEMETHYLENEDIOXY
2.060 (or dimethylmatairesinol) 25488-59-9

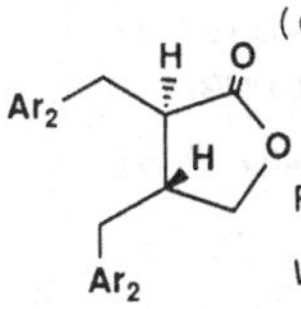

Ar2 H O H O Ar2

Phytochem.1983,22,1516

Virola sebifera

(-) - **HINOKININ** 26543-89-5
2.057 C20 H18 O6

Ar3 H O H O 8R,8'R Ar3

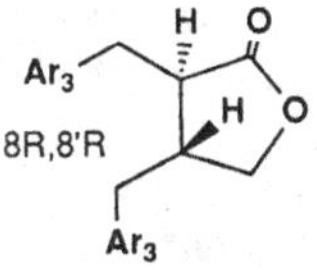

Phytochem.1983,22,1516

Virola sebifera & many other sources

HINOKININ 5 - METHOXY 112448-63-2
2.058 C21 H20 O7

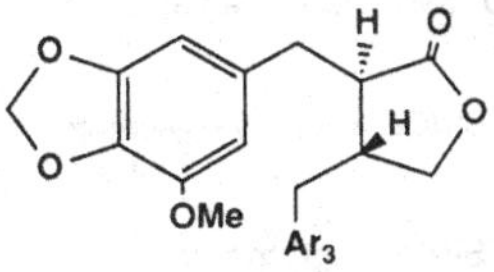

O O OMe H O H O Ar3

Phytochem.1987,26,2033

Piper cubeba

(-) - KUSUNOKININ 78684-83-0
2.061 C21 H22 O6

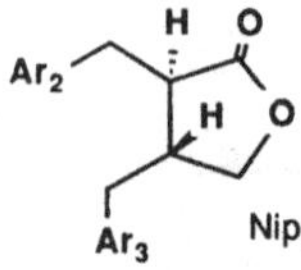

Nippon Kagaku Kaishi,1975,2192 (PMR)

Cinnamomum camphora
Haplophyllum vulcanicum

(+) - MATAIRESINOL DI - O - METHYL
2.064 (ent - 2.063)

Biorg.Khim.1980,6,1415
Chaerophyllum maculatum

MATAIRESINOL 4' - O - METHYL
2.067 (arctigenin) 7770-78-7
C21 H24 O6

J.Pharm.Soc.Japan,1929,49,565
Arctium lappa L.
with 4 - O -ß -D -glucoside (arctiin)

MATAIRESINOL 4 - O - ß - glucoside
34446-06-5

MATAIRESINOL diglucoside

(-) - MATAIRESINOL 580-72-3
2.062 C20 H22 O6

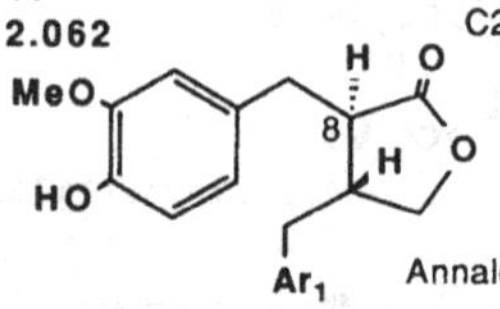

Annalen,1983,1202 (isol)

Podocarpus spicatus
Heliopsis sp.

(-) - MATAIRESINOL DI-O-METHYL
2.063 (maculatin) C22 H22 O7

Yakugaku Zasshi,1936,56,982

Anthriscus nemerosa

MATAIRESINOL 7' (R) - HYDROXY
2.065 20268-71-7
C20 H22 O7

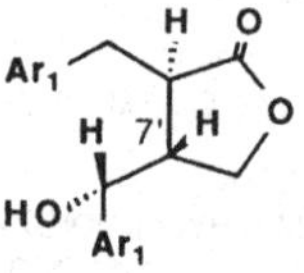

MATAIRESINOL 7' (S) - HYDROXY
2.066
Chem.Ber.1957,90,2857
Chem.Pharm.Bull.1980,28,850 (CMR)

MATAIRESINOL 7-OXO
2.066A C20 H20 O7

J.Chinese Chem.Soc.1985,32,75
(isol, PMR)

Tsuga chinensis pritz (Hay)

(-) - PARABENZLACTONE 27675-77-0

2.068 C20 H18 O7

H O H H O HO R

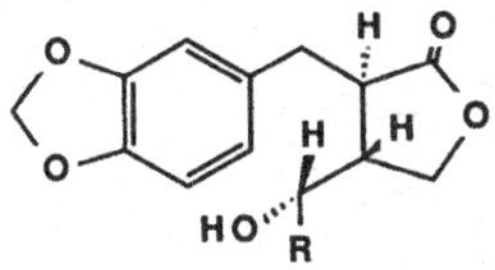

Tetrahedron Lett.1970,2017

Parabenzoin trilobum Nakai

(-) - PLUVIATOLIDE 28115-68-6

2.070 C20 H20 O6

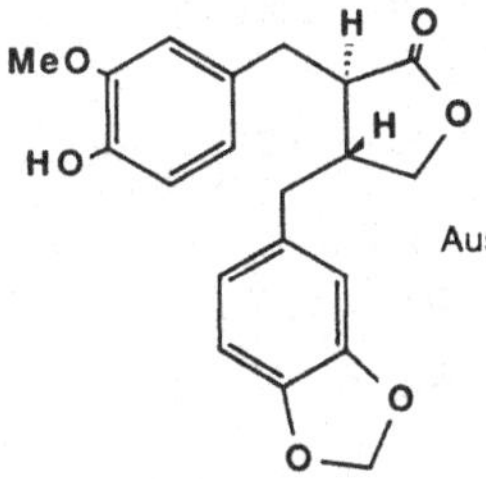

MeO HO H O H O O O

Aust.J.Chem.1970, 23,133

Zanthoxylum pluviatile Hartley

(-)-PRESTEGANE A 87605-56-9

2.075 C21 H24 O6

(R_1 = H, R_2 = Me) cf arctigenin

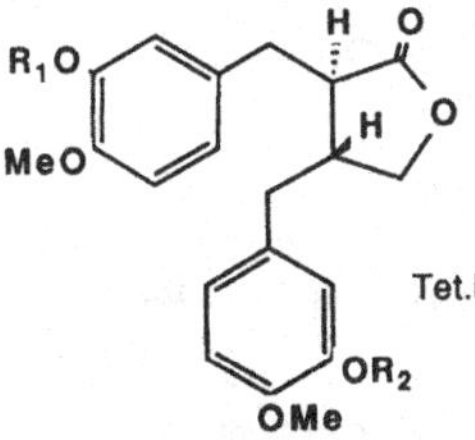

R_1O MeO H O H O OR_2 OMe

Tet.Lett.1983,24,3237
1984,25,4127

(-) - PRESTEGANE B 93376-04-6

2.076 ($R_1 = R_2 = H$) C20 H22 O6

Steganotaenia araliaceae Hochst.

PINOPALUSTRIN 34444-37-6

2.069 C20 H22 O7

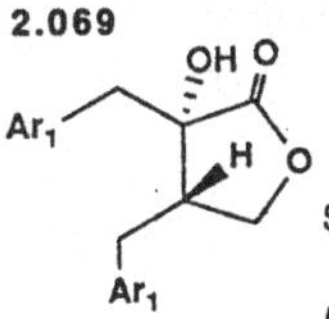

OH O Ar_1 H O Ar_1

Svensk.Kem.Tidskr. 1959,71,440

Pinus palustris Mill.

PODORHIZOL 17187-78-9

2.071 C22 H24 O8

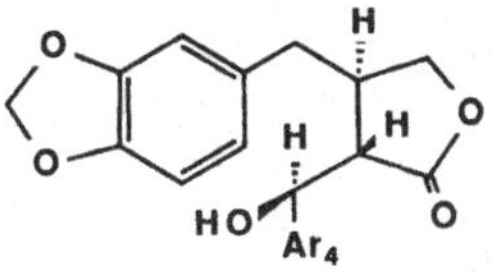

H O O H H O HO Ar_4 O

Helv.Chim.Acta,1967,50,1546
Tetrahedron,1982,38,3667 (synth)

(-) - PODORHIZONE 70191-79-6

2.071A C22 H22 O8

7-keto derivative of 2.071

Tetrahedron Lett.1978,4687 (synth)

PODORHIZOL 5' - METHOXY

2.072 C23 H26 O9

Heterocycles,1985,23,309

Hernandia cordigera

(-) - PODORHIZOL 7-DEOXY C22 H24 O7

2.073

J.Pharm.Sci.1972,61,1992* (M/S)

(-) - PODORHIZOL 5-DEMETHOXY-7-DEOXY

2.074 C21 H22 O6

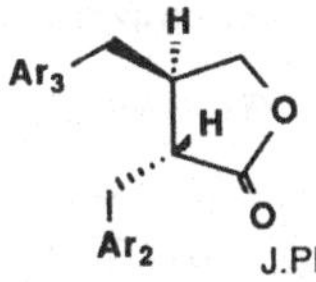

H Ar_3 H O O Ar_2

J.Pharm.Sci.1972,61,1992*

* the absolute config. of 2.073 & 2.074 needs to be checked

SVENTENIN 28168-95-8
2.077 C20 H18 O7

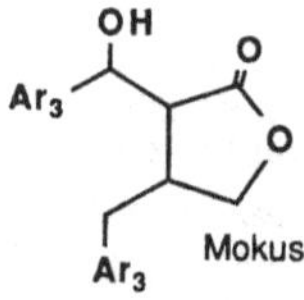

OH O Ar3 O Ar3

Mokusai Gakkaishi,1974,20,558

Taiwania cryptomerioides

THUJASTANDIN 19473-16-6
2.083 C20 H22 O8

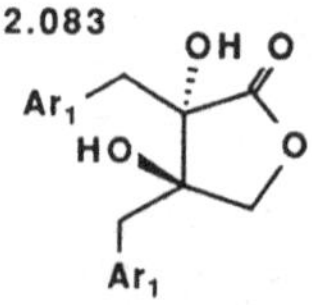

OH O Ar1 HO O Ar1

Mokusai Gakkaishi,1967,13,265 (PMR)

Thuja standishii

(+) - **TRACHELOGENIN** 34209-69-3
2.084 C21 H24 O7

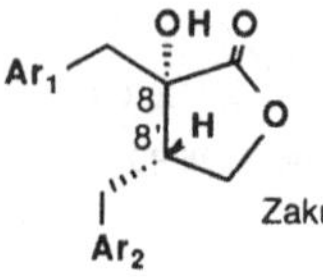

OH O Ar1 8 8' H O Ar2

8S,8'S

Zakugaku Zasshi,1973,93,541

Arctium lappa

(+) - **TRACHELOGENIN NOR**
2.085 615211-74-2
C20 H22 O7

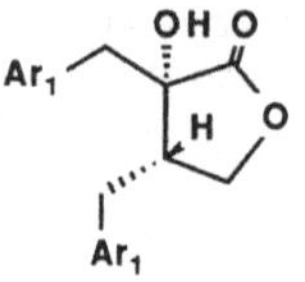

OH O Ar1 H O Ar1

J.Nat.Prod.1979,42,159

Wikstroemia indica

(-) - TRACHELOGENIN NOR 34444-37-6
2.086 C20 H22 O7

Phytochem.1971,10,2232

THUJAPLICATIN 6512-66-9
2.078 (R = H) C20 H22 O7

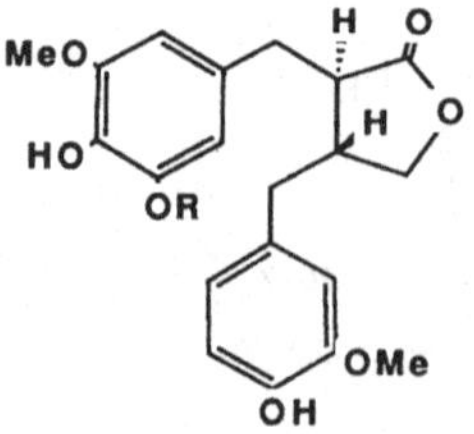

MeO HO OR H O H O OMe OH

THUJAPLICATIN 3-O-METHYL (R = Me)
2.079 C21 H24 O8

Canad.J.Chem.1967,45; 305 ,739

Thuja plicata Donn

THUJAPLICATIN 8 - HYDROXY - 3 - O-METHYL
2.080

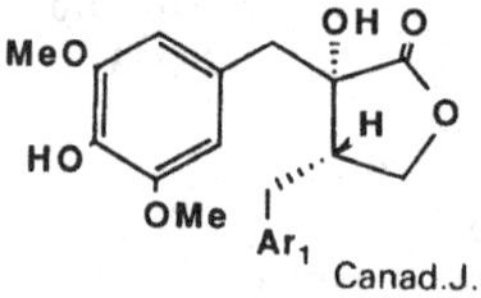

MeO HO OMe OH O H O Ar1

Canad.J.Chem.1966,44,1827

T.plicata

THUJAPLICATIN 8,8' - DIHYDROXY
2.081 R = H C20 H22 O9

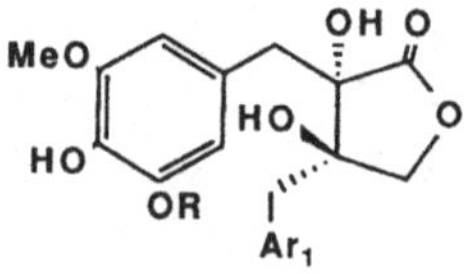

MeO HO OR OH O HO O Ar1

also the 3 - METHYL ether R = Me
2.082

Canad.J.Chem.1967,45,305,739

T. plicata

(-) - **TRACHELOSIASIDE** 106647 -12-5

2.087 C26 H32 O11

(Matairesinol 3-C- ß -D-glucoside

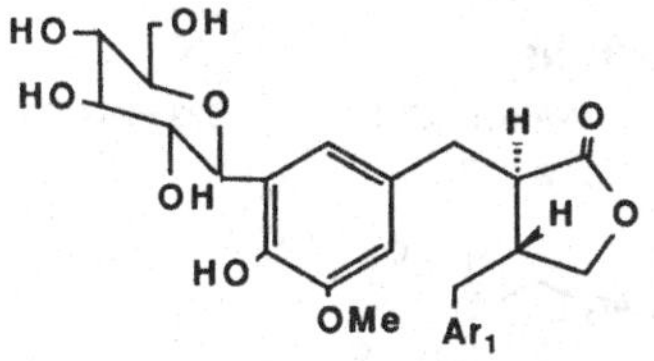

Chem.Pharm.Bull. 1986,34,4340 (CMR)

Trachelospermum asiaticum

(-) - **YATEIN** ISO 101751-72-8

2.089 C22 H24 O7

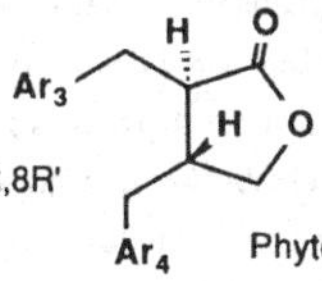

8R,8R'

Phytochem.1986,25,487

Piper cubeba

YATEIN ISO - 5 - METHOXY 101751-71-7

2.090 [(-) - cubebinone] C23 H26 08

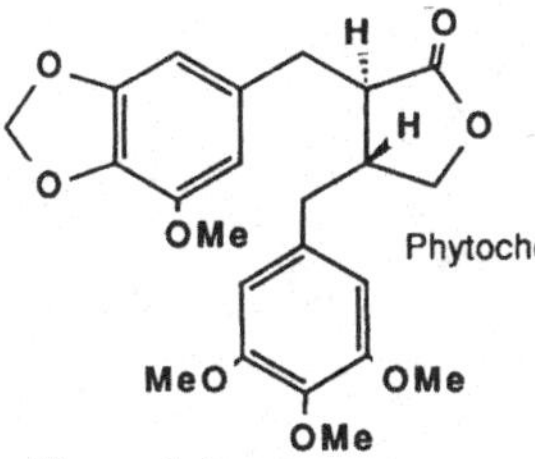

Phytochem.1986,25,487

Piper cubeba

(+)-**WIKSTROMOL** 61521-74-2

C20 H22 07

equiv [(+) - Nortrachelogenin, 2.085]

Planta Med.1984,50,264

Passerina vulgaris

(-) - **YATEIN** * 40456-50-6

2.088 C22 H24 07

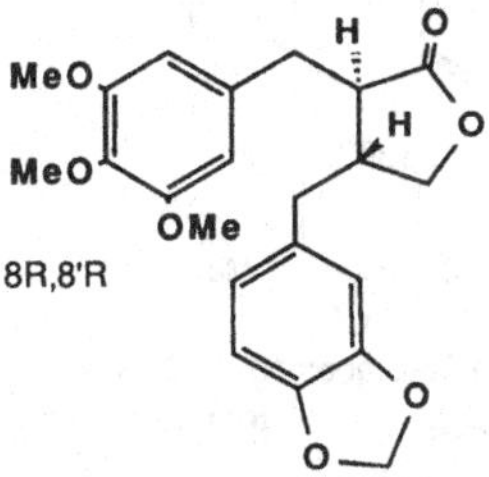

8R,8'R

Phytochem.1979,18,1495

Heterocycles,1985,23,309

Hernandia cordigera *Librocedrus yateensis*

* (-) - Yatein = (-) -7-deoxypodorhizol

by CD : Coll.Czech.Chem.Comm.1982,47,644

UNNAMED LIGNANS

LIGNANOLIDE 1 C23 H28 O7

2.091

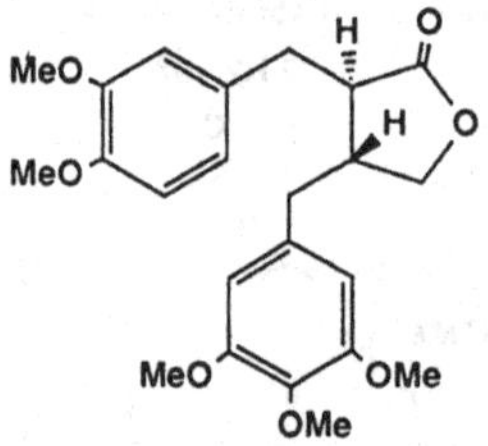

Bull.Chem.Soc.Japan,1977,50,2821

Cinnamomum camphora Sieb. (PMR, CD)

2.093

C21 H22 O6

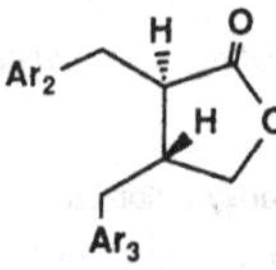

J.Nat.Prod.1988,51,44 (isol,CMR)

3,4-dimethoxy-3',4'-methylenedioxylignanolide

Phyllanthus niruri

LIGNANOLIDE 2 C22 H26 O8

2.092

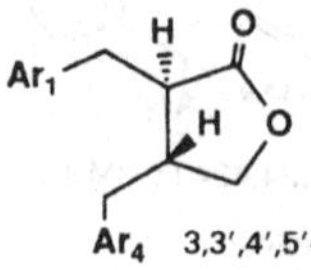

4-hydroxy-3,3',4',5'-tetramethoxylignanolide

'Nat. Prod. of Woody Plants Extraneous to the Lignocellulosic Cell Wall', Springer, 1989

Two laevorotatory lignans of this class were reported by P.B.Mc Doniel & J.R.Cole in J.Pharm.Sci.,1972,61,1992-4. There is some confusion about their absolute configuration but one appears to be (-)-yatein (or deoxy-podorhizone) and the other is (-)-O-methyl-pluviatolide (cf 2.070 & 2.093).

UNSATURATED DIARYLBUTYROLACTONES

ACANTHOTOXIN 63831-09-4
2.095 C20 H18 O6

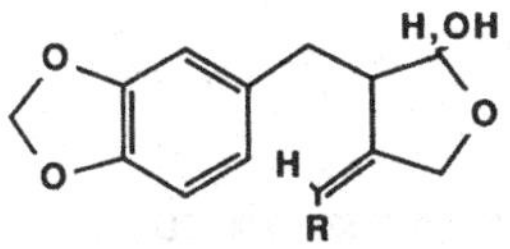

H,OH
H
R

Chem. & Ind. 1977,6,231

Zanthoxylum acanthapodium DC

(+) - CALOCEDRIN 98891-33-9
2.096 C20 H16 O7

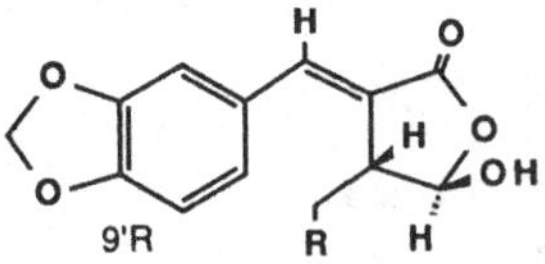

9'R R H

Phytochem.1985,24,1863

Calocedrus formosana

ANHYDROPODORHIZOL
2.098 cf NEMEROSIN 2.102

(-) - CHAEROPHYLLIN 75590-33-9
2.097 (var Kaerophyllin) C21 H20 O6

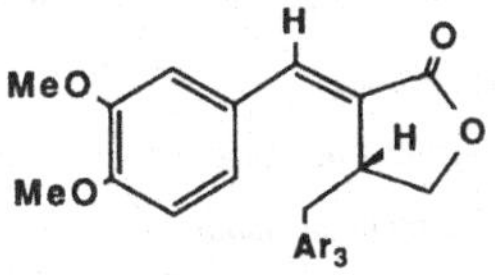

Ar3

Planta Med.1981,43,378

Chaerophyllum maculatum

(-) - HIBALACTONE 493-95-8
C20 H16 O6
2.099 (Savinin, Tawanin B)

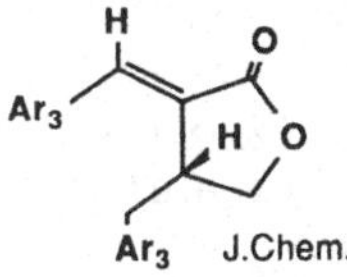

J.Amer.Chem.Soc.1955, 77,1906

J.Chem.Soc.(C),1969,693 (CD)

8'-(R)-(7E)

Chamaecyparis obtusa Sieb. & Zucc.

(+) - GADAIN 15914-41-7
2.098 C20 H16 O6

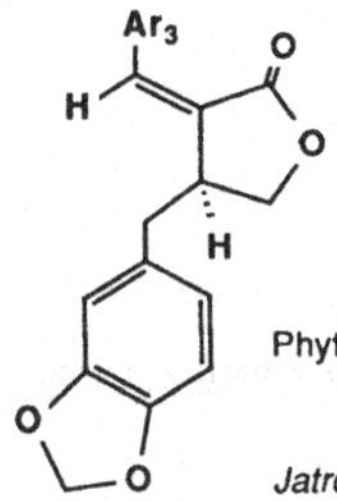

8' (S) - (7Z)*

Phytochem.1984,23,2329 (isol,PMR)

Jatropha gossypifolia

JATROPHAN 50816-74-5
2.100 C21 H20 O6

The 7 E isomer of gadain q.v.

* In the 7 E isomer the PMR spectrum shows a vinyl proton resonance in the region of the aromatic peaks - 6.6 - 7.1 d. In the 7 Z-isomer the vinyl peak lies at lower field typically 7.3 - 7.5d

7,8-DEHYDROHELIOBUPHTHALMIN
2.101 66547-92-0
$C_{22}H_{20}O_8$

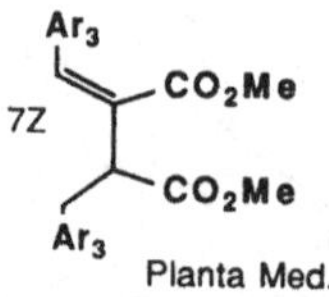

Ar_3 CO_2Me 7Z CO_2Me Ar_3

Planta Med.1986,52,18
Planta Med.1981,43,378 (MS)

Heliopsis buphthalmoides

PODOTOXIN 73191-25-0
2.104 $C_{21}H_{22}O_6$

As 2.095 but with R = Ar_2

Chem. & Ind.,1979,667 (isol,pharm)

Zanthoxylum accananthopodium

SUCHILACTONE 50816-74-5
2.105 $C_{21}H_{20}O_6$

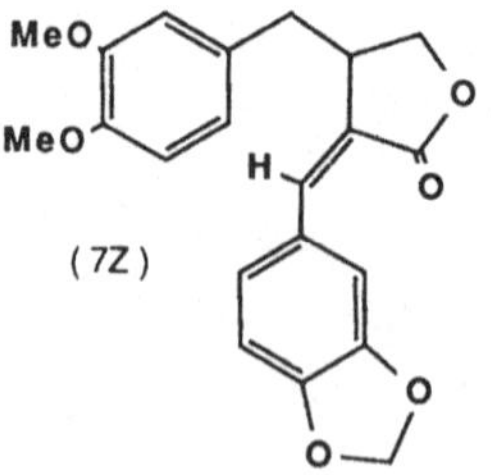

MeO MeO H O O (7Z) O O

Phytochem.1973,12,2550
J.Indian Chem.Soc.1978,55,1201 (synth)

Polygala chinensis

NEMEROSIN 17187-79-0
2.102 $C_{22}H_{22}O_7$

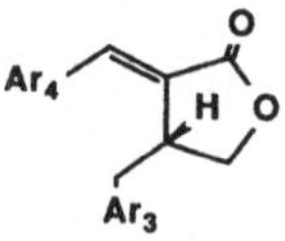

O Ar_4 H O Ar_3

Soviet J. Bio-org.Chem.1982,8,374 (CD)

Anthriscus nemerosa Bieb

PHEBALARIN 56136-37-9
2.103 $C_{26}H_{30}O_{10}$

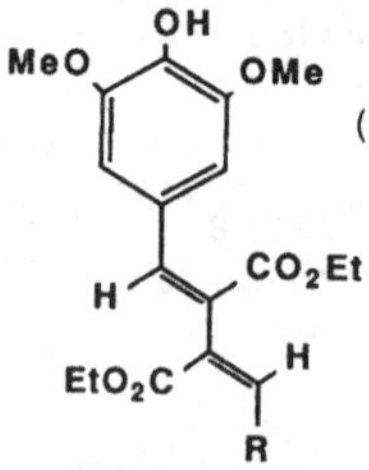

OH MeO OMe (7 Z, 7'Z) H CO_2Et EtO_2C H R

Tetrahedron,1958,2,256

Phebalium nudum Hook

SUCHILACTONE ISO 72598-33-5
$C_{21}H_{20}O_6$

2.106

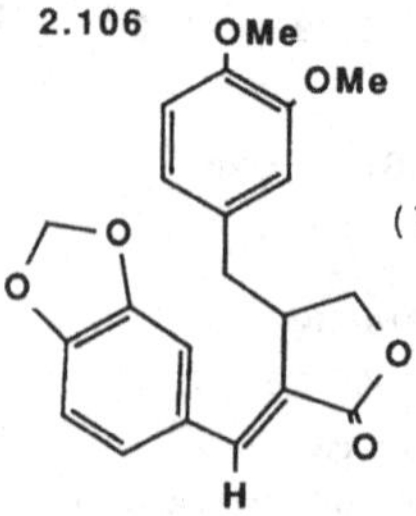

OMe OMe (7E) O O O O H

J.Indian Chem.Soc.1978,55,1201
Synthetic material

CHISULACTONE 50294-59-9
2.107 C21 H22 O6

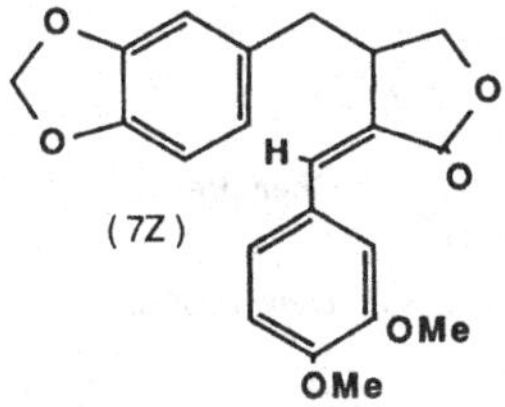

(7Z)

OMe
OMe

Phytochem.1974,13,1933
Polygala chinensis

TAIWANIN A 4650-68-4
2.108 C20 H14 O6

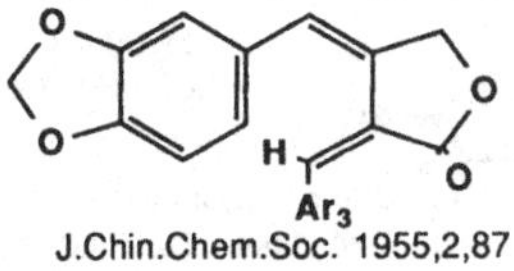

Ar_3

J.Chin.Chem.Soc. 1955,2,87

Tawania cryptomeriodes Hayata

THUJAPLICATENE 29725-59-5
2.109 C20 H20 O7

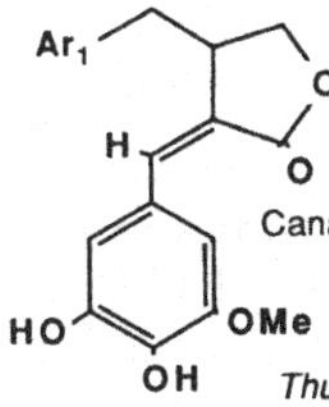

Ar_1 (7Z)

HO OMe
OH

Canad.J.Chem.1970,48,3144

Thuja plicata Donn.

FURANS

9,9' - EPOXYLIGNANS

(-) - **BRASSILIGNAN** 62624-76-4
2.110 C22 H28 O5

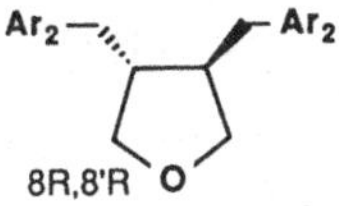

Ar_2 Ar_2

8R,8'R

Aust.J.Chem.1977,30,451
Flindersia brassii
Hartley & Hyland

BURSERAN 23284-23-3
2.111 C22 H26 O6

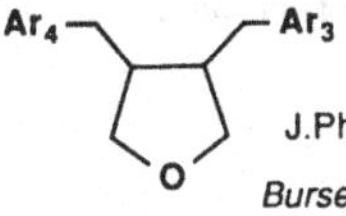

Ar_4 Ar_3

J.Pharm.Sci.1969,58,175
Bursera microphylla A.Gray

cis - isomer 73465-38-0

(-) - DIVANILLYL TETRAHYDROFURAN
2.112 34730-78-4
$C_{20}H_{24}O_5$

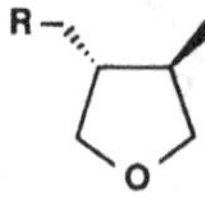

R, Ar_1, O

Chem.Nat.Compd.1971,7,802
Phytochem.1985,24,626 (synth)

Larix dahurica; L. sibirica

LIOVIL 484-39-9
2.113 $C_{20}H_{24}O_7$

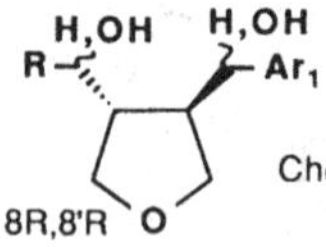

H,OH H,OH
R, Ar_1, O

8R,8'R

Chem.Ber.1957,90,2857

Picea excelsa; *Larix desidua*

2.114 $C_{20}H_{24}O_6$

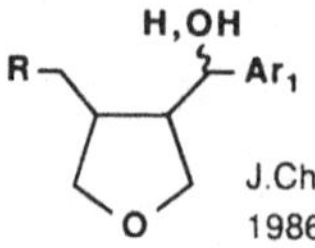

H,OH
R, Ar_1, O

J.Chem.Soc.Perkin Trans.1, 1986,1181 (pharm)

Tinospora cordifolia Miers

4,4'-Dihydroxy-3,3'-dimethoxy-9,9'-epoxylignan-7 -ol

7,9' - EPOXYLIGNANS

ACUMINATIN 41744-39-2
2.116 $C_{20}H_{22}O_6$

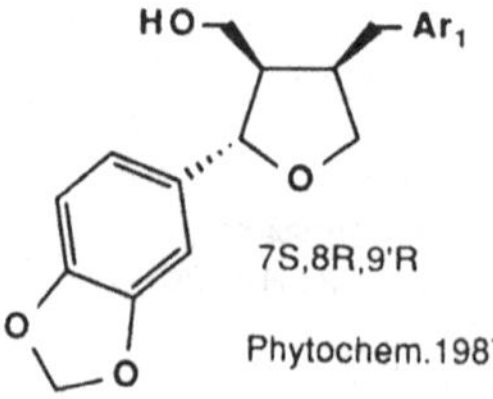

HO, Ar_1, O, O, O

7S,8R,9'R

Phytochem.1987,26,803 (NOE)

Helichrysum sp.

ARBORONE 108069-03-0
2.117 $C_{20}H_{18}O_8$

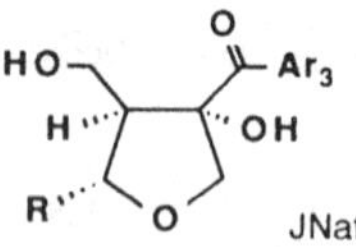

O
HO, Ar_3, H, OH, R, O

JNat.Prod.1986,49,1061 (chromatog, PMR)

Gmelina arborea

DIHYDROGMELINOL-7'-OXO
2.118 $C_{22}H_{26}O_8$

As 2.117 but R = Ar_2

(+) - LARICIRESINOL 27003-73-2
2.119 C20 H24 O6

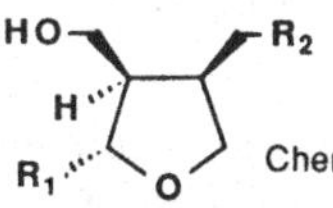

$R_1 = R_2 = Ar_1$

Chem.Rev.1955,55,987

(7S) Mokuzai Gakkaishi,1979,25,665

Larix decidua; L. Leptolepis; Picea excelsa

(-) - LARICIRESINOL 83327-19-9
2.121 C20 H24 O6

Lariciresinol - 9 - glucoside: Chem.Pharm.Bull. 1978,26,1619

Lariciresinol - 4 - glucoside: Phytochem. 1986,25,2581

Lariciresinol - bisglucoside: Agric.Biol.Chem. 1982,46,853

MAGNOLENIN C 66779-67-7
2.124 (X = H) C28 H36 O14

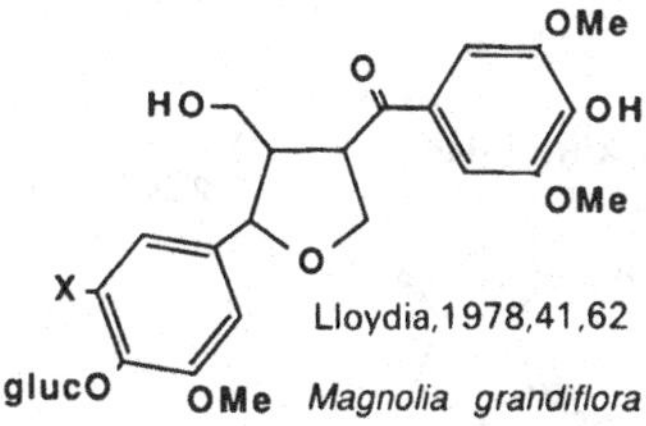

Lloydia,1978,41,62 *Magnolia grandiflora*

2.125
The related aglucone to 2.124 with X = OMe
C22 H26 O9

Planta Med.1979,37,32

Berberis chilensis Gillies et Hock

LARICIRESINOL O - DIMETHYL
2.120 C22 H28 O6

As 2.119 but $R_1 = R_2 = Ar_2$

Turraea nilitica; Monechma ciliatum

(+) - LARICIRESINOL O - METHYL
2.122 C21 H26 O6

As 2.119 but $R_1 = Ar_1$, $R_2 = Ar_2$

Int.J.Crude Drug Res.1987,25,15 (pharm)

Phytochem.1979,18,1703 (CMR)

(+) - LARICIRESINOL O-METHYL
2.123 C21 H26 O6

As 2.119 but $R_1 = Ar_2$, $R_2 = Ar_1$

J.Nat.Prod.1984,47,875

T.nilitica

MAGNOSTELLIN A 87562-70-7
2.126 $R_1 = R_2 = Ar_2$ C22 H28 O6

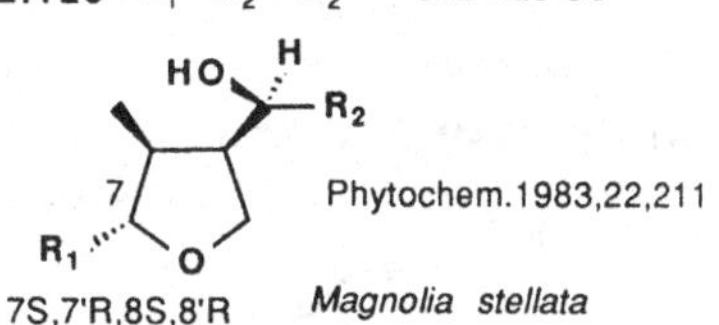

Phytochem.1983,22,211

7S,7'R,8S,8'R *Magnolia stellata*

MAGNOSTELLIN C 101205-03-2
2.127 C21 H24 O6

As 2.126 but $R_1 = Ar_3$; $R_2 = Ar_2$

Phytochem.1986,25,279

Virola elongata

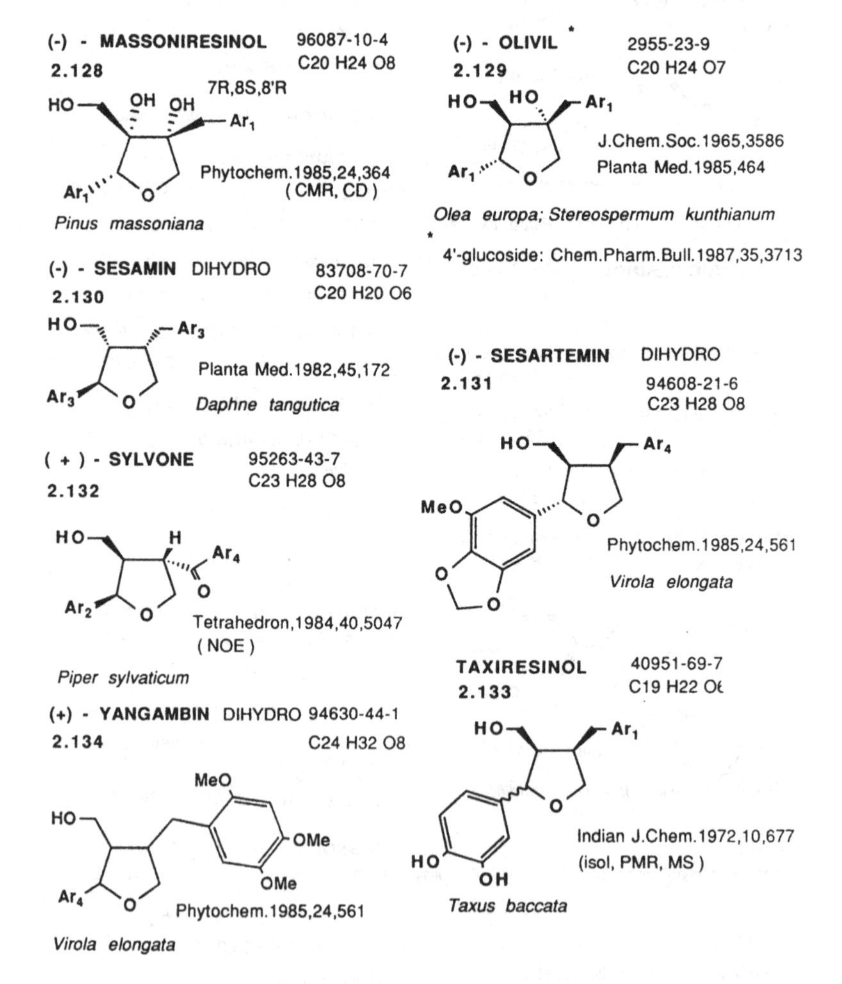
(-) - MASSONIRESINOL 96087-10-4
2.128 C20 H24 O8
7R,8S,8'R
HO
OH OH
Ar1
Ar1
O
Phytochem.1985,24,364
(CMR, CD)
Pinus massoniana
(-) - OLIVIL * 2955-23-9
2.129 C20 H24 O7
HO HO Ar1
Ar1
O
J.Chem.Soc.1965,3586
Planta Med.1985,464
Olea europa; Stereospermum kunthianum
* 4'-glucoside: Chem.Pharm.Bull.1987,35,3713
(-) - SESAMIN DIHYDRO 83708-70-7
2.130 C20 H20 O6
HO Ar3
Ar3
O
Planta Med.1982,45,172
Daphne tangutica
(-) - SESARTEMIN DIHYDRO
2.131 94608-21-6
C23 H28 O8
HO Ar4
MeO
O
O
O
Phytochem.1985,24,561
Virola elongata
(+) - SYLVONE 95263-43-7
2.132 C23 H28 O8
HO H Ar4
Ar2
O
O
Tetrahedron,1984,40,5047
(NOE)
Piper sylvaticum
TAXIRESINOL 40951-69-7
2.133 C19 H22 O6
HO Ar1
O
HO OH
Indian J.Chem.1972,10,677
(isol, PMR, MS)
Taxus baccata
(+) - YANGAMBIN DIHYDRO 94630-44-1
2.134 C24 H32 O8
MeO
HO
OMe
OMe
Ar4
O
Phytochem.1985,24,561
Virola elongata

7,8' - EPOXYLIGNAN

MAGNOSALICIN *

2.135

93376-03-5
C24 H32 O7

MeO OMe OMe 9 R 7 O 8'

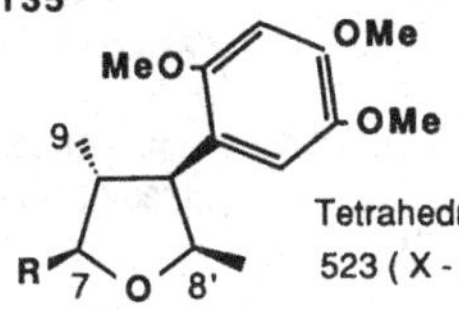

Tetrahedron 1986,42, 523 (X - ray)

7R,7'S,8R,8'R

Magnolia salicifolia

*Anti - allergic agent

UNNAMED LIGNANS

LIGNAN A

2.136

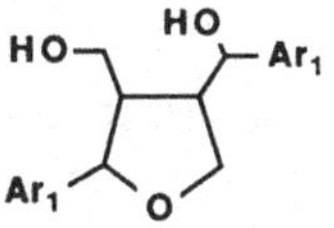

HO HO Ar1 Ar1 O

4,4'-dihydroxy-3,3'-dimethoxy-7,9'-epoxy-lignan-7',9-diol

LIGNAN B

2.137

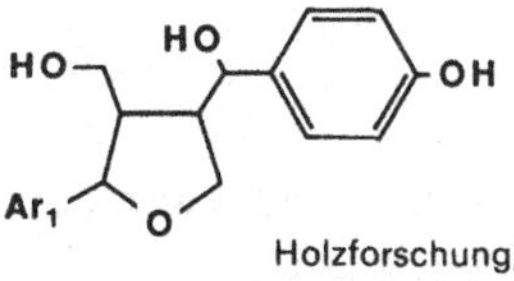

HO HO OH Ar1 O

Holzforschung,1976,30, 79 (GLC)

Picea abies (L). H.Karst

7,7' - EPOXYLIGNANS

(+) - ARISTOLIGNIN

2.138

110268-34-3
C21 H26 O5

7R,7'S,8S,8'S

Ar1 7' 7 O Ar2

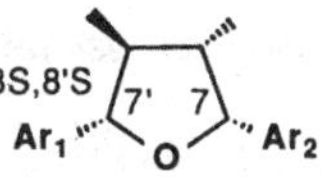

Phytochem.1987,26,1509 (NOE)

Aristolochia chilensis

meso - **CALOPHYLLIN**
2.140 C22 H28 O4

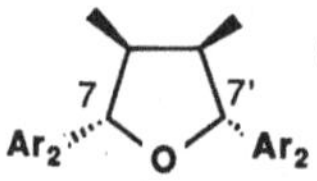

7 7'
Ar2 O Ar2

Chem.Abs.1988,108,183594

7S,7'R,8S,8'R *Virola calophylla*

(+) - **CALOPIPTIN** 19950-67-5
2.141 C21 H24 O5

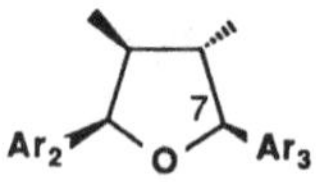

7
Ar2 O Ar3

7S Aust.J.Chem.1968,21,2095 (MS)

Magnolia acuminata; M.liliflora
Piptocalyx moorei Oliv.

(+) - **CHICANINE** 78918-28-5
2.142 C20 H22 O5

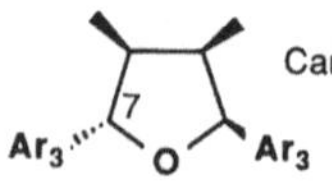

7
Ar3 O Ar3

Canad.J.Chem.1981,59,1680 (absol. stereo.)

7S,7'S,8S,8'R
Schizandra sp.

(+) -**AUSTROBAILIGNAN 7**
2.139 55890-25-0 C20 H22 O5

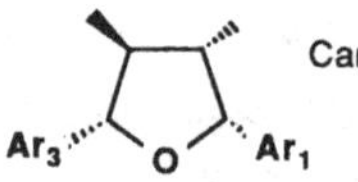

Ar3 O Ar1

Canad.J.Chem.1981,59, 1680 (CD)

Aust.J.Chem.1968,21,2095 (MS)
Schizandra sp.
Austrobaileya scandens

EN SHI ZHI SHU 82869-20-3
2.143 C20 H22 O6

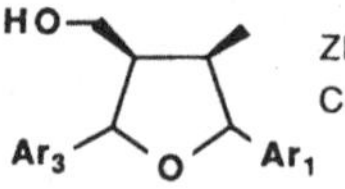

HO
Ar3 O Ar1

Zhongcaoyao,19822,13,22
Chem.Abs.1987,97,107029

Schizandra henryi Clarke

EPIGALBACIN 84709-25-1
2.144 C20 H20 O5

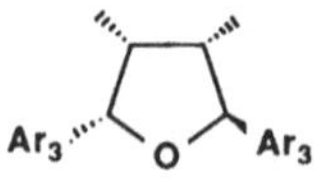

Ar3 O Ar3

Chem.Abs.1984,101,3913

7S,7'S,8R,8'S

Schizandra spenanthera Rehd. & Wils.

The FRAGRANSIN group

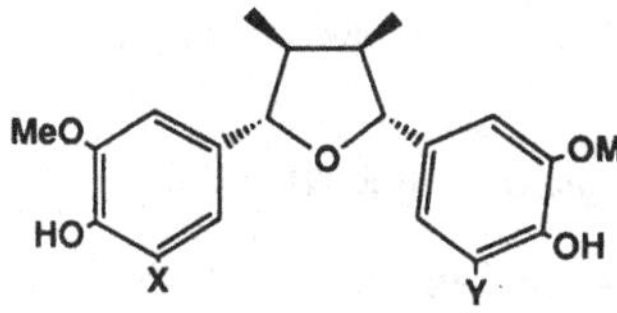

MeO HO X O OMe OH Y (1)

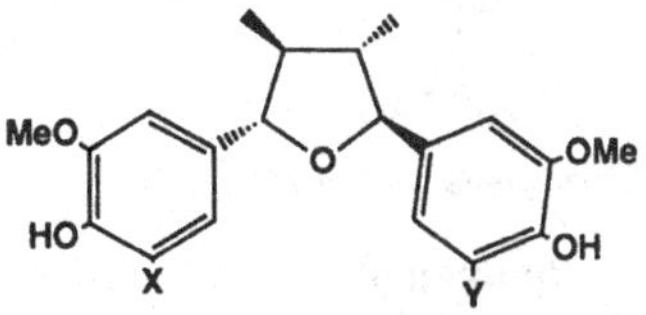

MeO HO X O OMe OH Y (3)

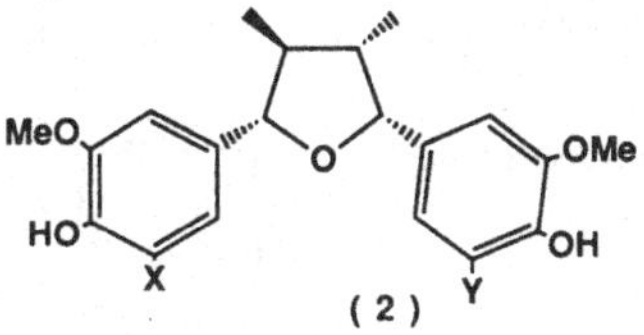

MeO HO X O OMe OH Y (2)

Structures (1)

2.145 FRAGRANSIN B_1
X = Y = OMe

2.146 FRAGRANSIN C_1
X = OMe, Y = H

Myristica fragrans

Structures (2)

2.147 FRAGRANSIN A_2
X =Y = H

2.148 FRAGRANSIN B_2
X = Y= OMe

Chem.Pharm.Bull.1987,35,3315
(chromatog. NOE)

Structures (3)

2.149 FRAGRANSIN B_3
X = Y= 0Me

2.150 FRAGRANSIN 3a
X = OMe; Y = H

2.151 FRAGRANSIN 3_B
X = H; Y = OMe

2.151A FRAGRANSIN C_2
X = OMe; Y = H

(-) - GALBACIN * 528-64-3
2.152 C20 H20 O5

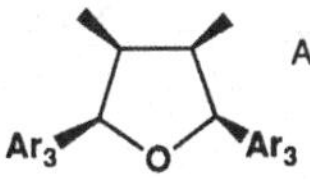

Aust.J.Chem.1954,7,104 (isol)

Ar_3 O Ar_3

Himantandra baccata Bail.

(+) -GALBACIN 61891-31-4
2.153

* Active against Mycobacterium tuberculosis

(-) - GALBELGIN 10569-12-7
2.154 C22 H28 O5

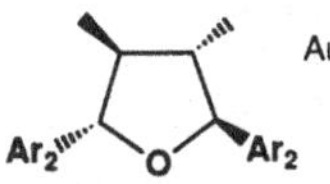

Aust.J.Chem.1954,7,104

Ar_2 O Ar_2

Himantandra belgaveana F.Muell

(+) - GALBELGIN 61949-24-4
2.155

meso - **GALGRAVIN** 528-63-2
2.156 C22 H28 O5

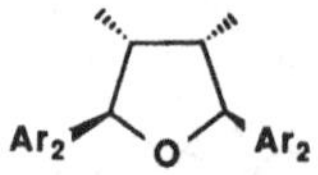

J.Chem.Soc.1958,4471

Ar2 O Ar2

Himantandra belgraveana

(+) - HENRICINE 107783-46-0
2.158 C22 H26 O6

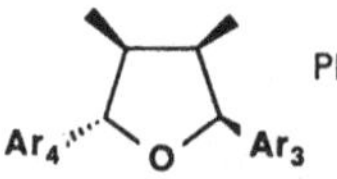

Planta Med.1986,52,493

Ar4 O Ar3

Schizandra henryi

(-)- MACHILUSIN 61989-57-9
2.164 C21 H24 O5

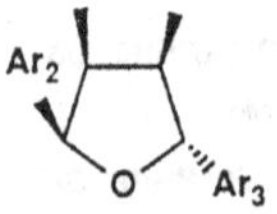

Ar2 O Ar3

Bull.Chem.Soc.Japan,1976, 49,3564 (isol. PMR)

Machilus japonica

NEOOLIVIL (as the tetra-acetate)
2.166 77790-55-7
C20 H24 O7

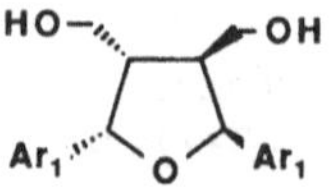

HO OH Ar1 O Ar1

Phytochem.1981,20,181

Thymus longiflorus

(-) - GRANDISIN 53250-50-3
2.157 C24 H32 O7

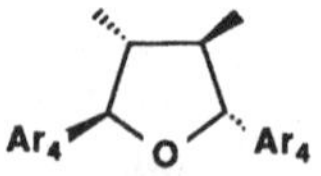

Phytochem.1974,13,1232

Ar4 O Ar4

Litsea grandis (Wall.) Hook

The **MACHILINS**

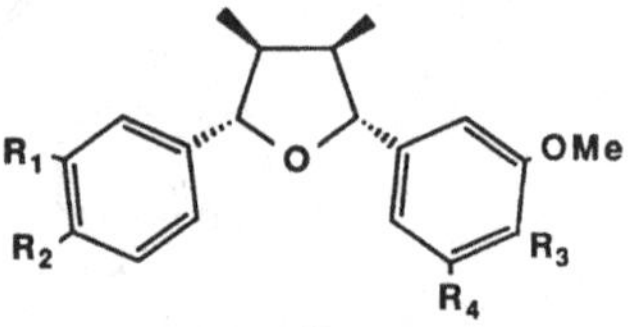

R1 R2 O OMe R3 R4

	R_1	R_2	R_3	R_4
MACHILIN F 2.159	$O-CH_2-O$		OH	H
MACHILIN G 2.160	$O-CH_2-O$		OMe	H
MACHILIN H 2.161	OMe	OH	OH	OMe

MACHILIN I
2.162

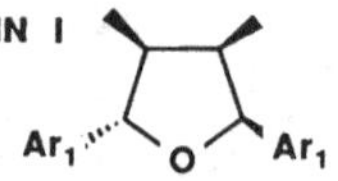

Ar1 O Ar1

with **NECTANDRIN A 2.163** 74683-15-1
as machilin general structure C21 H26 O5
R_2 = OH, R_4 = H, R_1 = OMe, R_3 = OMe
Phytochem.1988,27,634
Machilus thunbergii

NECTANDRIN B 2.165 74683-16-2
as machilin general structure C20 H24 O5
R_2 = OH, R_4 = H, R_1 = OMe, R_3 = OH
Chem.Pharm.Bull. 1987,35,3315
Myristica fragrans

TETRAHYDROFUROGUAIACIN B
2.167 C20 H24 O5

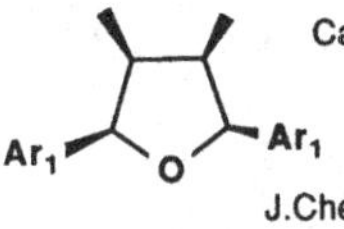

Ar1 O Ar1

Canad.J.Chem.1979,57,441 (CMR)

J.Chem.Soc.1964,4011(isol.)

(Isol. as O-dimethyl)

(+) -VERAGUENSIN 19950-55-1
2.168 C22 H28 O5

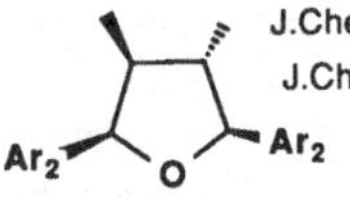

Ar2 O Ar2

J.Chem.Soc.1962,1459 (isol)
J.Chem.Soc.Perkin Trans.1, 1978,1147 (synth)

Ocotea veraguensis Mez
Magnolia liliflora

VERRUCOSIN 83198-63-4
2.169 (cf fragransin B_3) C20 H24 O5

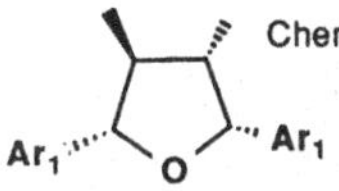

Ar1 O Ar1

Chem.Pharm.Bull.1987,35,3315

Myristica fragrans

ZUHONIN A 79120-58-4
2.170 C20 H20 O5
R = 8B-Ar_3; 8'a-Ar_3

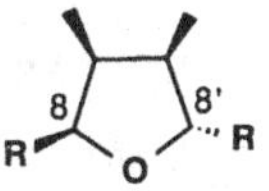

8 8' R O R

UNNAMED LIGNANS C18 H16 O3
2.173 (cf epigalbacin 2.144)

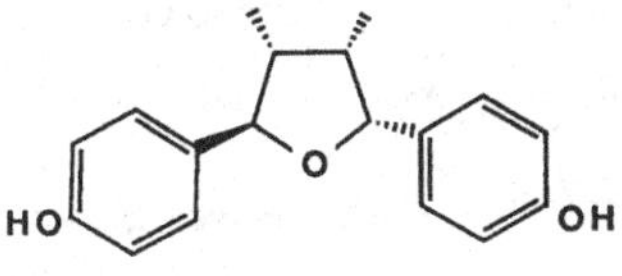

HO O OH

Nat.Prod.Rept.1987,500
Krameria triandra Ruiz

(7R,7'R,8R,8'S)-7,7'-EPOXYLIGNAN-4,4'-DIOL

ZUHONIN B 79120-59-5
2.171
As 2.170 but with
R = 8α-Ar_3; 8'α-Ar_3

ZUHONIN C 79120-60-8
2.172
As 2.170 but with
R = 8B-Ar_3; 8'a-Ar_2

Chem.Abs.1981,95, 129348 (isol)

Phytochem.1987,26,1509

Machilus zuihoensis

OTHER 7,7'- TETRAHYDROFURANS

C18 H16 O3

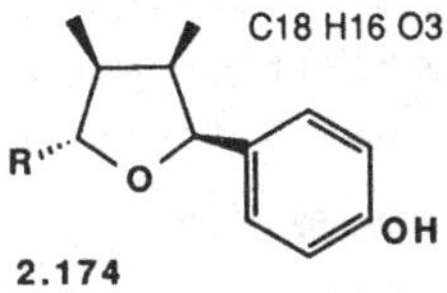

R O OH

2.174
7S,7'S,8S,8'R

C19 H18 O4

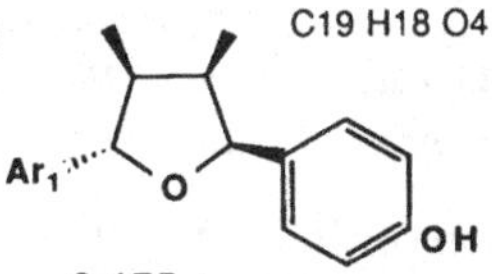

Ar1 O OH

2.175
7S,7'S,8S,8'R

C19 H18 O4

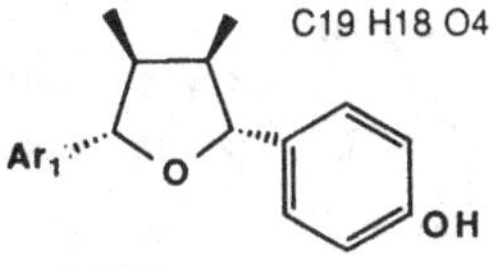

Ar1 O OH

2.176
7S,7'R,8S,8'R

Phytochem.1987,26,1159 (isol, CD, CMR)
Krameria cystoides with CHICANINE 2.142

FURANOID LIGNANS

FUROGUAIACIN 10035-27-5
2.177 $C_{20}H_{20}O_5$

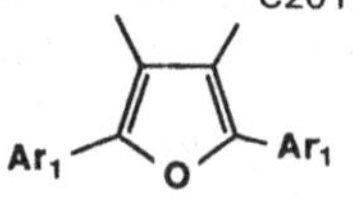

Ar_1 — furan — Ar_1

J.Chem.Soc.1964,4011 (isol,UV)
from *Guaiacum officinale*
as their methyl ethers

FUROGUAIACIN METHYL $C_{21}H_{22}O_5$
2.178

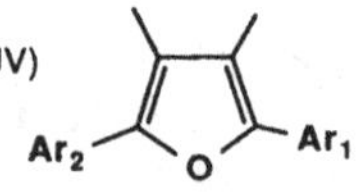

Ar_2 — furan — Ar_1

FUROFURANS
(7,9',7',9-BISEPOXYLIGNANS)

COMPOUNDS WITHOUT SIDE-CHAIN SUBSTITUTION

ASARININ see EPISESAMIN

ASCHANTIN 13060-15-6
2.179A $C_{22}H_{24}O_7$

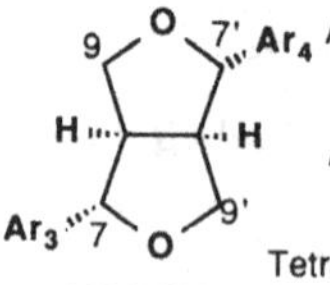

9, O, 7', Ar_4, H, H, Ar_3, 7, O, 9'

Arch.Pharm.1961,294,699 (isol)
Arch.Pharm.1969,302,940 (PMR)
Tetrahedron,1981,37,1181 (CD)

7S,7'S8R,8'R
Piper guineense

The BUDDLENOLS

R_2, OH, OMe, O, H, H, MeO, O, O, OMe, OH, OH, R_1, OH

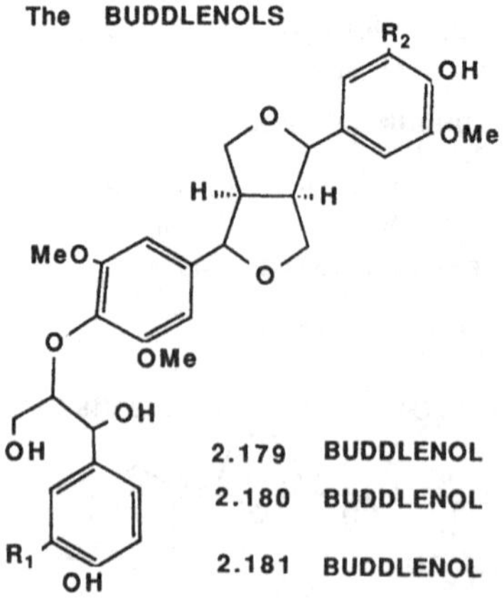

		R_1	R_2	Reg. no.	
2.179	BUDDLENOL C	H	OMe	97465-73-1	$C_{32}H_{38}O_{12}$
2.180	BUDDLENOL D	OMe	OMe	97465-75-3	$C_{33}H_{40}O_{13}$
2.181	BUDDLENOL E	H	H	97399-82-1	$C_{31}H_{36}O_{11}$
2.182	BUDDLENOL F	OMe	H	97399-83-2	$C_{32}H_{38}O_{12}$

Stereochemistry unknown
Buddleja davidii

Phytochem.1985,24,819 (isol, MS)
J.Ethnopharmacol.1984,11,293 (pharm)

(+) - DIAEUDESMIN * 16499-02-8
2.183 C22 H26 O6

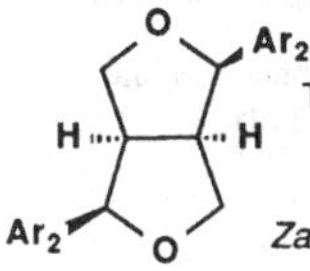

Ar2 O H H Ar2 O

Tetrahedron,1976,32,2783 (CMR)

Zanthoxylum acanthapodium

DIASESAMIN (EPIASARININ)
2.184 551-30-4
as 2.183 both Ar = Ar_3 C20 H18 O6

Tetrahedron,1976,32,2783 (CMR)
J.Chem.Soc.1967,2229 (isol)

Piper peepuloides

DIASESARTEMIN 77449-33-3
2.185 C23 H26 O8

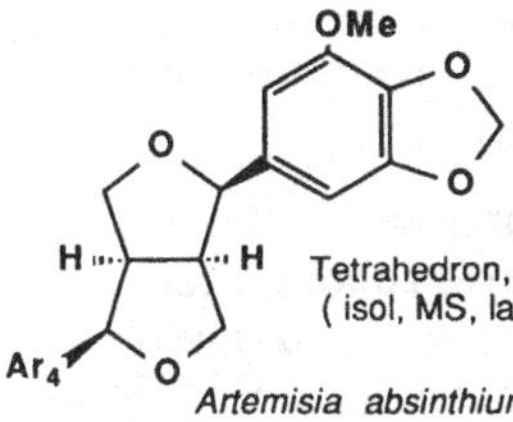

OMe O O O H H Ar_4 O

Tetrahedron,1980,36,3551 (isol, MS, lanthanide shift)

Artemisia absinthium

(+) - DIAYANGAMBIN 21453-68-9
2.186 C24 H30 O8
(LIRIORESINOL C DIMETHYL)
As 2.183 both Ar = Ar_4

Tetrahedron,1980,36,3551
J.Chem.Soc.1968,3042 (isol)
Phytochem.1973,12,1799 (MS)

Eremophilia glabra

EPIASCHANTIN 41689-50-3
2.187 C22 H24 O7

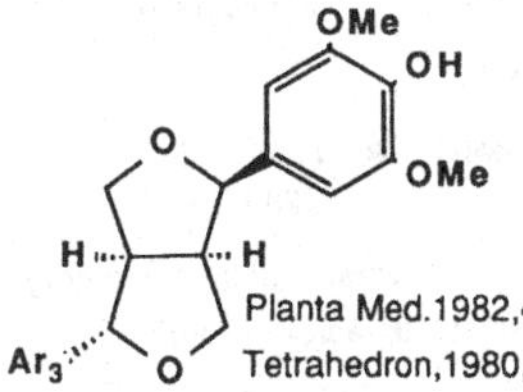

OMe OH OMe O H H Ar_3 O

Planta Med.1982,46,119 (isol)
Tetrahedron,1980,36,3551 (PMR)

Hernandia peltata

(+)- EPIEUDESMIN ** 60102-89-8
2.188 C22 H26 O6

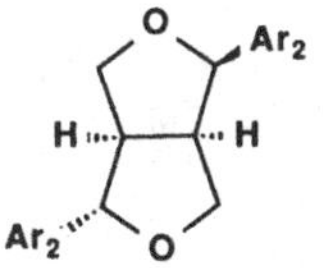

O Ar_2 H H Ar_2 O

Phytochem.1984,23,2021 (CD, ORD); p2647 (CMR)
Phytochem.1982,21,673

Virola sebifera *Magnolia kobus*

(-) - EPIEUDESMIN 4375-03-5
2.189 C22 H26 O6

* Diaxially substituted compounds may be isomerised during chromatography:
J.Chem.Soc.(C),1968,3042

** Sodium D line correlations are useful when CD results are not available:

eudesmin +60 epi- +120 dia +300

(+) - EPIEXCELSIN 51020-09-8
2.190 C22 H22 O8

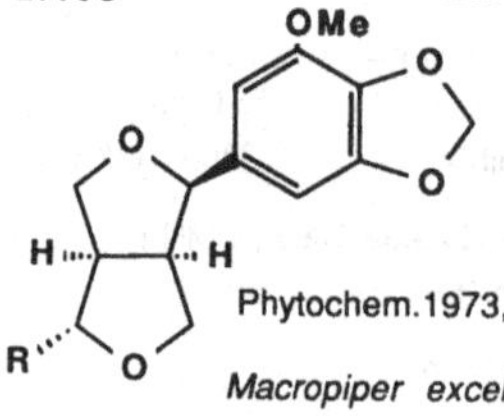

Phytochem.1973,12,1799 (PMR)
MS

Macropiper excelsum

(+) - EPIMAGNOLIN 41689-51-4
2.192 C22 H26 O10

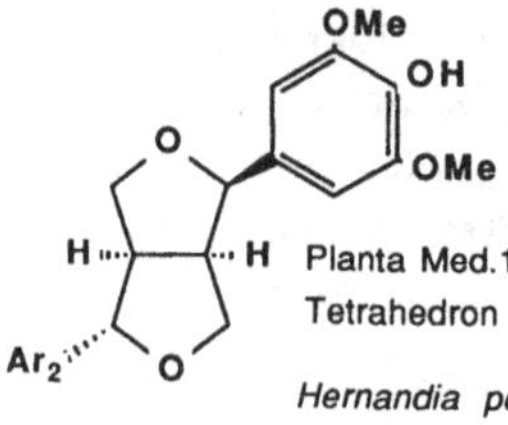

Planta Med.1982,46,119
Tetrahedron 1980,36,3558

Hernandia peltata

(+)- EPISESAMIN 133-03-9
2.194 (cf ASARININ) C20 H18 O6

as 2.188 both aryl = Ar_3

Tetrahedron,1976,32,2783 (isol, CMR)

Asarum sp.
Justicia simplex D. Don

(-)- EPISESAMIN 133-04-0
2.195 C20 H18 O6
Asarum sieboldi

EPIFARGESIN *
2.191

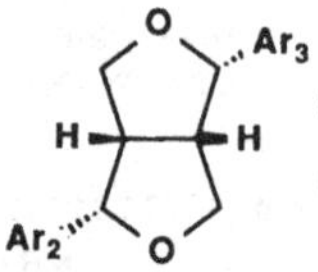

Chem.Ber.1977,110,301

Bull.Chem.Soc. Japan
1970,3631

Magnolia fargesii

EPIPINORESINOL 24404-50-0
2.193 C26 H22 O6
as 2.188 both aryl = Ar_1

Chem.Ber.1961,94,851 (isol)
Chem.Pharm Bull.1984,32,4653 (CMR)

Picea excelsa

(-)- EPIPINORESINOL 10061-38-8
see SYMPLOCOSIGENOL

EPIPINORESINOL O - METHYL
193A 487-39-8
C21 H24 O6

EPISESARTEMIN A 77449-31-1
2.196 C23 H26 O8

as 2.190 R = Ar_4

EPISESARTEMIN B 77449-32-2
2.197 C23 H26 O8

as 2.185 but with inversion of the 3-methoxy-4,5-methylenedioxy group

Tetrahedron,1980,36,3551 (lanthanide shift)
Phytochem.1985,24,561
Artemisia absinthium
Virola elongata

* Absolute stereochemistry uncertain

(+) - EPISYRINGARESINOL
2.198 (LIRIORESINOL A;SYMPLICOSIGENOL) 21453-71-4
C22 H26 O8

As 2.192 but both aryl
4-hydroxy-3,5-dimethoxyphenyl

J.Org.Chem.1958,23,179 (isol)
J.Chem.Soc.(C),1968,3042 (PMR)

Liriodendron tulipfera L.

(-) - EPISYRINGARESINOL 6216-82-6
Nandina domestica

(+) - EXCELSIN 50886-57-2
2.201 C22 H22 O8

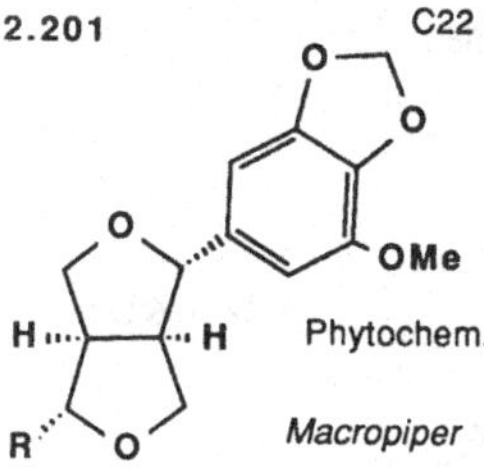

Phytochem.1973,12,1799

Macropiper excelsum

(+) - EXCELSIN DEMETHOXY
2.202 50696-38-3
C21 H20 O7

as 2.201 but R = Ar_3

(+) - HORSFIELDIN 85922-40-3
2.204 C20 H20 O6

as 2.203 but with Ar_1 for Ar_2
the stereochemistry needs confirmation
Phytochem.1982,21,2719

Horsfieldia iryaghedhi

(+) - EPIYANGAMBIN 24192-61-1
2.199 (LIRIORESINOL A O - DIMETHYL) C24 H30 O8

As 2.192 but both aryl = Ar_4

Tetrahedron,1980,36,3551 (PMR)
J.Pharm.Sci.1973,62,1561 (pharm)

Virola peruviana

(+) - EUDESMIN 29106-36-3
2.200 C22 H26 O6

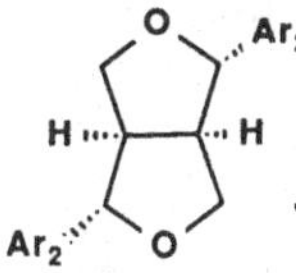

C.R.Acad.Sci.Paris (C),1968, 266,1284 (X-ray)

J.Chem.Soc.Perkin Tr.1,1982, 175 (synth)

Acta Chem.Scand.1956,10, 445 (isol)

Araucaria angustifolia

(-) - form from *Eucalyptus hemiphloia*
Amer.J.Pharm,1896,679 526-06-7

FAGARAL see SESAMIN 2.223

(+) - FARGESIN 68296-27-5
2.203 (PLUVIATILOL O- METHYL) C21 H22 O6

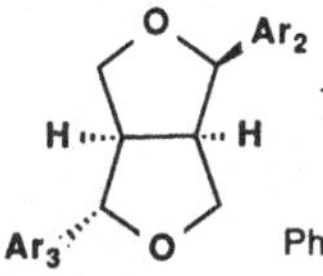

Tetrahedron,1976,32,2783 (CMR)

Phytochem.1972,11,2289 (isol, pharm, MS)

Magnolia fargesii

JUSTISOLIN 74061-79-3
2.205 C20 H18 O7

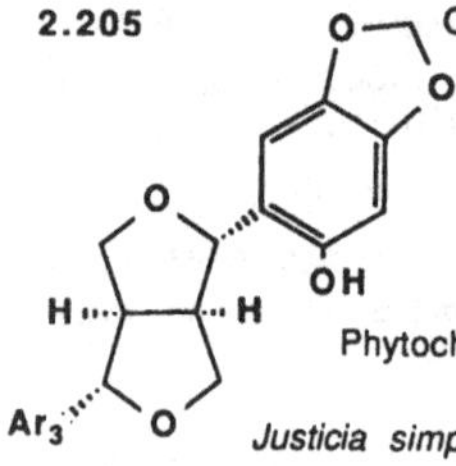

H, H, OH, Ar_3, O

Phytochem.1980,19,332

Justicia simplex

(+) - KOBUSIN 36150-23-9
2.206 C21 H22 O6

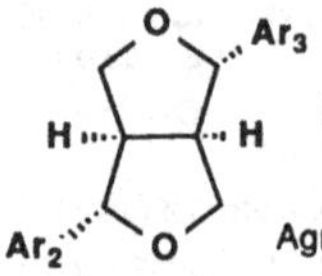

Ar_3, H, H, Ar_2, O

Phytochem.1983,22,211
also 21,673 (PMR)
Agric.Biol.Chem.1975,39,
833 (pharm, PMR)

Magnolia stellata; M.kobus

(-) - LIGBALLINOL 55332-75-7
2.207 C18 H18 O4

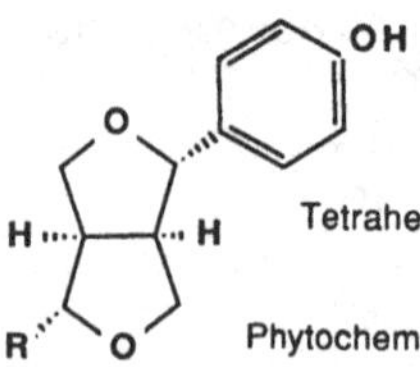

OH, H, H, R, O

Tetrahedron,1974,30,3309
(isol)
Phytochem.1983,22,1257
(PMR, MS)

Absolute configuration doubtful cf 2.279
Ecbalium elaterum

LIRIORESINOL A 21453-71-4
(see EPISYRINGARESINOL) C22 H26 O8
2.198
J.Org.Chem.1958,23,179

(-) - form 6216-82-6
2.208

(+) - LIRIORESINOL B 21453-69-0
2.209 C22 H26 O8
(SYRINGARESINOL)

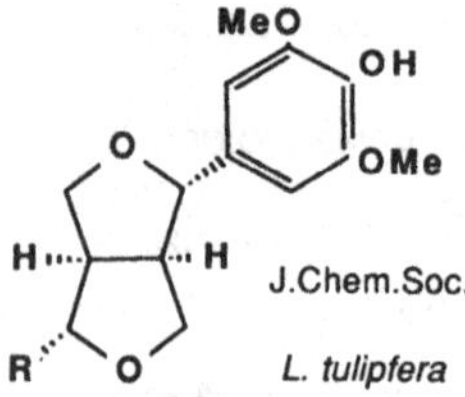

MeO, OH, OMe, H, H, R, O

J.Chem.Soc.(C),1968,3043

L. tulipfera

(-) - form 6216-81-5
2.210

(+) - MAGNOLIN 31008-18-1
2.213 C23 H28 O7

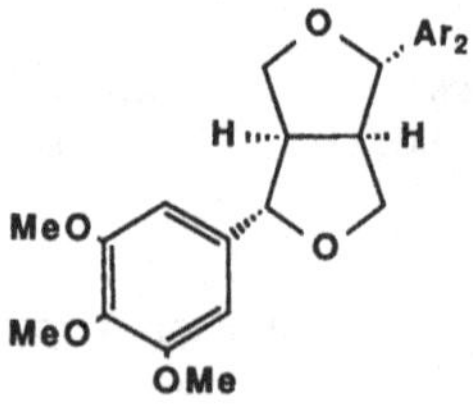

Ar_2, H, H, MeO, MeO, OMe, O

Phytochem.1987,26,319
Licaria armeniaca

(+) - LIRIORESINOL C 551-29-1
2.211 C22 H26 O8
(O - DIMETHYL equiv DIAYANGAMBIN)
as 2.209 but both aryl ß- (axial)
Chem.Ber.1961,94,851 (abs. config.)
Macropiper excelsum

2.212 (-) - form 19046-36-7

(+) - MEDIORESINOL 40957-99-1
2.214 C21 H24 O7

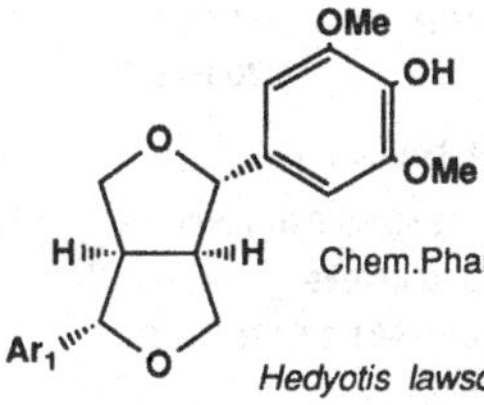

OMe, OH, OMe, O, H, H, Ar_1, O

Chem.Pharm.Bull.1985,33,1444

Hedyotis lawsoniae

(-) - form from *Dirca occidentalis* A.Gray
2.215 J.Pharm.Sci.1983,72,1285 (pharm)

Diglucoside from *Eucommia ulmoides*
Chem.Pharm.Bull.1983,31,2993

(+) - PIPERITOL (cf SIMPLEXOSIDE)
2.219 52151-92-5
C20 H20 O6

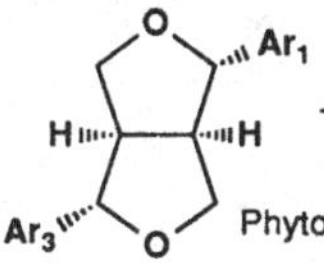

O, Ar_1, H, H, Ar_3, O

Tetrahedron Lett.1976,2221 (isol)
Phytochem.1987,26,265 (isol)

Aptosimum spinescens
Nectandra turbacensis

(-) - form 54983-96-9
2.220 Chem.Pharm.Bull.1974,22,2650 (PMR)

Xanthoxylum piperitum

(+) - PIPERITOL O - METHYL C21 H22 O6
2.221 (equiv. demethoxyaschantin)

Arch.Pharm.1972,305,33 (isol)
J.Chem.Soc.Perkin Trans.1,1985,587 (synth)

Sassafras sp.

(+) - PHILLYGENOL (PHILLYGENIN, FORSYTHIGENOL)
2.216
487-39-8
C21 H24 O6

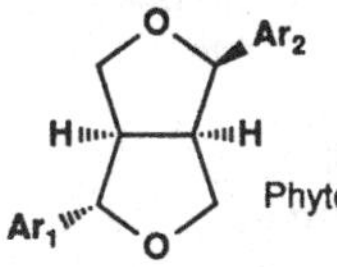

O, Ar_2, H, H, Ar_1, O

Phytochem.1984,23,1635 (isol)

Forsythia sp. cf Lantibeside 2.270
Phyllyrea sp.

(+) - PINORESINOL 487-36-5
2.217 C20 H22 O6

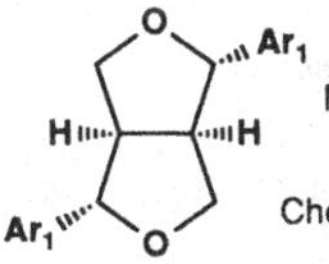

O, Ar_1, H, H, Ar_1, O

Monatsch.1894,15,505 (isol)
Chem.Ber.1953,86,1157 (synth)

Chem.Pharm.Bull.1984,32,1612, 4482 (pharm)
J.Chem.Soc.Perkin 1,1984,1159 (PMR)

Pinus laricio

(+) - PINORESINOL O - DIMETHYL
see EUDESMIN 2.200
J.Indian Chem.Soc.1978,55,1204
Machilus glaucescens

(-) - PINORESINOL 81446-29-9
2.218 Chem.Pharm.Bull.1982,30,1525

Balanophora japonica

(+) - PIPERITOL 3-METHOXY
2.219A Phytochem.1987,26,265

(-) - PLUVIATILOL 28115-67-5

2.222 C20 H20 O6

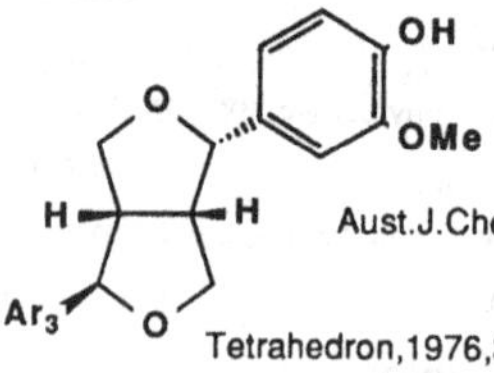

OH

OMe

H H

Ar_3 O

Aust.J.Chem.1970,23,133 (isol)

Tetrahedron,1976,32,2783 (PMR)

J.Chem.Soc.Perkin Trans.1,1985,587 (synth)

Zanthoxylum pluviatile

3 - methylbut-2-enyl ether

Yaoxue Xuebao 1984,19,268; Chem.Abs. 103,19810

X. podocarpum Hemsl.

PLUVIATILOL O - METHYL

equiv **FARGESIN 2.203**

SESARTEMIN 77394-27-5

2.226 C22 H22 O8

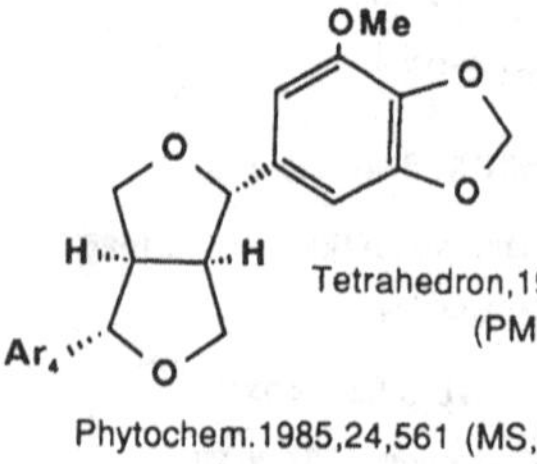

OMe

H H

Ar_4 O

Tetrahedron,1980,36,3551 (PMR,lanthanide)

Phytochem.1985,24,561 (MS,lanthanide shift)

J.Ethnopharmacol.1984,12,75 (pharm)

Virola elongata *Artemisia absinthum*

(+) - SESAMIN 607-80-7

2.223 C20 H18 O6

As 2.219 but both aryl = $Ar_{3\ (equat)}$

Chem.Rev.1955,55,996 (isol)

Curr.Sci.1978,48,949 (pharm)

Tetrahedron 1981,37,1181 (CD)

Sesamum sp.

(-) - form 13079-95-3

2.224 C20 H18 O6

Chem.Abs.1985,103,19810

Zanthoxylum podocarpum Hemsl.

(+) - SESANGOLIN 100268-57-3

2.225 C21 H20 O7

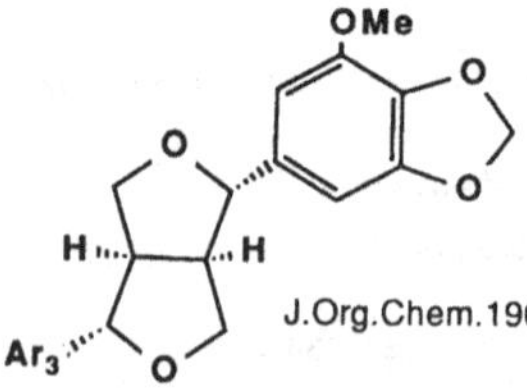

OMe

H H

Ar_3 O

J.Org.Chem.1962,27,3232 (isol)

Sesamum angolense

SPINESCEN 36150-23-9
2.227 C21 H22 O6

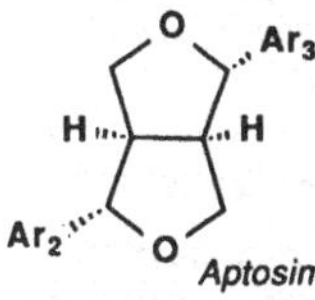

Tetrahedron Lett.1976,2221 (IR, MS)

Aptosimum spinescens Thornby

(+) - SYLVATESMIN 487-39-8
2.228 C21 H24 O6

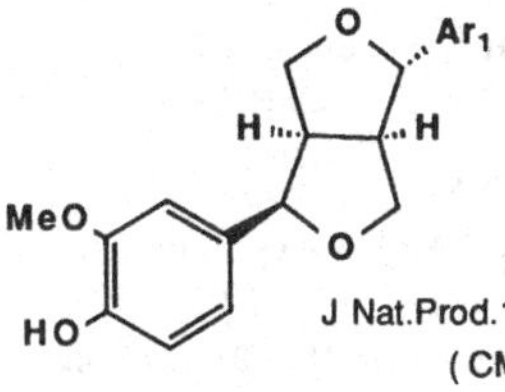

J Nat.Prod.1982,45,672 (CMR, SFORD)

Piper sylvaticum Roxb.

SYMPLOCOSIGENOL 10061-38-8
equiv EPIPINORESINOL **2.193**
C20 H22 O6

(+) - YANGAMBIN 13060-14-5
2.232 C24 H30 O8

SYRINGARESINOL DI - O - METHYL

As 2.227 but both Ar = Ar_4

Aust.J.Chem.1961,14,175 (isol)
J.Ethnopharmacol.1984,12,75 (pharm)

Eremophila glabra

(+) - SYRINGARESINOL 21453-69-0
2.229 C22 H26 O8

As 2.227 but both aryl =

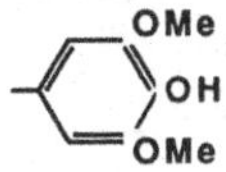

Chem.Pharm.Bull.1985,33,1444 (CMR)
J.Pharm.Sci.1983,72,1285 (sepn, pharm)
Aust.J.Chem.1987,40,1913 (X-ray)
J.Chem.Soc. Perkin 2,1976,341 (X-ray)

Hedyotis lawsoniae

(-) - form 6216-81-5
2.230 C22 H26 O8
Indian J.Chem.1981,20B,85

Guazuma tomentosum Kunth.

(-) - XANTHOXYLOL 54983-95-8
2.231 C20 H20 O6

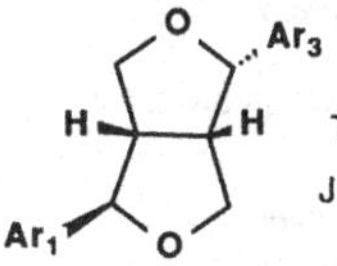

Tetrahedron,1976,32,2783
J.Chem.Soc.Perkin 1,1985, 587 (synth)
Chem.Pharm.Bull.1974,22,2650 (isol)

Xanthoxylum acanthapodium

2.233 C18 H18 O4

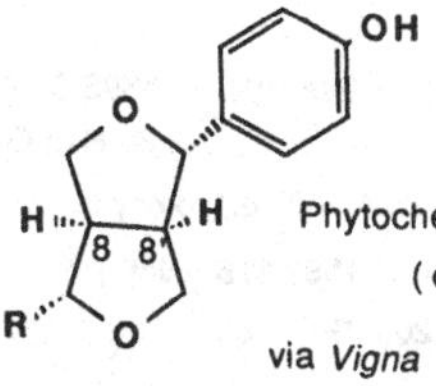

Phytochem.1983,22,1257 (chromatog, PMR)

via *Vigna angularis*

(+)-(7S,8R,7'S,8'R)-4,4'-dihydroxy-7,9',7',9-diepoxylignane

OXYFURO FURANS

(+) - ACETOXYPINORESINOL
2.235 81426-14-4
C22 H24 O8

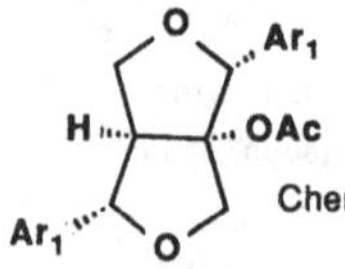

Ar1, H, OAc, Ar1, O

Chem.Pharm.Bull.1979,27,2868

Olea europa

(+) - APTOSIMOL 61254-18-0
2.236 C20 H18 O7

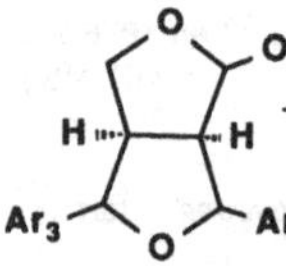

OH, H, H, Ar3, Ar3, O

Tetrahedron Lett.1976,2221

This structure is probably incorrect see-

Tetrahedron,1979,35,861

Aptosimum spinescens

(+) - ARBOREOL 58868-71-2
2.237 C20 H18 O8

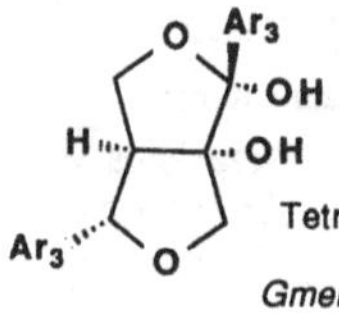

Ar3, OH, H, OH, Ar3, O

Tetrahedron,1975,31,1277

Gmelina arborea

ARBOREOL ISO 54286-94-7
2.238

as 2.237 but *trans*- 7,8-dihydroxy

2' - BROMOISOARBOREOL
2.239 Tetrahedron,1977,33,133

Gmelina arborea

(-) - DIHYDROXYSESAMIN 63398-39-0
2.241 C20 H18 O8

as sesamin 2.223 but 9,9'- dihydroxy

J.Chem.Soc.Perkin 1,1982,175 (synth)

Phytochem.1981,20,2271 (isol)

Kigelia pinnata

CINNAMONOL 58262-61-6
2.240 C20 H18-O7

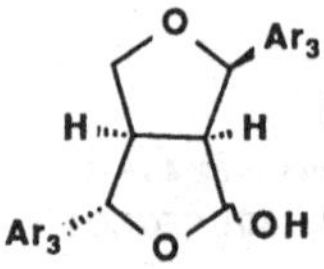

Ar3, H, H, Ar3, OH

Nippon Kagaku Kaishi 1975,2192

Chem.Abs.1976,84,71488

Cinnamomum camphora

(-) - DIHYDROXY-O-DIMETHYL-PINORESINOL C22 H26 O8
2.242

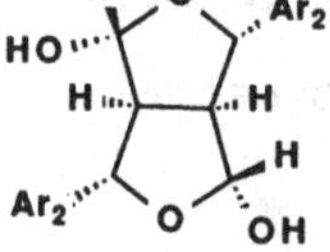

H, HO, Ar2, H, H, H, Ar2, O, OH

Aust.J.Chem.1987,40,405 (isol; X-ray)

Eremophila dalyana

EPIPHRYMAROL $C_{23}H_{24}O_{10}$
2.243

Chem. of Lignans,1978,p245
Phryma leptostachya

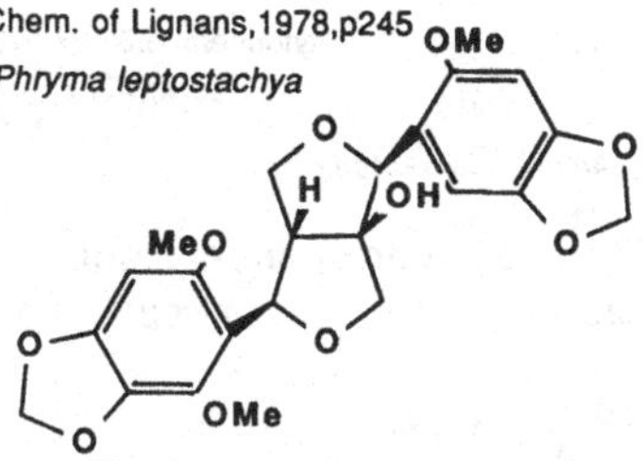

The absolute configuration is doubtful

(+) - GMELINOL 469-28-3
2.245 $C_{22}H_{26}O_7$

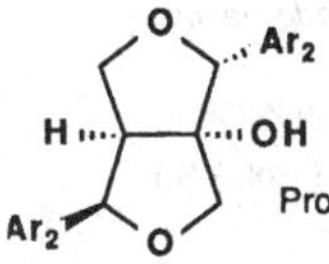

Proc.Roy.Soc.N.S.Wales 1912,46,187 (isol)

Gmelina leichardtii

(+) - GMELINOL ISO
2.246 597-01-3
$C_{22}H_{20}O_7$

GMELINOL NEO
2.247 17669-45-3
$C_{22}H_{20}O_7$

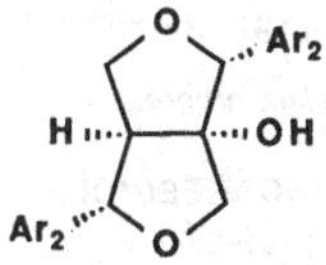

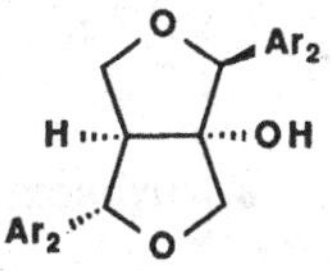

J.Chem.Soc.1967,1968 (absol stereo)
also Chapter 6, pp.210,214

(+) - FRAXIRESINOL $C_{21}H_{24}O_8$
2.244

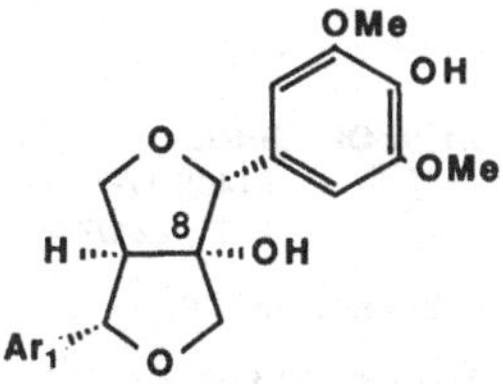

Chem.Pharm.Bull.1984,32,4482 (isol;CMR)

Fraxinus mandshurica var *japonica*

C-8-glucoside $C_{27}H_{34}O_{13}$
Chem.Pharm.Bull.1985,33,1232

(+) - GUMMADIOL 57684-87-4
2.248 $C_{20}H_{18}O_8$

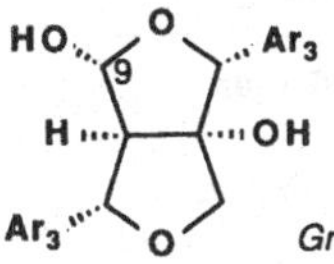

Tetrahedron Lett.1975, 1803

Gmelina arborea

(+) - 9 - EPIGLUCOSIDE $C_{26}H_{28}O_{13}$
2.249 Tetrahedron,1977,33,133

Gmelina arborea

(-) - GMELANONE 54826-95-8
2.250 C20 H16 O7

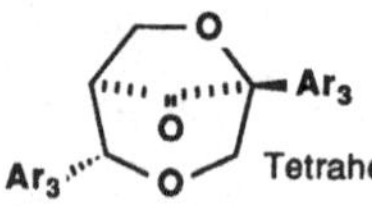

Tetrahedron,1975,31,1277

Gmelina arborea

8 - HYDROXYPINORESINOL
2.252 81426-17-7
C20 H22 O7

As 2.244 but both aryl = Ar_1

Chem.Pharm.Bull.1984,32,4482
Fraxinus mandshurica ; *Olea europa*

also see sambacolignoside

2.252 as diglucoside: Chem.Pharm.Bull.1985, 33,3651

Eucommia ulmoides

(+) - 8 - HYDROXYPINORESINOL -
2.253 4-O-ß-D-GLUCOSIDE

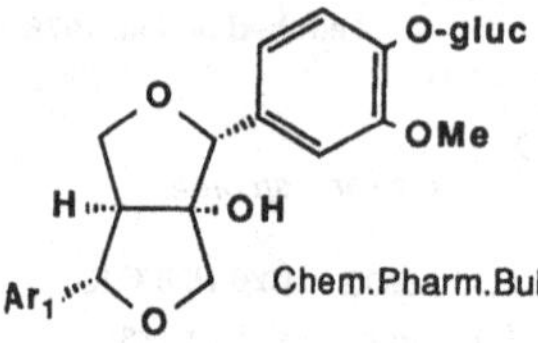

Chem.Pharm.Bull.1986,34,523 (CMR)
Chem.Pharm.Bull.1984,32,4482 (isol)

Eucommia ulmoides Oliv.
Fraxinus mandshurica

(+)-4'-O-glucoside
2.254 Chem.Pharm.Bull.1986,34,523

9 - HYDROXYMEDIORESINOL
2.251 C21 H24 O8

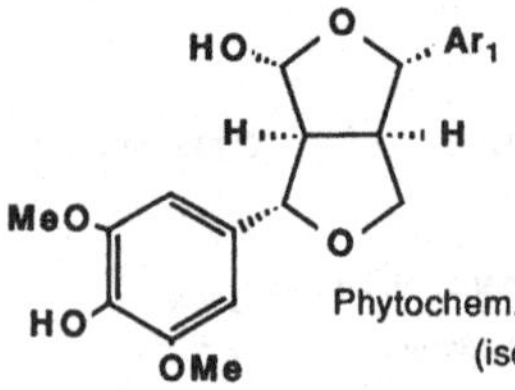

Phytochem.1988,27,575 (isol;NOE)

Allamanda nerifolia

(+) - 9-HYDROXYPINORESINOL
2.255 C21 H24 O8

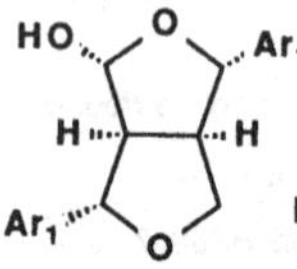

Tetrahedron,1977,33,133 (CMR, MS)
Phytochem.1988,27,575 (isol, CMR)

also see 2.235 *Allamendu neriifolia*

2.255A
The 9-ß-hydroxy isomer of 2.255
Phytochem.1985,24,628 (isol, MS)
Lonicera hypoleuca

9 - HYDROXYSESAMIN C20 H18 O7
2.256

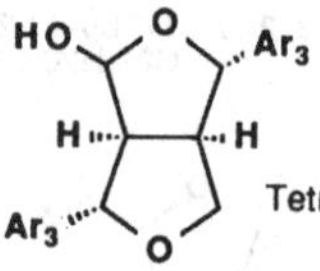

Tetrahedron,1977,33,,133

Gmelina arborea

(+) - 8 - HYDROXYSYRINGARESINOL
2.257 89199-95-1
C22 H26 O9

as 2.253 but both aryl = (4-hydroxy-3,5-dimethoxyphenyl: OMe, OH, OMe)

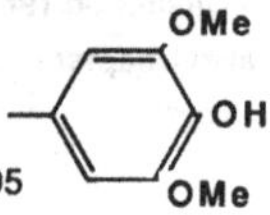

Chem.Abs.1983,100,117805
Fraxinus mandshurica

ISOPAULOWNIN 10590-41-7
2.258 C20 H18 O7

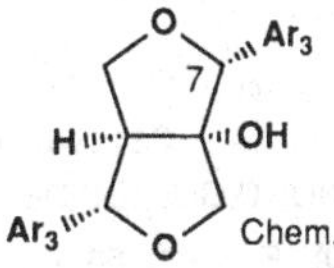

Chem.Pharm.Bull.1966,14,641

In vitro product

LEPTOSTACHYOL ACETATE
2.260 C26 H28 O12

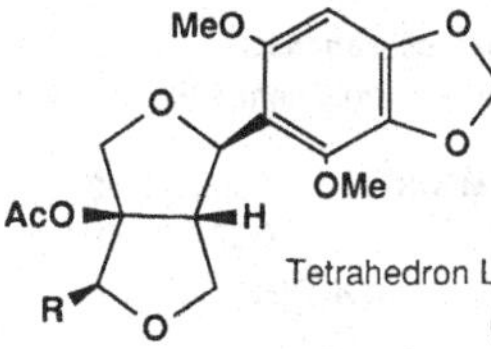

Tetrahedron Lett.1972,653
(isol; PMR)

Absolute stereochemistry unknown
Phryma leptostachya

LEPTOSTACHYOL 35942-44-0
C24 H26 O11

(-) - PRINSEPIOL 82667-99-0
2.262 C20 H22 O8

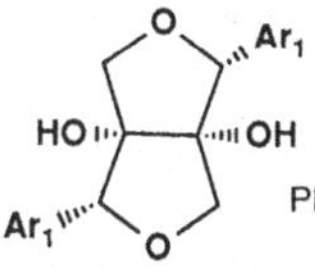

Phytochem.1982,21,796

Prinsepia utilis Royle
The oil of this plant has been used for the treatment of rheumatism.

ISOPHRYMAROL
2.259

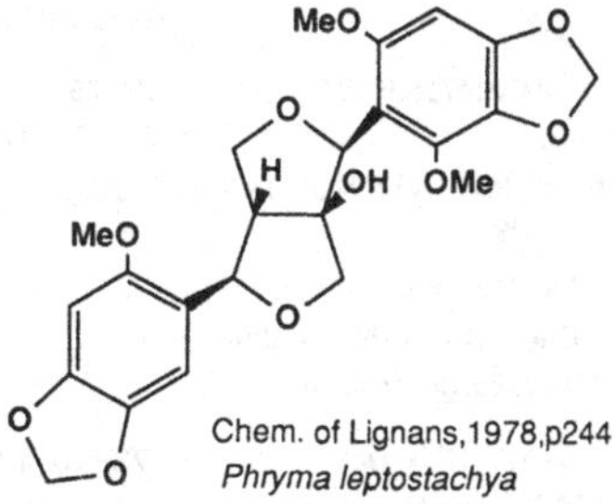

Chem. of Lignans,1978,p244
Phryma leptostachya
The absolute stereochemistry is doubtful
cf 2.243

KIGELIOL C20 H18 O8
2.259A
SESAMIN DIHYDROXY

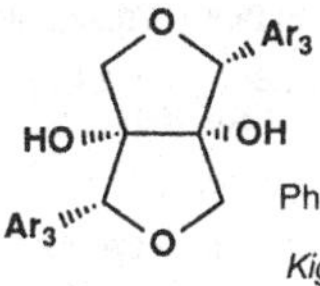

Phytochem.1981,20,2271
Kigelia pinnata

PAULOWNIN 13040-46-5
2.261 C20 H18 O7
As 2.258 but epimeric at C-7
Chem.Pharm.Bull.1979,27,2868 (CD,CMR)
Pharmazie,1980,35,122 (isol)
Paulonia tomentosa
Dolichandrone crispa Seem.

(-) - WODESHIOL 58007-97-9
2.263 C20 H18 O8

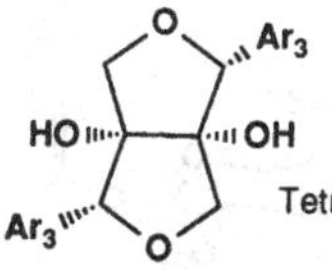

Tetrahedron,1981,37,3641
(isol; CMR; MS)

Cleistanthus collinus

FUROFURAN GLYCOSIDES

(-) - ACANTHOSIDE B 7374-79-0
2.265 C28 H36 O13
or SYRINGARESINOL MONOGLUCOSIDE 2.229

Lloydia,1978,41,56
Chem.Abs.1966,64,8290
Magnolia grandiflora

(+)-AFRICANAL* 73340-43-9
2.267 8-acetoxypinoresinol-4-ß-D-glucoside C28 H35 O13

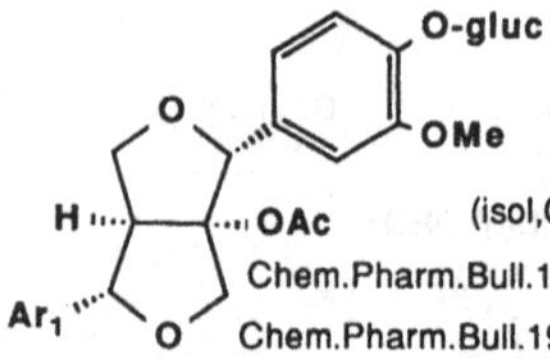

O-gluc, OMe, O, H, OAc, Ar$_1$, O

(isol,CMR,MS,CD)
Chem.Pharm.Bull.1979,27,2868
Chem.Pharm.Bull.1985,33,1232 (isol,CMR,pharm)

Olea africana Mill *O. europa* L.

* This name was also given to the tetralin 2.310

AFRICANAL 4'-O-METHYL
2.268

As 2.267 with Ar$_2$ for Ar$_1$

Olea europa L

LANTIBESIDE
2.270 A diglycoside of (+)-phillygenol cf 2.273

Zhiwu Xuebao,1985,27,402

Lancan tibetica Hook

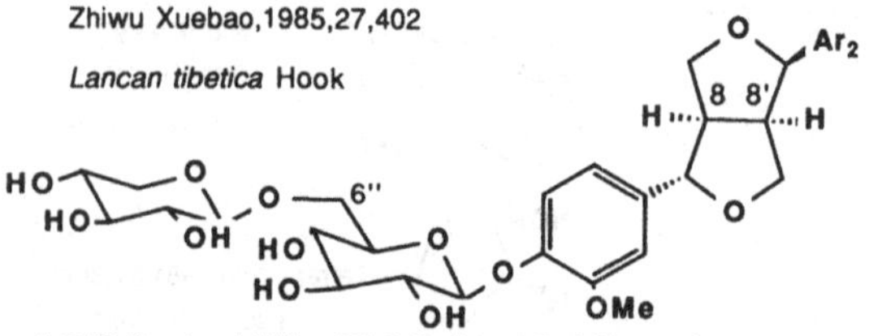

4-O-[ß-D-xylosyl-(1'''→ 6'')-ß-D-glucosyl]-phillygenol

(-) - ACANTHOSIDE D 6276-79-1
2.266 C34 H46 O18
(-) - SYRINGARESINOL DIGLUCOSIDE
or ELEUTHEROSIDE E 39432-56-9

Arch.Pharm.Res.1981,4,59
Saengyak Hakhoechi,1985,16,151
Chem.Abs.1986,104,192983

Acanthopanax chiisanensis
A.koreanum - J.Org.Chem.1980,45,1327

(+) - EUCOMMIN A 99633-12-2
2.269 C27 H34 O12
MEDIORESINOL MONOGLUCOSIDE

OMe, O-gluc, OMe, O, H, H, Ar$_1$, O

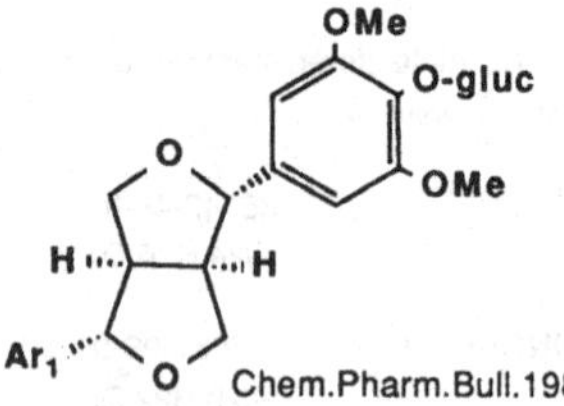

Chem.Pharm.Bull.1985,33,3651 (CMR)

Eucommia ulmoides

(+) - LIRIODENDRIN 573-44-4
2.271 C34 H46 O8

(+)-SYRINGARESINOL DIGLUCOSIDE

J.Org.Chem1980,45,1327(isol; pharm)
Phytochem.1986,25,1907(isol)
Helv.Chim.Acta.1981,64,3(CMR; pharm)

Globularia alypum L.
E.ulmoides ; *Desfontania spinosa*

(+) - PHILLYRIN 481-41-2
2.273 C27 H34 O11

PHILLYGENOL GLUCOSIDE

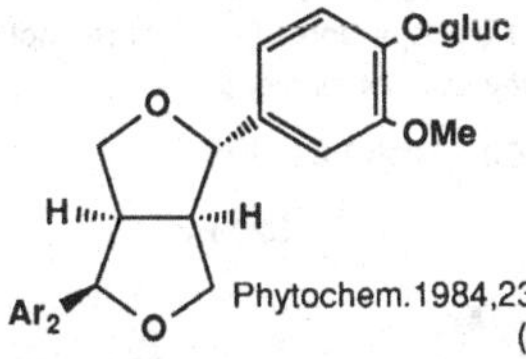

O-gluc
OMe
H
H
Ar_2
O
O

Phytochem.1984,23,1635 (isol)
Holzforschung,1977,31,41(biochem)
Phytochem.1980,19,335(CMR)
Forsythia sp.

(+) - SAMBACOLIGNOSIDE 114449-12-6
2.274 C43 H54 O22

cf 8-hydroxypinoresinol 2.252

Chem.Pharm.Bull.1987,35,5032(COSY; CMR)

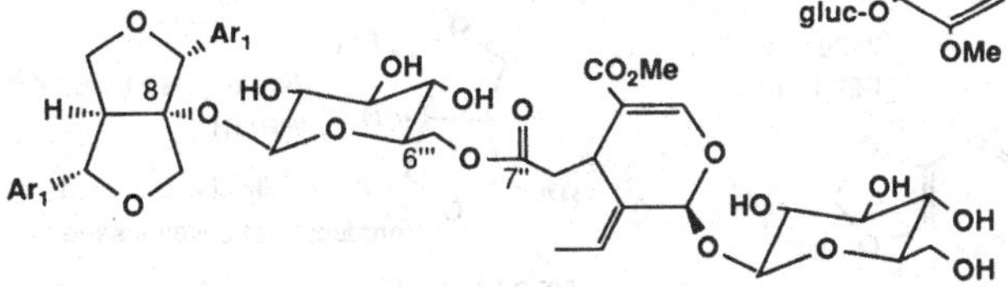

O, Ar_1, 8, H, O, HO, OH, OH, O, 6''', O, 7'', CO_2Me, O, Ar_1, O, HO, OH, OH, O, OH

7-O-[(+)-8-hydroxypinoresinol-ß-D- glucoside-(7,6''')]-oleoside-11-methyl ester

MAGNOLENIN C 66779-67-7
2.272 C28 H36 O14

Magnolia grandiflora

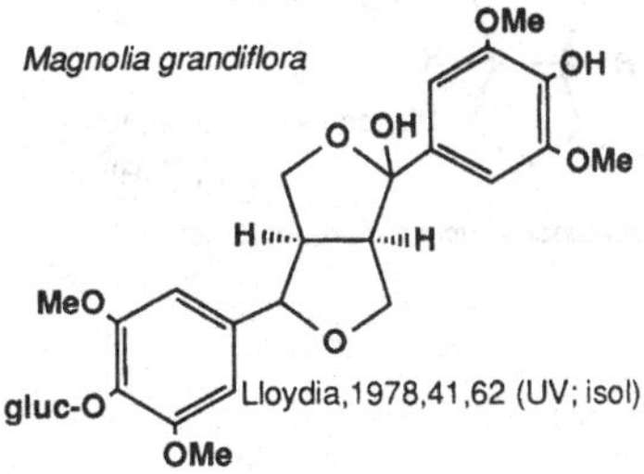

OMe, OH, O, OH, OMe, H, H, MeO, O, gluc-O, OMe

Lloydia,1978,41,62 (UV; isol)

(-) - SIMPLEXOSIDE 74061-78-2
2.275 C26 H30 O11

PIPERITOL GLUCOSIDE
2.219

Phytochem.1980,19,332(isol,UV,PMR,pharm)
Justicia simplex D. Don

(+) - EPIPINORESINOL GLUCOSIDE
2.276 cf 2.277

Phytochem.1980,19,335
Pharmazie,1968,23,519

Forsythia sp.

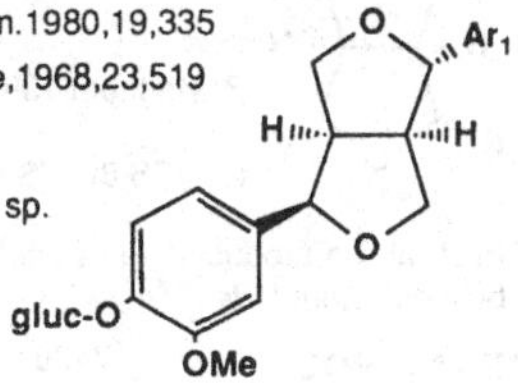

O, Ar_1, H, H, O, gluc-O, OMe

(-) - SIMPLOCOSIN 11042-30-1
2.277 C26 H32 O11
(-)-epipinoresinol glucoside

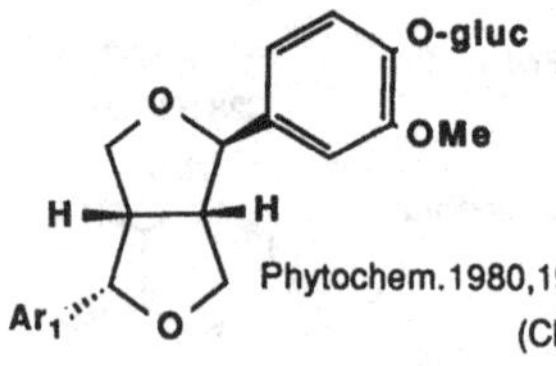

O-gluc
OMe
H H
Ar_1 O

Phytochem.1980,19,335
(CMR)
Simplocos lucida Sieb. et Zucc.

VERSICOSIDE 101391-02-0
2.278 EPIPINORESINOL C32 H42 O15
2.278 DIGLUCOSIDE
Chem.Nat.Compds.1985,21,584
(CMR; isol)

where R =

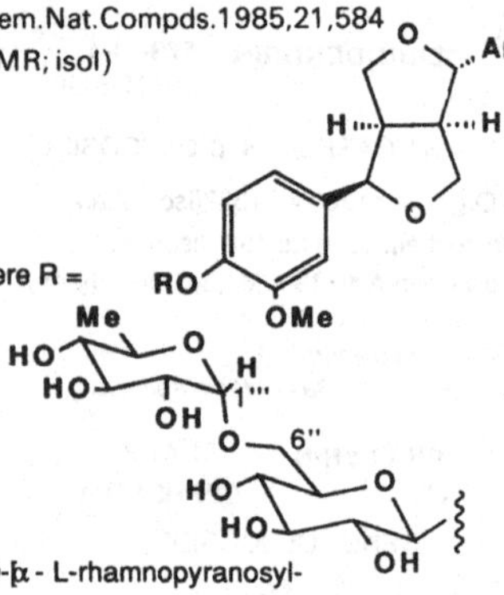

4-O-[α - L-rhamnopyranosyl-
(1'''--6'')-β-D-glucopyranosyl]-epipinoresinol
Haplophyllum versicolor

UNNAMED FUROFURANS AND RELATED COMPOUNDS

2.279 C18 H18 O4
cf, (-) - ligballinol

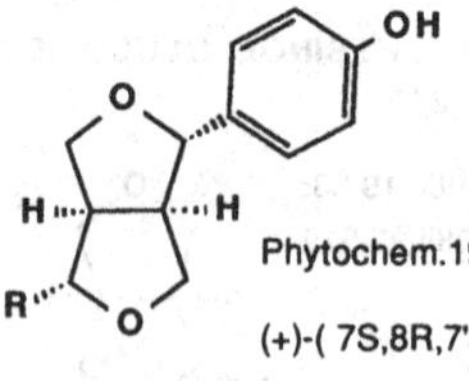

Phytochem.1983,22,1257
(+)-(7S,8R,7'S,8'R)

Phytoalexin produced in cell culture by *Vigna angularis*

2.280 C20 H22 O7

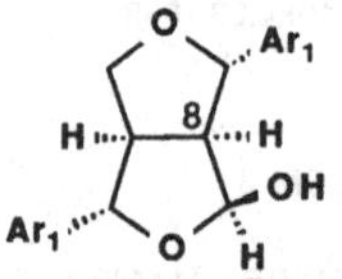

Phytochem.1985,24,628
(7S,8R,7'S,8'R,9S)
Lonicera hypoleuca

4,4'-dihydroxy-3,3'-dimethoxy-(7,9',7',9)-bis-epoxylignan-9-ol

SESAMINOL 74061-79-3
2.281 C20 H18 O7

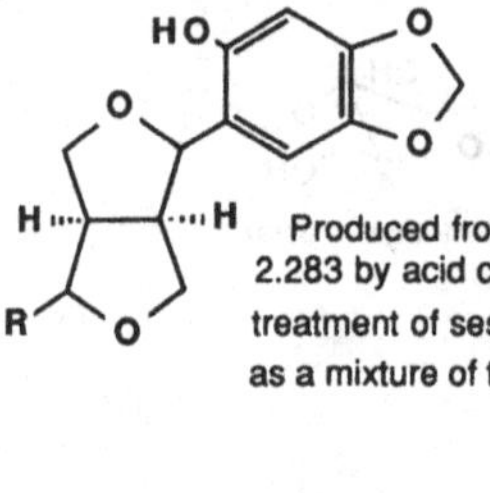

Produced from sesamolin 2.283 by acid catalysis during treatment of sesame seed oil as a mixture of two isomers.

SAMIN C13 H14 O5
2.282

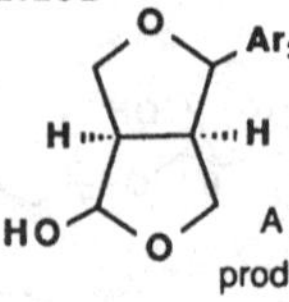

Heterocycles,1986,24,923
(HPLC)

A partially characterised product from sesame seed oil

SESAMOLIN 526-07-8
2.283 C20 H18 O7
The 3,4-methylenedioxyphenyl ether of 2.282
Heterocycles,1986,24,923

SESAMOLINOL 100016-94-2
2.284 C20 H20 O7

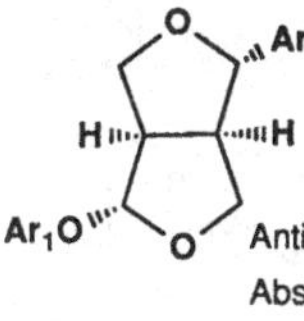

Agric.Biol.Chem.1985,49, 3351 (PMR; X-ray)

Anti-oxidant from sesame seed
Absolute configuration unknown

SIMPLEXOLIN 71328-57-9
2.284A C20 H18 O8

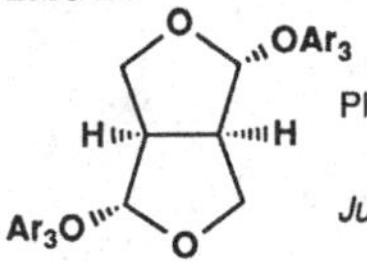

Phytochem.1979,18,503

Justicia simplex

(+) - PHRYMAROLIN 1 38303-95-6
2.286 C24 H24 O11

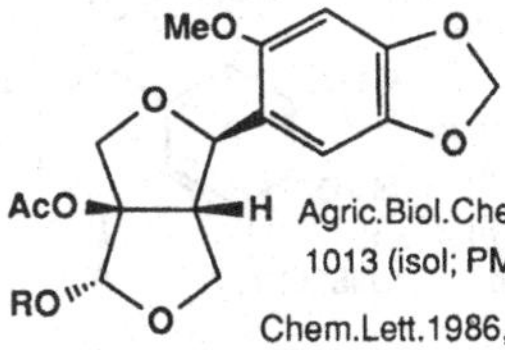

Agric.Biol.Chem.1972,36, 1013 (isol; PMR)
Chem.Lett.1986,1771 (synth)

Phryma leptostachya L.

(+) - APTOSIMON 61254-17-9
2.288A C20 H16 O7

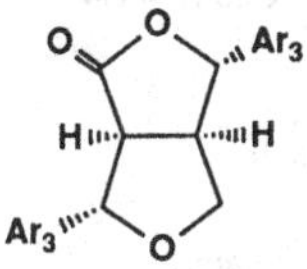

Tetrahedron Lett.1976,2221 (isol); 1986,4629 (synth)

(+) - APTOSIMOL 61254-18-0
2.288B C20 H18 O7

The ketal formed by reduction of 2.288A

Aptosimum spinescens

2.285 A X=H C21 H22 O8
2.285 B X=OMe

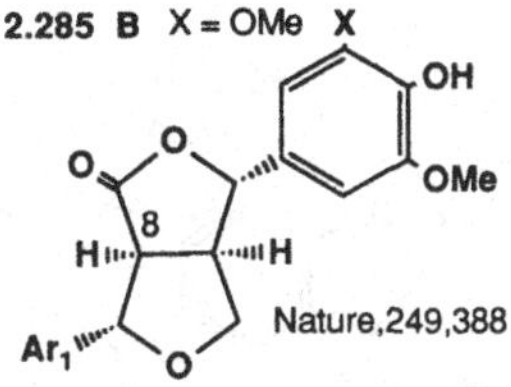

Nature,249,388 (isol)

Tetrahedron,1979,35,861 (synth, X-ray)

Experientia,1980,36,662 (pharm)

Aegilops ovata L.

4,4'-dihydroxy-3,3'5'-trimethoxy-(7,9',7',9)-bis-epoxylignan-9-olide

(+) - PHRYMAROLIN 2 23720-86-7
2.287 C23 H22 O10

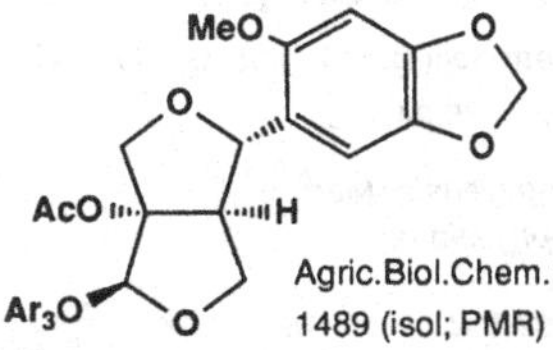

Agric.Biol.Chem.1972,36, 1489 (isol; PMR)

In this paper the absolute configuration of (+)-gmelinol is written incorrectly and the absolute configuration of the (+)-phrymarolins is therefore uncertain.

STYRAXIN 69742-32-1
2.289 C20 H18 O7

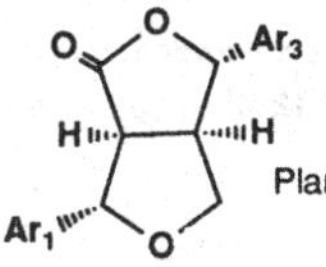

Planta Med.1978,34,403 (isol, pharm)

Tetrahedron Lett.1986,4629 (synth, NOE)

Styrax officinalis

ARYLTETRAHYDRONAPHTHALENES (Tetralin lignans)
9,9' - DIMETHYL COMPOUNDS

ATTENUOL 66761-07-7
2.290 $C_{19} H_{20} O_3$

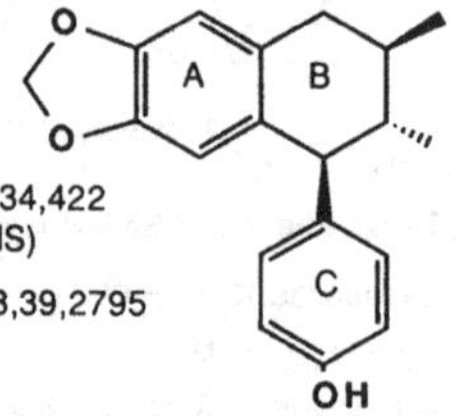

Experientia,1978,34,422
(isol; CMR; MS)

Tetrahedron,1983,39,2795
(synth)

Knema attenuata Wall.

AUSTROBAILIGNAN 3 55890-22-7
2.291 $C_{21} H_{22} O_5$

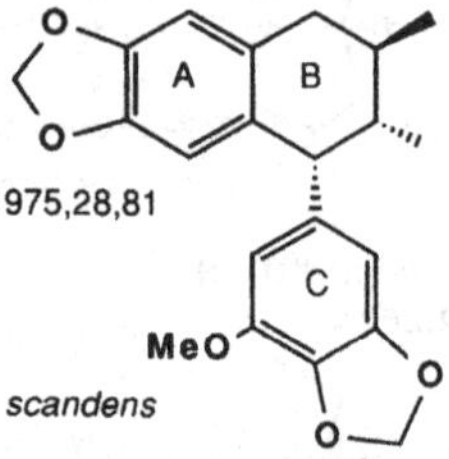

Aust.J.Chem.1975,28,81
(isol)

Austrobaileya scandens

(-) - CAGAYANIN 99096-51-2
2.293 $C_{20} H_{20} O_4$

As 2.290 but with ring C = Ar_3

J.Chin.Chem.Soc(Tapei),1985,32,177
Phytochem.1985,24,1867

Myristica cagayensis Merr.
Virola calophylloidea

(-) - GALCATIN 4892-34-6
2.294 $C_{21} H_{24} O_4$

As 2.290 but with ring C = Ar_2

Phytochem.1976,15,773
Tetrahedron Lett.1980,21,4827 (synth)
Aust.J.Chem.1954,7,104 (isol)

Virola carinata *Himantandra baccata*
The (+) - form has been synthesised-
Canad.J.Chem.1981,59,1291

AUSTROBAILIGNAN 4 55924-17-9
2.292 $C_{22} H_{26} O_5$

As 2.291 but with ring C = Ar_4

(-) - GALBULIN 521-54-0
2.295 $C_{22} H_{28} O_4$

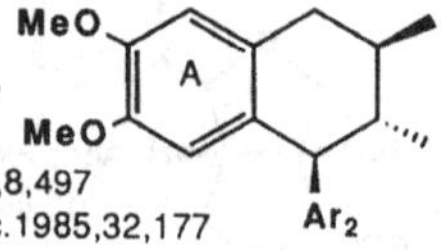

Phytochem.1979,
18,1703 (CMR)
Phytochem.1963,8,497
J.Chin.Chem.Soc.1985,32,177
(synth)

Araucaria angustifolia

(+) - GUAIACIN 36531-08-5
2.296 $C_{20} H_{24} O_4$

As 2.291 but with ring C = Ar_1

J.Chem.Soc.1964,4011 (isol)
Phytochem.1972,11,811
Phytochem.1985,24,1051 (pharm)

Guaiacum officinale

ISOGALBULIN 521-55-1
2.297 C22 H28 O4

J.Chem.Soc.1958,4471

In vitro product

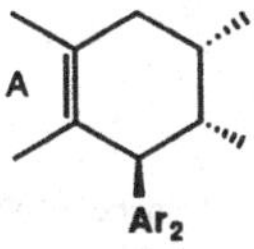

A

Ar_2

ISOGALCATIN 24150-38-7
2.298 C21 H24 O4

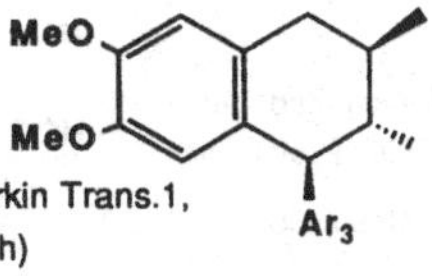

MeO
MeO

J.Chem.Soc. Perkin Trans.1, 1981,1681 (synth)

Ar_3

ISOGUAIACIN 78341-26-1
2.299 (EPIGUAIACIN) C20 H24 O4

J.Chem.Soc.1964,4011

Guaiacum officinale

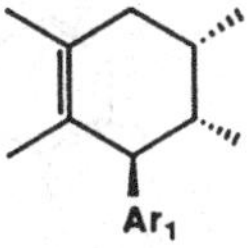

Ar_1

ISOGUAIACIN NOR 50376-42-6
2.300 C19 H22 O4

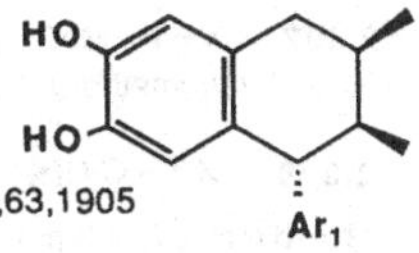

HO
HO

J.Pharm.Sci.1974,63,1905
(isol; pharm)

Larrea divaricata

Ar_1

ISOOTOBAIN 24150-38-7
2.302 C21 H24 O4

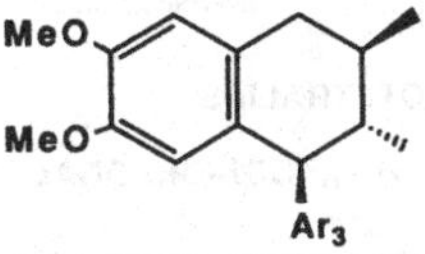

MeO
MeO

Ar_3

J.Amer.Chem.Soc.1921,43,199 (isol)
J.Chem.Soc.1963,1445
Myristica otoba

ISOGUAIACIN 3'-DEMETHOXY
2.301 C19 H22 O3

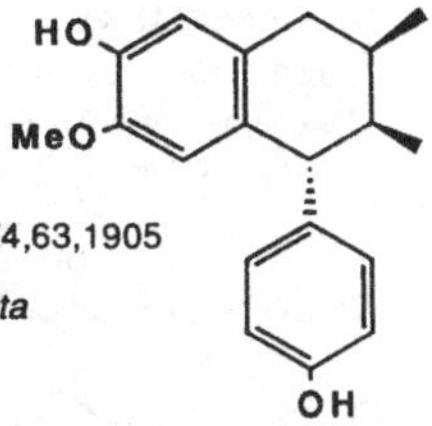

HO
MeO

OH

J.Pharm.Sci.1974,63,1905

Larrea divaricata

(-)-ISOOTOBAPHENOL 59985-35-2
2.303 C20 H22 O4

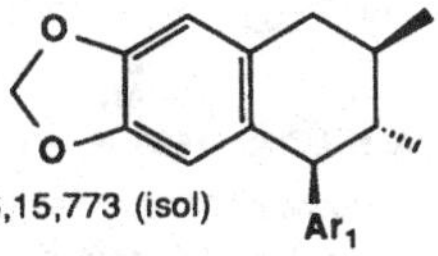

O
O

Phytochem.1976,15,773 (isol)

Ar_1

Phytochem.1985,24,1051 (isol; pharm)

Virola carinata
Myristica otoba

OTOBAIN 3738-01-0
2.304 C20 H20 O4

J.Chem.Soc.1962, 1780 (isol)

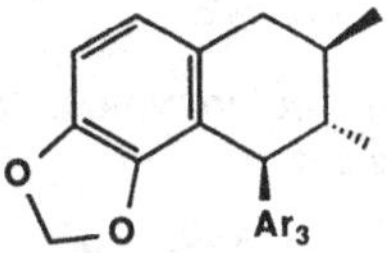

O
O

J.Chem.Soc.1963, 1445 (PMR)

Ar_3

J.Chin.Chem.Soc.1971,18,45 (isol)
Chem.& Ind.1962,270 (pharm)

M.otoba M.cagayensis Merr.

OTOBAPHENOL 10240-16-1
2.305 C20 H22 O4

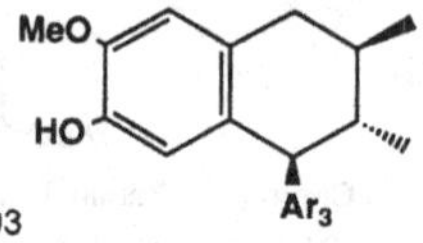

MeO, HO, Ar_3

Planta Med.1984, 50,53 (isol; CMR)

J.Chem.Soc.1966,893

Osteophloeum platyspermum

2.306 C18 H19 O2
4,4'-dihydroxy-2,7'-cyclolignan

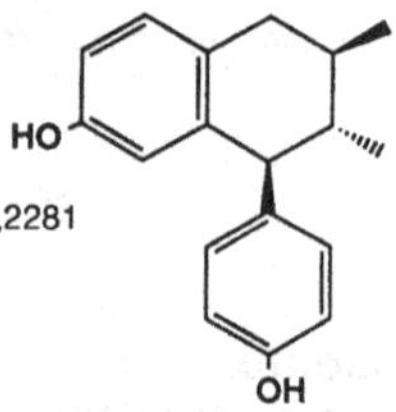

HO, OH

Phytochem.1983,22,2281 (isol; MS; CD)

Iryanthera grandis

2.307 X = Y = MeO; Z = H C22 H27 O4
3,3',4,4'-tetramethoxy-2,7'-cyclolignan

2.308 X,Y = O-CH_2-O; Z = H C21 H23 O4
3',4'-dimethoxy-4,5-methylenedioxy-2,7'-cyclolignan

2.309 X = H; Y,Z = O-CH_2-O C21 H23 O4
3',4'-dimethoxy-3,4-methylenedioxy-2,7'-cyclolignan

X, Y, Z, Ar_2

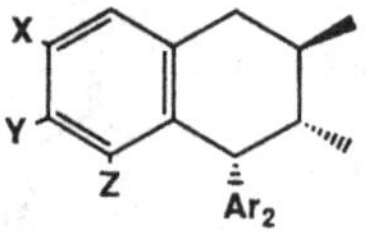

Phytochem.1986,25,959

Myristica otoba

OXY- and OXOTETRALINS

(+) - AFRICANAL 73340-43-9
2.310 C20 H22 O7
This name was later also given to the furofuran glucoside 2.267

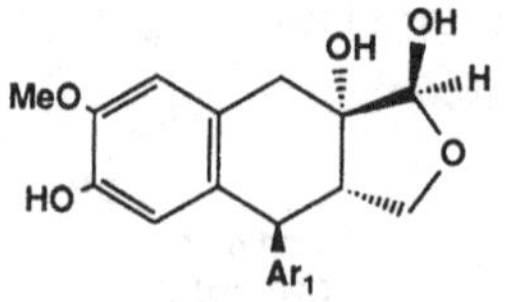

MeO, HO, OH, OH, H, O, Ar_1

Tetrahedron Lett.1979,3773

Olea africana

(+) -α -CONIDENDRAL O-METHYL
2.311
TSUGACETAL 71724-99-7
101312-79-2 C21 H24 O6

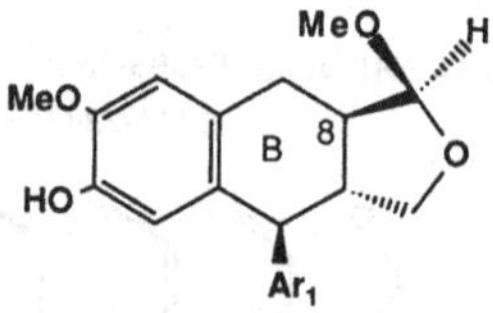

MeO, H, MeO, HO, B, 8, O, Ar_1

Aust.J.Chem.1985,38,1631 (isol; CMR)

Phytochem.1985,24,1363 (X-ray)

J.Chin.Chem.Soc.1985,32,377 (isol)

Dacrydium intermedium Kirk
Juniperus formosana

ß - CONIDENDRAL O-METHYL
2.312 71772-19-5
C21 H24 O6

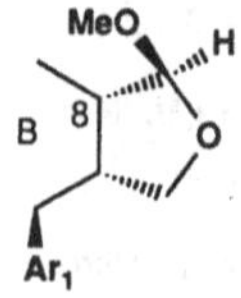

MeO, H, B, 8, O, Ar_1

Tetrahedron,1986,42,2005 (synth)

(+) - CYCLOLARICIRESINOL
2.313 (isolariciresinol) 548-29-8
C20 H24 O6

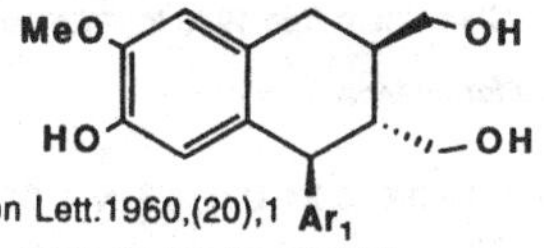

MeO HO OH OH Ar1

Tetrahedron Lett.1960,(20),1
Phytochem.1978,17,499 (isol; CMR)
Mokusai Gakkaishi,1980,26,759
J.Org.Chem.1986,51,3490 (synth)

Araucaria angustifolia *Abies sachalinensis*

9'-ß-D-XYLOSIDE (Schizandriside)
2.314 Chem.Pharm.Bull.1979,27,1422

Schizandra nigra Max.

(+) - CYCLOLARICIRESINOL 4'-O-METHYL
2.316

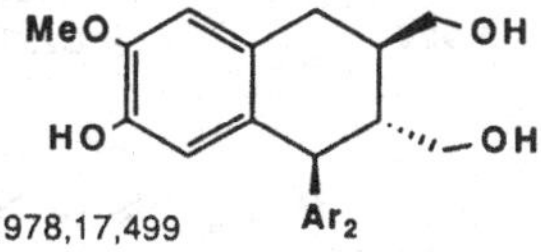

MeO HO OH OH Ar2

Phytochem.1978,17,499
(isol; CMR)

Araucaria angustifolia

The 4-O-methyl and 4,4'-O-dimethyl derivative (21966-92-7) were synthesised.

(+) - CYCLOOLIVIL 3064-05-9
2.318 (isoolivil) C20 H24 O7

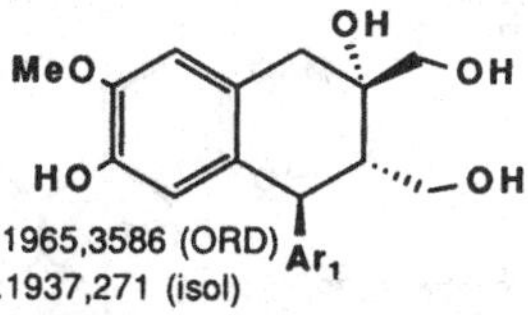

OH MeO HO OH OH Ar1

J.Chem.Soc.1965,3586 (ORD)
J.Chem.Soc.1937,271 (isol)
Chem.Pharm.Bull.1981,29,2082 (synth; CD)

Olea cunninghamii

(-) -CYCLOLARICIRESINOL 5'-METHOXY
2.315 as 9'-ß-D-XYLOSIDE
also see LYONIRESINOL

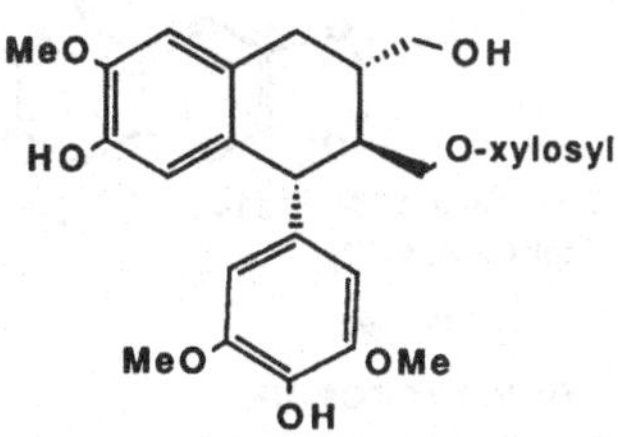

MeO HO OH O-xylosyl MeO OMe OH

Phytochem.1979,18,1847 (isol; CD; CMR)

Cinnamosma madagascariensis

(-) - CYCLOLARICIRESINOL
2.317 as 9'-ß-D-XYLOSIDE

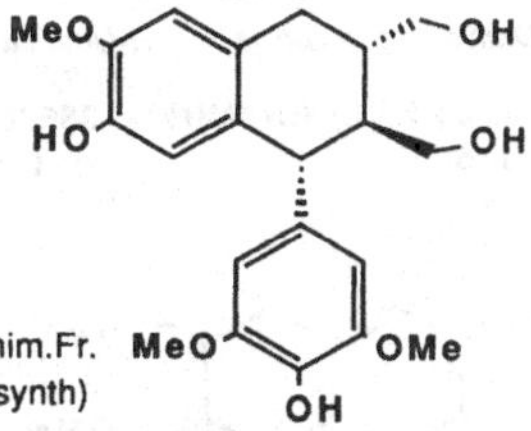

MeO HO OH OH MeO OMe OH

Bull.Soc.Chim.Fr.
1985,871 (synth)

Cinnamosma madagascariensis

(+) - CYCLOTAXIRESINOL 26194-57-0
2.319 (isotaxiresinol) C19 H22 O6
(X = H)

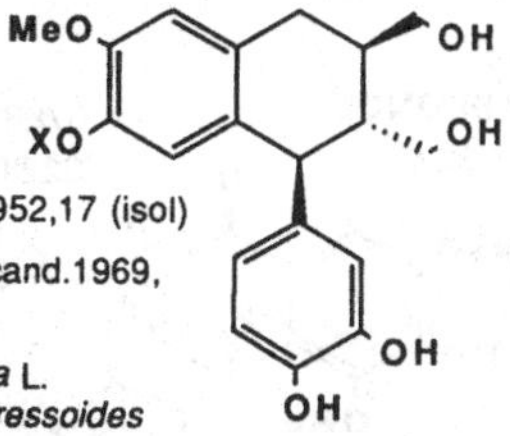

MeO XO OH OH OH OH

J.Chem.Soc.1952,17 (isol)
Acta Chem.Scand.1969,
23,2021 (isol)
Taxus baccata L.
Fitzroya cuppressoides

CYCLOTAXIRESINOL 4-O-METHYL
2.320 as 2.319 but X = Me 23141-17-5
C20 H24 06

Acta Chem.Scand.1969,23,2021 (PMR)
Phytochem.1969,8,931 (isol)
Taxus baccata

(-) - ENSHICINE 93710-74-8
2.321 C20 H20 O5
see SCHIZANDRONE

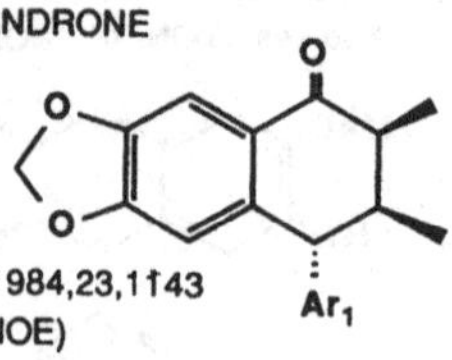

O, O, O, Ar_1

Phytochem.1984,23,1143
(isol; CMR; NOE)

Schisandra henryi

HYDROXYOTOBAIN C20 H20 O5
2.322 as 2.323 but 7-deoxy

Planta Med.1984,50,53 (isol)
Phytochem.1982,21,751 (CD)
J.Chem.Soc.1963,1445 (isol; PMR)

Osteophloeum platyspermum
Dialyanthera otoba *Myristica otoba*

FORMOSANOL 101312-79-2
C21 H24 O6
equiv. TSUGACETAL **2.311**

J.Chin.Chem.Soc.1985,32,377 (isol; CMR)

J.formosana

HYDROXYOXOOTOBAIN C20 H18 O6
2.323

Phytochem.1982,21,751
(isol; CD)

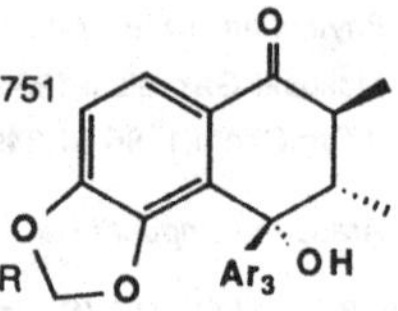

O, O, O, Ar_3, OH

8S,8'S,7'R

Virola sebifera

ent **HYDROXYOXOOTOBAIN**
2.324 Phytochem.1982,21,751

V.sebifera

HYPOPHYLLANTHIN 33676-00-5
2.325 C24 H30 O7

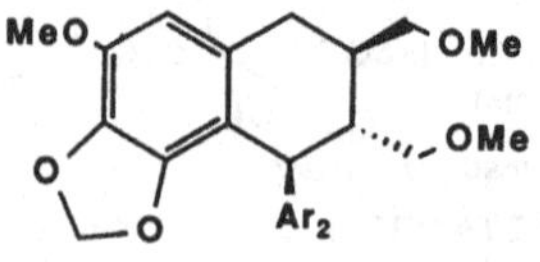

MeO, OMe, OMe, O, O, Ar_2

LINTETRALIN 73231-44-4
2.326 C23 H28 O6

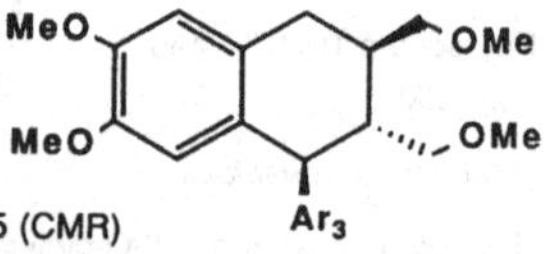

MeO, OMe, MeO, OMe, Ar_3

Tetrahedron,1973,29,1291

Tetrahedron Lett.1979,3045 (CMR)
J.Chem.Soc.Perkin Trans.1,1982,999 (synth, pharm)

Phyllanthus niruri

LIRIONOL 67881-22-5
2.327 C22 H24 O8

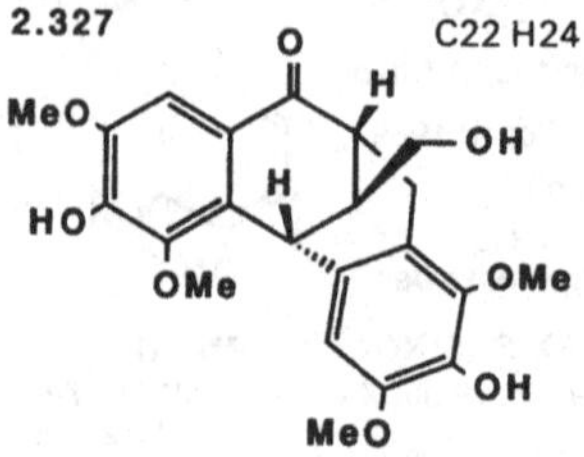

O, H, MeO, OH, H, HO, OMe, OMe, OH, MeO

Phytochem.1978,17,779 (isol; PMR)
Liriodendron tulipfera

(rac) - LYONIRESINOL 23212-64-8
2.328 see 2.334 C22 H28 O8
(R = H)

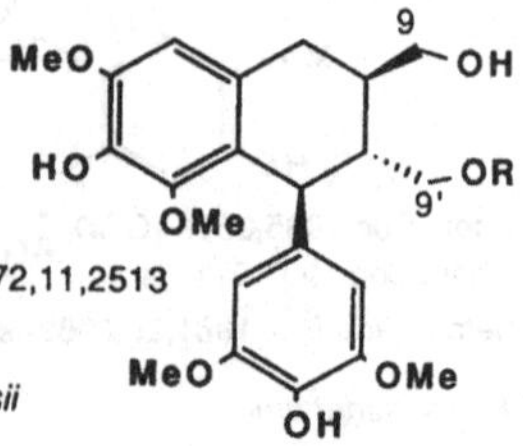

9, MeO, OH, HO, OR, 9', OMe, MeO, OMe, OH

Phytochem.1972,11,2513

Ulmus thomasii

also the (+)-rhamnoside 2.328 , R = α-rhamnosyl

(+) - LYONIRESINOL 14464-90-5
2.329 $C_{22}H_{28}O_8$

As shown for 2.328 with R = H
Chem.Ber.1961,94,2522
Alnus glutinosa

LYONISIDE 34425-25-7
2.330 $C_{27}H_{36}O_{12}$

The 9'-xyloside of 2.329
Canad.J.Chem.1962,40,1118
A.glutinosa *Sorbus desora* Sarg.

PHYLTETRALIN 50656-79-6
2.333 $C_{24}H_{32}O_6$

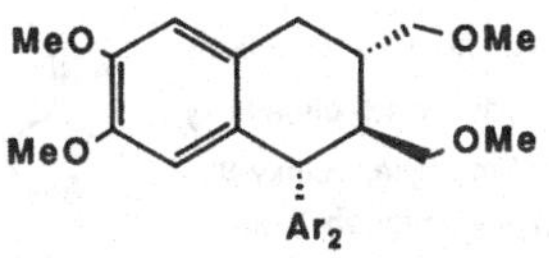

MeO, MeO, OMe, OMe, Ar_2

J.Chem.Soc.Perkin Trans.1,1981,1681 (synth)
J.Chem.Soc.Chem.Commun.1982,430 (synth)
Tetrahedron,1973,29,1291 (isol)
Phyllanthus niruri

SCHISANDRISIDE 71222-06-5
$C_{25}H_{32}O_{10}$

The 9'-ß-D-xyloside of CYCLOLARICIRESINOL 2.313

SCHISANDRONE 98619-25-1
2.336 $C_{21}H_{24}O_5$

As ENSHICINE 2.321 but with ring A 4-methoxy-5-hydroxysubstituted
Planta Med.1985,51,217 (NOE)
Schisandra sphenanthera

NIRTETRALIN 50656-78-5
2.328 A $C_{24}H_{30}O_7$

As 2.325 but with 3-methoxy-4,5-methylenedioxy substitution in ring A.
J.Chem.Soc.Perkin Trans.1,1982,999 (synth)
Tetrahedron Lett.1979,3043 (pharm)
Phyllanthus niruri

NUDIPOSIDE 62058-46-2
2.331 $C_{27}H_{36}O_{12}$

The 9'-xyloside of (-) - LYONIRESINOL
Chem.Pharm.Bull.1976,24,2102 (isol; CD)

OTOBANONE 34426-79-4
2.332 $C_{20}H_{18}O_5$

J.Chin.Chem.Soc.1985, 32,177 (PMR) also 1971,18,45.

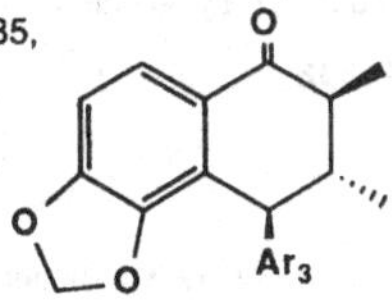

O, O, O, Ar_3

Myristica cagayensis *M.otoba*

PYGEORESINOL 31768-94-2
2.334 $C_{22}H_{28}O_8$

The (-) - form of 2.328
Indian J.Chem.1980,19B,279
Pygeum acuminatum

PYGEOSIDE 75587-29-0
2.335 $C_{27}H_{36}O_{12}$
also occurs in this plant.
It is the 9-ß-xyloside cf (-)-lyoniresinol

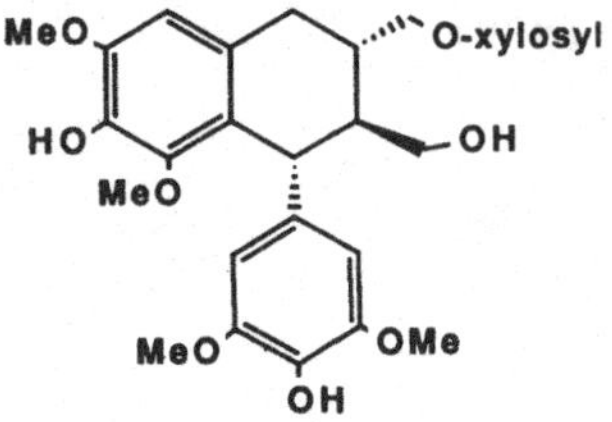

MeO, O-xylosyl, HO, OH, MeO, MeO, OMe, OH

UNNAMED TETRALINS & RELATED KETONES

2.337 C21 H22 O7

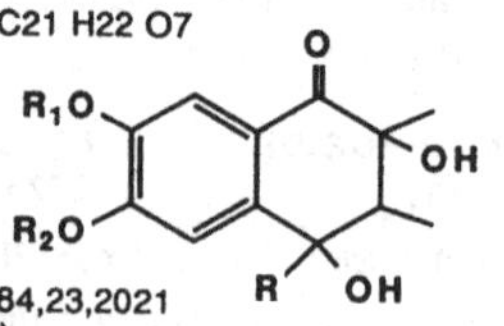

Phytochem.1984,23,2021
(isol, CMR, MS)

$R_1 = R_2 = Me; R = Ar_3$
7',8-dihydroxy- 4,5-dimethoxy-3'4'-methylene-dioxy-2,7'-cyclolignan-7-one

2.338 C21 H22 O7

As above but with $R_1, R_2 = -CH_2-$, $R = Ar_2$

7',8-dihydroxy-3',4'-dimethoxy-4,5-methylene-dioxy-2,7'-cyclolignan-7-one

Virola sebifera

2.339 C21 H20 O6

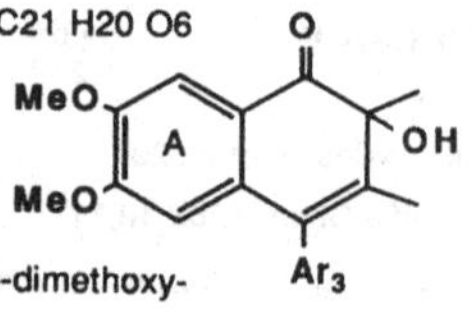

8-hydroxy-4,5-dimethoxy-3',4'-methylenedioxy-7'ene-2,7'-cyclolignan-7-one

2.340 C21 H20 O6

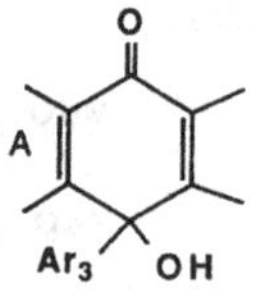

Phytochem.1984,23,2021

7'-hydroxy-4,5-dimethoxy-3',4'-methylenedioxy-8'-ene-2,7'-cyclolignan-7-one

V.sebifera

2.341

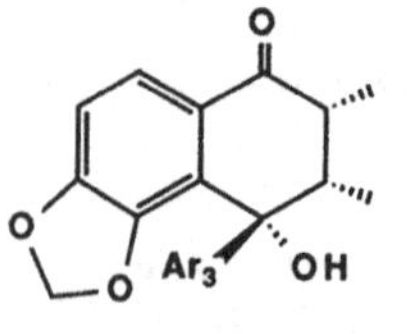

(7'R,8R,8'S)-7'-hydroxy-3,4,3',4'-bismethylenedioxy-2,7'-cyclolignan-7-one

Virola sebifera

2.342

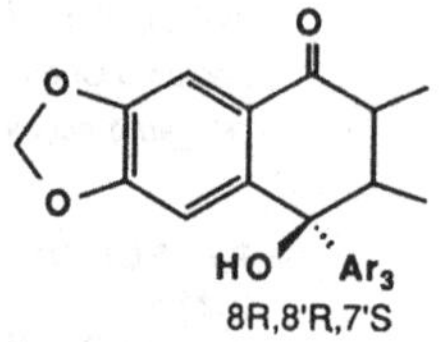

8R,8'R,7'S

Phytochem.1982,21,751
(isol, CD)

2.343

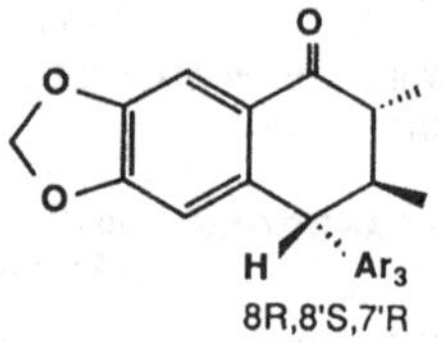

8R,8'S,7'R

see otobanone 2.332

(-) - ARISTOCHILONE 113952-98-0
2.343 A R_1 = Me, R_2 = H C21 H24 O5

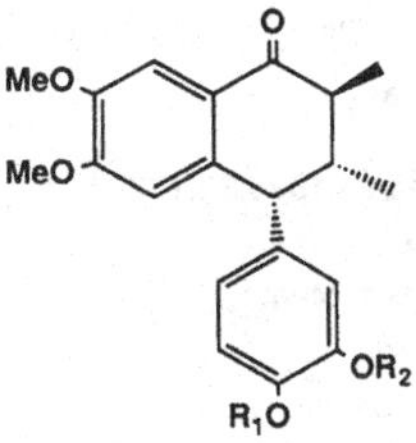

MeO
MeO
O
OR₂
R₁O

(-) - ARISTOLIGONE 114029-72-0
2.343 B R_1 = R_2 = Me C22 H26 O5

(-) - ARISTOSYNONE 114029-73-1
2.343 C C22 H26 O5

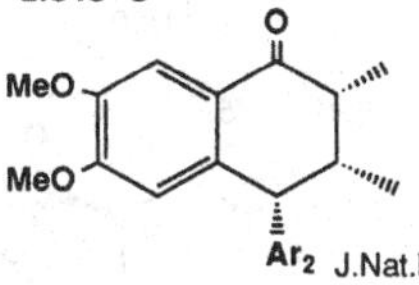

MeO
MeO
O
Ar_2 J.Nat.Prod.1988,51,117 ((isol, NOE, CD)

TODOLACTOL C20 H22 O6
2.346

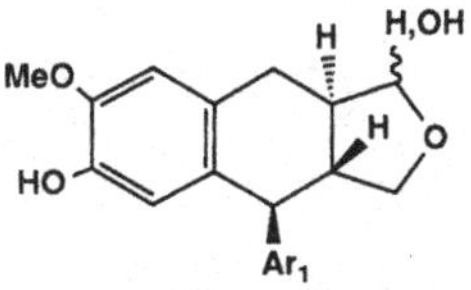

MeO
HO
H H,OH
H O
Ar_1

Mokusai Gakkaishi,1987,33,747 (isol,PMR,COSY)

Abies sachaliensis Masters

(-) - ARISTOTETRALONE 111188-75-1
2.343 D X = H C21 H22 O5

MeO
MeO
O
X
O
O

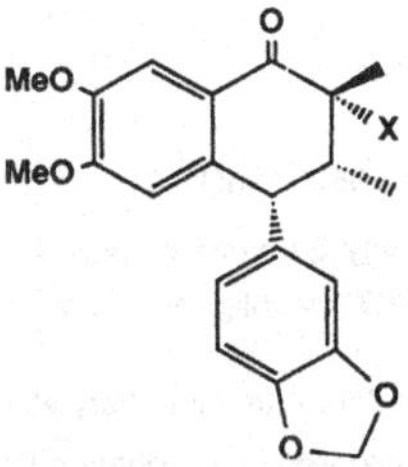

Phytochem.1987,26,2414 (isol, NOE)
Aristolochia chilensis

(-) - ARISTOTETRALONE
2.343 E 8 - HYDROXY
as above but with X = OH

(-) - ARISTOTETRALONE
2.343 F 8 - ACETOXY
as above but with X = OAc

J.Nat.Prod.1988,51,117 (isol, NOE, CD)

ARISTOTETRALOL 113952-99-1
2.343 G C21 H24 O5

The naturally occurring reduction product of 2.343 B

2.344

Phytochem.1985,24,1867* (isol,PMR)

(7'R,8S,8'S)-7'-hydroxy-5-methoxy-3,4,3',4'-bismethylenedioxy-2,7'-cyclolignan-7-one

2.345

J.Chin.Chem.Soc.(Taipei),1985,32,177

Virola calophylloidea
Myristica cagayensis Merr.

* An alkene derived from these tertiary alcohols was considered to be an artefact, this is also likely for otobaene(Phytochem.1969,8,497)

2,7' - CYCLOLIGNAN - 9 - OLIDES

(-)-α -CONIDENDRIN 518-55-8
2.348 C20 H20 O6

Chem.Rev.1955,55,1031
J.Chem.Soc.1910,97,1028; 1935,1576(isol)
Wood Sci.Technol.1970,4,122(isol)
Chem.Pharm.Bull.1981,29,2082(stereo)

Picea ajanensis *Tsuga heterophylla*
Podocarpus spicatus *Abies sachalinensis*

ß-CONIDENDRIN 5474-93-1
2.349 C20 H20 O6

Tetrahedron,1986,42,2005 (synth)

In vitro product

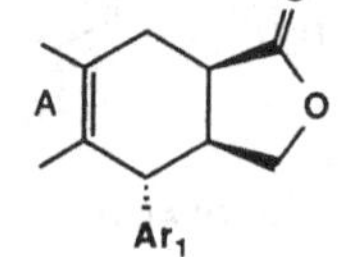

2,7' - CYCLOLIGNAN - 9' - OLIDES

ANTHRICIN see PODOPHYLLOTOXIN - DEOXY

AUSTROBAILIGNAN 1 55955-07-2
2.350 C21 H18 O7

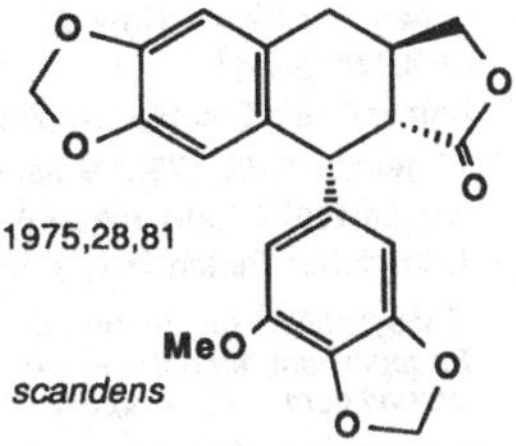

Aust.J.Chem.1975,28,81
(isol)

Austrobaileya scandens

MeO O O

AUSTROBAILIGNAN 2 55890-20-5
2.351 C21 H18 O7

As 2.350 but with *cis* -lactone

Probably an *in vitro* product

HERNANDION see PODOPHYLLOTOXIN - DEOXY

J.Chem.Soc.Japan,1942,63,1540

Hernandia ovigera

PICROPODOPHYLLIN 477-47-4
2.352 C22 H22 O8

J.Org.Chem.1966,31,4004
(synth)

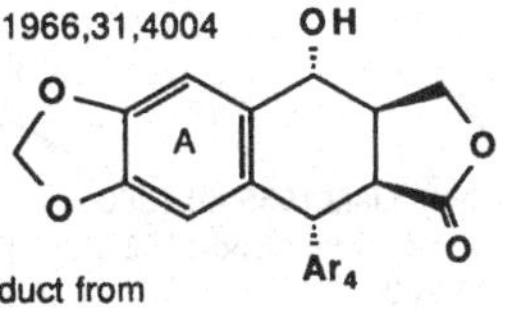

OH A Ar_4

in vitro product from podophyllotoxin 2.357

PICROPOLYGAMAIN 98819-09-1
2.353 C20 H16 O6

see POLYGAMAIN 2.365

ß-PELTATIN 518-29-6
2.355 C22 H22 O8

as 2.354 but with R = Me

Phytochem.1980,19,2479 (isol)

Chem.Pharm.Bull.1984,33 1754(X-ray)

Tetrahedron Lett.1982,23,949(synth)

Podophyllum peltatum
Thujopsis dolabrata

7'-ß-analogues Tetrahedron Lett.1982,23,949 (synth)

α - PELTATIN 568-53-6
2.354 R = H C21 H20 O8

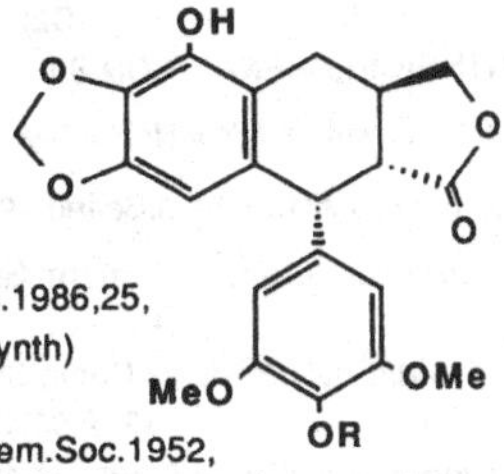

OH MeO OMe OR

Phytochem.1986,25, 2089 (biosynth)

J.Amer.Chem.Soc.1952, 74,4470 (isol)

Drugs Exp.Clin.Res.1981,7,711 (pharm)

PLICATIN 16462-67-2
2.356 $C_{20}H_{20}O_9$

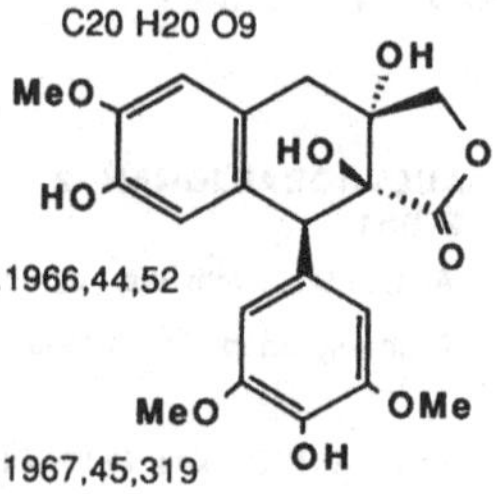

Canad.J.Chem.1966,44,52
(isol)

Canad.J.Chem.1967,45,319
(ORD,CD)

Thuja plicata D.Don

PODOPHYLLOTOXIN EPI 4375-07-9
2.359 $C_{22}H_{22}O_8$

As 2.357 but with 7 ß-OH
Precursor of the glycosides Etoposide and Teniposide, see Chapter 8
in vitro product

PODOPHYLLOTOXIN DEOXY
2.360 19186-35-7
$C_{22}H_{22}O_7$

(ANTHRICIN,HERNANDION,SILICOLIN

2.358 but with X = Y = H; R = Me

P.peltatum Acta Chem.Scand.1955,9,346
Anthriscus sylvestris J.Pharm.Soc.Japan 1940,60,629
Hernandia ovigera L. J.Chem.Soc.Japan 1942,63,1540
Juniperus silicola J.Amer.Chem.Soc.1953, 75,2138
P.pleianthum J.Pharm.Soc.Japan,1962 82,777

PODOPHYLLOTOXIN $C_{22}H_{22}O_8$
2.357 518-28-5

7R,7'R,8R,8'R

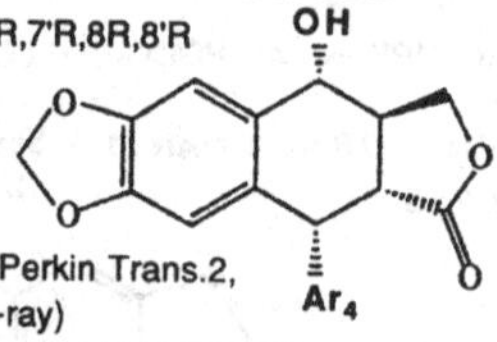

J.Chem.Soc.Perkin Trans.2, 1973,288 (X-ray)
J.Amer.Chem.Soc.1951,73,2909 (struct)
J.Chem.Soc.1898,73,209 (isol)
Lloydia,1967,30,291 (pharm)
J.Chem.Soc.Perkin Tr.1,1972,1343 (PMR)

For synthesis see chapter 8
Podophyllum emodi (P. hexandrum)
P. peltatum P. versipelle

PODOPHYLLOTOXIN 4'-DEMETHYL
2.358 $C_{21}H_{20}O_8$

X = H, Y = OH, R = H

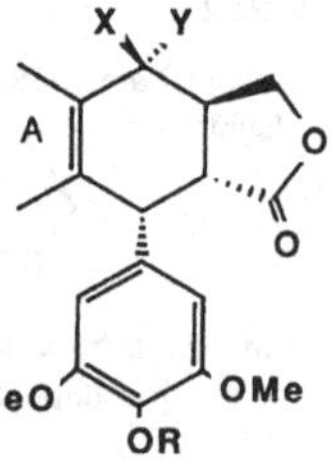

Chem.Listy,1955,49,1550
J.Amer.Chem.Soc.1952, 74,280

PODOPHYLLOTOXIN EPIISO 55568-79-1
2.361 X = 7ß-OH $C_{22}H_{22}O_8$

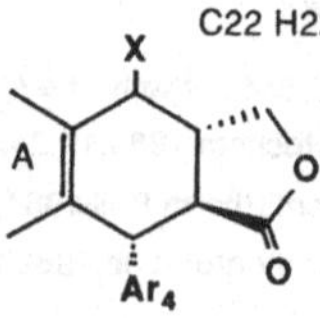

J.Org.Chem.1977, 42,246.

PODOPHYLLOTOXIN ISODEOXY
2.362 X = H The term -iso is ambiguous and isopodophyllotoxin is described as: (7'R,8S,8'S)-3',4'5'-trimethoxy-4,5-methylenedioxy-2,7'-cyclolignan-9,9'-olide

PODOPHYLLOTOXIN NEO 1456-54-8
2.363 C22 H22 O8

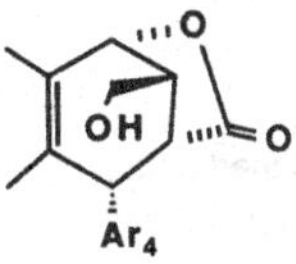

Experientia,1963,19,391
(synth)

In vitro product

OH
Ar4

PODOPHYLLOTOXONE 477-49-6
2.364 C22 H20 O8

As 2.358 but with X,Y = oxo
Phytochem.1975,14,1440 (isol)
Phytochem.1982,20,2277 (isol)

P.hexandrum; P.peltatum; P.pleianthum

POLYGAMAIN 71640-49-8
2.365 C20 H16 O6

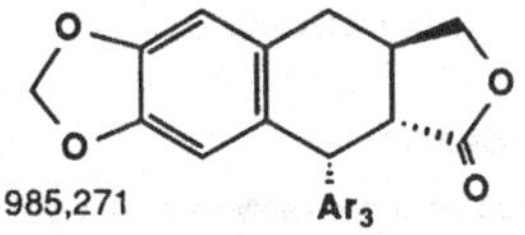

Ar3

Planta Med.1985,271

Commiphora incisa Chiov

The C8'-epimer picropolygamain was also isolated but it is probably an artefact.

POLYGAMATIN 71640-48-7
2.366 C20 H20 O6

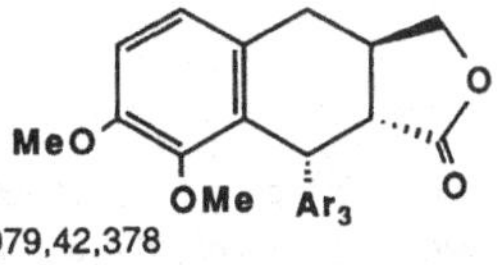

MeO
OMe Ar3

J.Nat.Prod.1979,42,378
(isol, pharm)

Polygala polygama

GLYCOSIDES of the PODOPHYLLOTOXIN GROUP

Their occurrence is reviewed by Cole J.R. & Weidhopf R.M. in "Chemistry of Lignans", Andhra Univ. Press, 1978. For pharmacological aspects see this work chapter 4 also:

Minocha A. & Long B.H.(1984),Biochem.Biophys.Res. Commun.123,165(pharm)

Markkanen T.,Makinen M.L.,Maunuksela E. & Himamen P.(1981),Drugs Exp.Clin.Res,711(pharm)

Schwabe K.& Tschiersch B.(1979),Chem.Abs.,91,175684(synth)

Saito H.,Yoshikawa H.,Nishimura Y.,Kondo S.,Takeuchi T. & Umezawa H.(1986), Chem.Pharm.Bull.,34,3733(synth)

Stringfellow D.A. & Schurig J.E.(1987),Cancer Treat. Rev.,14,291(pharm)

OTHER TETRALIN LIGNANS

CARPANONE 26430-30-8
2.367 C20 H18CO6

Tetrahedron Lett.1969,5159 (isol);
1981,22,4437 (synth)

J.Amer.Chem.Soc.1971,93,
6696(synth)

Cinnamomum laura
The carpano tree

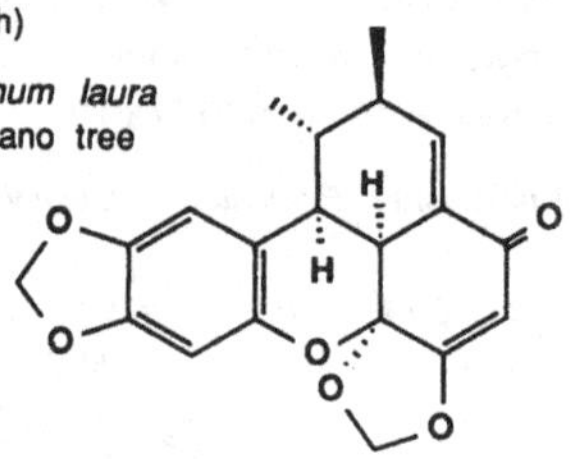

PLICATIC ACID 16462-65-0
2.368 C20 H22 O10

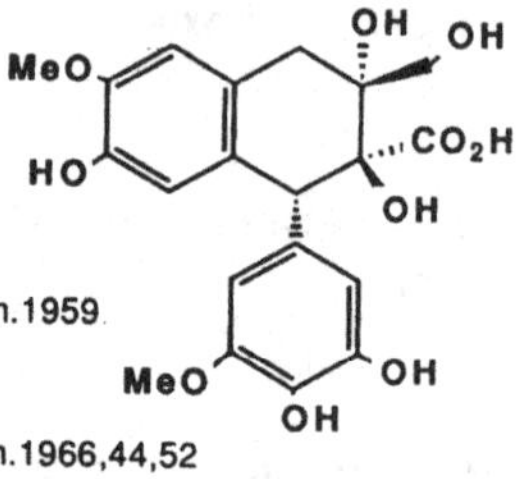

Canad.J.Chem.1959
37,1703(isol)

Canad.J.Chem.1966,44,52
(X-ray); 1967,45,319(ORD)

Western red cedar

APOCOMPOUNDS

The prefix "apo" refers to structures which are or could conceivably be obtained by the elimination of a ring B hydroxyl group with loss of a water molecule.

α - **APOPICROPODOPHYLLIN**
2.371 518-32-1
C22 H20 O7

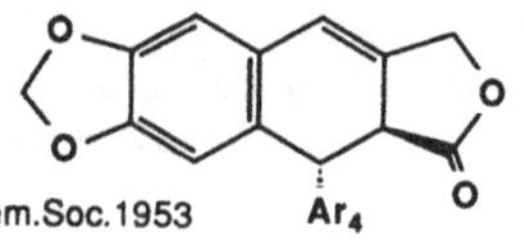

J.Amer.Chem.Soc.1953
75,5916

In vitro product ,the ß-isomer or 8-ene is a potent cytotoxic agent - Indian J.Chem.1985 24B,505.

The 7'-ene is also known as a synthetic product - J.Amer.Chem.Soc.1952,74,5676

β - apopicropodophyllin 477-52-1
γ - apopicropodophyllin 668-01-9

ß-APOPLICATITOXIN 39993-02-7
2.372 C20 H18 O7

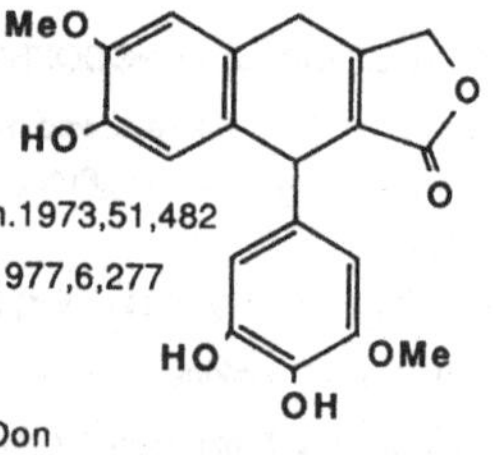

Canad.J.Chem.1973,51,482
Heterocycles,1977,6,277
(synth)

Thuja plicata Don
The first apotetralin to be isolated from a natural source

COLLINUSIN 17990-72-6
2.373 C20 H18 O6

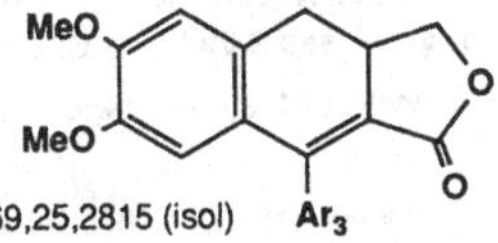

MeO, MeO, O, O, Ar_3

Tetrahedron 1969,25,2815 (isol)
J.Org.Chem.1971,36,3453 (synth)

Cleistanthus collinus

CYCLOGALGRAVIN 73366-01-5
2.374 C22 H26 O4

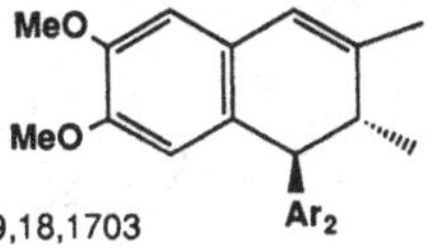

MeO, MeO, Ar_2

Phytochem.1979,18,1703
(isol, CMR)

Araucaria angustifolia

DIHYDROTAIWANIN C C20 H14 06
2.375

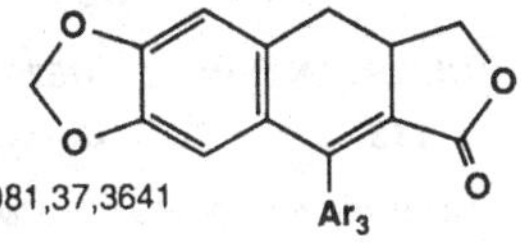

O, O, O, O, Ar_3

Tetrahedron,1981,37,3641

Cleistanthus collinus

(rac) - KONYANIN 90902-18-4
2.376 C20 H16 O6

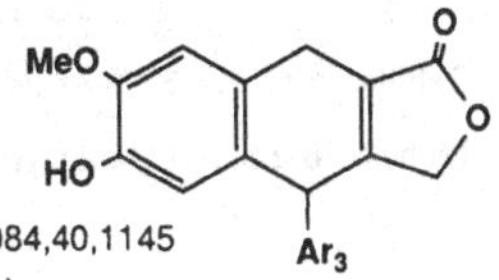

O, MeO, O, HO, Ar_3

Tetrahedron,1984,40,1145
(isol, CD, NOE)

Haplophyllum vulcanicum

(rac) - MAGNOSHININ 86702-02-5
2.377 C24 H30 O6

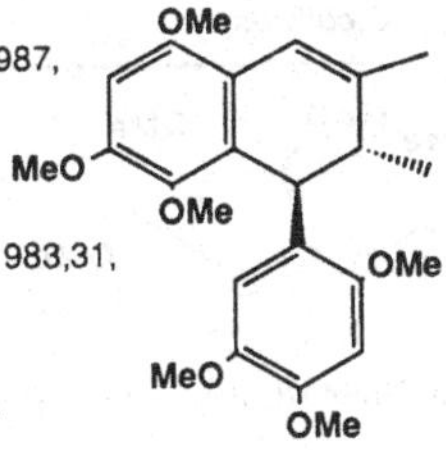

OMe, MeO, OMe, OMe, MeO, OMe

Tetrahedron Lett.1987,
28,2857 (pharm)

Chem.Pharm.Bull.1983,31,
1112 (isol)

Magnolia salicifolia

THOMASIC ACID 21880-64-8
2.378 C22 H24 O9

X = - CH_2OH

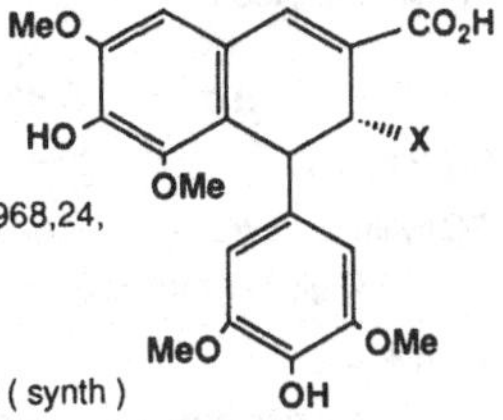

MeO, CO_2H, HO, X, OMe, MeO, OMe, OH

Tetrahedron,1968,24,
1475 (isol)

Tetrahedron
1973,29,3753 (synth)

OTOBAENE* 10240-18-3
2.380 C20 H18 O4

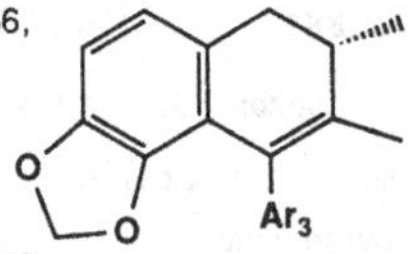

J.Chem.Soc.(C),1966,
1775

O, O, Ar_3

Phytochem.1969,8,497
Virola cuspidata

THOMASIDOIC ACID 26350-60-7
2.379 C22 H22 O10

X = $-CO_2H$
Tetrahedron,1969,25,2325 (isol)

* possibly an artefact cf 2.322

ARYLNAPHTHALENES

CHINENSIN 31888-76-3
2.382 X = H C21 H16 O6

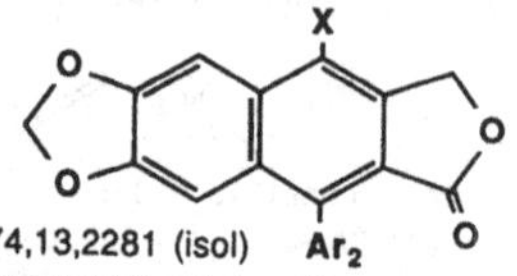

X
O O O O Ar$_2$

Phytochem.1974,13,2281 (isol)
Indian J.Chem.1985,24B,151 (synth)
Polygala chinensis
Bupleurum frutiscens L.

CLEISTANTHIN D 81417-78-9
2.386 C29 H30 O11

Tetrahedron,1981,37,3641 (isol, CMR, MS)

Diphyllin-ß-2,3-5-tri-O-methyl-D-xyloside

CLEISTANTHIN A 25047-48-7
2.383 see 2.394 C28 H28 O11
DIPHYLLIN 3,4-DI-O-METHYL-D-XYLOSIDE

Phytochem.1975,14,1875 (isol)
Tetrahedron,1981,37,3641
Cleistanthus collinus

CLEISTANTHIN B 30021-77-3
2.384 C27 H26 O12
DIPHYLLIN GLUCOSIDE

CLEISTANTHIN C 57576-58-6
2.385 C34 H38 O16

Tetrahedron,1981,37,3641 (CMR, MS)
Phytochem.1975,14,1875 (isol)

Diphyllin-ß-2,3-di-O-methyl-D-xylopyran-
osyl-(1- 4)-ß-D-glucopyranoside
C.collinus

CLEISTANTHIN E 81417-77-8
2.387 C42 H52 O20

Tetrahedron,1981,37,3641
(isol, CMR, MS)

MeO OMe MeO OMe
OH
HO O O O OMe
DIPHYLLIN O O O O
OH

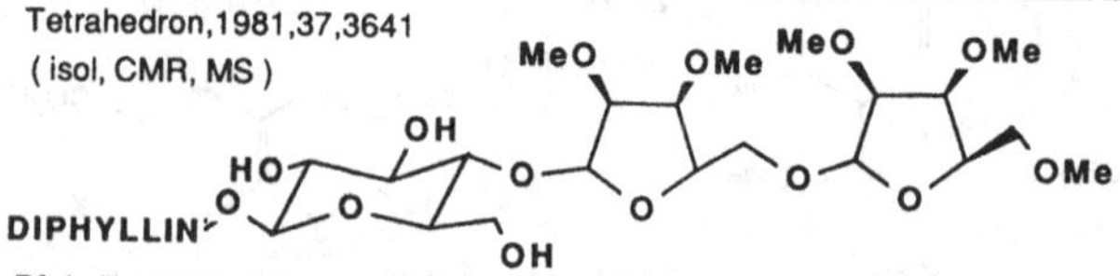

D[phyllin-2,3,5-tri-O-methyl-D-xylofuranosyl-2,3-di-O-methyl-D-xylofuranosyl-
(1- 4)-D-glucopyranoside

C. collinus

CLEISTANTHOSIDE A 86402-39-3
2.388 C34 H38 O16

DIPHYLLIN-4-O-[ß-D-glucopyranosyl(1- 2)]-
ß-3,4-di-O-methyl-D-xylopyranoside

Planta Med.1983,47,227 (isol,chromatog)
Cleistanthus patulus Muell.

CLEISTANTHOSIDE B 109145-66-6
2.389 C27 H26 O11

Phytochem.1987,26,1153

DIPHYLLIN-4-O-ß-D-4"-O-methylxylo-
pyranoside
Cleistanthus patulus

DAURINOL 79862-78-5
2.390 $C_{20}H_{14}O_6$

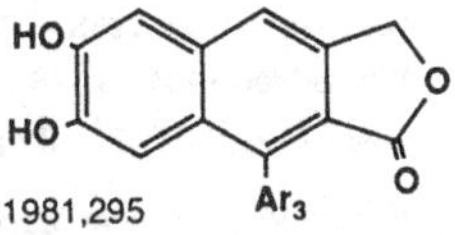

Khim.Prir.Soedin,1981,295
Chem.Nat.Compd.1987,23,63 (CMR)

Haplophyllum dauricum

DEHYDROCONIDENDRIN $C_{20}H_{16}O_6$
2.392

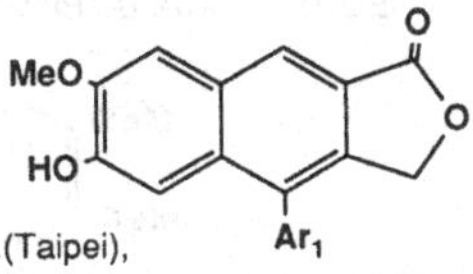

J.Chin.Chem.Soc.(Taipei), 1982,29,213 (MS); Chem Abs.1982,97,212698

Juniperus formosana

DIPHYLLIN 22055-22-7
2.394 X = OH $C_{21}H_{16}O_7$

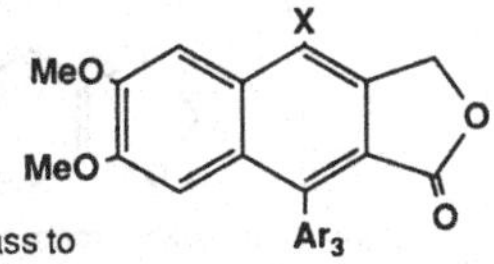

The first of this class to be isolated - J.Pharm.Soc.Japan, 1961,81,1596.
Chem.Pharm.Bull.1969,17,1878 (structure)
Planta Med.1979,36,200 (pharm)
J.Nat.Prod.1986,49,348 (pharm)

Diphylleia grayi *Justicia procumbens*
Haplophyllum hispanicum
Tawania cryptomeriodes

GLUCOSIDE (CLEISTANTHIN B, 2.384)
Phytochem.1987,26,1153

APIOSIDE - Planta Med.1985,217
see TUBERCULATIN 2.419

DEHYDROGUAIARETIC ACID
2.391 20601-86-9
$C_{20}H_{20}O_4$

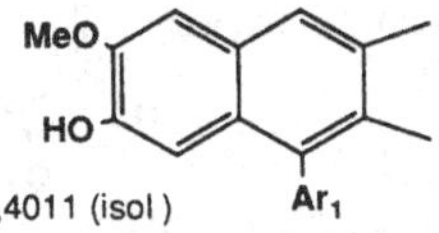

J.Chem.Soc.1964,4011 (isol)
Taxus baccata

DEHYDROPODOPHYLLOTOXIN
2.393 $C_{22}H_{20}O_8$

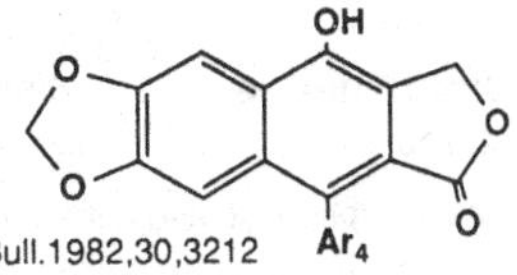

Chem.Pharm.Bull.1982,30,3212 (isol)
Chem.Pharm.Bull.1987,35,4162 (CMR, synth)

Hemandia ovigera

DIPHYLLIN 7 - DEOXY 17951-19-8
2.395 X = H $C_{21}H_{16}O_6$
(JUSTICIDIN B)
J.Org.Chem.1971,36,3453 (synth)
Tetrahedron Lett.1970,923 (isol)

Justicia procumbens

DIPHYLLIN ISO 53965-06-3
2.396 $C_{21}H_{16}O_7$
(7 - HYDROXY CHINENSIN)
Anales Quim.1975,71,109 (isol)
Tetrahedron,1978,34,1011 (synth)

DIPHYLLIN ISO - 7 - DEOXY
(CHINENSIN, 2.382) ref. as for 2.396

HAPLOMYRTIN 105037-92-1
2.397 $C_{20}H_{14}O_7$
(4 - DEMETHYL DIPHYLLIN)
Phytochem.1986,25,1949 (isol, NOE)
Haplophyllum myrtifolium

HELIOXANTHIN 18920-47-3
2.398 C20 H12 O6

J.Amer.Chem.Soc.
1951,73,100 (isol)

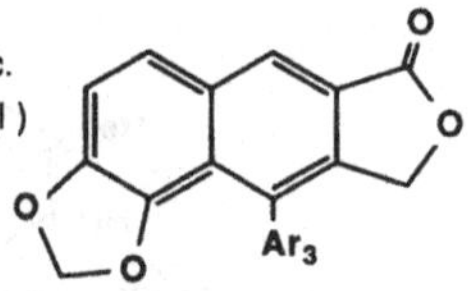

O O Ar3 O O

J.Chem.Soc.(C),1969,693 (PMR)
Indian J.Chem.1979,18B,391 (synth)
Planta Med.1982,44,154 (CMR)

Heliopsis scabra *Justicia simplex*
J. flava

JUSTICIDIN B 17951-19-8
2.400 C21 H16 O6

As for diphylin (2.394) but X = H

J.Chem.Soc.Chem.Commun.1980,995 (synth)
Fitoterapia,1981,52,37 (isol)
J.Nat.Prod.1986,49,1175 (pharm)

H. tuberculatum H. popovii
Boenninghausenia albiflora
Sesbania drummondii

JUSTICIDIN D 27041-98-1
2.402 C21 H14 O7
(NEOJUSTICIDIN A)
4,5-methylenedioxy-4,5-demethoxy - justicidin C
Tetrahedron,1970,26,4301 (isol, PMR)

J. procumbens L.

JUSTICIDIN F 30403-00-0
2.404 C21 H14 O7

O - METHYL TAWANIN E (2.418)

Chem.Pharm. Bull.1977,25,1803 (synth, isol)
J.Org.Chem.1981,46,3881 (synth)

J. procumbens L.

JUSTICIDIN A 25001-57-4
2.399 see 2.394 C22 H18 O7
(O - METHYL DIPHYLLIN)

Tetrahedron Lett.1965,4167 (isol)
Planta Med.1981,43,148 (PMR, lanthanide)
J.Nat.Prod.1986,49,348 (pharm)

Justicia hayatai var *decumbens*
Haplophyllum tuberculatum

JUSTICIDIN C 17803-12-2
2.401 C22 H18 O7
(NEOJUSTICIDIN B, RETROJUSTICIDIN A)

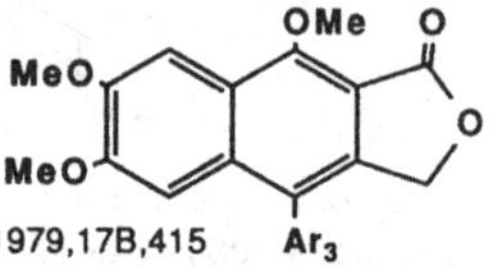

OMe O MeO MeO O Ar3

Indian J.Chem.1979,17B,415

Justicia simplex *J. procumbens*

JUSTICIDIN E 27792-97-8
2.403 C20 H12 O6

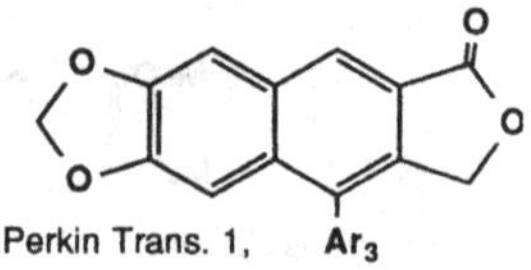

O O O O Ar3

J.Chem.Soc.Perkin Trans. 1, 1973,1266 (synth)
Tetrahedron Lett.1986,27,365 (synth)

J. simplex

JUSTICIDIN P 95585-91-4
2.405 $C_{23}H_{20}O_8$

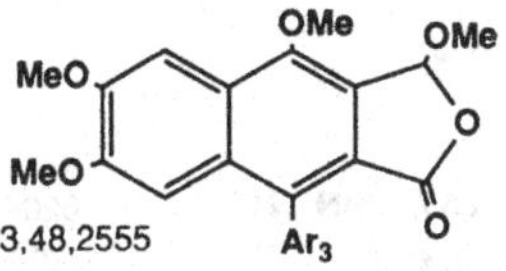

J.Org.Chem.1983,48,2555
(isol, synth)
Justicia extensa

OROSUNOL 82012-44-0
2.407 R = Me $C_{21}H_{16}O_7$

Planta Med.1982,44,154
(isol,UV,MS,PMR)
Justicia flava

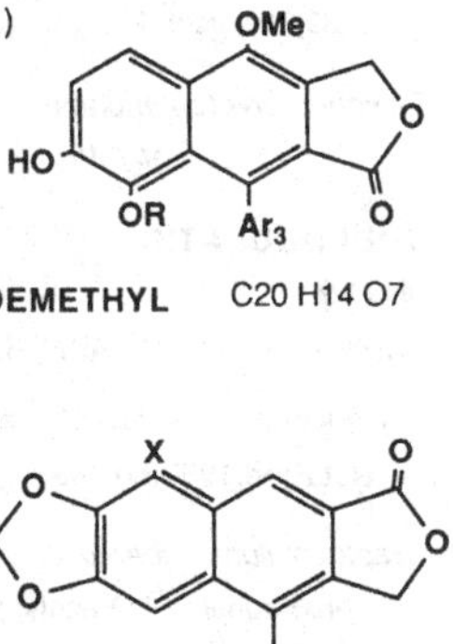

OROSUNOL 3-DEMETHYL $C_{20}H_{14}O_7$
2.408 R = H

PROSTALIDINS

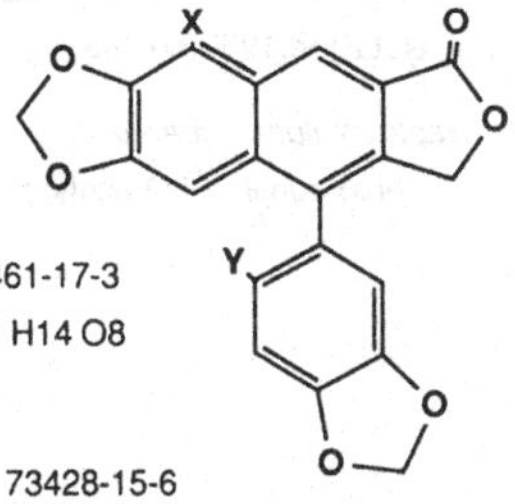

A
2.411 73461-17-3
X = OMe; $C_{21}H_{14}O_8$
Y = OH

B
2.412 73428-15-6
$C_{22}H_{16}O_8$
X = Y = OMe

C
2.413 73428-14-5
X = H; Y = OH $C_{20}H_{12}O_7$

Chem. & Ind.1979,854 (isol, MS, pharm)
Justicia prostrata

JUSTICINOL 75340-41-9
2.406 $C_{20}H_{12}O_7$

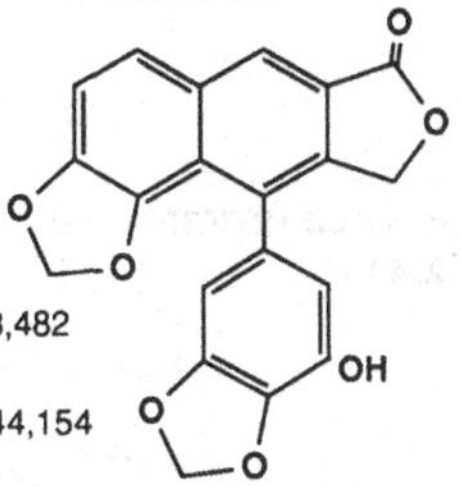

J Nat.Prod.1980,43,482
(isol, PMR)
Planta.Med.1982,44,154
(CMR)
Justicia flava

PLICATINAPHTHALENE 26560-82-7
2.409 X = H, Y= OH $C_{20}H_{16}O_7$

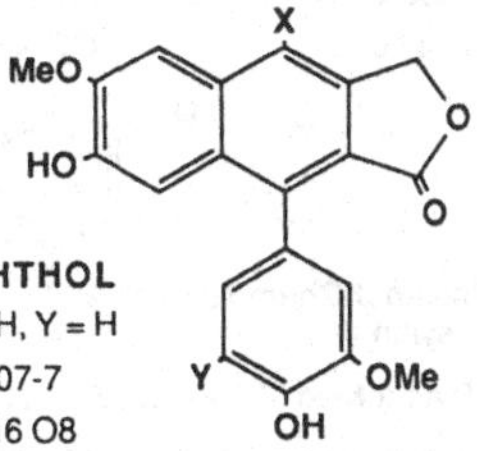

PLICATINAPHTHOL
2.410 X = OH, Y = H
22127-07-7
$C_{20}H_{16}O_8$

Canad.J.Chem.1969,47,457,4495
(chromatog, isol)
Thuja plicata Donn

RETROCHINENSIN 5707-96-0
2.414 $C_{21}H_{16}O_6$

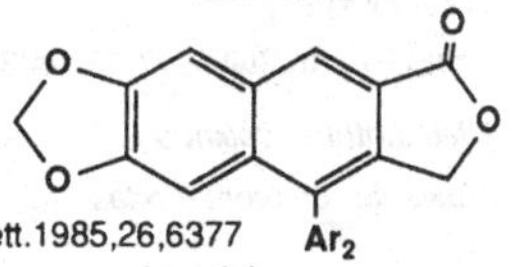

Tetrahedron Lett.1985,26,6377
(synth)
J.Chem.Soc.Chem.Commun.1979,165 (synth)
Chem & Ind.1979,854 (pharm)
Justicia prostrata

RETROJUSTICIDIN B 82001-16-9
2.415 C21 H16 O 6

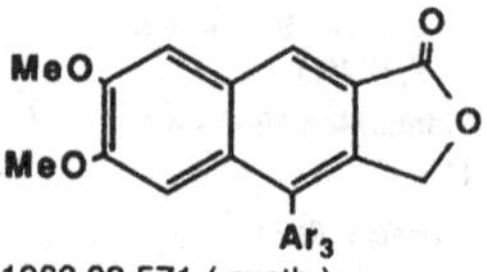

MeO, MeO, O, O, Ar_3

Tetrahedron Lett.1982,23,571 (synth)

TAWANIN C 14944-34-4
2.416 C20 H12 O6

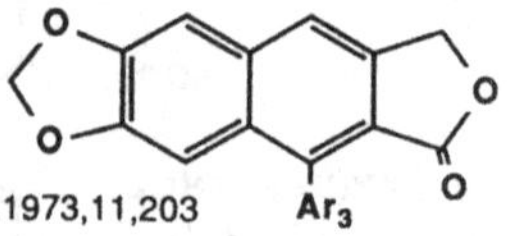

O, O, O, O, Ar_3

Indian J.Chem.1973,11,203
(synth)

Planta Med.1983,47,227 (isol)

Tetrahedron,1981,37,3641 (MS)

TAWANIN E 22743-05-1
2.418 C20 H12 O7

7 - HYDROXY TAWANIN C

Tetrahedron Lett.1967,849; 1969,1079
(isol, PMR)

Chem.Pharm.Bull.1977,25,1803 (synth)

Cleistanthus collinus

Tawania cryptomeriodes Hayata

TAWANIN H 102040-04-0
2.417 C20 H16 O7

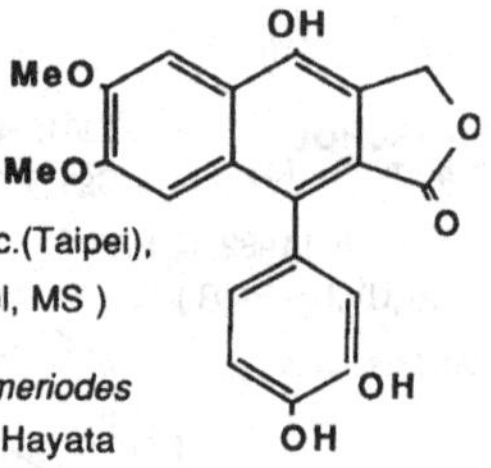

OH, MeO, MeO, O, O, OH, OH

J.Chin.Chem.Soc.(Taipei),
1985,32,381 (isol, MS)

Tawania cryptomeriodes
Hayata

TUBERCULATIN 90706-10-8
2.419 C26 H24 O11

DIPHYLLIN - ß - D - APIOFURANOSIDE

Phytochem.1984,23,151 (isol, MS)

J. Nat.Prod.1987,50,748

Haplophyllum tuberculatum

H. buxbaumii (O-acetyl)

DIBENZOCYCLOOCTADIENES

(-) - **ARALIANGINE** 85915-63-5
2.420 R = angeloyl C26 H26 O8

Tetrahedron Lett.1986,27,2871
(PMR)

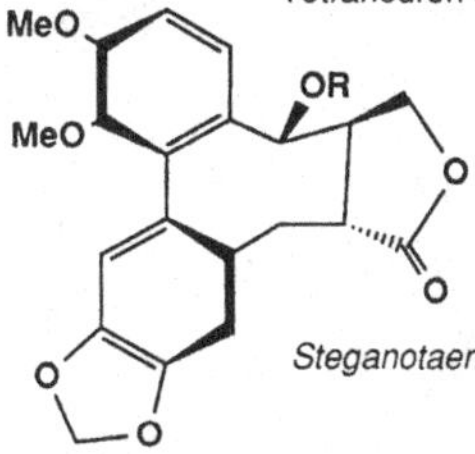

MeO, MeO, OR, O, O, O, O

Steganotaenia araliacea

(+) - **DEOXYSCHIZANDRIN** 61281-38-7
2.422 C24 H32 O6
DIMETHYL GOMISIN J

Tetrahedron Lett.1986,27,5377
(synth)
Chem.Pharm.Bull.
1985,33,3599 (synth)
Chem.Pharm.Bull.
1975,23,3296 (isol)

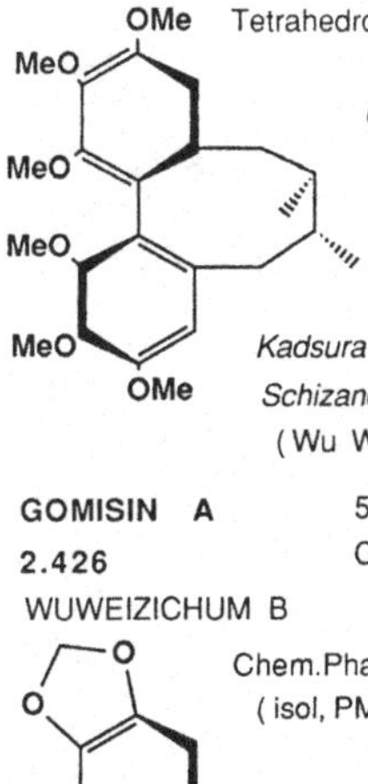

OMe, MeO, MeO, MeO, MeO, OMe

Kadsura coccinea
Schizandra chinensis
(Wu Wei Zi)

GOMISIN A 58546-54-6
2.426 C23 H28 O7
WUWEIZICHUM B

Chem.Pharm.Bull.1979,27,1383
(isol, PMR, CD, NOE, pharm)

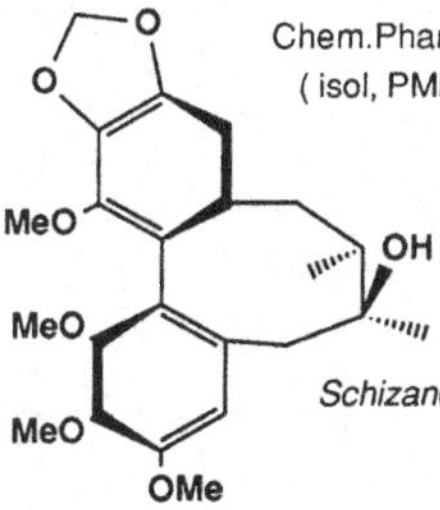

O, O, MeO, MeO, MeO, OMe, OH

Schizandra chinensis Baill

BINANKADSURIN A 77165-79-8
2.421 C22 H26 O7
R = acetyl, or angeloyl or caproyl

Chem.Pharm.Bull.1981,29,123
(isol, CMR)

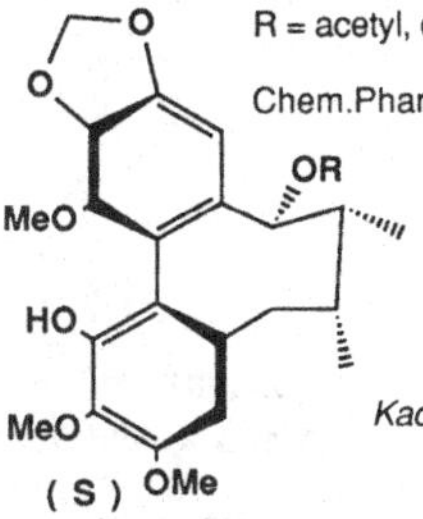

O, O, OR, MeO, HO, MeO, OMe

Kadsura japonica

(S)

EUPODIENONE 1 R = Ac 75539-66-1
2.423 C25 H32 O8

EUPODIENONE 2 R = H 39350-63-5
2.424 C23 H30 O7

EUPODIENONE 3 R = Ph.CO 39350-66-8
2.425 C30 H36 O8

Aust.J.Chem.1980,33,1823
(PMR, lanthanide)

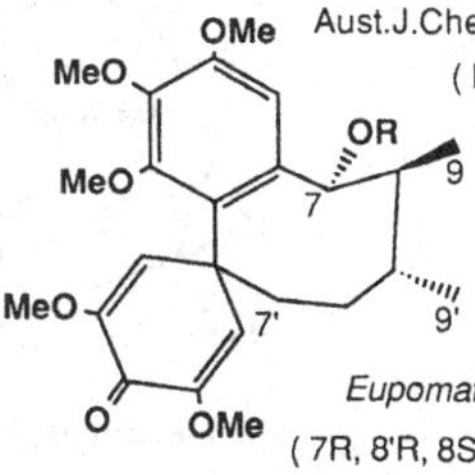

OMe, MeO, MeO, OR, 7, 9, MeO, 7', 9', O, OMe

Eupomatia laurina
(7R, 8'R, 8S)

GOMISIN B 58546-55-7 C28 H34 O9
2.427 SCHIZANTHERIN C R = angeloyl

GOMISIN C
2.428 SCHIZANTHERIN A R = benzoyl
58546 56-8 C30 H32 O9

Chem Pharm.Bull.1979,27,1383
Planta.Med.1984,50,213
(pharm)

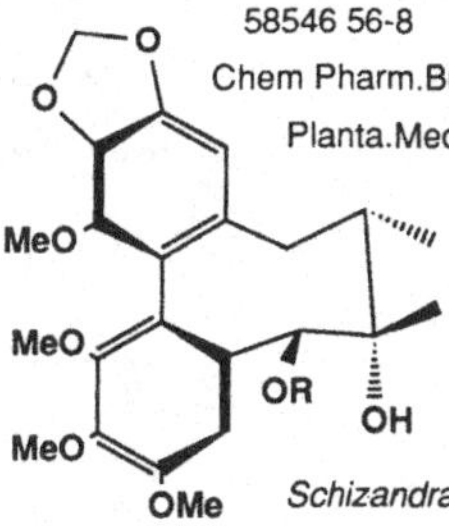

O, O, MeO, MeO, MeO, OMe, OR, OH

Schizandra chinensis

GOMISIN D 60546-10-3

2.429 X = OH; Y = Me C28 H34 O10

GOMISIN E 72960-21-5

2.430 X = Me; Y = H C28 H34 O9

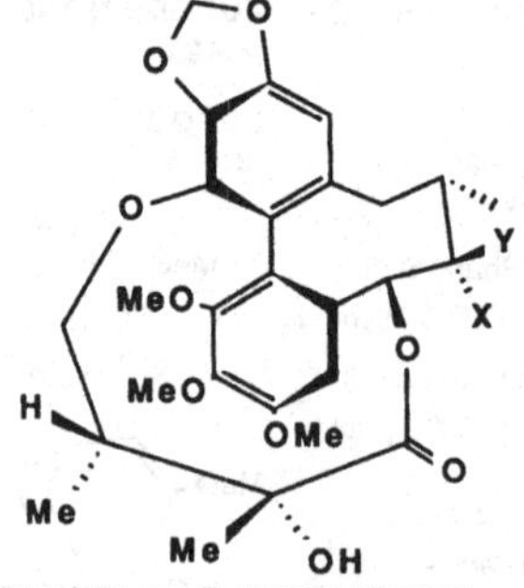

Chem.Pharm.Bull.1979,27,1395
(isol,MS,PMR)

Planta Med.1985,51,297

S. chinensis *Kadsura coccinea*

GOMISIN F 62956-47-2
2.431 R = angeloyl C28 H34 O9

GOMISIN G
2.432 R = benzoyl 62596 -48-3
C30 H32 O9

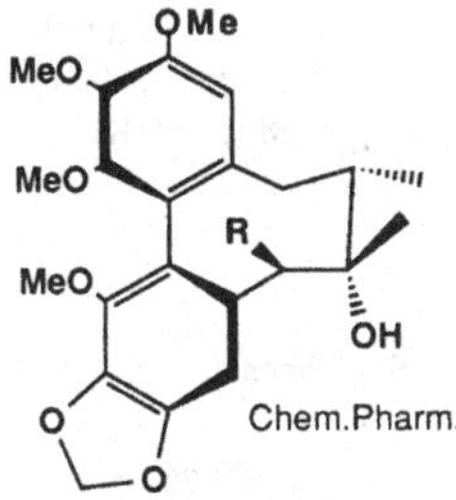

OMe, MeO, MeO, R, MeO, OH, O, O

Chem.Pharm.Bull.1979,27,1383 (isol, NOE)

Schizandra chinensis Baill

(-) - **GOMISIN K$_1$** 75629-20-8
2.435 C23 H30 O6

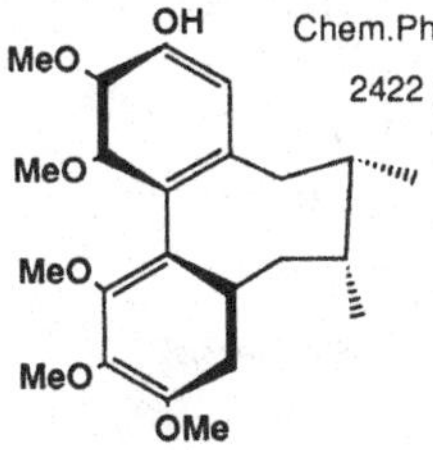

OH, MeO, MeO, MeO, MeO, OMe

Chem.Pharm.Bull.1980,28, 2422 (isol, NOE, CMR)

(+) - **GOMISIN K$_2$** 75684-44-5
2.436 R = H, X = OMe C23 H30 O6
(+) - **GOMISIN K$_3$** 69363-14-0
2.437 R = Me, X = OH C23 H30 O6

Literature as 2.435

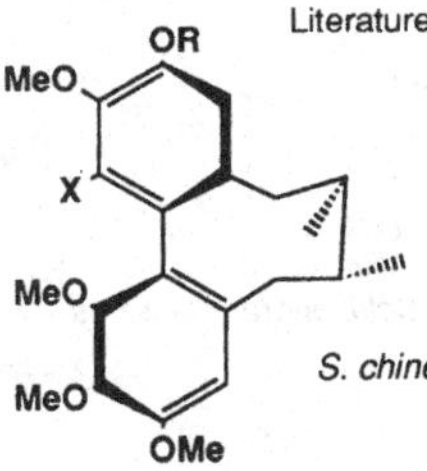

OR, MeO, X, MeO, MeO, OMe

S. chinensis

GOMISIN H 66056-20-0
2.433 (NORSCHIZANDRIN) C23 H30 O7

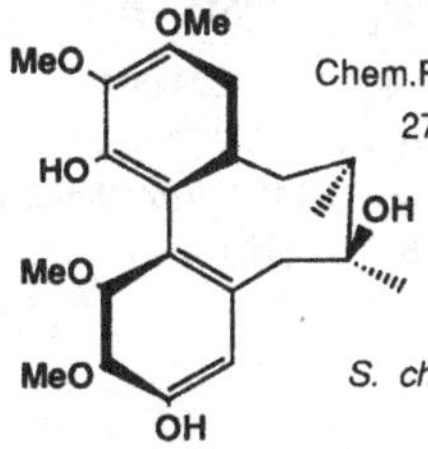

OMe, MeO, HO, OH, MeO, MeO, OH

Chem.Pharm.Bull.1979, 27,1576 (isol, PMR)

S. chinensis

GOMISIN J 66280-25-9
2.434 (see gomisin N) C22 H28 O6

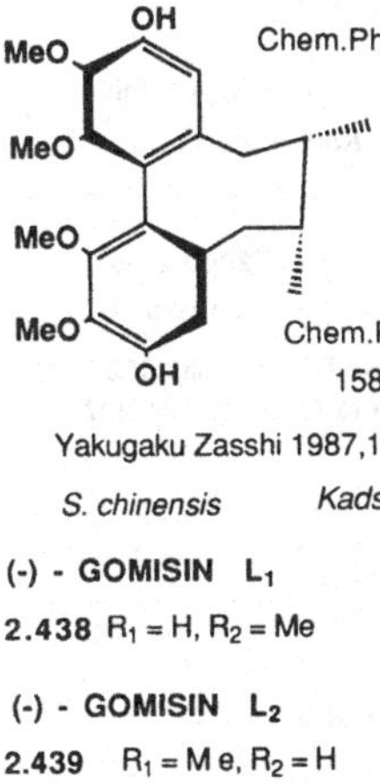

OH, MeO, MeO, MeO, MeO, OH

Chem.Pharm.Bull.1978,26, 682 (isol, CD)

Chem.Pharm.Bull.1979,27, 1583 (NOE, biosynth)

Yakugaku Zasshi 1987,107,720 (pharm)

S. chinensis *Kadsura coccinea*

(-) - **GOMISIN L$_1$** 82425-43-2
2.438 R_1 = H, R_2 = Me C22 H26 O6

(-) - **GOMISIN L$_2$** 82425-44-3
2.439 R_1 = Me, R_2 = H C22 H26 O6

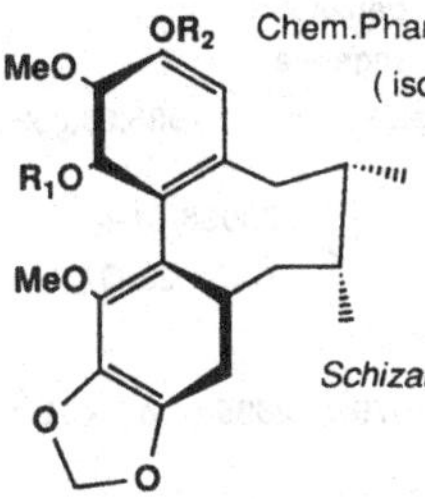

OR_2, MeO, R_1O, MeO, O, O

Chem.Pharm.Bull.1982,30,132 (isol, CMR, HPLC)

Schizandra chinensis

(rac) - GOMISIN M_1 82467-50-3
2.440 R_1 = H, R_2 = Me C22 H26 O6
R- (+)- GOMISIN M_1 95152-95-7

(+) - GOMISIN M_2 82425-45-4
2.441 R_1 = Me, R_2 = H C22 H26 O6
Chem.Pharm.Bull.1982,30,132 (isol,CMR,CD)
Planta Med.1985,51,297 (isol, MS)

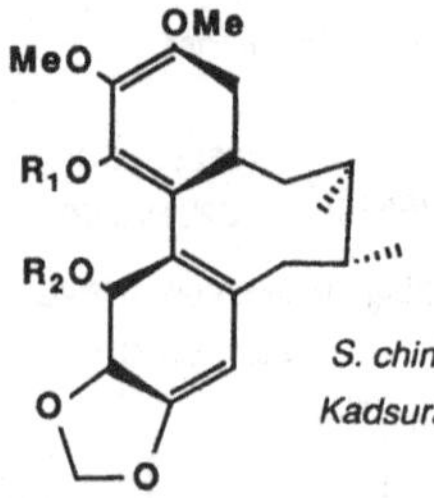

MeO, OMe, R_1O, R_2O, O, O

S. chinensis Baill
Kadsura coccinea

GOMISIN O EPI 72960-22-6
2.443 C23 H28 O7

Chem.Pharm.Bull.1982,30,3202
Planta Med.1985,51,297

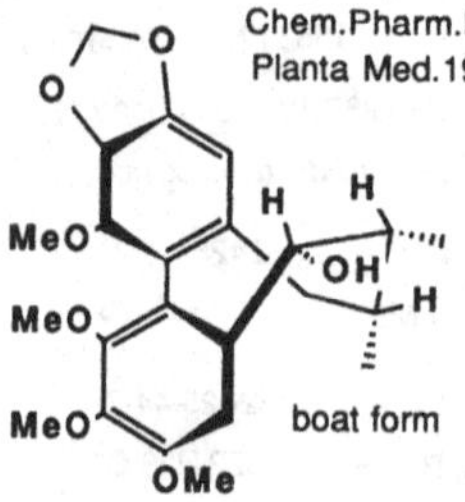

O, O, MeO, MeO, MeO, OMe, H, H, OH, H

boat form

Kadsura coccinea benzoate
S. chinensis angelate
S. rubriflora (Acta Pharm.Sin.1985,20,832)

GOMISIN O 73036-31-4
2.444 C23 H28 O7

The 7-ß-epimer of 2.443
Chem.Pharm.Bull.1979,27,2695 (isol, NOE)
S. chinensis

GOMISIN N 69176-52-9
2.442 (see gomisin J) C23 H28 O6

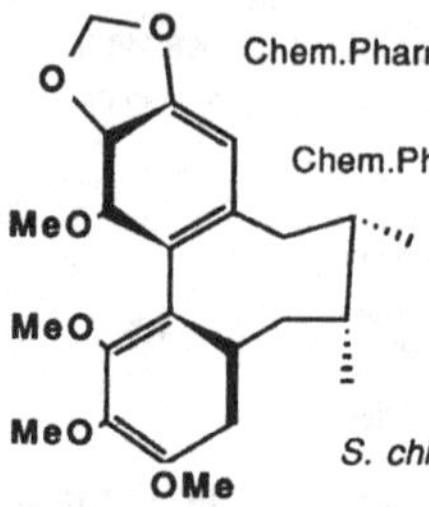

Chem.Pharm.Bull.1978,26,3257 (isol)
Chem.Pharm.Bull.1979,27, 2695 (PMR)

O, O, MeO, MeO, MeO, OMe

S. chinensis

GOMISIN P 69980-01-4
2.445 C23 H28 O8

As 7-angelate with 7-tiglate
Chem.Pharm.Bull.1980,28,3357 (isol, NOE, GLC)

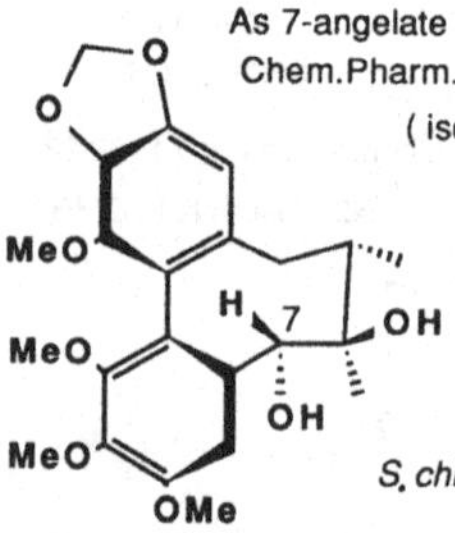

O, O, MeO, MeO, MeO, OMe, H, 7, OH, OH

S. chinensis

GOMISIN Q 66096-74-0
2.446 C24 H32 O8

As 7-angelate

Chem.Pharm.Bull.1979,27, 2536

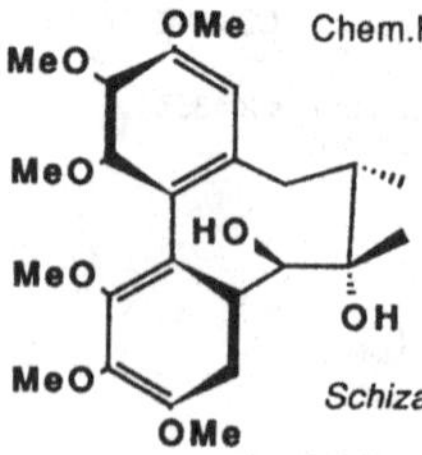

OMe, MeO, MeO, MeO, MeO, OMe, HO, OH

Schizandra chinensis

GOMISIN R 83864-72-6
2.447 $C22\ H24\ O7$

as 2.473, boat with subst. 7-ß-OH

Chem.Pharm.Bull.1982,30,3207 (isol, CMR)
Schizandra chinensis Baill.

(-) - **KADSURARIN** 51670-41-8
2.451 $C30\ H36\ O11$

Bull.Chem.Soc.Japan,1977, 50,1824 (isol , NOE)

OAc
MeO
MeO
OH
O-angeloyl
MeO
OMe

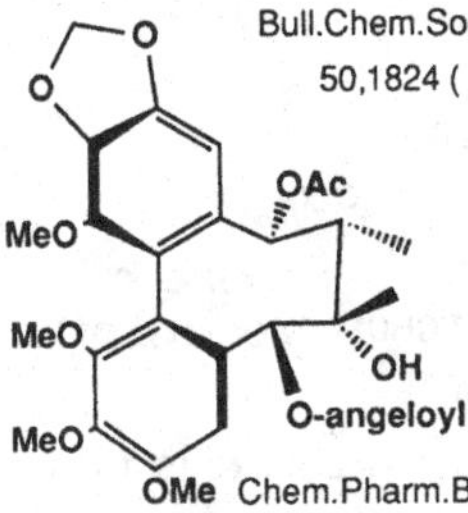

Chem.Pharm.Bull.1979,27,1383; 1981,29,123

Kadsura japonica Dunal *S. chinensis*

(+) - **KADSUTHERIN** 99481-39-7
2.453 R = angeloyl $C27\ H32\ O7$

Planta Med.1985,51,297 (isol, CD, MS)

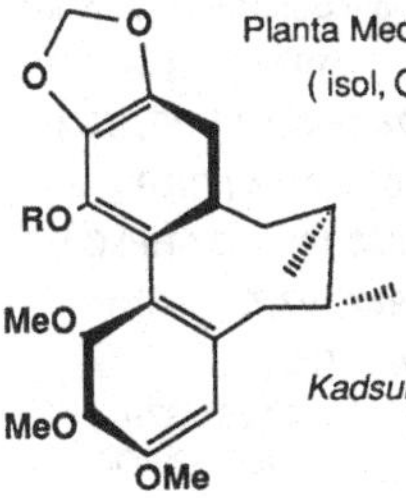

RO
MeO
MeO
OMe

Kadsura coccinea

KADSURANIN $C23\ H28\ O6$
2.448 as 2.449; R_1,R_2 = $-CH_2-$; R_3,R_4 = Me

KADSURANIN ISO 82467-52-5
2.449 R_1,R_2 = Me; R_3,R_4 = $-CH_2-$ $C23\ H28\ O6$

OR_4
Planta Med.1985,51,297 (isol)

R_3O
MeO
MeO
R_2O
OR_1

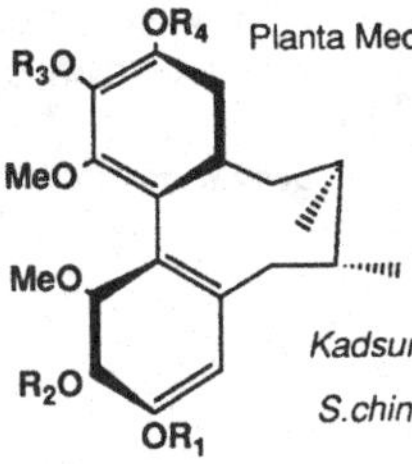

Kadsura coccinea
S.chinensis

KADSURANIN NEO 115181-68-5
2.450 $C23\ H26\ O7$

Planta Med.1988,54,45 (isol, CD,NOE)

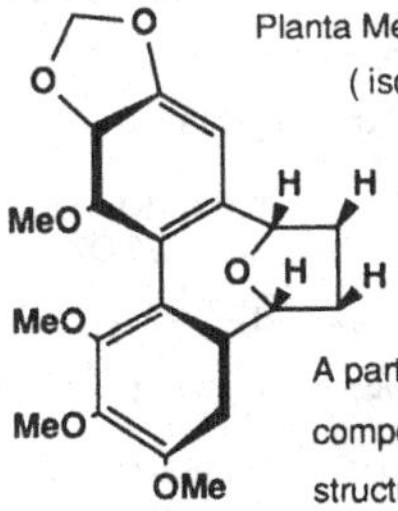

MeO
H H
O H H
MeO
MeO
OMe

Kadsura coccinea

A partially characterised compound of this possible structure isolated from-

Clerodendron inerme Phytochem.1981,20,2757

(-) - **KADSURIN** 51670-40-7
2.452 $C25\ H30\ O8$

Tetrahedron Lett.1973, 4257 (pharm)

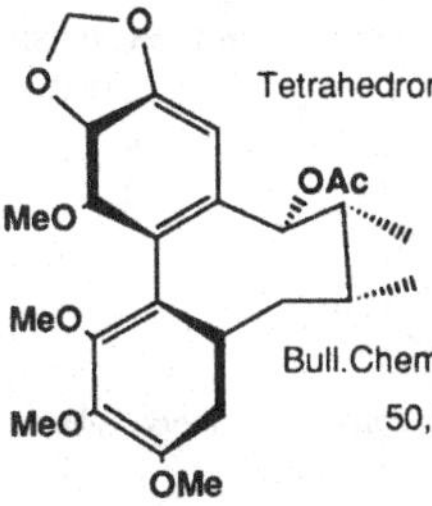

OAc
MeO
MeO
MeO
OMe

Bull.Chem.Soc.Japan,1977, 50,1824 (isol, NOE)

Kadsura japonica Dunal
S. chinensis

(-) - RUBSCHIZANDRIN 82467-51-4 WITH **(-) - RUBSCHIZANTHERIN** 102637-03-6

2.454 (see gomisin N) C23 H28 O6 **2.455** C25 H30 O8

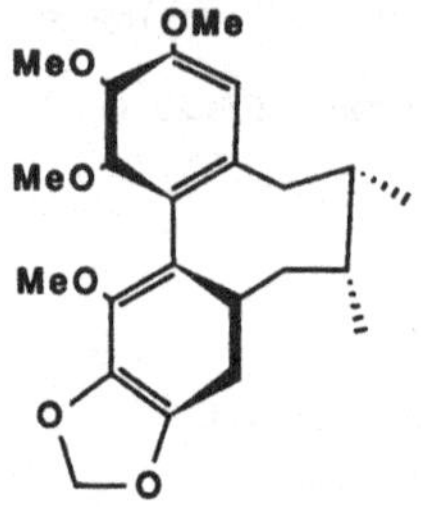

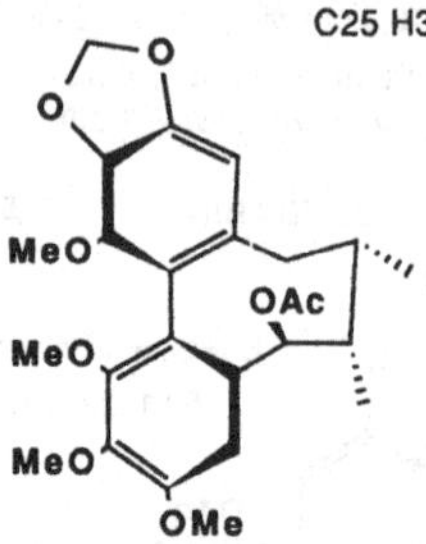

MeO, OMe, MeO, MeO, O, O

Acta Pharm.Sin.1985,20,832
(isol, CMR, CD,MS)

Schizandra rubrifolia Rhed et Wils

O, O, MeO, OAc, MeO, MeO, OMe

SCHIZANDRIN 7432-28-2

2.456 (WUWEIZICHUM A) C24 H32 O7

$X = OH, R_1, R_2 = Me$

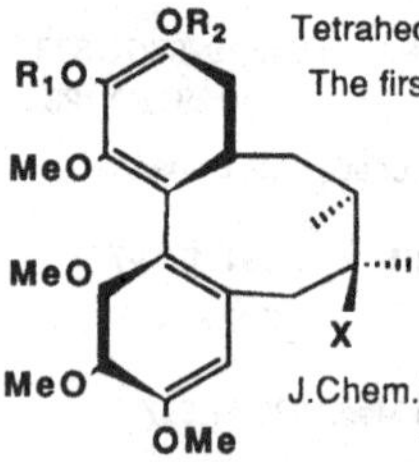

OR_2, R_1O, MeO, MeO, MeO, OMe, X

Tetrahedron Lett.1961,730
The first known example of this class

J.Chem.Soc.Chem.Commun. 1978,480 (synth)

Chem.Pharm.Bull.1978,26,328 (NOE, CD)
1980,28,2414 (CMR)

Yakugaku Zasshi ,1983,103,743 (HPLC)
1987,107,720 (pharm)

Schizandra chinensis Baill

(+) - SCHIZANDRIN DEOXY 61281-38-7

2.458 C24 H32 O6

As 2.456 $X = H, R_1 = R_2 = Me$

Tetrahedron Lett.1962,361 (synth)

SCHIZANDROL B 58546-54-6

2.459 C23 H28 O7

As 2.456 $X = OH, R_1, R_2 = -CH_2-$

Planta Med.1985,51,297 (isol, X-ray)

(+) - SCHIZANHENOL 69363-14-0

2.460 $R_1, R_3, R_4 = Me$ $R_2 = H$ C23 H30 O6

Chem.Abs.1979,90,103711 (isol)

(+) - SCHIZANHENOL B C22 H26 O6

2.461 $R_1 = H; R_2 = Me, R_3, R_4 = -CH_2-$

Acta Pharm.Sin.1985,20,832
(CMR, CD)

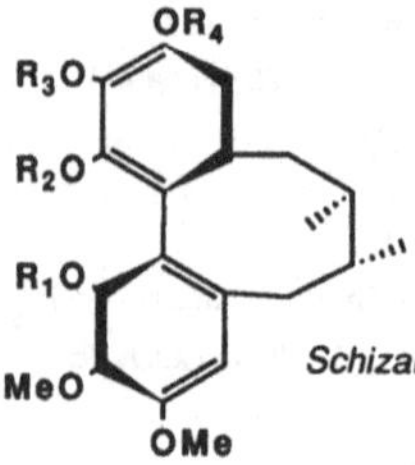

OR_4, R_3O, R_2O, R_1O, MeO, OMe

Schizandra henryi Clarke

γ - SCHIZANDRIN 61281-37-6

2.457 (WUWEIZISU B) C23 H28 O6

$X = H, R_1, R_2 = -CH_2-$

Izv.Akad.Nauk SSSR,1964,1036 (isol)

Chem.Pharm.Bull.1982,30,132 (CMR)

Sci.Sin.1976,19,276 (isol as wuweisu B)

SCHIZANTHERIN A 58546-56-8
(GOMISIN C 2.428) C30 H32 O9

SCHIZANTHERIN B 58546-55-7
(GOMISIN B 2.427) C28 H34 O9

SCHIZANTHERIN C 64938-51-8
2.462 C28 H34 O9
R = angeloyl; R_1,R_2 = Me; R_3,R_4 = $-CH_2-$

SCHIZANTHERIN D 64917-82-4
2.463 C29 H28 O9
R = benzoyl; R_1,R_2, = R_3,R_4 = $-CH_2-$

SCHIZANTHERIN E 64917-83-5
2.464 C30 H34 O9
R = benzoyl; R_1,R_2,R_3 = Me; R_4 = H

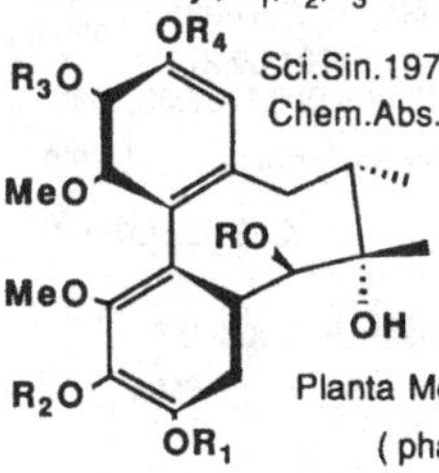

OR_4 R_3O MeO RO MeO OH R_2O OR_1

Sci.Sin.1978,21,483 (isol,NOE)
Chem.Abs.1982,99,019670 (pharm)
Planta Med.1984,50,213 (pharm, review)

Schizanthera spenanthera *S. chinensis*
S. rubrifolia *S. lancifolia*

(+) - **STEGANE** NEOISO** 87084-98-8
2.469 C23 H26 O7

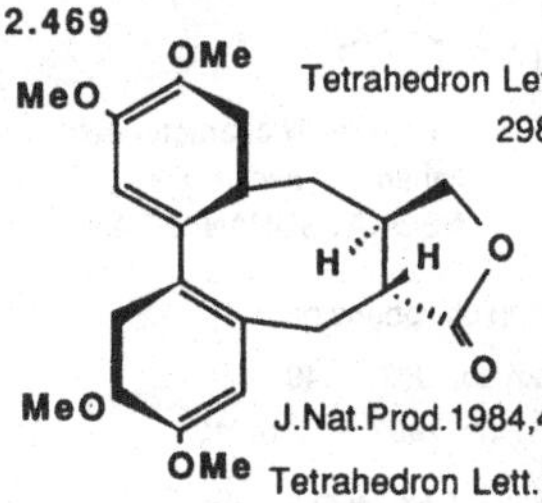

OMe MeO H H O (R) O MeO OMe

Tetrahedron Lett.1983,24, 2987 (isol)
J.Nat.Prod.1984,47,600 (CD)
Tetrahedron Lett.1986,27,1785 (synth)

S. araliacea

(-) - **STEGANACIN** 41451-68-7
2.465 R = Ac C24 H24 O9

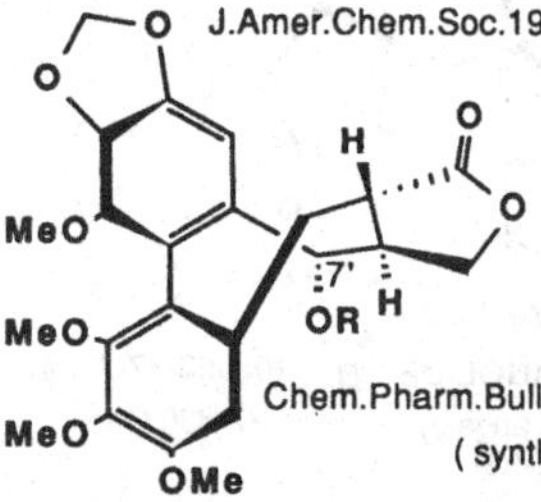

O O MeO H O O MeO 7' OR H MeO OMe

J.Amer.Chem.Soc.1973,95,1335* (isol, pharm)
Chem.Pharm.Bull.1985,33,609 (synth. NOE)
Tetrahedron Lett.1980,21,2973 (synth, + form)
J.Chem.Soc.Perkin Trans. 1,1982,521 (synth)
J Amer.Chem.Soc.1987,109,5446 (synth)

(-) - **STEGANANGIN** 41451-69-8
2.466 R = angeloyl C27 H28 O9

(-) - **STEGANOL** 41451-71-2
2.467 R = H C22 H22 O8

(-) - **STEGANONE** 41451-70-1
2.468 C-7' = carbonyl

Steganotaenia araliacea

* This original report gave the incorrect absolute stereochemistry

** Neoisostegane is the first reported natural stegane.
For details of synthetic (S)- NEOSTEGANE and related compounds see:
Tetrahedron Lett.1978,3613
1979,1409
J.Nat.Prod.1984,47,600

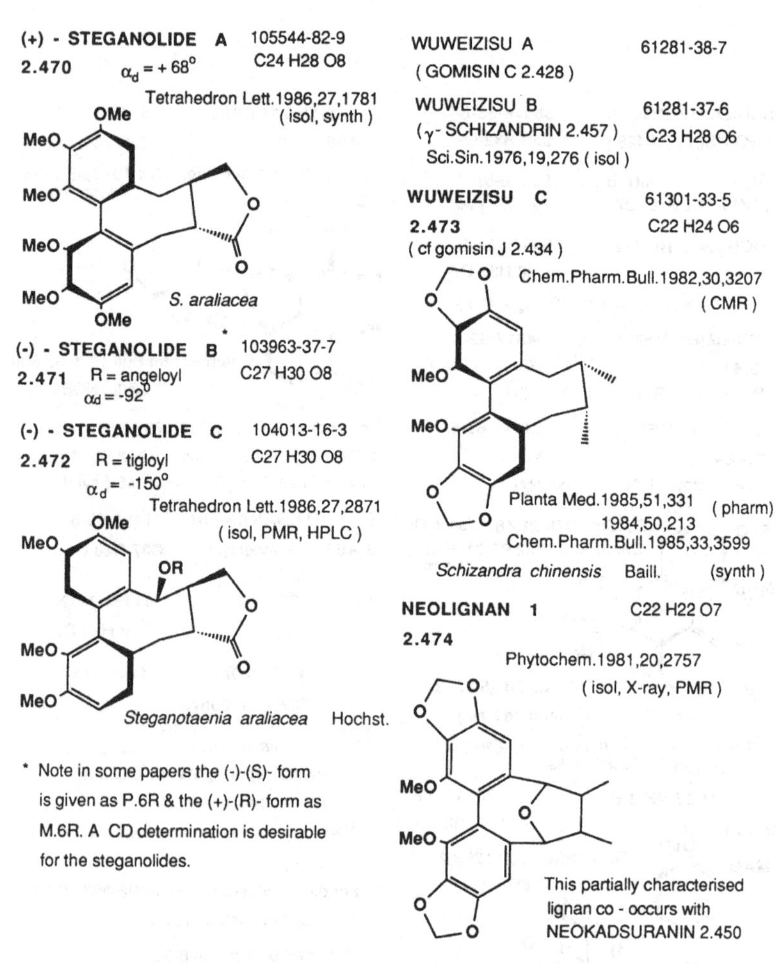

A review of the pharmacological activity of this group of lignans has been published: Song W. and Peigen X. Zhongcaoyao,1982,13,40 - 48
Chem.Abs. 1982, 96,214219

For a review of sources of lignans and neolignans see Gottlieb,O.R. and Yoshida,M in " Natural Products Extraneous to the Lignocellulosic Cell Wall of Woody Plants, ed Rowe,J.W. and Kirk,C.H., Springer, 1989.

3

Biological and clinical properties of podophyllotoxin and other lignans

Introduction

This chapter is divided into four sections. The first is a synopsis of the use of lignans and podophyllotoxin (**3.1**; X = OH) in ancient and modern clinical medicine. The second section summarises the known mechanism of action of podophyllotoxin in a variety of eukaryotic cell systems. The third summarises recent literature on the identification of naturally occurring lignans in man. The last section reviews recent studies describing new biological or physiological properties of plant lignans.

Use of lignans in folk and modern medicine

Lignans have a long and fascinating history beginning with their use as folk remedies by many diverse cultures. The earliest recorded medicinal use of lignans dates back over 1000 years (Kelly and Hartwell, 1954). According to the Leech book of BALD, an early English medical book (Cockayne, 1961), the roots of wild Chervil (which contains several lignans, including deoxypodophyllotoxin (**3.1**; X = H) were used in a salve for treating cancer. About 400–600 years ago, both the natives of the Himalayas and the American Penobscot Indians of Maine discovered independently that the resin produced from an alcoholic extract of the roots and rhizomes of *Podophyllum* perennials was a cathartic and poison. The American Indians believed that the roots possessed effective venom antidotes and applied the resin topically to treat poisonous snake bites, or

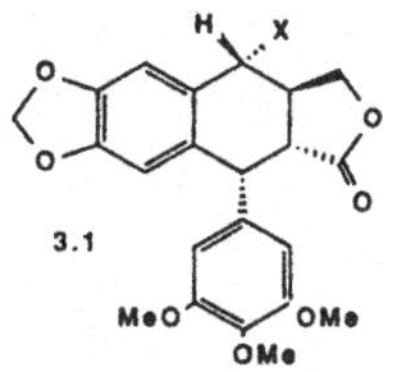

H X O O O O 3.1 MeO OMe OMe

used the roots as a suicide agent and poison (Kelly and Hartwell, 1954). The American colonists also used the extracts of the roots and rhizomes of May apple or American Mandrake as purgative, anthelminthic, chlorotic and vesicant agents (Kelly and Hartwell, 1954; Hartwell and Schrecker, 1958).

Lignan-rich plant products were also active ingredients in the treatment of disease in Chinese and Japanese folk medicine. Unfortunately, many of the active ingredients of these plant products have not been scientifically tested as therapeutic agents. For example, *Kadsura coccinea* [Lem. A.C. Smith (Schizandraceae)] is a climbing plant widely distributed in the southern part of China. The dried roots and stems of this plant are listed in the Chinese Pharmacopoeia for the treatment of rheumatoid arthritis, gastric and duodenal ulcers (Pharmacopoeia of the People's Republic of China, 1977). These roots contain at least ten dibenzocylo-octadiene lignans which have recently been isolated and identified structurally (Lian-niang *et al.*, 1985). Typical examples are shown in Scheme 3.1 (structures **3.2–3.4**) and the others are listed in the registry (**2.422–2.474**). Another example is the use of the dried bark of *Fraxinus mandshurica* (which contains at least eight different lignans) as a substitute for the Chinese crude drug 'qin pi' (Wu *et al.*, 1983). Tsukamoto *et al.* (1984a) isolated five known lignans and three new lignans from the bark of *Fraxinus mandschurica* Rupr. var *japonica* Maxim (oleaceae) (Scheme 3.2, structures **3.5–3.7**). The dried bark of *F. japonica* Blume has been used for

Scheme 3.1 Lignans of *Kadsura coccinea*

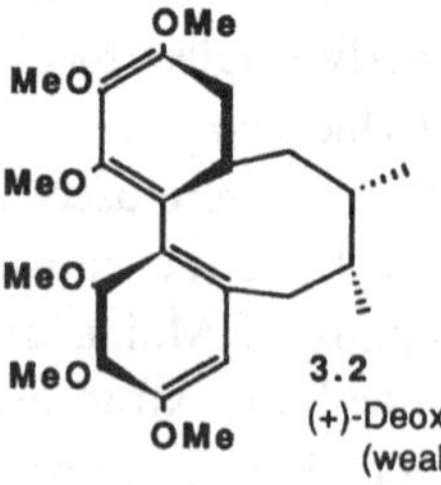

3.2 (+)-Deoxyschizandrin (weakly active)

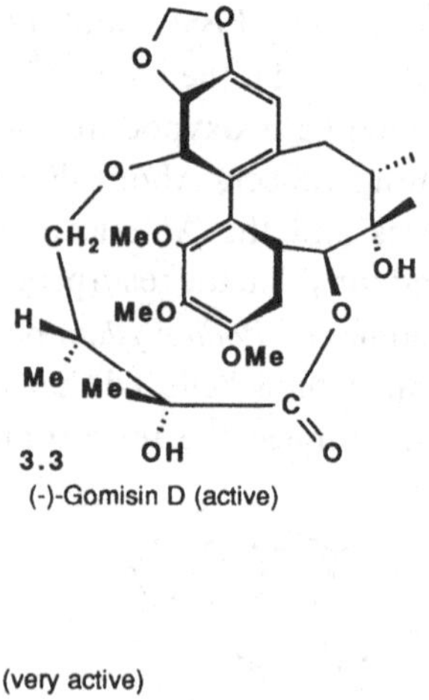

3.3 (-)-Gomisin D (active)

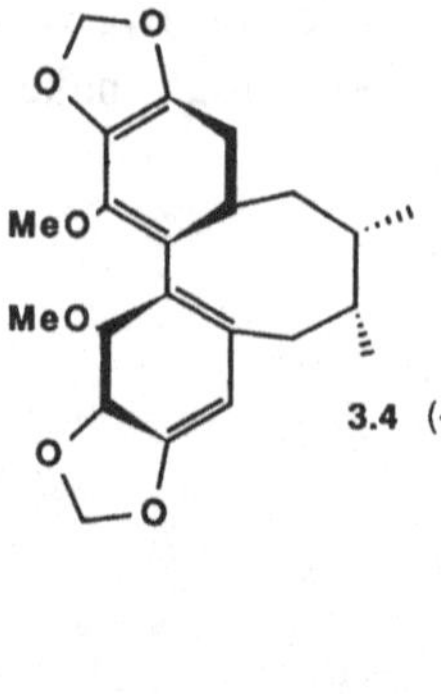

3.4 (+)-Wuweizisu C (very active)

Scheme 3.2 Lignans from *Fraxinus mandshurica*

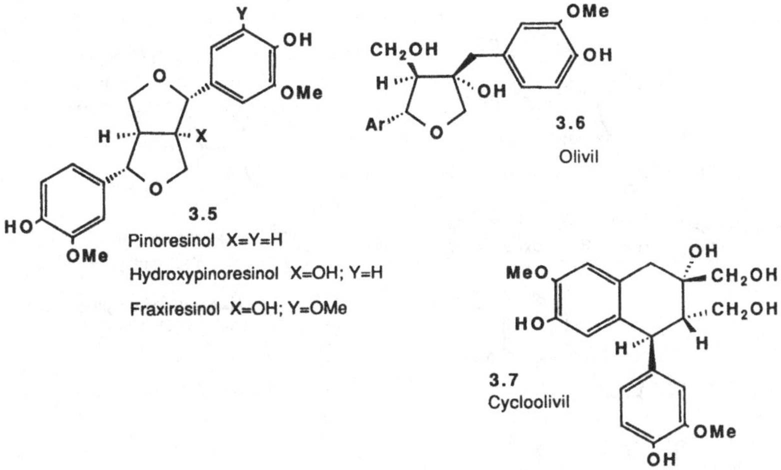

3.5
Pinoresinol X=Y=H
Hydroxypinoresinol X=OH; Y=H
Fraxiresinol X=OH; Y=OMe

3.6
Olivil

3.7
Cycloolivil

hundreds of years in Japan as a diuretic, an antipyretic, an analgesic and an antirheumatic agent (Kariyone and Kimura, 1976). Several of these lignans have been isolated (Kariyone and Kimura, 1976) and have demonstrated inhibitory activity *in vitro* against cyclic adenosine monophosphate phosphodiesterase and may possess other biological properties as well.

The bark of *Olea europaea* L. has been used since olden times in the Orient as an antipyretic, an antirheumatic, a tonic and a remedy for scrofula. Scrofula is a tuberculous-like disease of the lymph glands generally localised in the neck. Tsukamoto *et al.* (1984c) have isolated a group of new lignans (Scheme 3.3) from the bark of *Olea europaea* L. which they believe may be responsible for the bark's therapeutic properties.

The widespread use of several of these lignans as folk medicines suggests that lignans are potential parent compounds for development of new classes of chemotherapeutic agents. Thus, the isolation and chemical identification of these lignans coupled with an understanding of their biological activity will be a fruitful area for future research.

(i) *History of podophyllotoxin*

The May apple or American mandrake (*Podophyllum peltatum*) is a member of the Berberidaceae. Podophyllin is the alcoholic extract of the roots and rhizome and contains the active principle, podophyllotoxin (**3.1**; X = OH), and other lignans. Podophyllin was such a popular cathartic and cholagogue in America that it was included in the U.S. Pharmacopeia from 1820 until 1942. In 1942, it was removed from U.S. Pharmacopeia

Scheme 3.3 New lignans from *OLea europa*

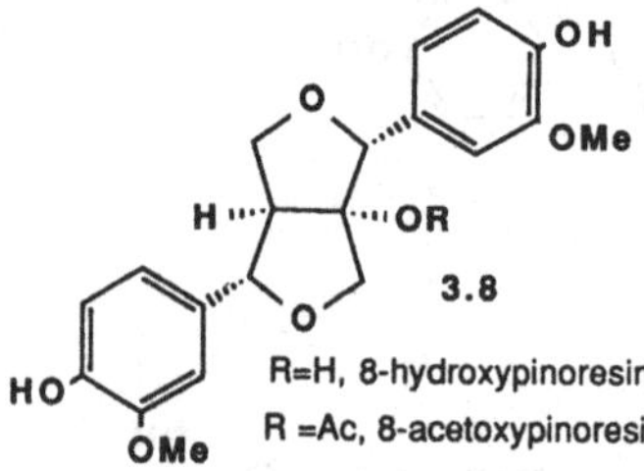

3.8
R=H, 8-hydroxypinoresinol
R =Ac, 8-acetoxypinoresinol

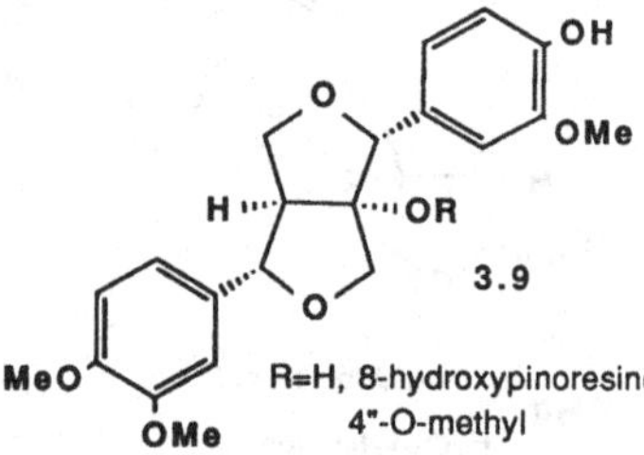

3.9
R=H, 8-hydroxypinoresinol 4"-O-methyl
R=Ac, 8-acetoxypinoresinol 4"-O-methyl

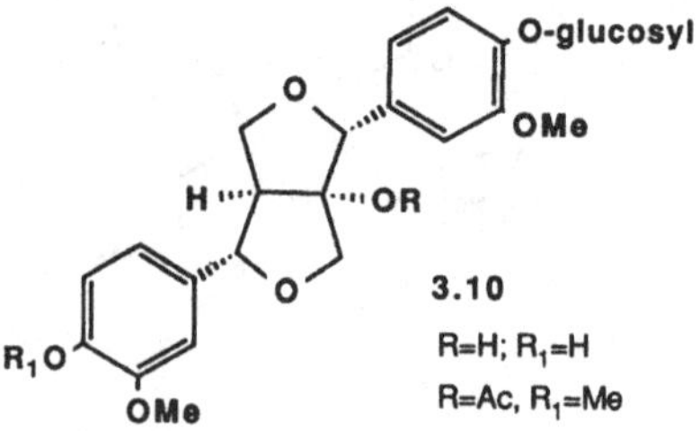

3.10
R=H; R_1=H
R=Ac, R_1=Me

because of its severe gastrointestinal toxicity. Ironically, that same year, Kaplan (1942) reported that topical application of podophyllin was an effective treatment for venereal warts. Currently, podophyllin remains as an effective therapy for anogenital warts (Perez-Figarado and Baden, 1976) and may also be effective in the treatment of nasal papillomas (Bennett and Grist, 1985). Interestingly, other members of the Berberidaceae family, in particular *Berberis* and *Epimedium* have been used in the treatment of warts and solid tumours in China and India from at least the second century A.D. (Pettit, 1977).

Modern treatment of venereal warts involves the application of 25% podophyllin in tincture of benzoin to the wart area. However, podophyllin is extremely toxic (Cassidy *et al.*, 1982; Fisher, 1981) and its ingestion or absorption from the mucosal surfaces in toxic doses will produce nausea, vomiting, and diarrhoea, followed by delirium, stupor and coma. Renal impairment and myelosuppression with a decrease in leukocytes and platelets are other side effects of podophyllotoxin. In fact, the fatal dose of 25% podophyllin has been estimated to be between 0.3 and 0.6 g or about half a teaspoon.

Von Krogh (1981) considered the toxicity of crude podophyllin and questioned its use in dermatology when purified podophyllotoxin (**3.1**; X = OH) is readily available. In fact, Von Krogh (1981) demonstrated that the quantity of active ingredients in podophyllin can vary from 3 to 8% in different podophyllin preparations. In addition, there are two varieties of podophyllin, the American and the Indian, the latter being more toxic and irritating to tissues.

The variability of active ingredients in different podophyllin preparations and the potential toxicity of podophyllin led Von Krogh to treat venereal warts with purified podophyllotoxin. He found that two applications with 8% podophyllotoxin, separated by a 72-h interval, was most effective: 54% of the patients treated became wart-free and rarely exhibited cutaneous irritation. Thus, the use of podophyllotoxin instead of podophyllin resin in the treatment of venereal warts should be investigated further for it could lower the risk of systemic toxicity and decrease the frequency of severe adverse inflammatory reactions.

Condyloma acuminatum, the most common form of venereal wart is caused by a papilloma virus; these are small double-stranded DNA viruses. The effect of podophyllotoxin on venereal, planters and nasal warts is thought to be mediated by its antiviral properties of which little is known. There is however, one report describing the antiviral properties of podophyllotoxin and some of its derivatives. Markkanen *et al.* (1981) tested the effects of 21 podophyllotoxin lignans on the replication of herpes simplex 1 virus (a DNA virus) in primary amnion cell cultures. The concentrations of podophyllotoxin and their derivatives which have effective antiviral activities were remarkably lower than concentrations which exhibit cytotoxic or cytostatic properties in amnion cell cultures. Markkanen *et al.* (1981) found that of all the natural lignans tested, podophyllotoxin exhibited the highest chemotherapeutic index (6000). The chemotherapeutic index is defined as the ratio of maximum tolerated concentration divided by minimum inhibitory concentration. For podophyllotoxin this index corresponds to a concentration of 10 ng/ml of drug. At this concentration, podophyllotoxin inhibits viral replication and is non-toxic to normal cells. In fact, amnion cell cultures could tolerate doses of up to 60 000 ng/ml in an overnight incubation.

The successful application of podophyllotoxin in the treatment of other infections caused by DNA viruses or RNA viruses remains largely unexplored. The effective treatment of warts caused by venereal papilloma with podophyllotoxin and the recent progress made in monitoring and maintaining papilloma infections in cell cultures, indicate the need to investigate further the mechanism of action of podophyllotoxin as an antiviral agent.

There is another application of podophyllotoxin in dermatology. A recent study by Lassus and Rosen (1986) tested the effects of podophyllotoxin on 152 patients suffering from stable psoriasis vulgaris. One daily dose of podophyllotoxin was applied during a 16-week double blind study. After two weeks of treatment with podophyllotoxin, three-quarters of the patients showed good to excellent response in those psoriatic lesions. This is the first study to suggest that podophyllotoxin may be effective in treating psoriasis. Furthermore, this study complements earlier clinical reports in which podophyllin was shown to possess anti-psoriatic properties.

Biological properties and mechanism of action of podophyllotoxin

(i) *Microtubules*

While there are numerous biological properties of podophyllotoxin, the action of podophyllotoxin as a microtubule inhibitor is understood best. Microtubules are tubular polymers whose protomeric unit, consisting of α- and β-tubulin, forms a heterodimer (see McKeithan and Rosenbaum, 1984, for a review) and are the dynamic constituents of the cytoskeleton. The cytoplasm of eukaryotic cells contains a soluble pool of unpolymerised tubulin protomers as well as an organised array of microtubules. Microtubules can be rapidly assembled or disassembled in response to various stimuli with little or no net change in the total tubulin content of the cell. Cytoplasmic microtubules have been implicated in such diverse processes as cellular motility, intracellular transport, secretion, organisation of the cytoplasm, organisation of protein in the membrane, and growth factor signalling.

Podophyllotoxin is a well established spindle poison which disrupts the dynamic equilibrium of assembled and disassembled microtubules *in vitro* and *in vivo*. *In vitro*, podophyllotoxin inhibits the assembly of tubulin into microtubules in a dose-dependent fashion. *In vivo*, podophyllotoxin inhibits microtubule assembly and disrupts the dynamic equilibrium between microtubules and tubulin, thereby inducing the disassembly of microtubules into tubulin. The net result is destruction of the cytoskeletal framework in the cytoplasm and in the spindle fibres. The most pronounced cytotoxic effect of podophyllotoxin is its inhibition of cell division in metaphase by destroying the spindle fibres and preventing the duplicated chromosomes from separating. This results in the arrest of cell division at the mitotic stage of the cell cycle.

Podophyllotoxin is not the only drug which disrupts the assembly and function of microtubules. Other microtubule inhibitors include colchicine, steganacin, taxol, and the vincristine-type alkaloids. The mechanism of podophyllotoxin and colchicine disruption of microtubules is similar. Wilson and Friedkin (1967) first showed that podophyllotoxin competes for the colchicine-binding site on grasshopper embryo tubulin. However,

Table 3.1. *The effects of podophyllotoxin derivatives on microtubule assembly*

	R_1	R_2	R_3	A[a] MK_1(μM)	B[b] ID_{50}(μM)
Podophyllotoxin (PT)	OH	O	C=O	0.51	0.6
Deoxypodophyllotoxin (DPT)	H	O	C=O	1.2	0.5
PT cyclic ether	OH	O	H_2	5.2	1
DPT cyclic ether	H	O	H_2	—	0.8
DPT-cyclopentane	H	H_2	H_2	—	5
DPT-cyclopentanone	H	C=O	H_2	—	5
PT cyclic sulphide	OH	S	H_2	—	10
DPT cyclic sulphide	H	S	H_2	—	10
PT cyclic sulphone	OH	SO_2	H_2	—	c
DPT cyclic sulphone	H	SO_2	H_2	—	c
Podophyllic acid 2-ethyl hydrazide	OH	3.11A	3.11A	4.5	—
Picropodophyllin	OH	3.11B	3.11B	10	30

[a] Inhibition constants (K_1) were determined for the inhibition of [^{3}H]colchicine binding to mouse brain tubulin. From Kelleher (1977).

[b] Chicken brain tubulin (1.0 mg/ml) was incubated at 37°C in a standard reaction mixture with various concentrations of drug to determine the ID_{50} or 50% inhibition dose. From Loike *et al.* (1978). Reprinted with permission: copyright Cancer Research Journal.

[c] No inhibition of microtubule assembly was seen at 100 μM.

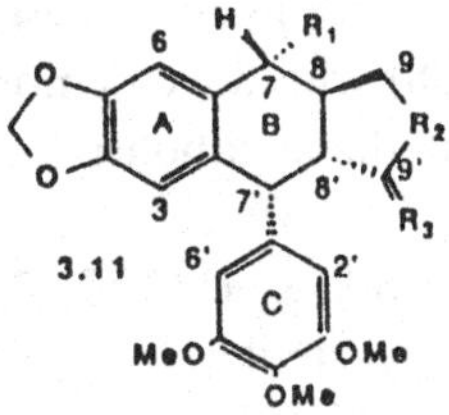

3.11 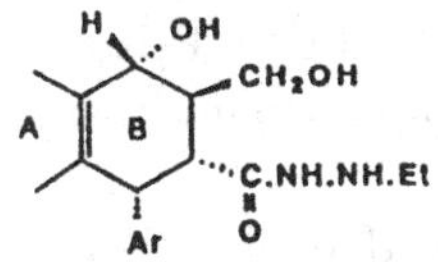 3.11A 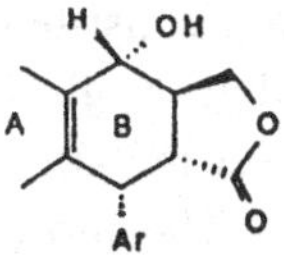 3.11B

podophyllotoxin binding to tubulin does differ from that of colchicine. For example, podophyllotoxin binding to tubulin is rapid and readily reversible; in contrast, colchicine displays slow rates of binding to tubulin and the process is irreversible. Podophyllotoxin binding to tubulin is less temperature-dependent than the binding of colchicine to tubulin. These differences have led to speculation that podophyllotoxin and colchicine occupy overlapping rather than identical sites on tubulin (Bhattacharyya and Wolff, 1974).

There are several structure–activity relationship studies (Brewer *et al.*, 1979; Kelleher, 1977; Loike *et al.*, 1978) (see Table 3.1) which have examined the interaction of podophyllotoxin with tubulin and its effect on microtubule assembly *in vitro*. The conclusion from structure–activity relationship studies of Loike *et al.* (1978), Brewer *et al.* (1979), and Kelleher *et al.* (1977) suggest that the B- and lactone rings of these compounds are involved in their interaction with tubulin. Configuration, size and/or hydrophobic character of substituents at the C-7 position in the B-ring and steric constraint at the R_2 position of the lactone ring dramatically affect the activity of podophyllotoxin analogues on tubulin. In fact, Etoposide and Teniposide (glucoside derivatives of 4′-demethylepipodophyllotoxin, see Chapter 4) are not inhibitors of microtubule assembly *in vitro* or *in vivo* because the sugar moieties at the C-7 position of the B-ring are thought to block sterically the interaction of the drug with the tubulin binding site. Brewer *et al.* (1979) showed by NMR analysis of Etoposide that the glucoside moiety preferentially occupies a position over the lactone ring of the molecule and may interfere with the interaction of this ring with tubulin.

Can cells metabolise the glucoside derivatives of podophyllotoxin such as Etoposide to a form which could act as an inhibitor of microtubule assembly? Several studies (Pelsor *et al.*, 1978; Phaire-Washington *et al.*, 1980) have suggested that Etoposide is not metabolised by cells to the non-glucoside derivative, 4′-demethylepipodophyllotoxin. Phaire-Washington *et al.* (1980) showed that murine macrophages treated with Etoposide continue to express assembled microtubules as measured by immunofluorescent staining using anti-tubulin antibodies. In contrast, cells treated with podophyllotoxin do not exhibit intracellular microtubules. These studies are consistent with those by Pelsor *et al.* (1978) who have demonstrated that the glucoside moiety remains intact during the metabolism of Etoposide in man.

The morpholino derivatives of benzyl-benzodioxole (**3.12**, **3.13**) are interesting compounds because of their structural similarity to podophyllotoxin, steganacin and colchicine (Batra *et al.*, 1986). The benzyl ring (B-ring of these compounds) appears analogous to the trimethoxylbenzene ring (C-ring) of podophyllotoxin. These compounds do not appear to possess antitumour activities *in vivo*, yet they are potent inhibitors of tubulin polymerisation and compete with the colchicine binding site on tubulin. The morpholino compounds which are chemically similar to the lactone ring of podophyllotoxin were the least active in their ability to inhibit tubulin assembly into microtubules (Batra *et al.*, 1986). A single methoxy group at position C-4′ on the B-ring of the morpholino compounds yielded maximum activity, while additional methoxy groups at the C-3′ and C-5′ led to significantly reduced activity. In particular, the 2′,3′,4′-trimethoxybenzene derivative exhibits a greater inhibitory effect than the

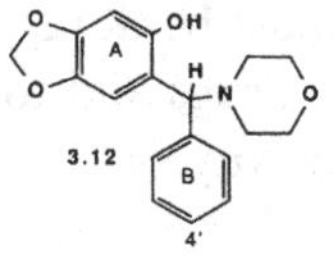

3′,4′,5′-trimethoxybenzene form. The authors speculated that it is the C-4′-methoxy group in the benzyl-benzodioxole derivatives which corresponds to the C-5′-methoxy group of podophyllotoxin.

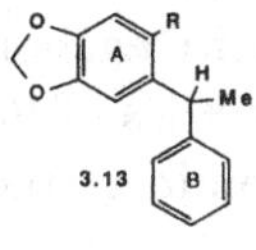

(ii) *Antiviral properties*

Markkanen *et al.* (1981) examined several podophyllotoxin derivatives as antiviral agents but did not speculate on the mechanism of the antiviral action. They found no clear correlation between the capacity of a podophyllotoxin derivative to inhibit microtubules and its antiviral property. Several podophyllotoxin derivatives possess high antimicrotubule inhibitory effects but do not possess antiviral activity. For example, epipodophyllotoxin is less potent than picropodophyllotoxin as an antiviral but it is much more potent as a microtubule inhibitor. Furthermore, 4′-demethylpodophyllotoxin exhibits pronounced antiviral properties whereas 4′-demethylepipodophyllotoxin appears to be inactive. Yet both derivatives are inhibitors of microtubule assembly. In all cases tested, the glucoside derivatives of podophyllotoxin exhibited decreased antiviral activity. For example, glucoside derivatives of β-peltatin, α-peltatin and podophyllotoxin exhibit much less antiviral activity than their respective aglucones.

There are a few reports describing the antimicrobial properties of lignans. For example, the norlignan pachypostaudin B (**3.14**; X = H) and its dihydroderivative pachypostaudin A were isolated (Ngadjui *et al.*, 1987) from an extract of the plant *Pachypodanthium staudtii* and found to be active against the polio virus. The structurally related apolignan magnoshinin (**3.14**; X = Me) is an anti-inflammatory agent of comparable activity to hydrocortisone acetate (Kadota *et al.*, 1987). Polygamain (**2.365**) is structurally very similar to deoxypodophyllotoxin and was shown (Sheriha *et al.*, 1987) to be strongly antibacterial in its action against *Staphylococus aureus*, and *Escherichia coli.*

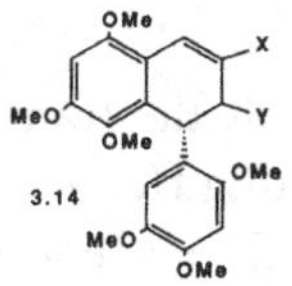

Clearly, much more structural activity analysis must be done in order to characterise the antimicrobial properties of these drugs and to understand their mechanism of antiviral or antibacterial activity.

(iii) *Nucleoside transport*

Although mammalian cells do not require exogenous nucleosides for cell growth, they possess specific plasma membrane transport components which mediate the movement of nucleosides across the plasma membrane (see Plagemann and Richey, 1974; Young and Jarvis, 1983, for reviews). Most cells transport nucleosides into their cytoplasm by a process described as facilitated diffusion which is a saturable carrier-mediated mechanism. A small amount of nucleosides enters the cell via a non-saturable process which is thought to be diffusion. Once the nucleoside traverses the cell membrane, specific kinases phosphorylate them to their respective 5′-mono-, 5′-di-, and 5′-triphosphate forms.

Convincing evidence for a single mammalian nucleoside transport system in some cells like the lymphocytic S49 cell line stems from the generation and characterisation of nucleoside transport-deficient lymphoma cells (Aronow *et al.*, 1985; Cohen *et al.*, 1979). These nucleoside transport-deficient cells fail to transport and incorporate virtually all nucleosides and are resistant to growth inhibition by a spectrum of cytotoxin nucleosides and their analogues. However, some mammalian cells may express multiple nucleoside carrier systems. Moreover, cells differ widely in their sensitivities towards two inhibitors of nucleoside transport: 4-nitrobenzylthioinosine (NBMPR) (Young and Jarvis, 1983) and dipyridamole (Plagemann and Wohlhueter, 1984). While nucleoside transport in S49 cells is sensitive to nanomolar concentrations of NBMPR, Novikoff hepatoma cells and Walker 256 carcinosarcoma cells are resistant to its action (Plagemann and Wohlhueter, 1984; Belt and Noel, 1985). Thus, there exist NBMPR sensitive and insensitive nucleoside transporters.

The nucleoside transport system also plays a major role in the uptake of certain anti-nucleoside analogues by tumour cells. For example, cytosine-β-D-arabinofuranoside (ara-C) is the most effective antimetabolite for treatment of acute myelogenous leukaemia and current evidence indicates that ara-C is a substrate for the nucleoside transporter. Experimental tumours or mutant cells that lack nucleoside transport activity are highly resistant to ara-C and other cytotoxic nucleosides (Cohen *et al.*, 1979; Rathmell *et al.*, 1984).

Podophyllotoxin and many of its derivatives inhibit the transport of nucleosides in animal cells. Within 5 min after incubating HeLa cells with 100 μM podophyllotoxin, the uptake of thymidine, uridine, adenosine and guanine into the acid soluble fraction of the cell is inhibited by more than 80%. Loike and Horwitz (1976) showed that these drugs inhibited the saturable component of nucleoside transport without affecting either the non-saturable (diffusional) component or the phosphorylation of nucleosides into nucleotides. Recently, Yalowich and Goldman (1984) confirmed that podophyllotoxin does not directly affect the rates of thymidine triphosphate formation in Ehrlich ascites tumour cells. While most podophyllotoxin-like compounds inhibit nucleoside transport activity, the chemical sites of these compounds which interact with the nucleoside transporter have not been identified.

(iv) *Antitumour effects*

The antitumour effects of podophyllotoxin and its congeners have been extensively investigated from the 1940s through the 1970s and are clearly reviewed in a number of papers (Kelly and Hartwell, 1954; Hartwell and Schrecker, 1958). Current wisdom suggests that those podophyllotoxin derivatives which are the best inhibitors of microtubule assembly possess the greatest antitumour properties. However, since these compounds attack both normal and cancerous cells, the toxic side effects of these lignans have limited their application as drugs in cancer chemotherapy. Only the discovery of a new mechanism of action of some of the podophyllotoxin derivatives has led to their use as antitumour agents (see Chapter 4). These derivatives, Etoposide and Teniposide, do not exhibit any effect on intracellular microtubules but induce breaks in single- and double-stranded DNA.

Mammalian lignans

(i) *Structure*

A number of lignans have recently been identified in humans and in several animals. Both *trans*-8,8′-bis(3-hydroxybenzyl)-γ-butyrolactone and 8,8′-bis(3-hydroxybenzyl)-butane-9,9′-diol, known respectively as enterolactone and enterodiol (**3.15**, **3.16**) have recently been described as the major lignans present in serum, urine, bile, and seminal fluids of humans and animals (Setchell *et al.*, 1980, 1981a,b). The mammalian-derived lignans differ from plant-derived lignans in possessing phenolic hydroxyl groups only in the *meta* position of the aromatic rings.

The dietary precursors for these lignans are thought to be secoisolariciresinol (Setchell *et al.*, 1980) and matairesinol (Borriello *et al.*, 1985). Moreover, it is believed that the Clostridia group of gut microorganisms are responsible for converting plant fibre to mammalian lignans (Setchell *et al.*, 1981a,b). Both enterolactone and enterodiol are produced from secoisolariciresinol by facultative faecal bacteria, whereas enterolactone is also produced from matairesinol (Borriello *et al.*, 1985). A proposed scheme for the formation of enterodiol and enterolactone from the dietary precursor secoisolariciresinol glycoside is presented in Scheme 3.4.

Generally, in man the amount of lignan excreted correlates positively with the amount of fibre ingested. Thus, vegetarians secrete a large amount of lignans. The concentrations of lignans in the urine of vegetarians can reach more than 800 times those of total urinary estrogens (Adlercreutz *et al.*, 1982) which is comparable to that of many steroid metabolites. The lignans excreted in bile and urine are mainly conjugated glucuronides (Axelson and Setchell, 1980, 1981) whereas the faeces contain the unconjugated forms of lignans (Setchell *et al.*, 1981b) that enter the enterohepatic circulation. Axelson *et al.* (1982) measured the amounts of lignan excreted by rats fed on different diets. Lignan excretion in rat urine was measured over a 24-h period. Rye, buckwheat, millet, soya, oats and barley are the prime sources of mammalian lignans in diets. Rats fed these lignan precursors generate about 2–6 μg of lignans per g of meal added. Interestingly, linseed was found to be the most abundant source of lignan in rats, producing 800 μg of lignan per g of meal added.

Lignan excretion does not always correlate with fibre intake. Young vegetarians have much higher lignan excretion than their older counter-

Scheme 3.4 Production of enterodiol and enterolactone by faecal flora

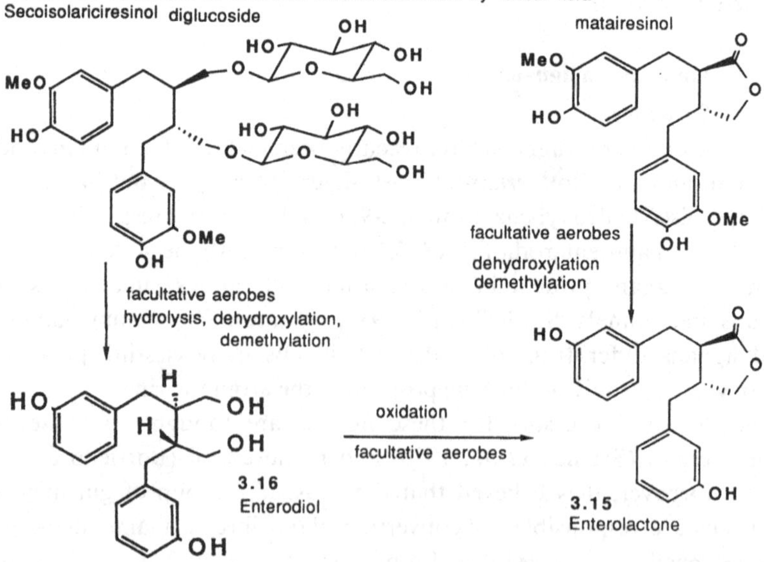

Secoisolariciresinol diglucoside

facultative aerobes
hydrolysis, dehydroxylation,
demethylation

matairesinol

facultative aerobes
dehydroxylation
demethylation

oxidation
facultative aerobes

3.16
Enterodiol

3.15
Enterolactone

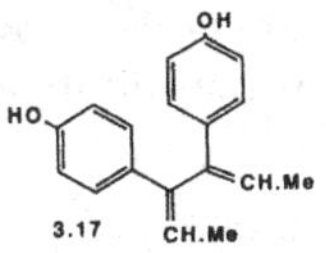

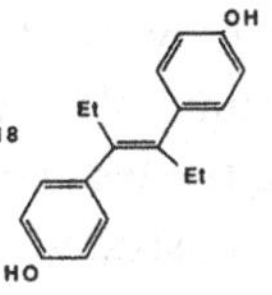

parts, although they eat similar amounts of fibre (Adlercreutz *et al.*, 1981). Furthermore, there was no significant difference in enterolactone excretion between the postmenopausal omnivorous and vegetarian groups despite significant difference in the fibre intake (Adlercreutz *et al.*, 1981).

(ii) *Biological properties*

The diphenolic structure of lignans shares similarities with several estrogen-like substances such as dienoestrol (**3.17**), diethylstilboestrol (**3.18**), and its dihydro derivative hexoestrol. The structural similarity between lignans and oestrogens has led to the idea that lignans may serve a biological role in man. In this regard, enterolactone has been shown to depress oestrogen-stimulated rat uterine RNA synthesis (Waters and Knowler, 1982). In women, peak lignan excretion is observed in the luteal phase of the menstrual cycle, and high values are seen in early pregnancy (Setchell *et al.*, 1980). The physiological relevance of these observations remains to be investigated since enterodiol and enterolactone bind weakly to rat uterine cytosol oestrogen receptor (Erb *et al.*, 1982) and exhibit no oestrogenic activity *in vivo* in mice (Setchell *et al.*, 1981).

Studies by Adlercreutz *et al.* (1982) have shown that lignan excretion is lower in women who have breast cancer than in healthy women with no history of breast cancer. The authors speculate that the low rate of breast cancer in some women may be in part due to the presence of lignan precursors in their fibre-rich diets. It is possible that the lignans can serve in some anti-oestrogen or anti-proliferative capacity to suppress the growth of human malignancies. The current evidence which demonstrates the protective role of fibre-rich diets in certain cancers is very limited. Further investigations are required to characterise the role of lignans in this process before the biological role of these compounds in man can be defined.

There are several other studies describing new and interesting biological properties of mammalian lignans. For example, Fagoo *et al.* (1986) demonstrated that enterolactone can displace ^{3}H-ouabain from its binding sites on cardiac digitalis receptor. In addition, it inhibits the Na^{+}, K^{+}-ATPase activity of human and guinea-pig heart in a dose-dependent manner. These authors speculate that the γ-butyrolactone of enterolactone is common to ascorbic acid and cardiac glycosides, agents which also inhibit Na^+, K^+-ATPase activity. Furthermore, Fagoo *et al.* suggest that the existence of lignans in mammals may account for the putative digitalis-like activity found in their tissues and fluids.

Sanghvi *et al.* (1984) demonstrated that enterolactone and enterodiol exhibit significant inhibitory properties against cholesterol 7α-hydroxylase activity *in vitro*. Cholesterol 7α-hydroxylase is the rate-limiting enzyme in the formation of primary bile acids. The gut bacteria then convert these primary bile acids to secondary bile acids like deoxycholic acid and release these substances into the colon. Sanghvi *et al.* (1984) speculate that lignans may provide some protection against colonic cancer, based on the correlation between colorectal cancer risk and faecal levels of the secondary bile acid, deoxycholic acid. Thus, inhibition of cholesterol 7α-hydroxylase by lignans would decrease primary bile acids and in turn, prevent accumulation of deoxycholic acid in the colon.

Furthermore, synthetic enterolactone and other lignans are cytotoxic to human lymphoid cells (Setchell *et al.*, 1981a,b) and have demonstrated effective action against animal tumours (Cairnes *et al.*, 1980; Tonance *et al.*, 1979).

A paper by Wahala (1986) reviews the mammalian lignans and their medical uses and describes a method for labelling these compounds with tritium.

Structure and biological activity of other lignans

This section provides examples of lignans which exhibit a wide range of biological activities and potential medical applications. A review of the biological activity of lignans was published recently by MacRae and Towers (1984) and useful summaries have been compiled by Pelter (1986) and Braquet and Godfroid (1986).

(i) *Acyclic lignans*

Nordihydroguiaretic 'acid' (NDHGA, **3.19**; Shroeter *et al.*, 1918) has been used commercially as an antioxidant (Elakovich and Stevens, 1985) and had limited use in the treatment of Parkinson's disease. NDHGA has been called 'the penicillin of hydroquinones' and has been shown (Waravdekar *et al.*, 1955) to inhibit cytochrome oxidase in oxidative phosphorylation. The inhibition of tumour growth by the more complex lignans of the podophyllotoxin group has also been established (Cole *et al.*, 1969), and is thought to be mediated by the inhibition of cytochrome oxidase. Anti-leukaemic action has also been reported (Tandon and Rastogi, 1976) for the hydroxylated lactone wikstromol (**3.20**) isolated from *Wikstroemia viridiflora*.

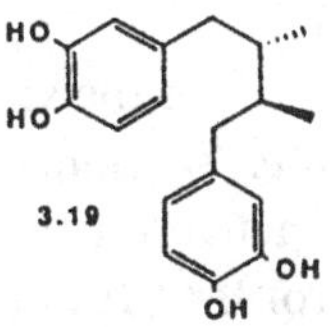

3.19

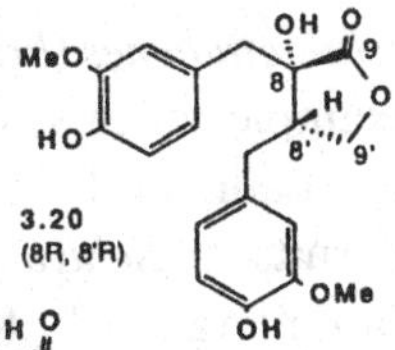

3.20 (8R, 8'R)

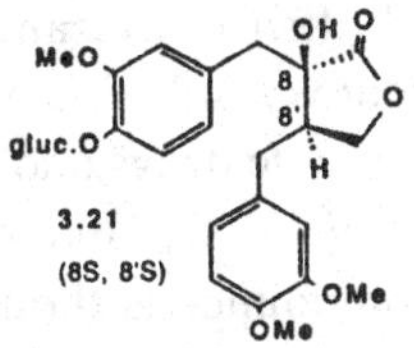

3.21 (8S, 8'S)

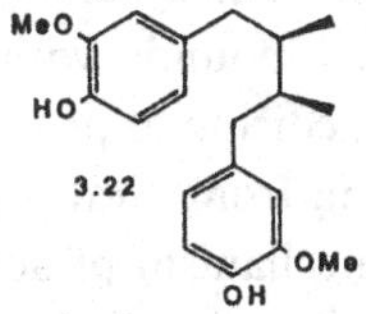

3.22

The Formosan variety *Wikstroemia indica* is used as a folk remedy for whooping cough and arthritis and the extract is said to be active against the central nervous system (Kato *et al.*, 1979). This activity could not be confirmed in (+)-nortrachelogenin (equiv. to **3.20**) following its isolation (Ji-Xian *et al.*, 1984) from the stems of *Passerina vulgaris*. The laevorotatory enantiomer of wikstromol has been isolated (Nishibe *et al.*, 1973) and the glucoside of its monomethyl ether (**3.21**) has the same structure as the cathartic component identified in safflower meal (Palter *et al.*, 1972). Antifungal and antimicrobial action are exhibited by some lignans in this group, notably by NDHGA (**3.19**) and dihydroguaiaretic 'acid' (**3.22**; Elakovich and Stevens, 1985).

A steam volatile dibenzylbutane (**2.015**) from the arils of *Myristica fragrans* has been shown (Shin and Woo, 1986) to modify hepatic drug metabolism. Flavonolignans, such as silybin, have antihepatotoxic activity and drugs based upon them have been used for treatment of liver disease (Kurkin, 1987).

(ii) *Arylnaphthalenes*

Compounds of this subclass are strongly fluorescent and have been identified as plant inclusion products. Thus, diphyllin (**3.23**) was the first to be isolated in 1961 by Murakami and Matsushima from *Justicia hayatai* although it was not until 1969 that the structural details were correctly assigned (Horii *et al.*, 1986). A number of these compounds are fish poisons (Ghosal *et al.*, 1979; Hui *et al.*, 1986) with toxicity comparable to that of rotenone, and others, such as prostalidin A (**3.24**) are mild depressants (Ghosal *et al.*, 1979).

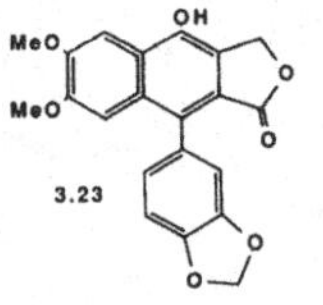

3.23

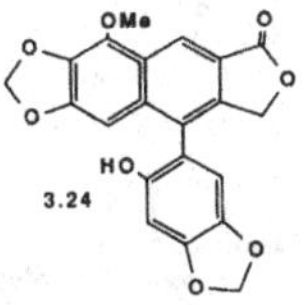

3.24

(iii) *Dibenzocyclo-octadienes (bridged biphenyls)*

The first examples of this class were isolated from the seed oil of *Schizandra chinensis* Baill. in 1961 by Kochetkov *et al.* Some lignans obtained from the fruits of *Schizandra* species are active on the central nervous system (Yan-Yong *et al.*, 1976). Schizandrol B (**3.25**) and other compounds with the methylenedioxy substituent are used in Chinese medicine to protect the liver from toxic drugs and to treat patients suffering from viral hepatitis (Fang *et al.*, 1981; Hikino and Kiso, 1988). The mechanism of action of these compounds is thought to be inhibition of mixed function oxidase (Casida, 1970), this activity requires a methylenedioxy group on these compounds.

Of the 34 biphenyl lignans isolated from plant sources 15 were examined (Suekawa *et al.*, 1987) for inhibition of arterial contraction in dogs. Several of the gomisins and Schizandrin (**3.26**) were effective in increasing blood flow: gomisin J (**2.434**) was almost as effective as the drug Diltiazem. Schizandrin (**3.26**) and isoschizandrin (**3.27**) both exhibited anti-ulcer activity in stress induced rats (Ikeya *et al.*, 1988).

Steganotaenia species are a source of anticancer lignans; the first of this group, steganone (**3.28**), was first isolated by Kupchan *et al.* in 1973.

(iv) *Lignans acting on cyclic adenosine-3′,5′-monophosphate (cAMP)*

Some of the dibenzylbutyrolactones and furofurans inhibit the action of cAMP phosphodiesterase. cAMP was identified as a second messenger within cells by Sutherland and Ral (1958) – the action of the phosphodiesterase is to catalyse the irreversible hydrolysis of the cyclic phosphate (**3.29**) to the 5′-phosphate dianion (**3.30**), thereby terminating the intracellular effects of the cyclic nucleotide.

Examination of 222 extracts of plants used in oriental medicine revealed 22 which showed reproducible inhibition of beef heart phosphodiesterase and exhibited potential use as drugs with pronounced activity on the central nervous system (Nikaido *et al.*, 1981). Hence, substances

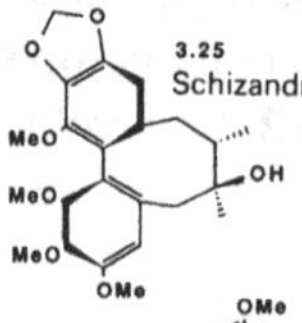

3.25 Schizandrol B

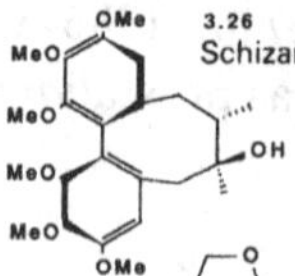

3.26 Schizandrin

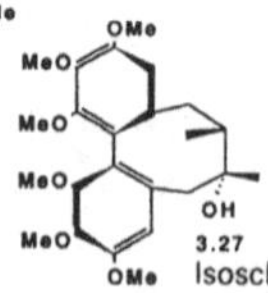

3.27 Isoschizandrin

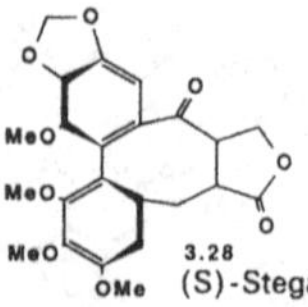

3.28 (S)-Steganone

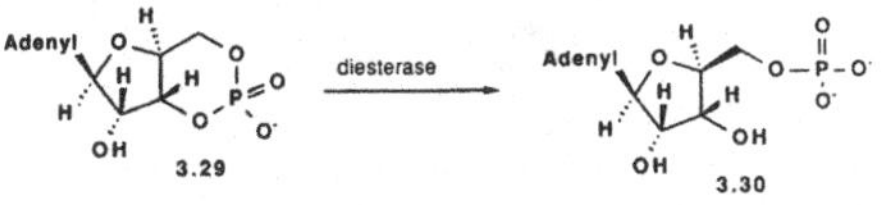

which inhibit cAMP phosphodiesterase may be of use as CNS-active drugs. It should be noted that some anti-parasitic agents are also active against cAMP phosphodiesterase (Amer and Kreighbaum, 1975).

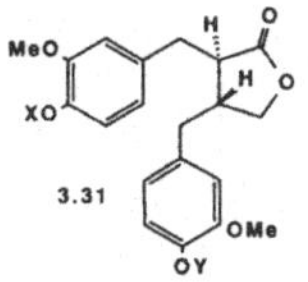

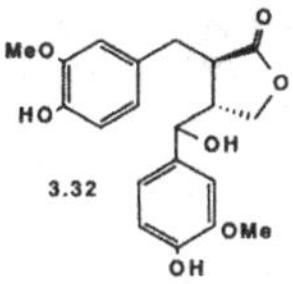

The acyclic, furan and furofuran classes of lignans also include compounds which act on cAMP metabolism. Matairesinol (**3.31**; X, Y = H) is active; arctigenin (**3.31**; X = H, Y = Me) with only one phenolic group is also active but at a lower level. The structure–activity relationship is complex (Cole *et al.*, 1969), because although activity is lost on glucosylation of arctigenin (**3.31**; X = glucosyl, Y = Me), it is retained in the diglucoside of matairesinol (**3.31**; X, Y = glucosyl). Regiosensitivity is also shown, since activity is retained on hydroxylation at C-8 as in wikstromol (**3.20**) but is lost in 7′-hydroxymatairesinol (**3.32**); cyclisation to the corresponding phenylnaphthalane, α-conidentrin (**2.348**), also leads to loss of activity.

Of the furofuran lignans, pinoresinol (**3.33**; X = H, Y = H, Z = H) is an effective inhibitor of cAMP (Tsukamoto, 1984b). The level of activity is highest with this configuration and as observed for matairesinol (**3.31**; X, Y = H), activity is reduced in the monoglucoside and restored in the diglucoside. In the lignans isolated from *Eucommia* bark, Deyama *et al.* (1988) reported that the diglucoside > aglycone > monoglucoside in activity and suggested that the lignan diglucosides derived from aglucones may exhibit marked inhibitory activity of cAMP phosphodiesterase. In fact, (+)-8-acetyloxypinoresinol 4′,4″-di-*O*-β-D-glucopyranoside (**3.33**; X = glucosyl, Z = AcO-, Y = H) exhibited an IC_{50} of 1.1×10^{-5} M, which represented the most potent inhibitor of cAMP phosphodiesterase tested in this group of lignans (Deyama *et al.*, 1988).

There are several other physiological properties of these lignans which may be related to their inhibitory effects on cAMP metabolism. For example, pinoresinol monoglucoside is an antihypertensive agent (Kitagawa *et al.*, 1984) whilst the diglucoside of syringaresinol (**3.33**; X = glucosyl, Y = OMe, Z = H) and episesartemin B (**3.34**) reduce aggression in

3.33

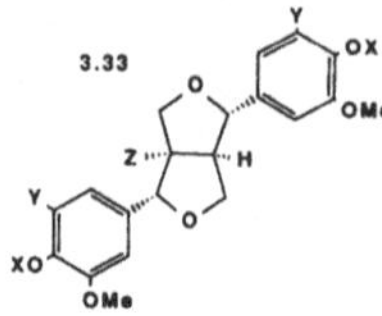

3.34

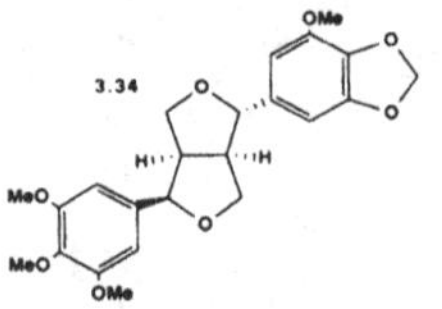

3.35

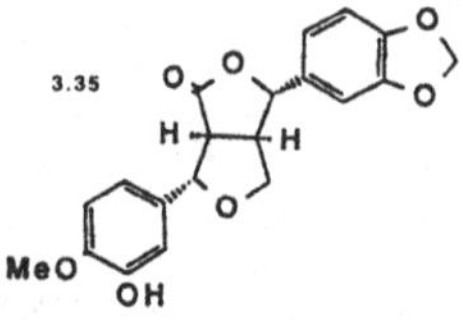

3.36

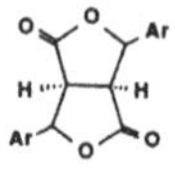

experimental animals. The latter (**3.34**) is one of four component lignans in the resin of *Virola elongata*, which has been used as an arrow poison. Interestingly, the resin was found (MacRae and Towers, 1984) not to be toxic to mice following intraperitoneal injection. Liriodendrin (**3.33**; X = glucosyl, Y = OMe, Z = H) is a cytotoxic agent (Jolad *et al.*, 1980) as also is the related lactone styraxin (**3.35**; Ululeben *et al.*, 1978) which shows antitumour activity of 141% at a level of 10 mg/kg against the PS tumour line system. The dilactone (**3.36**; Ar = 3,4-dihydroxyphenyl-) is an inhibitor of cAMP phosphodiesterase (Kumada *et al.*, 1978); as a fungal metabolite it may be derived from a lignan precursor. Dilactones of the same general type are obtained *in vitro* by the oxidative coupling of cinnamic acids (p. 304).

(v) *Lignans acting on platelet activating factor (PAF)*

This bioactive phospholipid, PAF, has been identified as 1-*O*-hexadecyl/octadecyl-2-acetyl-*sn*-glyceryl-3-phosphorylcholine (**3.37**; Snyder, 1985), and it has been shown to be linked to various haematological responses including aggregation and degranulation of platelets and neutrophils and to be an important mediator of inflammation and asthma. This has led to a search for PAF inhibitors for use as anti-asthma drugs.

One example is the neolignan kadsurenone (**3.38**), which inhibits the

3.39

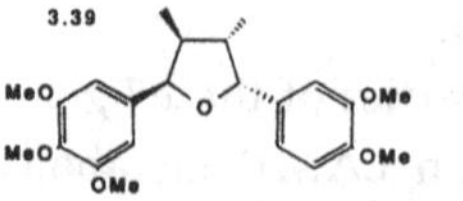

3.37

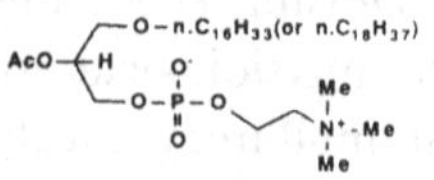

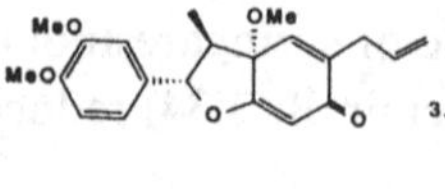

3.38

action of PAF in aggregating human neutrophils. The neolignan also counteracts induced cardiovascular changes in rabbits and endotoxic shock in rats. (+)-Dihydroguaiaretic acid (**3.22**; 8*S*,8′*S*-) is also antagonistic to PAF (Han *et al.*, 1986) as are the related presteganes A and B (**2.075**, **2.076**) and methoxymatairesinols (**2.064**, **2.067**; Braquet, 1986).

The 9,9′-epoxylignan, burseran (**2.111**) is a moderate inhibitor of PAF (Braquet *et al.*, 1986) but attention has centred recently on tetrahydrofuran lignans with the 7,7′-epoxy structure. The synthetic *cis-trans-cis* compound **3.39** proved to be most active with an IC_{50} of 0.02 μM (Biftu *et al.*, 1986; Biftu and Stevenson, 1987). Naturally occurring compounds of this group such as (+)-veraguensin (**2.168**), (+)-galbelgin (**2.155**) and *meso*-galgravin (**2.156**) are effective at concentrations of the order of 1–3 μM. The analogous lignan, magnosalicin (**2.134**), has been shown to be an anti-allergic compound (Mori *et al.*, 1986).

Antagonism to PAF has also been detected (Pan *et al.*, 1987) in those lignans of the furofuran class which have been isolated from flowers of *Magnolia biondii*. This source is used in Chinese medicine for the treatment of headache and nasal empyema. Of the lignan constituents the most active against PAF were fargesin (**2.203**; ED_{50} = 1.3 μM) and dimethylpinoresinol **2.217**; ED_{50} = 1.7 μM); that is at a level of about one-tenth that of kadsurenone (**3.38**). Other lignans from *Magnolia kobus* showed activity against contact dermatitis.

(vi) *Biological effects of lignans on plants and insects*

Some acyclic and furofuran lignans with methylenedioxy-phenyl groups have a synergistic effect on a range of insecticides (Haller *et al.*, 1942). This biological property may be due to an association between the phenyl group of the lignans and inhibition of mixed function oxidase, although this association is not specific because matairesinol (**3.31**; X, Y = H) is also an insecticide synergist (Kerr, 1951). A tetrahydrofuran lignan (**2.114**) isolated (Hanuman *et al.*, 1986) from *Tinospora cordifolia* is synergistic with pyrethrum and also acts as a stimulant of the central nervous system.

Lignans also play a role in the regulation of plant growth. In common with other phenolics, including cinnamic and *ortho*- and *para*-coumaric acids, the inhibition they exert on the germination of barley and lettuce is the result of interference with the transport of amino acids and of protein formation in the seeds (Van Sumere *et al.*, 1972). Activity is also linked to the methylenedioxy group (Bhiravamurty *et al.*, 1978), because fargesin (**3.40**) strongly inhibits germination of the peanut (*Arachis hypogaea* L.) and sesamin (**3.41**) is also active but eudesmin (**2.200**) is not. The complexity of this mechanism is well illustrated by Lavie's work (Gutterman *et al.*, 1980) with the lignan (**3.42**) isolated from *Aegilops ovata*. This compound inhibits the germination of lettuce in the light but *not*, as is usual, in the dark; and the activity depends on the presence of both the precursor (**3.42**) and its photo-product. This type of monolactone (**3.42**) is also of interest in that it was the first to be isolated from a plant source (cf. 185).

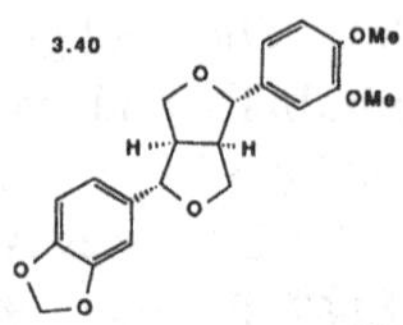

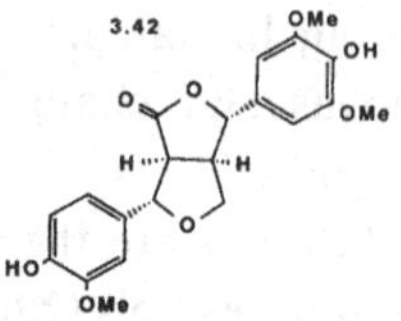

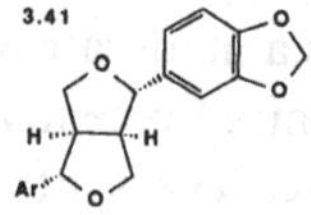

There is evidence that fungal attack in wood leads to lignan synthesis and that further degradation is prevented. The action of *Fomes annosus* on spruce leads to the production of matairesinol (**3.31**; X, Y = H) and 7'-hydroxymatairesinol (**3.32**) which then limits further fungal growth (Shain and Hillis, 1971). Cyclo-olivil (**3.7**) is formed (Hasegawa, 1959) by fungal attack of *Prunus* wood and injury to the wood leads to the production of syringaresinol (**3.33**; X = H, Y = OMe, Z = H) (Chen *et al.*, 1977). One indicator of a possible catabolic path is the co-occurrence in *Helichrysum* sp. (Jakupovic *et al.*, 1987) of acuminatolide (**3.43**) and the 7,9'-epoxy-lignan acuminatin (**3.44**). Thus, oxidative cleavage of the phenolic ring of the latter, followed by lactonisation, would lead to **3.43** with loss of one aryl group.

In the last of a series of papers on the phenols of discoloured Sugi wood, Takahashi and Ogiyama (1986) have proposed a sequence of catabolic changes to account for the accumulation of the lignan products yateresinol (**3.45**; scheme 3.5) and hinokiresinol (**3.46**) in the sapwood. In this system the 1,4-diarylbutadiene (**3.47**) and agatharesinol (**3.48**) with its

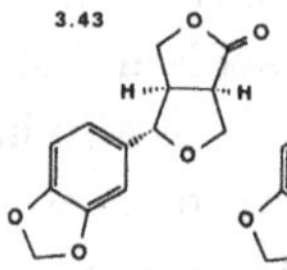

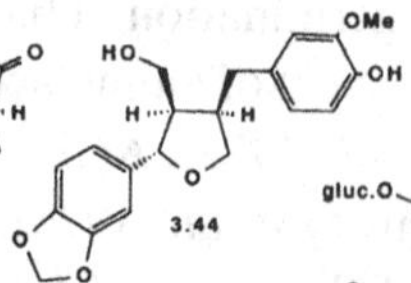

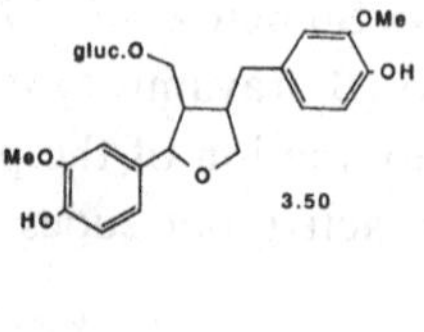

Scheme 3.5 Catabolic products in sugi wood

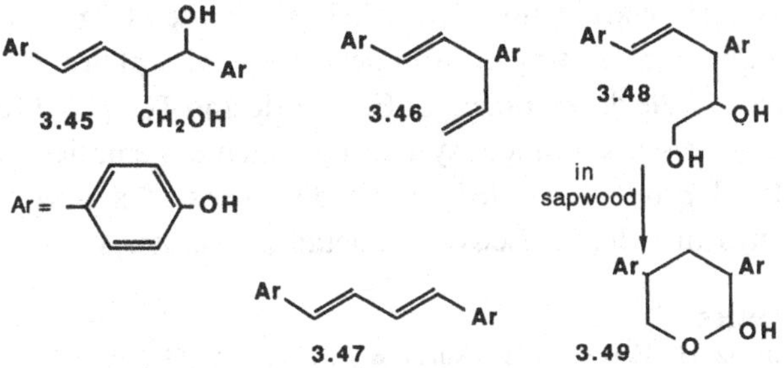

rearranged structure are also formed. This latter diol is further metabolised in the sapwood to products such as sugiresinol (**3.49**), which then accumulates in the heartwood. In contrast, yateresinol and hinokiresinol are stable and accumulate in the sapwood.

A stress compound (**3.50**) of unknown stereochemistry has been isolated (Yoshihara *et al.*, 1982) from the roots and stolons of potatoes infected with the nematode *Globodera rostochiensisis*.

The modification of lignans and their synthesis from other natural products by the action of moulds and fungi in culture is an area of research which merits further investigation.

Conclusions

Lignans are found throughout the plant kingdom (Cole and Wiedhopf, 1978) although the neolignans are restricted in their occurrence to Magnoliales and the related Piperales. The biological functions of plant lignans have not been identified. However, the antifungal and insecticidal properties of several lignans suggest that these compounds may serve in the plant host defence system. Lignans may also play a role in the regulation of plant growth. Lignan precursors are thought to be byproducts and/or components of the pathway of cinnamate biosynthesis leading to the formation of lignins. This is discussed in Chapter 7.

Clearly lignans possess a diverse spectrum of biological properties. The variety of biological properties described for lignans suggest that these compounds possess a variety of mechanisms of action. Therefore, extensive investigations correlating structure to activity must be carried out to understand which chemical sites on these lignan molecules are responsible for the specific biological properties. In this way it may be possible to develop new and effective pharmacological agents from the lignan family.

Podophyllotoxin, Etoposide and Teniposide are the major lignans which have defined applications in clinical medicine. The role of lignans in diet, their antiviral properties and their presumed protective roles against certain cancers await clarification. Nonetheless, the use of lignans in folk medicine has offered interesting leads in developing new pharmacological agents. The story of the development of Etoposide and Teniposide as antitumour agents represents a model system by which systematic modification of a natural product has led to the discovery of a new class of antitumour drugs, and this is discussed in detail in Chapter 4.

References

Adlercreutz, H., Fotsis, T., Heikkinen, R., Dwyer, J.T., Goldin, B.R., Gorbach, S.L., Lawson, A.M. and Setchell, K.D. (1981). Diet and urinary excretion of lignans in female subjects. *Med-Biol.* **59**, 259–61.

Adlercreutz, H., Fotsis, T., Heikkinen, R., Dwyer, J.T., Woods, M., Goldin, B.R. and Gorbach, S.L. (1982). Excretion of the lignans enterolactone and enterodiol and of equol in omnivorous and vegetarian postmenopausal women and in women with breast cancer. *Lancet* **ii**, 1295–9.

Amer, M.S. and Kreighbaum, W.E. (1975). Cyclic nucleotide phosphodiesterases: properties, activators, inhibitors, structure activity relationships & possible role in drug development. *J. Pharm. Sci.* **64**, 1–37.

Aronow, B., Allen, K., Patrick, J. and Ullman, B. (1985). Altered nucleoside transporters in mammalian cells selected for resistance to the physiological effects of inhibitors of nucleoside transport. *J. Biol. Chem.* **260**, 6226–33.

Axelson, M. and Setchell, K.D. (1980). Conjugation of lignans in human urine. *FEBS Lett.* **122**, 49–53.

Axelson, M. and Setchell, K.D. (1981). The excretion of lignans in rats – evidence for an intestinal bacterial source for this new group of compounds. *FEBS Lett.* **123**, 337–42.

Axelson, M., Sjovall, J., Gustafsson, B.E. and Setchell, K.D. (1982). Origin of lignans in mammals and identification of a precursor from plants. *Nature* **298**, 659–60.

Batra, J.K., Jurd, L. and Hamel, E. (1986). Morpholino derivatives of benzyl-benzodioxole, a study of structural requirements for drug interactions at the colchicine/podophyllotoxin binding site of tubulin. *Biochem. Pharmacol.* **35**, 4013–18.

Belt, J.A. and Noel, L.D. (1985). Nucleoside transport in Walker 256 rat carcinosarcoma and S49 mouse lymphoma cells. Differences in sensitivity to nitrobenzylthioinosine and thiol reagents. *Biochem. J.* **232**, 681–8.

Bennett, R.G. and Grist, W.J. (1985). Nasal papillomas: Successful treatment with podophyllin. *Southern Med. J.* **78**, 224–5.

Bhattacharyya, B. and Wolff, J. (1974). Promotion of fluorescence upon binding of colchicine to tubulin. *Proc. Natl. Acad. Sci. USA* **71**, 2627–31.

Bhiravamurty, P.V., Kanakala, R.D., Rao, E.V. and Sastry, K.V. (1978). Effect of some furofuranoid lignans on three species of seeds. *Current Sci.* **48**, 949–50.

Biftu, T., Gamble, N.F., Doebber, T., Hwang, S-B., Shen, T-Y., Snyder, J., Springer, J.P. and Stevenson, R. (1986). Conformation & activity of tetrahydrofuran lignans & analogues as specific platelet activating factor antagonists. *J. Med. Chem.* **29**, 1917–21.

Biftu, T. and Stevenson, R. (1987). Natural 2,5-bisaryl-tetrahydrofuran lignans: platelet-activating factor antagonists. *Phytother. Res.* **1**, 97–106.

Borriello, S.P., Setchell, K.D., Axelson, M. and Lawson, A.M. (1985). Production and metabolism of lignans by the human faecal flora. *J. Appl. Bacteriol.* **58**, 37–43.

Braquet, P. and Godfroid, J.J. (1986). PAF-acether specific binding sites. Part 2. Design of specific antagonists. *Trends in Pharm Sci.*, 397–9.

Brewer, C.F., Loike, J.D., Horwitz, S.B., Sternlicht, H. and Gensler, W.J. (1979). Conformational analysis of podophyllotoxin and its congeners. Structure activity relationship in microtubule assembly. *J. Med. Chem.* **22**, 215–21.

Cairnes, D.A., Ekundayo, O. and Kingston, D.G. (1980). Plant anticancer agents X. Lignans from *Juniperus phoenica. J. Natural Products* **43**, 495–7.

Casida, J.E. (1970). Mixed function oxidase involvement in the biochemistry of insecticide synergists. *J. Agric. Food Chem.* **18**, 753–72.

Cassidy, D.E., Drewry, J. and Fanning J.P. (1982). Podophyllum toxicity: a report of a fatal case and a review of the literature. *J. Toxicol. Clin. Toxicol.* **19**, 35–44.

Chen, C-L., Chang, H.M., Chang, C-Y., Huang, H. and Cowling, E.B. (1977). Chromophores & phytoalexins formed in response to injury of sapwood in *Liriodendrin tulipfera* L. *Proc. Amer. Phytopath. Soc.* **4**, 135.

Cockayne, T.O. (1961). Leech book of Bald: In: Lecchdom, Wortcunnings and Starcraft of Early England, vol II. The Holland Press, London, p. 313.

Cohen, A., Ullman, B. and Martin, D.W. Jr. (1979). Characterization of a mutant mouse lymphoma cell with deficient transport of purine and pyrimidine nucleosides. *J. Biol. Chem.* **254**, 112–16.

Cole, J.R., Bianchi, E. & Trumbull, E.R. (1969). Anti-tumour agents from *Bursera microphylla* (Burseraceae). Part 2. Isolation of a new lignan, burseran. *J. Pharm. Sci.* **58**, 175.

Cole, J.R. and Wiedhopf, R.M. (1978). Distribution. In: *Chemistry of Lignans*, ed. Rao, C.B.S. Andhra Univ. Press, pp. 39–64.

Elakovich, S.D. and Stevens, K.L. (1985). Phytotoxic properties of nordihydroguaiaretic acid, a lignan of *Larrea divaricata* (creosote bush). *J. Chem. Ecol.* **11**, 27–33.

Erb, L., Lasley, B.I., Czekda, N.M., Monfort, S.L. and Bercovitz, A.B. (1982). A dual radioimmunoassay and cytosol receptor binding assay for the measurement of estrogenic compounds applied to urine, faecal and plasma samples. *Steroids* **39**, 33–46.

Fagoo, M., Braquet, P., Robin, J.P., Esanu, A. and Godfraind, T. (1986). Evidence that mammalian lignans show endogenous digitalis-like activities. *Biochem. Biophys. Res. Commun.* **134**, 1064–70.

Fang, S.D., Xu, R.S. and Gao, Y-S. (1981). Some recent advances in the chemical studies of Chinese herbal medicine. *Amer. J. Botany* **68**, 300–3.

Fisher, A.A. (1981). Severe systemic and local reactions to topical podophyllum resin. *Cutis* **28**, 233.

Ghosal, S., Banerjee, S. and Frahm, A.W. (1979). Prostalidins A, B, C & retrochinensin: a new anti-depressant: 4-aryl-2,3-naphthalide lignans from *Justicia prostrata. Chem. & Ind.* pp. 854–5.

Gupta, R.S. and Chenchaiah, P.C. (1987). Synthesis & biological activities of the C-4 esters of 4′-demethylepipodophyllotoxin. *Anticancer Drug Des.* **2**, 13–23.

Gutterman, Y., Evenari, M., Cooper, R., Levy, E.C. and Lavie, D. (1980). Germination inhibition activity of naturally occurring lignans from *Aegilops ovata* L. in green & infrared light. *Experientia* **36**, 662.

Haller, H.L., La Forge, F.B. and Sullivan, W.N. (1942). Some compounds related to sesamin: their structures & their synergistic effect with pyrethrum insecticides. *J. Org. Chem.* **7**, 185–8.

Han, G-Q., Chang, M.N. and Hwang, S-B. (1986). The investigation of lignans of *Sargentodoxa cuneata* (Oliv) Rehd & Wils. *Yaoxue Xuebao* **21**, 68–70.

Hanuman, J.B., Mishra, A.K. and Sabata, B. (1986). A natural lignan from *Tinospora cordifolia* Miers. *J. Chem. Soc. Perkin Trans.* **1**, 1181–5.

Hartwell, J.L. and Schrecker, A.W. (1958). The chemistry of podophyllum. *Progress in the Chemistry of Organic Natural Products* **15**, ed. L. Zechmeister, pp. 83–166. Springer, Berlin.

Hasegawa, M. and Shirato, T. (1959). Abnormal constituents of *Prunus* wood: isoolivil from the wood of *Prunus jamasakura*. *J. Jap. Forest Soc.* **41**, 1–3.

Hikino, H. and Kiso, Y. (1988). Natural products for liver diseases. In: *Econ. & Med. Plant Res.* **2**, ed. Wagner, H., Hikino, H. and Farnsworth, N.R. pp. 39–72. Academic Press, London.

Horii, Z., Ohkawa, K., Kims, S-W. and Momose, T. (1969). Synthetic studies on lignans & related compounds. Part 2. Synthesis of 1-hydroxy-2-hydroxymethyl-6,7-dimethoxy-4-(3,4-methylenedioxyphenyl)-2-naphthoic acid-γ-lactone & its non-identity with diphyllin. *Chem. Pharm. Bull. Tokyo* **17**, 1878–82.

Hui, Y.H., Chang, C.J., Mclaughlin, J.L. and Powell, R.G. (1986). Justicidin B, a bioactive trace lignan from the seeds of *Sesbania drummondii*. *J. Nat. Prod.* **49**, 1175–6.

Ikeya, Y., Taguchi, H., Mitsuhashi, H., Takeda, S., Kase, Y. and Aburada, M. (1988). Isoschizandrin a lignan from *Schizandra chinensis* (*Schizandraceae*), *Phytochemistry* **27**, 569.

Jakupovic, J., Pathak, U.P., Bohlman, F., King, R.M. and Robinson, H. (1987). Obliquin derivatives & other constituents from Australian *Helichrysum* sp. *Phytochemistry* **26**, 803–7.

Ji-Xian, G., Handa, S.S., Pezzuto, J.M., Kinghorn, A.D. and Farnsworth, N.R. (1964). Plant anti-cancer agents. Part 33. Constituents of *Passerina vulgaris*. *Planta Med.* **50**, 264.

Jolad, S.D., Hoffman, J.J., Cole, J.R., Tempesta, M.S. and Bates, R.B. (1980). Cytotoxic agent from *Penstemon deustus* (*Scrophuliacea*). Isolation & stereochemistry of liriodendrin, a symmetrically substituted furofuranoid lignan diglucoside. *J. Org. Chem.* **45**, 1327–9.

Kadota, S., Tsubono, K., Makino, K., Takeshita, M. and Kikuchi, T. (1987). Convenient synthesis of magnoshinin, an anti-inflammatory neolignan. *Tetrahedron Lett.* **28**, 2857–60.

Kaplan, I.W. (1942). *Condyloma acuminata. New Orleans Med. Surg. J.* **94**, 388.

Kariyone, T. and Kimurta, Y. (1976). Saishin Wakanyakuyoshokubutsu. *Hirokawas Shoten, Tokyo*, p. 105.

Kato, A., Hashimoto, Y. and Kidokuro, M. (1979). (+)-Nortrachelogenin, a new pharmacologically active lignan from *Wikstroemia indica*. *J. Nat. Prod.* **42**, 159–62.

Kelleher, J.K. (1977). Tubulin binding affinities of podophyllotoxin and colchicine analogues. *Mol. Pharmacol.* **13**, 232–41.

Kelly, M.G. and Hartwell, J.L. (1954). The biological effects and the chemical composition of podophyllin: a review. *J. Nat. Canc. Inst.* **14**, 967–1010.

Kerr, R.W. (1951). *Australian C.S.I.R.O. Bulletin*, no. 261, 31.

Kitagawa, S., Hisada, S. and Nishibe, S. (1984). Phenolic compounds from *Forsythia* leaves. *Phytochemistry* **23**, 1635–6.

Kochetkov, N.K., Khorlin, A., Chizhov, O.S. and Sheinchenko, V.I. (1961). *Schizandrin* lignan of unusual structure. *Tetrahedron Lett.* 730–4.

Kumada, Y., Naganawa, H., Takeuchi, T., Umezawa, H., Yamashita, K. and

Watanabe, K. (1978). Biochemical activities of the derivatives of dehydrocaffeic acid lactone. *J. Antibiotics* **31**, 106–11.

Kupchan, S.M., Britton, R.W., Ziegler, M.F., Gilmore, C.J., Restivo, R.S. and Bryan, R.F. (1973). Steganacin & steganangin, novel antileukemic lignan lactones from *Steganotaenia araliaceae. J. Amer. Chem. Soc.* **95**, 1335–6.

Kurkin, V.A. and Zapesochnaya, G.C. (1987). Flavolignans & other natural lignoids: problems of structural analysis. *Chem. Nat. Compd.* **23**, 7–25.

Lassus, A. and Rosen, B. (1986). Response of solitary psoriatic plaques to experimental application of podophyllotoxin. *Dermatologica* **172**, 319–22.

Lian-niang, L., Hung, X. and Rui, T. (1985). Dibenzocyclooctadiene lignans from roots and stems of *Kadsura coccinea. Planta Medica* **51**, 297–300.

Loike, J.D. and Horwitz, S.B. (1976). Effects of podophyllotoxin and VP-16-213 on microtubule assembly *in vitro* and nucleoside transport in HeLa cells. *Biochemistry* **15**, 5435–43.

Loike, J.D., Brewer, C.F., Sternlicht, H., Gensler, W.J. and Horwitz, S.B. (1978). Structure-activity study of the inhibition of microtubule assembly *in vitro* by podophyllotoxin and its congeners. *Cancer Res.* **38**, 2688–93.

Lokantha Rai, K.M. and Anjanamurthy, C. (1986). Biological assay & antimitotic activity of synthetic derivatives of podophyllotoxin. *Curr. Sci.* **55**, 701–6.

MacRae, W.D. and Towers, G.H.N. (1984). An ethnopharmacological examination of *Virola elongata* bark: a South American arrow poison. *J. Ethnopharmacol.* **12**, 75–92.

Markkanen, T., Makinen, M.L., Maunuksela, E. and Himanen, P. (1981). Podophyllotoxin lignans under experimental antiviral research. *Drugs Exptl. Clin. Res.* **7**, 711–18.

McKeithan, T.W. and Rosenbaum, J.L. (1984). The biochemistry of microtubules. *Cell Muscle Motil.* **5**, 225–88.

Minocha, A. and Long, B.H. (1984). Inhibition of the DNA catenation activity of type II topoisomerase by VP16-213 & VM26. *Biochem. Biophys. Res. Commun.* **122**, 165–70.

Morgan, J.L. and Spooner, J.S. (1983). Immunological detection of microtubule poison induced conformational changes in tubulin. *J. Biol. Chem.* **258**, 13127–33.

Mori, K., Komatsu, M., Kido, M. and Nakagawa, K. (1986). A simple biogenetic-type synthesis of magnosalicin. A new neolignan with antiallergic activity isolated from *Magnolia salicifolia. Tetrahedron* **42**, 523–8.

Murakami, T. and Matsushima, A. (1961). Studies on the constituents of Japanese *Podophyllaceae* plants. Part 1. On the constituents of *Diphylleia grayii. J. Pharm. Soc. Japan* **81**, 1596–600.

Ngadjui, B.T., Ayafor, J.F. and Lontsl, D. (1987).Unusual norlignans & antiviral agents from *Pachypodanthium staudtii. Fitotherapia* **58**, 340–1.

Nikaido, T., Ohmoto, T., Noguchi, M., Kinoshita, T., Saitoh, H. and Sankawa, U. (1981). Inhibitors of cyclic AMP phosphodiesterase in medicinal plants. *Planta Med.* **43**, 18–23.

Nishibe, S., Hisada, S. and Inagaki, I. (1973). Lignans of *Trachelospermum asiaticum* var. *intermedium*. Part 2. Structures of tracheoloside & nortracheloside. *Chem. Pharm. Bull.* **21**, 1108–13.

Palter, R., Lundin, R.E. and Haddon, W.F. (1972). A cathartic lignan glycoside isolated from *Carthamus tinctorus. Phytochemistry* **11**, 2871–4.

Pan, J-X., Hensens, O.D., Zink, D.L., Chang, M.N. and Hwang, S.B. (1987). Lignans with platelet activating factor antagonist activity from *Magnolia biondii. Phytochemistry* **26**, 1377–9.

Pelsor, F.R., Allen, L.M. and Creaven, P.J. (1978). Multicompartment pharmacokinetic model of 4′-demethylepipodophyllotoxin 9-(4,6-0-ethylidene-beta-D-glucopyranoside) in humans. *J. Pharm. Sci.* **67**, 1106–8.

Pelter, A. (1986). Lignans: some properties & synthesis. *Recent Adv. Phytochemistry* **20**, 201–41.

Perez-Figarado, R.A. and Baden, H.P. (1976). The pharmacology of podophyllum. *Prog. Dermatol.* **10**, 1–4.

Pettit, G.R. (1977). *Biosynthetic products for Cancer Chemotherapy*, p. 97. Plenum Press, New York.

Phaire-Washington, L., Silverstein, S.C. and Wang, E. (1980). Phorbol myristate acetate stimulates microtubule and 10-nm filament extension and lysosome redistribution in mouse macrophages. *J. Cell Biol.* **86**, 641–55.

Plagemann, P.G.W. & Richey, D.P. (1974). Transport of nucleosides, nucleic acids, bases, choline and glucose by animal cells in culture. *Biochim. Biophys. Acta* **344**, 263–305.

Plagemann, P.G.W. & Wohlhueter, R.M. (1984). Nucleoside transport in cultured mammalian cells. Multiple forms with different sensitivity to inhibition by nitrobenzylthioinosine or hypoxanthine. *Biochim. Biophys. Acta* **773**, 39–52.

Pharmacopoeia of the People's Republic of China (*1977*), vol. 1, p. 594.

Rathmell, J.P., White, J.C. and Capizzi, R.L. (1984). Nucleoside transport as a determinant of cellular uptake of ara-C. *Proc. Am. Assoc. Cancer Res.* **25**, 346.

Sanghvi, A., Diven, W.F., Seltman, H., Warty, V., Rizk, M., Kritchevsky, D. and Setchell, K.D.R. (1984). Inhibition of rat liver cholesterol 7 hydroxlase and acyl CoA:cholesterol acyl transferase activities by enterodiol and enterolactone. In *Drugs affecting lipid metabolism, Vol. 8*, ed. Kritchevsky, D., Parletti, R. and Holmes, W.L., pp. 311–22. Plenum Press, New York.

Setchell, K.D., Lawson, A.M., Mitchell, F.L., Adlercreutz, H., Kirk, D.N. and Axelson, M. (1980). Lignans in man and in animal species. *Nature* **287**, 740–2.

Setchell, K.D., Lawson, A.M., Conway, E., Taylor, N.F., Kirk, D.N., Cooley, G., Farrant, R.D., Wynn, S. and Axelson, M. (1981a). The definitive identification of the lignans trans-2,3-bis(3-hydroxybenzyl)-gamma-butyrolactone and 2,3-bis(3-hydroxybenzyl)butane-1,4-diol in human and animal urine. *Biochem. J.* **197**, 447–58.

Setchell, K.D., Lawson, A.M., Borriello, S.P., Harkness, R., Gordon, H., Morgan, D.M., Kirk, D.N., Adlercreutz, H., Anderson, L.C. and Axelson, M. (1981b). Lignan formation in man–microbial involvement and possible roles in relation to cancer. *Lancet* **ii**, (8236) 4–7.

Sheriha, G.H., Abouamer, K., Elshtaiwi, B.Z., Ashour, A.,S., Abed, F.A. and Alhallaq, H.H. (1987). Quinoline alkaloids & cytotoxic lignans from *Haplophyllum tuberculatum. Phytochemistry* **26**, 3339–41.

Shain, L. and Hillis, W.E. (1971). Phenolic extractions in Norway spruce & their effects on *Fomes annosus, Phytopathology* **61**, 841–5.

Shin, K.H. and Woo, W.S. (1986). Hepatic drug metabolism modifier from arils of *Myristica fragrans. Saengyak Hakkoechi* **17**, 91–9; *Chem. Abs.* **105**, 146184.

Shroeter, G., Lichenstadt, L. and Ireneu, D. (1918). On the constitution of *Guaiacum* resin. *Chem. Ber.* **51**, 1587–613.

Snyder, F. (1985). Chemical & biochemical aspects of platelet activating factor. *Med. Res. Rev.* **5**, 107n.

Stringfellow, D.A. and Schurig, J.E. (1987). The search for more active & less toxic mitomycin & etoposide analogues. *Cancer Treat. Rev.* **14**, 291–5.

Suekawa, M., Shiga, T., Sone, H., Ikeya, Y., Taguchi, H., Aburada, M. and Hosoya, E. (1987). Effects of gomisin J & analogous lignan compounds in *Schizandra* fruits on isolated smooth muscles. *Yakugaku Zasshi* **107**, 720–6.

Sutherland, E.W. and Rall, T.W. (1958). Fractionation & characterisation of a cyclic adenine ribonucleotide formed by tissue particles. *J. Biol. Chem.* **232**, 1077–91.

Takahashi, K. and Ogiyama, K. (1986). Phenols of discoloured Sugi (*Cryptomeria japonica* D.Don) sapwood. Part 4. Norlignans of Ayasugi (cultivar) in the Kyushu region. *Mokuzai Gakkaishi* **32**, 457–60.

Tandon, S. and Rastogi, R.P. (1976). Wikstromol, a new lignan from *Wikstroemia viridflora*. *Phytochemistry* **15**, 1789–91.

Thurston, L.S., Irie, H., Tani, S., Han, F-S., Liu, X-C., Cheng, Y-C. and Lee, K-H. (1986). Antitumour agents. Part 78. Inhibition of human DNA topoisomerase II by podophyllotoxin & a-peltatin analogues. *J. Med. Chem.* **29**, 1547–50.

Tonance, S.J., Hoffman, J.J. & Cole, J.R. (1979). Wikstromol, antitumour lignan from *Wikstroemia foetida* var. *oahuensis* Gray & *Wikstroemia uva-ursi* Gray (Thymelaeceae). *J. Pharm. Sci.* **68**, 664–5.

Tsukamoto, H., Hisada, S. and Nishibe, S. (1984a). Lignans from the bark of *Fraxinus mandshurica* var. *japonica* & *F. japonica*. *Chem. Pharm. Bull.* **32**, 4482–9.

Tsukamoto, H., Hisada, S. and Nishibe, S. (1984b). Lignans from the bark of *Olea* plants. *Chem. Pharm. Bull.* **32**, 2730–5.

Tsukamoto, H., Hisada, S., Nishibe, S. and Roux, D.G. (1984c). (−)-Aristotetralone from *Aristolochia chilensis*. *Phytochemistry* **26**, 2414–15.

Ululeben, A., Saiki, Y., Lotter, H., Cari, V.M. and Wagner, H. (1978). Chemical components of *Styrax officinalis* L. Part 4. The structure of a new lignan, styraxin. *Planta Med.* **34**, 403–7.

Urzua, A., Freyer, A.J. and Shamma, M. (1987). (−)-Aristotetralone from *Aristolochia chilensis*. *Phytochemistry* **26**, 2414–15.

Van Sumere, C.F., Cottenie, J., De Greef, J. and Kint, J. (1972). Biochemical studies in relation to the possible germination regulatory role of naturally occurring coumarins & phenolics. *Recent Advances in Phytochemistry* **4**, 165–221.

Von Krogh, G. (1981). Podophyllotoxin for condylomata acuminata eradication. Clinical and experimental comparative studies on *Podophyllum lignans*, colchicine and 5-fluorouracil. *Acta Derm. Venereol.* [Suppl] (Stockh). **98**, 1–48.

Wahala, K., Makela, T., Backstrom, R., Brunow, G. and Hase, T. (1986). Synthesis of the [^{2}H]-labelled urinary lignans enterolactone & enterodiol and the phytooestrogen daidzein & its metabolites equol & demethylangolensin. *J. Chem. Soc. Perkin Trans.* **1**, 95–8.

Waller, C.W. and Gisvold, O. (1945). A phytochemical investigation of *Larrea divaricata* Cav. *J. Amer. Pharm. Assoc.* **34**, 78–81.

Waravdekar, V.S., Paradis, A.D. and Leiter, J. (1955). Enzyme changes induced in normal & malignant tissues with chemical agents. *J. Natl. Cancer Inst.* **16**, 31.

Waters, A.P. and Knowler, J.T. (1982). Effect of a lignan (HOMF) on RNA synthesis in the rat uterus. *J. Reprod. Fertil.* **66**, 379–81.

Wilson, L. and Friedkin, M. (1967).The biochemical events of mitosis. II. The *in vivo* and *in vitro* binding of colchicine in grasshopper embryos and its possible relation to inhibition of mitosis. *Biochemistry* **6**, 3126–35.

Wu, J-L., Shen, J. and Xie, Z-W. (1983). Pharmcognostical studies on the Chinese drug Qin pi (*Cortex fraxini*). Part 2. Identification of the drug & its adulterants. *Acta Pharmaceuitica Sinica* **18**, 377–83.

Yalowich, J.C. and Goldman, I.D. (1984). Analysis of the inhibitory effects of VP-16-213 (etoposide) and podophyllotoxin on thymidine transport and metabolism in Ehrlich ascites tumor cells *in vitro*. *Cancer Res.* **44**, 984–9.

Yan-Yong, C., Zeng-Bao, S. and Lian-Niang, L. (1976). *Scientia Sin.* **19**, 276–90.

Yoshihara, T., Yamaguchi, K. and Sakamura, S. (1982). A lignan-type stress compound in potato infected with nematode (*Globodera rostochiensis*). *Agric. Biol. Chem.* **46**, 853–4.

Young, J.D. and Jarvis, S.M. (1983). Nucleoside transport in animal cells. *Biosci. Reports* **3**, 309–22.

4

Etoposide and Teniposide

Introduction

Etoposide (VP-16-213; 4′-demethyl-7-[4,6-*O*-ethylidene-β-D-glucopyranosyl epipodophyllotoxin] **4.3**) and Teniposide (VM-26; 4′-demethyl-7-[4,6-*O*-thenilidene-β-D-glucopyranosyl epipodophyllotoxin] **4.2**) are two semi-synthetic derivatives of podophyllotoxin (**4.1**) which were developed by Sandoz in the early 1970s in an attempt to develop podophyllotoxin derivatives which do not possess the severe gastrointestinal toxicity of podophyllotoxin (Issell *et al.*, 1984; O'Dwyer *et al.*, 1985; Stähelin and von Wartburg, 1989). Using a murine leukaemia L-1210 tumour model, the scientists at Sandoz targeted these two podophyllotoxin derivatives as potential anti-tumour agents for clinical trials in man. In the 15 years since Etoposide and Teniposide were introduced in clinical trials, these drugs have played an important role in a wide variety of cancer chemotherapy protocols. The use of Etoposide represents a therapeutic advance in cancer treatment due to its ease of administration and substantial activity in small-cell lung cancer, testicular

4.1
Podophyllotoxin

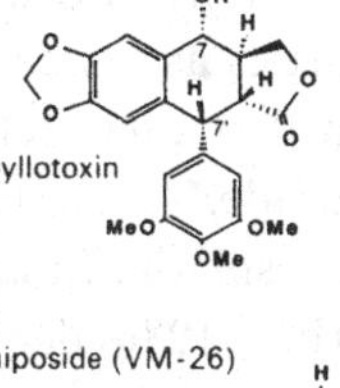

4.2
Teniposide (VM-26)

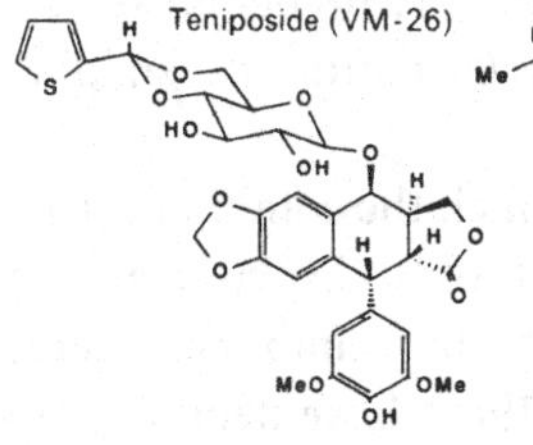

4.3
Etoposide (VP-16-213)

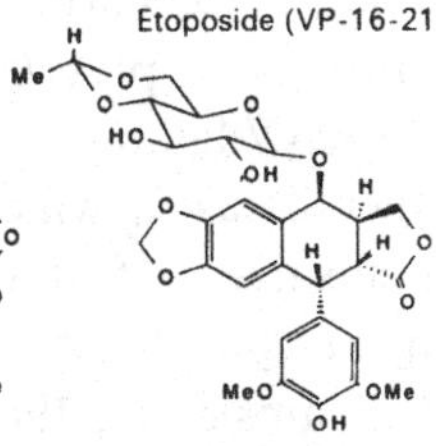

cancer, lymphoma and acute lymphocytic leukaemia (see Issell *et al.*, 1984; O'Dwyer *et al.*, 1985; Yarbro, 1986, for a detailed review of the clinical applications of Etoposide and Teniposide).

The successful development of anticancer agents from lignans represents a paradigm for targeting and developing new pharmacological agents from natural products (see previous chapter). In this chapter, recent studies on the mechanism of action, toxicology and pharmacokinetic properties of Etoposide and Teniposide will be reviewed. The clinical application of Etoposide and Teniposide will be restricted to recent reports involving phase II or III studies. The aim of phase II studies as defined by the Food and Drug Administration of the United States is to establish the general efficacy of a drug against a specific disease and to eliminate any drug which is inactive for that disease. Phase II trials also are meant to further characterise potential cytotoxic activity of a given drug. The aim of phase III studies is to further assess the efficacy of a drug even further by comparing its clinical activity with standard treatment modalities in a randomised control trial. Recent structure–activity studies which identify the chemical moieties of these compounds responsible for their biological properties are also summarised.

Clinical applications

The clinical use of Etoposide in treating cancer is entering a new phase. Etoposide is being incorporated into many combination chemotherapeutic regimens. Many of these combination chemotherapeutic protocols are associated with first-line regimens in the treatment of a variety of cancers, such as small cell lung cancer and germinal testicular cancer.

The results of Cisplatin (*cis*-diamminedichloroplatinum) plus Etoposide combination chemotherapy in small-cell cancer represents a highly productive first-line therapeutic synergism in human cancer chemotherapy (O'Dwyer *et al.*, 1985; Havemann *et al.*, 1987; Loehrer *et al.*, 1988). Furthermore, the combination of these drugs as a second line treatment regimen is due in part to the observation that small-cell lung cancer exhibits resistance to CAV (Cyclophosphamide, Adriamycin and Vincristine), but still responds to Etoposide and Cisplatin. In fact, several recent clinical studies (O'Dwyer *et al.*, 1985; Yarbro, 1986) demonstrated a 50% objective response using Etoposide plus Cisplatin for small-cell lung cancer as salvage therapy for CAV failure.

Testicular cancer is another disease in which the combination of Etoposide plus Cisplatin appears effective. In fact, only Cisplatin surpasses Etoposide as the most active agent against testicular tumours. Several studies (Hainsworth *et al.*, 1985; Williams *et al.*, 1987) have demonstrated significant activity of Cisplatin, Etoposide and Bleomycin as first-line therapy in far advanced germinal testicular cancer. Moreover, Hainsworth *et al.* (1986) reported that the combination of Cisplatin plus Etoposide resulted in a 25% cure rate as salvage therapy for patients not cured with the combination of Cisplatin, vinblastine and bleomycin.

The combination of Cisplatin plus Etoposide also appears to be the most effective non-surgical treatment for patients with non-small cell lung cancer: a recent trial (Finkelstein *et al.*, 1986) reported that this combination yields the highest 1-year survival rate (25%). Thus, from all of the current studies, it appears that the combination of Cisplatin and Etoposide will continue to play an important role in the treatment of certain cancers.

Encouraging results have also been obtained using Etoposide alone for certain forms of leukaemia (Pinkel, 1987) or in combination with ifosfamide, methotrexate and methyl-GAG in treating recurrent or refractory lymphoma (O'Donnell *et al.*, 1987). Studies incorporating Etoposide or Teniposide in the treatment of ovarian carcinoma (Piver, 1986), adenocarcinomas of the upper gastrointestinal tract (Kelsen *et al.*, 1987), refractory breast carcinoma (Cox *et al.*, 1989), and familial erythrophagocytic lymphohistiocytosis (Henter *et al.*, 1986) suggest partial responses but will require further clinical investigations.

Why is Etoposide such an effective agent for incorporation into multidrug treatment protocols? The answer may be related in part, to the fact that Etoposide is a well tolerated, active, anticancer agent. In addition, many cancers develop resistance to a first-line drug therapy. However, these drug-resistant cancers still respond to Etoposide. Thus, multidrug treatment protocols which include Etoposide would decrease the probability of tumours becoming resistant to the combination therapy. Finally, Etoposide and Teniposide exhibit unique mechanisms of action different from those observed in many of the first-line chemotherapy regimens, thereby providing a different mechanism for eradicating the tumour.

Although Teniposide is a more potent cytotoxic agent than Etoposide in L-1210 and HeLa tumour cell lines, its clinical application as an anticancer agent is still under investigation. Teniposide appears to have its greatest activity in haematological malignancies such as refractory lymphoma (Grossberg *et al.*, 1987) and acute lymphocytic leukaemia (Jacobs and Gale, 1984). There is also a report suggesting activity in non-small-cell lung cancer (Giaccone *et al.*, 1987).

An unusual clinical application of Etoposide was reported by Kushner *et al.* (1987). They used this drug as an agent for *ex vivo* bone marrow purging. Purging is the elimination of neoplastic cells from harvested bone marrow and re-administration of neoplastic-free bone marrow cells back into the patient. The use of Etoposide in purging is based on their *in vitro* studies which show that neoplastic lymphoid (SK-DHL2) or myeloid (HL-60) cell lines are more sensitive to the cytotoxic effects of Etoposide than normal human marrow cells. Mixtures of normal marrow cells with malignant cells were treated with 75 μmol/l Etoposide. At this concentration, Etoposide appeared to be cytotoxic to the neoplastic cell lines tested, but had little adverse effect on the proliferative capacity of marrow stromal progenitors (precursor cells which differentiate to form erythrocytes and leukocytes). Furthermore, using animal models, Stiff and Koester (1987) report *in vitro* chemoseparation of leukaemic cells from murine bone marrow using Etoposide. These types of studies suggest that Etoposide may be an effective agent for purging malignant cells from normal stem cells in bone marrow transplant protocols.

In their review article, O'Dwyer *et al.* (1985) emphasise that the 15 years of clinical trials of Etoposide underscore the importance of re-evaluating a new anticancer drug as the clinicians learn more about the pharmacokinetics and mechanism of action of that drug. The initial 'negative' results obtained in chemotherapeutic trials of Etoposide and Teniposide against other tumours may merit reconsideration. In a majority of these trials, the number of patients used for these studies was not sufficient to draw any conclusions. In addition, the pretreatment status of the patients was not stated; inclusion of patients who had relapses after several courses of prior chemotherapy may cause severe underestimation of the anti-tumour activity of Etoposide and Teniposide.

Structure and synthesis

The overall synthesis of Etoposide from podophyllotoxin is outlined in Scheme 4.1 (Issell *et al.*, 1984; Kuhn *et al.*, 1969). The structure of Etoposide has been analysed by NMR spectroscopy and confirmed by X-ray analysis. Brewer *et al.* (1979) used high field NMR (360 MHz) to assign the resonances of the aglycone portion of Etoposide. Jardine *et al.* (1982) published the assignments for the glucosidic part of Etoposide. These structural analyses emphasise that the *trans*-fused lactone portion of Etoposide (and many other podophyllotoxin derivatives) forces a rigid conformation of the A and B rings. Thus, the A and B ring systems are in one plane whereas the C-ring lies almost perpendicular to that plane. The rigidity of Etoposide may explain some of its mechanisms of action. It appears from both NMR (Brewer *et al.*, 1979) and X-ray analyses (Issell *et al.*, 1984) that the glycoside residue occupies a considerable amount of time above the lactone ring [C-9]. Since the lactone rings of podophyllo-

Scheme 4.1 Synthesis of Etoposide from podophyllotoxin

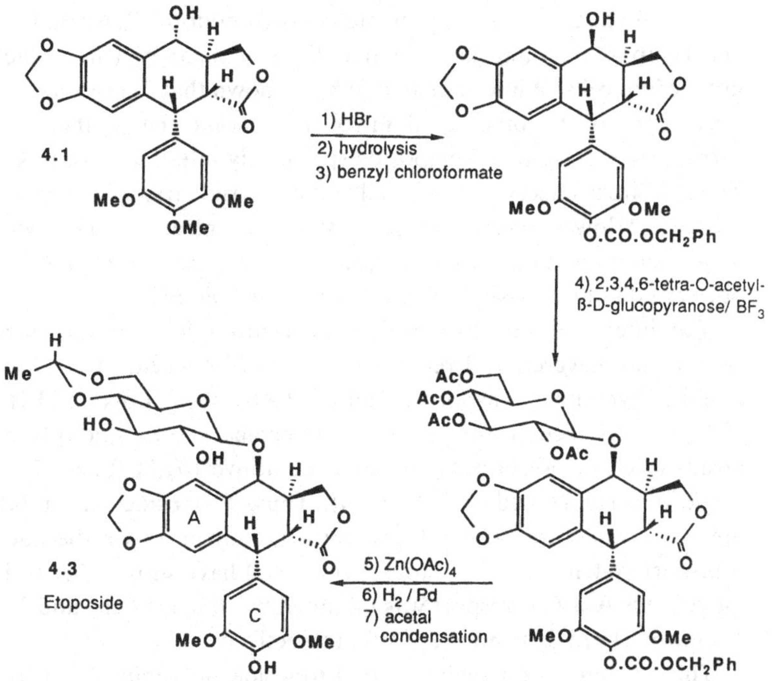

4.1

1) HBr
2) hydrolysis
3) benzyl chloroformate

4) 2,3,4,6-tetra-O-acetyl-ß-D-glucopyranose/ BF_3

5) $Zn(OAc)_4$
6) H_2 / Pd
7) acetal condensation

4.3
Etoposide

toxin derivatives are thought to interact with tubulin (see Chapter 3), the glycoside portion of Etoposide would then interfere sterically with the potential binding of Etoposide to tubulin.

Mechanism of action

The structural differences between Etoposide (and Teniposide) and podophyllotoxin confer new biological properties. In the mid 1970s Loike and Horwitz (1976a) demonstrated that Etoposide and Teniposide do not act as antimitotic poisons (see Chapter 3). Furthermore, it became clear that Etoposide and Teniposide exhibit a new biological property: they induce single- and double-stranded DNA breaks *in vitro* and *in vivo*. Published reports (Chen *et al.*, 1984; Liu, 1983; Long *et al.*, 1984; Ross *et al.*, 1984) in the last few years strongly support the idea that these DNA breaks occur through the interaction of Etoposide or Teniposide with topoisomerase II, a critical enzyme in DNA replication. However, Etoposide, and Teniposide do share one biological property with podophyllotoxin: they inhibit nucleoside transport in mammalian cells.

(i) *Inhibition of nucleoside transport*

As discussed in the previous chapter, podophyllotoxin, Etoposide and Teniposide inhibit the facilitated diffusional component of nucleoside uptake into cells. White *et al.* (1985) have shown that Teniposide competitively inhibits the binding of nitrobenzylthioinosine to the nucleoside carrier, suggesting that Teniposide specifically interacts with this carrier. The inhibitory effects of Teniposide on nucleoside transport are most significant at low extracellular nucleoside concentrations, i.e., where the major component of nucleoside uptake into the cell occurs via the transporter and not by passive diffusion across the lipid bilayer.

The interaction of Etoposide and Teniposide with the nucleoside carrier may have clinical relevance. At 10 μM concentration, Etoposide inhibits thymidine transport in HeLa cells by 50% (Loike and Horwitz, 1976a). This concentration is relevant to clinical studies since plasma Etoposide levels in cancer patients remained above 10 μM for 6–12 h after a single infusion (Hande *et al.*, 1984). Thus, Etoposide may inhibit the uptake of those anticancer drugs which are taken up via the nucleoside transport system. In fact, White *et al.* (1985) have shown that in Ehrlich ascites cells Ara-C transport was inhibited by 50% by 7 μM and 35 μM of Teniposide (VM-26) and Etoposide, respectively.

Thus, combination regimens of Etoposide or Teniposide with those anticancer agents which are transported into the cell via the nucleoside transport system might prove ineffective. This type of drug-drug counterbalance may be the biochemical explanation for the results obtained by Rivera *et al.* (1975) on the efficacy of Ara-C; VM-26 combination against L1210 leukaemia in mice. Only Ara-C; VM-26 combinations with the highest doses of Ara-C possessed the greatest antitumour activity. At lower Ara-C; VM-26 concentrations ratios, Teniposide blocked entry of Ara-C into the L1210 cells.

(ii) *DNA effects*

The first report that Etoposide may have an effect on DNA came from Huang *et al.* (1973) who showed that cells treated with Etoposide (VP-16-213) have chromosomal aberrations. Further studies by Grieder *et al.* (1974, 1977) demonstrated that Etoposide and Teniposide arrest cells in either late S or early G_2 phase of the cell cycle, suggesting that these drugs were not acting like inhibitors of microtubule assembly. Loike and Horwitz (1976a,b) first demonstrated that Etoposide and Teniposide did not interfere with microtubule assembly *in vitro* (see Chapter 3) but did induce single-stranded breaks in HeLa DNA. This effect was reversible and did not occur with purified DNA (from either calf thymus or adenovirus) treated with these agents. Roberts *et al.* (1980) confirmed the results of Loike and Horwitz (1976b) by demonstrating that these drugs caused single-stranded breaks in L1210 DNA which were repaired after drug removal.

In 1983, Gupta suggested that the cellular effects of Etoposide were similar to those of X-radiation. However, injury induced by X-ray radiation of DNA may not be a good model for the action of Etoposide on DNA. Smith *et al.* (1986) have shown that an SV-40 transformed fibroblast cell line originally derived from an ataxia telangiectasia patient is more sensitive to Etoposide than to its corresponding normal control cell line. Ataxia-telangiectasia is an autosomal recessive genetic disorder characterised by cerebellar ataxia, oculocutaneous telangiectasia and immunodeficiency. Many of these patients and cell lines derived from these patients exhibit a defect in DNA repair mechanisms and are more susceptible to irradiation injury and chromosomal breaks. Initially it would make sense that radiation sensitive cells (like the SV-40 transformed fibroblast line originally derived from an ataxia telangiectasia patient) should be more sensitive to the action of Etoposide. However, Smith *et al.* (1986) reported that the frequency of DNA lesions induced per lethal hit is the same for both the normal and the radiation sensitive cell lines, suggesting that normal and radiation sensitive cell lines do not differ in their intrinsic sensitivity to Etoposide. These authors (Smith *et al.*, 1986) also suggest that Etoposide treatment differs from X-ray irradiation in that a large percentage of Etoposide treated cells are in parasynchronous S-phase which is a property not observed in X-ray irradiated cells. Thus, it would appear that the cellular effects of Etoposide are not similar to X-ray irradiation.

In 1983, Wozniak and Ross confirmed that Etoposide did not induce breaks in purified DNA and demonstrated that Etoposide induces double-stranded DNA breaks and causes DNA–protein crosslinks in L1210 cells. Equally important, they reported that Etoposide induces both single- and double-stranded breaks in isolated nuclei from these cells and suggested that these drugs affected some protein involved in DNA synthesis or repair. Within a year of this report, Ross *et al.* (1984), Long *et al.* (1984) and Chen *et al.* (1984) demonstrated that Etoposide inhibited the action of topoisomerase II. Moreover, these drugs interact with type II topoisomerase *in vitro* (Ross *et al.*, 1984; Chen *et al.*, 1984). Wozniak and Ross (1983) and Long *et al.* (1984) presented a strong correlation between DNA break frequency and cytotoxicity in L1210 leukaemic cells or in lung adenocarcinoma cells treated with Etoposide. These groups also reported that Teniposide is approximately 10-fold more potent than Etoposide in causing DNA breaks; a finding that correlated well with Teniposide's increased cytotoxicity. Thus, the current evidence suggests that these lignan derivatives exert their cytotoxic action through an interaction and inhibition of topoisomerase II (Glisson & Ross, 1987).

Topoisomerases are enzymes found in both eukaryotes (animal cells) and prokaryotes (plant cells) which change the topology of DNA without altering the nucleotide sequence (see Liu, 1983, and Wang, 1987, for a review). It is now accepted that topoisomerase II is involved in a variety of biological properties, including (a) the elongation of DNA replication, (b) the resolution of intertwined pairs of newly replicated DNA molecules, (c) DNA transcription, (d) the segregation of chromosomal material at the end of replication, and (e) the nuclear matrix in chromosome scaffolds. Topoisomerase II also is thought to affect the changes in chromatin structure accompanying gene activation or inactivation (Glikin *et al.*, 1984; Riou *et al.*, 1986).

Topoisomerase II exists as a dimer with subunit molecular mass of 131–180 kilodaltons. This enzyme catalyses DNA topoform interconversions by introducing a transient enzyme-bridged double-strand break in crossing DNA segments. Liu (Liu, 1983; Rower *et al.*, 1984) and Long *et al.* (1985) have shown that topoisomerase II cuts DNA with a four base stagger and becomes covalently linked to the 5′ end of each broken strand via a phosphotyrosyl bond. In fact, this complex can be isolated and represents a key intermediate in the breaking–rejoining reactions. The energy for rejoining the broken DNA strands comes from an enzyme-associated ATPase activity. Thus, topoisomerase II requires Mg^{2+} and ATP, shows DNA-stimulated ATP hydrolysis, and requires the presence of a sulphydryl reagent for full activity. Work over the last four years has revealed that Etoposide and Teniposide stabilise the formation of a DNA–topoisomerase II intermediate resulting in increased DNA scission and a failure to rejoin the DNA. Thus, Etoposide traps topoisomerase II with DNA and effectively interferes with the DNA-breakage-reunion reactions normally catalysed by this enzyme.

The identification of DNA topoisomerase II as the intracellular target for Etoposide and Teniposide is consistent with the DNA-cutting properties of these drugs. Topoisomerase II can cause both single- and double-stranded DNA breaks, a reaction which is stimulated by but not dependent on ATP. The interaction of these drugs with topoisomerase II but not DNA, can also account for DNA protein crosslinks observed by alkaline elution. Interestingly, the podophyllotoxin ring system of Etoposide and Teniposide is similar to oxolinic acid which is a potent inhibitor of the ligase activity of the prokaryotic type II topoisomerase, gyrase (Cozzarelli, 1980; Ross *et al.*, 1984; Long *et al.*, 1985).

Inhibitors of topoisomerase II generate a variety of cellular effects including sister chromatid exchanges, chromosome aberrations, increased chromosome number and cell killing. Which of these effects are direct consequences of the inhibition of topoisomerase II? Moreover, what is the relationship (if any) between cell death (cytotoxicity) and the inhibition of topoisomerase II? These issues are addressed in a recent paper (Pommier *et al.*, 1988). Pommier *et al.* (1988) investigated the generation of sister chromatid exchange and chromosome aberrations in normal and in Etoposide resistant Chinese hamster cell lines treated with Etoposide or other inhibitors of topoisomerase II. They found that in a normal Chinese Hamster cell line the concentrations of Etoposide producing sister chromatid exchange and chromosomal aberrations are in the same concentration range as those producing protein-associated DNA strand breaks. In addition, cytotoxicity correlated well with sister chromatid exchange. These authors suggest that topoisomerase II inhibition leads to both drug-induced protein associated DNA strand breaks and chromosome aberrations. This hypothesis is supported by their studies with Etoposide resistant cells in which very few protein associated DNA strand breaks were found and no drug-induced sister chromatid exchange or chromosome aberrations were detected after treatment with Etoposide. Nonetheless, the biochemical pathway by which inhibition of topoisomerase II results in chromosomal rearrangements remains to be elucidated.

Most studies have examined the short term effects of Etoposide on DNA breakage. Recently, Jaxel *et al.* (1988) have reported that prolonged treatment of Etoposide with concanavalin A-stimulated splenocytes results in a secondary DNA fragmentation which is irreversible and yields small DNA fragments. In contrast, short term incubations of cells with Etoposide results in reversible DNA breaks. Therefore Jaxel *et al.* (1988) hypothesise that prolonged treatment (>20 h) of proliferating splenocytes by Etoposide induces DNA fragmentation by a mechanism (as yet to be defined) that may not directly involve topoisomerase II.

There are other drugs which interfere with topoisomerase II. *o*-AMSA, *m*-AMSA, adriamycin (daunorubicin), ellipticine and 2-methyl-9-hydroxyellipticinium are all intercalating agents which inhibit the enzymatic capacity of purified topoisomerase II. Furthermore, there are many drugs like Interleukin-2, which potentiate the cytotoxic activity of Etoposide presumably through their activation of topoisomerase II (Edwards *et al.*, 1987). There are two recent reports by Alexander *et al.* (1987a,b) that tumour necrosis factor (TNF) enhances the *in vitro* and *in vivo* efficacy of Etoposide. The mechanism by which TNF induces this enhanced Etoposide cytotoxicity and the role of topoisomerase II in this process merit further investigations. Epstein and Smith (1988) showed that oestrogen potentiates DNA damage and cytotoxicity in human breast cancer cells treated with Etoposide. Clearly, topoisomerase II is a multidrug target.

(iii) *Effects on oxidative phosphorylation*

There are a few reports in the literature that Teniposide inhibits oxidative phosphorylation. Mitochondrial abnormalities were early cytotoxic effects in chick embryo fibroblasts treated with Teniposide (Gotzos *et al.*, 1979). These effects are consistent with the observations of Gosalvez *et al.* (1972) that Teniposide inhibits NADH-linked respiration in Ehrlich ascites tumour cells and rat liver mitochondria. While it is generally believed that the cytotoxic effects of Teniposide are due to its interaction with topoisomerase II, the relationship between the toxic effects of this drug on oxidative metabolism and cytotoxicity remains to be further characterised.

Cell sensitivity and resistance towards Etoposide

Cell resistance towards antineoplastic agents is a frequent problem in chemotherapeutic management of cancer since tumours often become resistant to the administered drug and at times to other chemically unrelated anticancer drugs. Clinically, drug resistance leads to the limited duration or partial responses following chemotherapy. Understanding the mechanism of cell resistance to Etoposide and cross-resistance of Etoposide to other chemotherapeutic agents is vital in designing combination therapy protocols. In addition, comparing the cytotoxic effects of a drug in sensitive vs. resistant cell lines may yield insights into its mechanism of action and cell type specificity.

Resistance to antitumour agents is frequently associated with changes in the plasma membrane and decreased net drug accumulation into the cells. Roberts *et al.* (1987) have attributed resistance to Teniposide by Teniposide-selected resistant L-1210 cells in culture to a reduced accumulation of this drug. Moreover, Teniposide resistant L-1210 cells exhibited cross-resistance to vincristine, doxorubicin, actinomycin D, and amsacrine. This type of resistance is characteristic of multidrug resistance. Teniposide resistant cell lines also express a unique 22-kD protein not seen in drug-sensitive cells and in certain cell lines; this protein was associated with the plasma membrane. Thus, certain characteristics of multidrug resistance accompany the acquisition of resistance to Teniposide.

Long *et al.* (1986) used a unique approach to study the mechanism of cell resistance to Etoposide and Teniposide. They characterised primary human lung carcinoma cell lines for natural resistance to Etoposide and

Teniposide and identified the sites of resistance. They showed that the lethal effect of Etoposide may be related to the formation of double-stranded DNA breaks since a much better correlation exists between cytotoxicity and double-stranded DNA breakage than between cytotoxicity and single-stranded breaks.

Furthermore, in several of the naturally resistant cell lines the authors found lower levels of topoisomerase II than in parent cell lines. Thus, Long *et al.* provide indirect evidence that the concentration of topoisomerase II in human lung carcinoma cells may play an important role in conferring natural resistance or sensitivity to Etoposide.

Pommier *et al.* (1986a,b) studied the activity of topoisomerase II isolated from Chinese Hamster lung cells which were made resistant to Etoposide and other DNA intercalators. Those isolated cell lines which were made resistant to 9-hydroxyellipticine have reduced capacity to form protein-associated DNA strand breaks upon exposure to *m*-AMSA, 2-methyl-9-hydroxyellipticinium and Etoposide. Both whole cells and isolated nuclei from the resistant cells generated fewer protein-associated DNA breaks in response to Etoposide than their non-resistant control cells. The authors interpret these results to mean that the mechanism of resistance in Chinese Hamster lung cells appears not to be related to drug accumulation, but rather it corresponds to a reduction of topoisomerase II sensitivity to Etoposide. In addition, the authors present other evidence that resistance to Etoposide may be due to the presence of a modulating factor perhaps related to topoisomerase I. The presence of an associated modulating factor has also been suggested by Charcosset *et al.* (1988).

Further evidence that alterations in the topoisomerase II enzyme are associated with Etoposide resistance came from Glisson *et al.* (1986). They studied the mechanism of Etoposide resistance in a Chinese Hamster ovary cell line, Vpm^r-5, which was made resistant to Etoposide. From their results, they hypothesised that the Etoposide-resistant Vpm^r-5 line expressed a qualitative change in the type of topoisomerase II that altered the interaction of drug with the enzyme or enzyme–DNA complex. It is possible that the resistant cell line expressed a topoisomerase II with a mutation in the Etoposide-binding site. Danks *et al.* (1988) have also described a form of multidrug resistant phenotype of Teniposide-resistant human leukaemic cells with an alteration in the activity of DNA topoisomerase II or a modulator of this enzyme. The precise mechanism of resistance in these cell lines must await purification of topoisomerase II from the parent and resistant lines.

Sullivan *et al.* (1986) have been interested in understanding the effects of the cell cycle on Etoposide cytotoxicity by comparing Etoposide-induced DNA damage in log-phase cell growth versus plateau cells. They found that log-phase Chinese Hamster ovary cells exhibited dose-dependent drug-induced DNA breaks whereas cells in a plateau phase of growth were found to be resistant to the effects of Etoposide. The lack of sensitivity to Etoposide is thought to reflect an alteration in topoisomerase II activity in cells in a plateau phase of growth. In their studies, topoisomerase II activity was monitored by a variety of techniques (including the capacity to cleave pBR322 DNA or to precipitate DNA topoisomerase complexes) and found to be uniformly greater in log-phase Chinese Hamster ovary cells than it is in plateau cells. These studies are the first to suggest that the sensitivity of a cell to Etoposide may correlate with topoisomerase II activity during its cell cycle. However, in other cell types such as HeLa and L1210, Etoposide-induced DNA damage occurs in both log and plateau phases of cell growth.

In a recent study, Sullivan *et al.* (1987) extended their initial observations to examine the basis for the reduction in enzyme activity during quiescence in Chinese Hamster ovary cells. They demonstrated that loss of topoisomerase II activity during plateau phase is a function of reduction in enzyme content rather than post-translational modifications of the enzyme in Chinese Hamster ovary cells as determined by immunoblotting techniques with antibodies directed against the topoisomerase II. Thus, in some cell types topoisomerase II content is proliferation-dependent while in other malignant phenotypes different factors can decrease the sensitivity of the cell to Etoposide. These studies demonstrate that cells may differ in their regulation of topoisomerase II content as a function of their cell cycle. Equally important, Sullivan *et al.* (1987) demonstrate that L1210 cells exhibit different properties with respect to sensitivity to Etoposide. Sensitivity towards Etoposide-induced DNA breakage is equal in log and plateau phase in L1210 cells, yet plateau phase cells are more resistant to drug toxicity than log phase L1210 cells. The authors suggest that topoisomerase II content may not uniformly predict tumour sensitivity to Etoposide in the non-dividing state.

The results from the above mentioned studies suggest that there are multiple pathways by which tumour cells develop resistance or sensitivity to Etoposide. Yet, understanding how a particular cancer cell develops resistance to Etoposide or Teniposide will be particularly helpful in choosing cancer chemotherapy drug combinations which present the least cross-resistance. In addition, agents which potentiate topoisomerase II activity in tumour cells may in fact increase sensitivity towards Etoposide cytotoxicity. This is evident from studies by Edwards *et al.* (1987) who showed

Table 4.1. *Relative activities of some podophyllotoxin congeners*

Cytotoxicity
Teniposide, demethylepipodophyllotoxin benzylidine glucoside > 4′-demethylpodophyllotoxin > demethylpodophyllotoxin benzylidine glucoside > Etoposide > 4′-demethylepipodophyllotoxin > 4′-demethylepipodophyllotoxin glucoside > 4′-demethylpodophyllotoxin glucoside.

DNA breakage
Teniposide, demethylepipodophyllotoxin benzylidine glucoside > Etoposide > 4′-demethylepipodophyllotoxin > demethylpodophyllotoxin benzylidine glucoside > demethylpodophyllotoxin > demethylepipodophyllotoxin glucoside > demethylpodophyllotoxin glucoside.

that lymphocytes treated with mitogens like phytohaemagglutinin and immunomodulators like interleukin-2 become more sensitive to the cytotoxic actions of Etoposide. Thus, efforts to stimulate topoisomerase II content may improve the therapeutic efficacy of this drug.

Structure/activity studies – DNA effects

Earlier studies by Loike and Horwitz (1976a,b) demonstrated that all podophyllotoxin and Etoposide derivatives that contained a C-4′ hydroxyl group induced DNA breaks in HeLa cells. Later studies by Ross and colleagues (Wozniak and Ross, 1983; Ross *et al.*, 1984) and others (Long *et al.*, 1984; Chen *et al.*, 1984) confirmed that a C-4′ hydroxy group was necessary for an Etoposide derivative to induce DNA breaks both *in vitro* and *in vivo*. For example, Long *et al.* (1984) tested the effects of various podophyllotoxin congeners (structures **4.4–4.6**) on cytotoxicity and DNA breakage Table 4.1.

The structural requirements for a congener of Etoposide to induce DNA breaks are (a) demethylation at the C-4′ position, (b) epimerisation at the C-7 position to potentiate activity, (c) 6-*O* addition to the glucose moiety to potentiate activity; and (d) 7-D-glycosylation to decrease activity.

Thurston *et al.* (1986) investigated a series of podophyllotoxin and peltatin analogues with regard to their ability to inhibit DNA topoisomerase II activity. Using human DNA topoisomerase II from lymphoid cells, these authors confirmed previous studies in that (a) a free 4′-hydroxyl is required for activity, (b) a 7-β ether linkage enhances activity, and (c) a free hydroxyl group at the C-3′ position on the pendent C-ring also enhances activity. The additional hydroxyl group at the C-3′ position is thought to make an additional H-bonding site available. Thurston *et al.*

(1986) also demonstrated that cleavage of the methylenedioxy ring generated greater activity than α-peltatin. However, in their hands a number of compounds previously shown to be inactive *in vivo* were found to be active *in vitro*. For example, it is unclear why these authors found a 30% inhibition of topoisomerase II with epipodophyllotoxin (which possesses a 4′-methoxy group) while other investigators found no DNA cleavage either *in vitro* or *in vivo*. Furthermore, Thurston *et al.* (1986) found that 4′-demethylpodophyllotoxin was inactive whereas others reported that this compound did generate DNA breaks *in vivo*. It will be important to clarify why some derivatives induce DNA damage *in vitro* and not *in vivo*.

It is interesting that the podophyllotoxin ring system of both Etoposide and Teniposide is similar to that of oxolinic acid, a potent inhibitor of the ligase activity of prokaryote type II topoisomerase, gyrase (Gellert *et al.*, 1977; Synder and Drlica, 1979; Sugino *et al.*, 1977). In fact, oxolinic acid inhibits the ligase activity of gyrase *in vitro* and *in vivo* leading to the formation of protein-associated double-strand DNA breaks. These types of DNA break are similar to that observed in Etoposide-treated cells.

Saito *et al.* (1986a,b) have recently described an interesting method of modifying the sugar moiety of the podophyllotoxin glycosides. They synthesised numerous analogues of podophyllotoxin in which the sugar moiety of Etoposide was replaced with an aminosugar derivative. The antitumour activities of these synthesised compounds were evaluated using a leukaemia L-1210 tumour system of mice. The antitumour activities of these structures correlated with the nature of the sugar moiety. As compared to Etoposide, the cyclic acetals 1-*O*-(2-amino-2-deoxy-β-D-glucopyranosyl)-4′-*O*-demethyl-D-epipodophyllotoxin (**4.4**), 3-amino-3-deoxy-β-D-glucopyransosyl)-4′-*O*-demethyl-D-epipodophyllotoxin (**4.5**) and 1-*O*-(2-dimethylamino-2-deoxy-4:6-*O*-ethylidiene-β-D-glucopyranosyl)-4′-*O*-demethyl-L-epipodophyllotoxin (**4.6**) produced a significant increase in the survival time of mice injected with L-1210 tumour cells. In contrast, cyclic acetals of the corresponding α-D-glucosides and α- and β-glucosides did not show significant antitumour effects. These studies suggest that further understanding of the role of the sugar moiety in the anti-

4.4: $R_1 = OH, R_2 = NH_2$
4.5: $R_1 = NH_2, R_2 = OH$
4.6: $R_1 = OH, R2 = -NMe_2$

tumour activity of Etoposide or Teniposide may generate new derivatives with better clinical application.

Pharmacokinetics and pharmacology

(i) *Cellular uptake of Etoposide and Teniposide*

There are only a few studies which examined the membrane interaction and cellular uptake of Etoposide or Teniposide. The amphipathic nature of these compounds would allow for association either with membrane lipids or with hydrophobic domains on the membrane proteins. Electron-spin resonance studies by Wright and White (1987) demonstrate that Teniposide affects the acyl chain order of cellular and model membranes. Teniposide exhibited a three-fold increase in membrane effects as compared to the less toxic congener, Etoposide. This is probably related to the fact that the thiophene moiety of Teniposide renders it more hydrophilic than Etoposide. In addition, Wright and White (1987) showed that treating Ehrlich ascites tumour cells with Teniposide significantly reduced the mobility of membrane probes, further suggesting that Teniposide perturbs biological membranes. However, the cytotoxic effects resulting from the interaction of Teniposide with biological membranes remains to be clarified.

Both Allen *et al.* (1978) and Seeber *et al.* (1982) suggest that these drugs enter cells by passive diffusion, whereas the efflux of these drugs is thought to be mediated by an active energy-dependent pump system. The efflux of Etoposide was characterised by a single exponential decay process, with a small fraction of drug observed to be non-exchangeable. The role of the bound form of Etoposide (non-exchangeable) is presently unknown. Thus, the uptake and efflux of Etoposide are referred to a leak and pump mechanism of transport, which has also been suggested to operate in the cellular uptake of vinca alkaloids.

Interestingly, Yalowich *et al.* (1985a,b, 1987) found that cells treated with Verapamil (a calcium antagonist) were much more sensitive to Etoposide-induced DNA breaks. Furthermore, they found that Verapamil caused an elevation of cellular Etoposide levels which correlated with the enhancement of DNA damage and increased cytotoxicity. Studies of bidirectional Etoposide fluxes in and out of cells revealed that Verapamil lowered the rate of Etoposide efflux in the L-1210 leukaemic cell line. Further evidence that the potentiation effect of Verapamil is related to a membrane effect comes from experiments demonstrating that Verapamil does not potentiate Etoposide's DNA damage in isolated L-1210 nuclei. Thus, cells treated with Etoposide and Verapamil showed an overall increase in steady state accumulation of the Etoposide by inhibiting Etoposide efflux. Yalowich (1987) reported a similar effect with microtubule inhibitors. He found that Vinblastine and other microtubule inhibitors (including Vincristine and Maytansine but not Colchicine) elevated intracellular Etoposide levels (and in turn Etoposide-induced DNA breaks) by inhibiting Etoposide efflux of the exchangeable drug and by increasing the intracellular level of non-exchangeable drug.

(ii) *Pharmacokinetics*

The pharmacokinetics of Etoposide in humans are generally studied after i.v. administration or occasionally after oral administration. Most of the recent pharmacological studies of Etoposide and Teniposide utilise high-pressure liquid-chromatography assays and confirm earlier studies (Creaven, 1982; Issel *et al.*, 1984) which correlate well with the conventional compartmental approach for describing the pharmacokinetics of Etoposide.

The plasma levels of Etoposide after an i.v. bolus infusion follows a bi-exponential decay curve, with a distribution half-life of 0.6 to 1.3 h and elimination half-life time of four to eight hours. The distribution volume of the steady state ranges from 4.8 to 10.6 l/m^2 and the total body clearance for Etoposide ranges from 19.5 to 39.3 $ml/min/m^2$. Both Etoposide and Teniposide bind to plasma proteins. Several studies (Postmus *et al.*, 1984; Radice *et al.*, 1979; Stewart *et al.*, 1983) have measured Etoposide levels in cerebrospinal fluid and have suggested that Etoposide does disrupt the blood–brain barrier. In rats, both Etoposide and Teniposide are cleared mainly via the bile.

It is interesting that considerable interpatient variability in absorption and elimination of Etoposide has been described. Oral preparations have proven to be highly erratic in absorption and have not been recommended until further oral formulations are developed. The pharmacokinetics of Etoposide and Teniposide differ significantly, particularly in that the clearance of Teniposide occurs much more slowly than that observed for Etoposide.

A study by D'Incalci *et al.* (1986), investigated the pharmacokinetic properties of Etoposide in patients with abnormal renal and hepatic function. This is an important study since many cancer patients suffer from renal or hepatic insufficiency. It concludes that in patients with renal dysfunction, Etoposide clearance is reduced suggesting that the dose of Etoposide should be reduced in these patients. Furthermore, the authors stated that a good correlation was found between Etoposide and creatinine plasma clearance suggesting that the doses of Etoposide might be adjusted in relation to creatinine clearance values.

(iii) *Metabolism*

The metabolism of Etoposide has not been completely elucidated. Urinary excretion of Etoposide accounts for only 35–40% of drug disposition. The metabolism of the remaining 60% of an administered drug dose has not been fully determined. Earlier studies (Strife *et al.*, 1980; Evans *et al.*, 1982) have shown that Etoposide is metabolised to a hydroxyacid derivative which is formed as a result of hydrolysis of the *trans*-lactone ring. Other metabolites identified, include a *cis*-picro-lactone isomer (Evans *et al.*, 1982) which is found in low concentrations in plasma or urine and a glucuronic acid conjugate (Rossi *et al.*, 1982). Urinary excretion of Etoposide glucuronide accounted for roughly 9–17% of an administered Etoposide dose depending on the study. In fact, Arbuck *et al.* (1986) could not identify Etoposide conjugation in their study. Recently, Hande *et al.* (1988) found that etoposide glucuronide was a major metabolite in rabbits and rats and accounted for 30–40% of drug disposition. These authors speculated that the relative resistance of Etoposide glucuronide to some glucuronidase preparations may account for an underestimation of the Etoposide glucuronide formation in man.

There is also an interesting series of papers by Haim and Sinha's group (Haim *et al.*, 1986; Sinha *et al.*, 1983, 1985) demonstrating that Etoposide can undergo a cytochrome P-450 dependent *ortho*-demethylation to form the 3′,4′-dihydroxy derivative. This species is highly reactive and is thought to be oxidised to the *ortho*-quinone of Etoposide. Since the quinone derivatives are probably shortlived and would probably react rapidly with glutathione or other R-SH compounds, their identification in human fluids will be difficult. One can speculate that the *ortho*-quinone derivative might generate oxygen radicals and cause macromolecular damage. However, since Etoposide and Teniposide inhibit DNA topoisomerase II *in vitro* and inhibition of this enzyme is thought to induce DNA breaks, it is unlikely that metabolism of Etoposide to the *ortho*-quinones is required to generate DNA breaks. Nonetheless, Sinha *et al.* (1988) have recently reported that the *o*-dihydroxy metabolite to Etoposide is cytotoxic to human breast tumour cells. Furthermore, they found that Etoposide chelates iron and catalyses the formation of hydroxyl radicals from hydrogen peroxide and reduced glutathione. Under these conditions, the dihydroxy form of Etoposide can induce breaks in SV40 DNA. This is the first report that an Etoposide metabolite can induce DNA breaks in the absence of topoisomerase II. It is possible that the free radical formed could also inactivate topoisomerase II. Thus, these reactive metabolites of Etoposide and Teniposide may be responsible for other biological properties of these drugs.

(iv) *Toxicity*

The results from many studies highlight that the dose-limiting toxicity of Etoposide is leukopenia. Other side effects include anaemia, thrombocytopenia, alopaecia, peripheral neuropathy and mild gastro-intestinal toxicity. Bone marrow recovery is usually complete by 20 days after drug administration. Overall, one of the advantages of using Etoposide in chemotherapy is that it is well tolerated in patients. Moreover, the discovery that various growth factors such as granulocyte-monocyte colony stimulating factor or erythropoietin stimulate the proliferation of leukocytes and red blood cells, respectively, may offer new therapeutic ways of reducing the leukopenia and anaemia caused by Etoposide.

Conclusions and future studies

There has been a great deal of work done in the last five years on understanding the application of Etoposide and Teniposide to human cancer therapy strategies as well as elucidating the mechanism of action of these podophyllotoxin analogues. It is well established that Etoposide and Teniposide do not affect microtubules but induce both single-stranded and double-stranded DNA breaks through their interaction with topoisomerase II. What remains poorly understood, however, is the mechanism linking the accumulation of DNA breaks to the cytotoxic properties of Etoposide and Teniposide.

One future area of investigation is whether Etoposide and Teniposide induce DNA breaks at specific DNA loci? In fact, Fosse *et al.* (1988) have reported a pattern of recognition of DNA by topoisomerase II *in vitro.* Their studies suggest that topoisomerase II recognises specific DNA patterns and that these sequences may be the sites of Etoposide-induced DNA breaks.

Another area for future research is characterising which tumours are responsive to the cytotoxic effects of Etoposide. There are at least 14 tumours on which Etoposide does not appear to exert any therapeutic effect (Table 4.2). It would be interesting to examine the capacity of Etoposide to interact with topoisomerase II from these Etoposide-resistant tumours. Do Etoposide-resistant tumours exhibit an altered drug–topoisomerase II interaction, do they express reduced topoisomerase activity, or do they express altered transport properties to Etoposide?

In this regard, it is interesting to speculate about the relationship of

Table 4.2. *Tumour types insensitive to Etoposide*

Colorectal	Ovarian
Melanoma	Renal cell
Head and neck	Gastric
Soft-tissue sarcoma	Endometrium
Cervix	Bladder
Osteosarcoma	Esophageal

topoisomerase II activity to the susceptibility of malignant cells to Etoposide. Miskimins *et al.* (1983) have demonstrated that the exposure of human and mouse fibroblasts to epidermal growth factors results in increased topoisomerase II activity in both cytoplasm and nuclei. This raises the possibility that topoisomerase II activity may be greater in cancer cells than in normal cells, and may explain the selective cytotoxic effects of Etoposide towards tumour cells. Moreover, Potmesil *et al.* (1988) report that Etoposide exhibited diminished antitumour activity in chronic lymphocytic leukaemic cells and that these cells contain extremely low levels of topoisomerase II. In contrast, lymphocytes from other lymphoproliferative disorders are more sensitive to Etoposide and express high levels of topoisomerase II. The above studies suggest that the sensitivity of a cell type to Etoposide correlates with the level of topoisomerase II expression by that cell type.

Another area of future research is in studying the effects of Etoposide and Teniposide on oncogene expression. About 40 oncogenes have been characterised so far. While oncogenes are involved in human cancer, how they participate in cancer remains to be solved. In certain forms of cancer, in particular breast and lung, the number of oncogene copies in the tumours correlates positively with the severity of the disease. Those individuals with high DNA copy number of these oncogenes are less likely to respond to chemotherapy.

Topoisomerase II appears to be involved in oncogene activation. Riou *et al.* (1986) have shown that Teniposide induces non-random DNA breaks *in vivo* in the c-myc oncogene, within a region located 5′ (upstream) to the first c-myc exon. This region is thought to mediate the control of c-myc expression. It will be important to correlate the activity of Etoposide or Teniposide in a particular tumour with its capacity to induce non-random DNA breaks in the oncogene region of that tumour. It is possible that in future, chemotherapeutic decisions will depend on the oncogene profiles of the cancer.

Thus, a clearer picture of the regulation of topoisomerase II in the cell cycle, in oncogene activation, and in Etoposide-resistant cells will be an important area in evaluating future applications of Etoposide and Teniposide in cancer chemotherapy.

Etoposide is an established antitumour podophyllotoxin derivative with a new mechanism of action. Its development from the natural product, podophyllotoxin, represents a model system for developing new drugs from natural products. Structure–activity relationship studies have been and will be required for the rational development and design of new Etoposide derivatives with potentially greater antitumour activity and selectivity. These types of studies coupled with greater clinical experience in using Etoposide will provide a better understanding of the applications of Etoposide in medicine.

References

Alexander, R.B., Isaacs, J.T. and Coffey, D.S. (1987a). Tumor necrosis factor enhances *in vitro* and *in vivo* efficacy of chemotherapeutic drugs targeted at DNA topoisomerase II in treatment of murine bladder cancer. *J. Urol.* **138**, 427–9.

Alexander, R.B., Nelson, W.G. & Coffey, D.S. (1987b). Synergistic enhancement by tumor necrosis factor of *in vitro* cytotoxicity from chemotherapeutic drugs targeted at DNA topoisomerase II. *Cancer Res.* **47**, 2403–6.

Allen, L.M. (1978). Comparison of uptake and binding of two epipodophyllotoxin glycopyranosides, 4′-demethylepipodophyllotoxin ethylidene-β-D-glucopyranoside and 4′demethylepipodophyllotoxin-thenilidene-D-glucopyranoside in the L1210 leukemia cell. *Cancer Res.* **38**, 2549–54.

Arbuck, S.G., Douglas, H.O., Crom, W.L., Goodwin, P., Silk, Y., Cooper, C. and Evans, W.E. (1986). Etoposide pharmacokinetics in patients with normal and abnormal organ function. *J. Clin. Oncol.* **4**, 1690–5.

Brewer, C.F., Loike, J.D., Horwitz, S.B., Sternlicht, H. and Gensler, W.J. (1979). Conformational analysis of podophyllotoxin and its congeners. Structure–activity relationship in microtubule assembly. *J. Med. Chem.* **22**, 215–21.

Charcosset, J.Y., Saucier, J.M. and Jacquemin-Sablon, A. (1988). Reduced DNA topoisomerase II activity and drug-stimulated DNA cleavage in 9-hydroxyellipticine resistant cells. *Biochem. Pharmacol.* **37**, 2145–9.

Chen, G.L., Yang, L., Rowe, T.C., Halligan, B.D., Tewey, K.M. and Liu, L.F. (1984). Nonintercalative antitumor drugs interfere with the breakage–reunion reaction of mammalian DNA topoisomerase II. *J. Biol. Chem.* **259**, 13560–6.

Cox, E.B., Burton, G.V., Olsen, G.A. and Vugrin, D. (1989). Cisplatin and etoposide: an effective treatment for refractory breast carcinoma. *Am. J. Clin. Oncol.* **12**, 53–6.

Cozzarelli, N.R. (1980). DNA gyrase and the supercoiling of DNA. *Science* **207**, 953–60.

Creaven, P.J. (1982). The clinical pharmacology of VM-26 and VP16-213: a brief overview. *Cancer Chemother. Pharmacol.* **7**, 133–40.

Danks, M.K., Schmidt, C.A., Cirtain, M.C., Suttle, D.P. and Beck, W.T. (1988). Altered catalytic activity of the DNA cleavage by DNA topoisomerase II from human leukemic cells selected for resistance to VM-26. *Biochemistry* **27**, 8861–9.

D'Incalci, M., Rossi, M., Zucchetti, M., Urso, R., Cavalli, F., Mangioni, C., Willems, Y. and Sessa, C. (1986). Pharmacokinetics of Etoposide in patients with abnormal renal and hepatic functions. *Cancer Res.* **46**, 2566–71.

Edwards, C.M., Glisson, B.S., King, C.K., Smallwood-Kentro, S. and Ross, W.E. (1987). Etoposide-induced DNA cleavage in human leukemia. *Cancer Chemother. Pharmacol.* **20**, 162–8.

Epstein, R.J. and Smith, P.J. (1988). Estrogen-induced potentiation of DNA damage and cytotoxicity in human breast cancer cells treated with topoisomerase II-interactive antitumor drugs. *Cancer Res.* **48**, 297–303.

Evans, W.E., Sinkule, J.A., Crom, W.R., Dow, L., Look, A.T. and Rivera, G. (1982). Pharmacokinetics of Teniposide (VM-26) and Etoposide (VP-16-213) in children with cancer. *Cancer Chemother. Pharmacol.* **7**, 147–50.

Finkelstein, D.M., Ettinger, D.S. and Ruckdeschel, J.C. (1986). Long-term survivors in metastatic non-small-cell lung cancer: an Eastern Cooperative Oncology Group Study. *J. Clin. Oncol.* **4**, 702–9.

Fosse, P., Paoletti, C. and Saucier, J.M. (1988). Pattern of recognition of DNA by mammalian DNA topoisomerase II. *Biochem. Biophys. Res. Commun.* **151**, 1233–40.

Gellert, M., Mizuchi, K., O'Dea, M.H., Itoh, T. and Tomizawa, J.I. (1977). Nalidixic acid resistance: a second genetic character involved in DNA gyrase activity. *Proc. Natl. Acad. Sci. USA.* **74**, 4772–6.

Giaccone, G., Donadio, M., Ferrati, P., Bonardi, G., Cluffreda, L., Bagatella, M. and Calciati, A. (1987). Teniposide in the treatment of non-small cell lung carcinoma. *Cancer Treat. Rep.* **71**, 83–5.

Glikin, G.C., Ruberti, I. and Worcel, A. (1984). Chromatin assembly in Xenopus oocytes: *in vitro* studies. *Cell* **37**, 33–41.

Glisson, B., Gupta, R., Smallwood Kentro, S. and Ross, W. (1986). Characterization of acquired epipodophyllotoxin resistance in a Chinese hamster ovary cell line: loss of drug-stimulated DNA cleavage activity. *Cancer Res.* **46**, 1934–8.

Glisson, B. and Ross, W. (1987). DNA topoisomerase II: a primer on the enzyme and its unusual role as a multidrug target in cancer chemotherapy. *Pharmacol. Ther.* **32**, 89–106.

Gosalvez, M., Perez-Garcia, J. and Lopez, M. (1972). Inhibition of NAD-linked respiration with the anti-cancer agent 4′ demethyl-epipodophyllotoxin-β-D-thenylidene glucoside (VM-26). *Eur. J. Cancer* **8**, 471–3.

Gotzos, M., Cappelli-Gotzos, B. and Despond, J.M. (1979). A quantitative microdensitometric and autoradiographic study of the effects of 4′ demethyl-epipodophyllotoxin-β-D-thenylidene glucoside (VM-26) on the cell cycle of cultured fibroblasts. *Histochem. J.* **11**, 691–9.

Grieder, A., Maurer, R. and Stahelin, H. (1974). Effect of an epipodophyllotoxin derivative (VP 16-213) on macromolecular synthesis and mitosis in mastocytoma cells *in vitro*. *Cancer Res.* **34**, 1788–93.

Grieder, A., Maurer, R. and Stahelin, H. (1977). Comparative study of early effects of epipodophyllotoxin derivatives and other cytostatic agents on mastocytoma cultures. *Cancer Res.* **37**, 2998–3005.

Grossberg, H., Opfell, R., Glick, J., Bakemeier, R., Schnetzer, G-III and Muggia, F. (1987). Treatment of advanced refractory lymphoma with Teniposide and Lomustine. *Cancer Treat. Rep.* **71**, 215–16.

Gupta, R.S. (1983). Genetic, biochemical, and cross-resistance studies with mutants of Chinese hamster ovary cells resistant to the anticancer drugs, VM-26 and VP16-213. *Cancer Res.* **43**, 1568–74.

Haim, N., Roman, J., Nemec, J. and Sinha, B.K. (1986). Peroxidative free radical formation and O-demethylation of etoposide (VP-16) and teniposide (VM-26). *Biochem. Biophys. Res. Commun.* **135**, 215–20.

Hainsworth, J.D., Wiliams, S.D., Einhorn, L.H., Birch, R. and Greco, F.A. (1985). Successful treatment of resistant germinal neoplasms with VP-16 and cisplatin: results of a Southeastern Cancer Study Group trial. *J. Clin. Oncol.* **3**, 666–71.

Hainsworth, J.D., Porter, LL 3d, Johnson, D.H., Hande, K.R., Wolff, S.N., Birch, R., Enas, G. and Greco, F.A. (1986). Combination chemotherapy with vindesine, etoposide, and cisplatin in non-small cell lung cancer: a pilot study of the Southeastern Cancer Study Group. *Cancer Treat. Rep.* **70**, 339–41.

Hande, K., Anthony, L., Hamilton, R., Bennett, R., Sweetman, B. and Branch, R. (1988). Identification of Etoposide glucuronide as a major metabolite of Etoposide in the rat and rabbit. *Cancer Res.* **48**, 1829–34.

Hande, K.R., Wedlund, P.J., Noone, R.M., Wilkinson, G.R., Greco, F.A. and Wolff, S.N. (1984). Pharmacokinetics of high-dose etoposide (VP-16-213) administered to cancer patients. *Cancer Res.* **44**, 379–82.

Havemann, K., Wolf, M., Holle, R., Gropp, C., Drings, P., Manke, H.G., Hans, K., Schroeder, M., Heim, M., Victor, N. *et al.* (1987). Alternating versus sequential chemotherapy in small cell lung cancer. *Cancer* **59**, 1072–82.

Henter, J.I., Elinder, G., Finkel, Y. and Soder, O. (1986). Successful induction with chemotherapy including Teniposide in familial erythrophagocytic lymphohistiocytosis. *Lancet* **ii**, 1402.

Huang, C.C., Hou, Y. and Wang, J.J. (1973). Effects of a new antitumor agent, epipodophyllotoxin, on growth and chromosomes in human hematopoietic cell lines. *Cancer Res.* **33**, 3123–9.

Issell, B.F., Muggia, F.M., Carter, S.K., Schaefer, D. and Schurig, J. (1984). *Etoposide (VP-16): Current Status and New Developments.* Academic Press, New York.

Jacobs, A.D. and Gale, R.P. (1984). Recent advances in the biology and treatment of acute lymphoblastic leukemia in adults. *N. Engl. J. Med.* **311**, 1219–31.

Jardine, I., Strife, R.J. and Kozlowski, J. (1982). Synthesis, 470-MHz 1H NMR spectra, and activity of delactonized derivatives of the anticancer drug etoposide. *J. Med. Chem.* **25**, 1077–81.

Jaxel, C., Taudou, G., Portemer, C., Mirambeau, G., Panijel, J. and Duguet, M. (1988). Topoisomerase inhibitors induce irreversible fragmentation of replicated DNA in concanavalin A stimulated splenocytes. *Biochemistry* **27**, 95–9.

Kelsen, D.P., Buckner, J., Einzig, A., Magill, G., Heelan, R. and Vinciguerra, V. (1987). Phase II Trial of Cisplatin and Etoposide in adenocarcinomas of the upper gastrointestinal tract. *Cancer Treat. Rep.* **71**, 329–30.

Kuhn, M., Keller-Juslen, C. and von Wartburg, A. (1969). Partial synthesis of 4′-demethylepipodophyllotoxin. *Helv. Chim. Acta* **52**, 944–7.

Kushner, B.H., Kwon, J.H., Gulati, S.C. and Castro Malaspina, H. (1987). Preclinical assessment of purging with VP-16-213: key role for long-term marrow cultures. *Blood* **69**, 65–71.

Liu, L.F. (1983). DNA topoisomerases – enzymes that catalyse the breaking and rejoining of DNA. *CRC Crit. Rev. Biochem.* **15**, 1–24.

Loehrer, P.J. Sr, Einhorn, L.H. and Greco, F.A. (1988). Cisplatin plus Etoposide in small cell lung cancer. *Seminars in Oncol.* **15**(suppl. 3), 2–8.

Loike, J.D. and Horwitz, S.B. (1976a). Effects of podophyllotoxin and VP-16-213 on microtubule assembly *in vitro* and nucleoside transport in HeLa cells. *Biochemistry* **15**, 5435–43.

Loike, J.D. and Horwitz, S.B. (1976b). Effect of VP-16-213 on the intracellular degradation of DNA in HeLa cells. *Biochemistry* **15**, 5443–8.

Long, B.H., Musial, S.T. and Brattain, M.G. (1984). Comparison of cytotoxicity and DNA breakage activity of congeners of podophyllotoxin including VP16-213 and VM26: a quantitative structure–activity relationship. *Biochemistry* **23**, 1183–8.

Long, B.H., Musial, S.T. and Brattain, M.G. (1985). Single- and double-strand DNA breakage and repair in human lung adenocarcinoma cells exposed to etoposide and teniposide. *Cancer Res.* **45**, 3106–12.

Long, B.H., Musial, S.T. and Brattain, M.G. (1986). DNA breakage in human lung carcinoma cells and nuclei that are naturally sensitive or resistant to etoposide and teniposide. *Cancer Res.* **46**, 3809–16.

Miskimins, R., Miskimins, W.K., Bernstein, H. and Shimizu, N. (1983). Epidermal growth factor-induced topoisomerases. Intracellular translocation and relation to DNA synthesis. *Exp. Cell Res.* **146**, 53–62.

O'Donnell, M.R., Forman, S.J., Levine, A.M., Territo, M., Farbstein, M.J., Fahey, J.L., Gill, P., Lazar, G., Nademanee, A., Neely, S. and Snyder, D.S. (1987). Cytarabine, Cisplatin, and Etoposide chemotherapy for refractory non-Hodgkin's lymphoma. *Cancer Treat, Rep.* **71**, 187–9.

O'Dwyer, P.J., Leyland Jones, B., Alonso, M.T., Marsoni, S. and Wittes, R.E. (1985). Etoposide (VP-16-213). Current status of an active anticancer drug. *N. Engl. J. Med.* **312**, 692–700. (Review).

Pinkel, D. (1987). Curing children of leukemia. *Cancer* **59**, 1683–91.

Piver, M.S. (1986). Cisplatin plus Etoposide as second-line treatment in advanced ovarian carcinoma. *Cancer Treat. Rep.* **70**, 1466.

Pommier, Y., Kerrigan, D., Covey, J.M., Kao-Shan, C.S. and Whang-Peng, J. (1988). Sister chromatid exchanges, chromosomal aberrations, and cytotoxicity produced by antitumor topoisomerase II inhibitors in sensitive (DC3F) and resistant (DC3F/9-OHE) Chinese hamster cells. *Cancer Res.* **48**, 512–16.

Pommier, Y., Schwartz, R.E., Zwelling, L.A., Kerrigan, D., Mattern, M.R., Charcosset, J.Y., Jacquemin Sablon, A. and Kohn, K.W. (1986a). Reduced formation of protein-associated DNA strand breaks in Chinese hamster cells resistant to topoisomerase II inhibitors. *Cancer Res.* **46**, 611–16.

Pommier, Y., Kerrigan, D., Schwartz, R.E., Swack, J.A. and McCurdy, A. (1986b). Altered DNA topoisomerase II activity in Chinese hamster cells resistant to topoisomerase II inhibitors. *Cancer Res.* **46**, 3075–81.

Postmus, P.E., Holthuis, J.J., Haaxma-Reichge, H., Mulder, N.H., Vencken, L.M., van-Oort, W.J., Sleijfer, D.T. and Sluiter, H.J. (1984). Penetration of VP 16-213 into cerebrospinal fluid after high-dose intravenous administration. *J. Clin. Oncol.* **2**, 215–20.

Potmesil, M., Hsiang, Y-H., Liu, L.F., Bank, B., Grossberg, H., Kirschenbaum, S., Forlenzar, T.J., Penziner, A., Kanganis, D., Knowles, D., Traganos, F. and Silber, O. (1988). Resistance of human leukemic and normal lymphocytes to drug-induced DNA cleavage and low levels of DNA topoisomerase II. *Cancer Res.* **48**, 3537–43.

Radice, P.A., Bunn, P.A. Jr and Ihde, D.C. (1979). Therapeutic trials with VP-16-213 and VM-26: active agents in small cell lung cancer, non-Hodgkin's lymphomas, and other malignancies. *Cancer Treat. Rep.* **63**, 1231–9.

Riou, J.F., Multon, E., Vilarem, M.J., Larsen, C.J. and Riou, G. (1986). *In vivo* stimulation by antitumor drugs of the topoisomerase II induced cleavage sites in c-myc protooncogene. *Biochem. Biophys. Res. Comm.* **137**, 154–62.

Rivera, G., Avery, T.L. and Roberts, D. (1975). Response of L1210 to combinations of cytosine arabinoside and VM-26 or VP-16-213. *Eur. J. Cancer* **11**, 639–47.

Roberts, D., Hilliard, S. and Peck, C. (1980). Sedimentation of DNA from L1210 cells after treatment with 4'demethylepipodophyllotoxin-9[4,6-O-thenilidene-D-glucopyranoside or 1-beta-D-arabinofuranosylcytosine or both drugs. *Cancer Res.* **40**, 4225–31.

Roberts, D., Lee, T., Parganas, E., Wiggins, L., Yalowich, J. and Ashmun, R. (1987). Expressions of resistance and cross-resistance in teniposide-resistant L1210 cells. *Cancer Chemother. Pharmacol.* **19**, 123–30.

Ross, W., Rowe, T., Glisson, B., Yalowich, J. and Liu, L. (1984). Role of topoisomerase II in mediating epipodophyllotoxin-induced DNA cleavage. *Cancer Res.* **44**, 5857–60.

Rossi, C., Zucchetti, M., Sessa, C., Urso, R., Mangioni, C. and D'Incalci, M. (1982). Pharmacokinetic study of VM-26 given as a prolonged IV infusion to ovarian cancer patients. *Cancer Chemother. Pharmacol.* **13**, 211–14.

Rowe, T.C., Tewey, K.M. and Liu, L.F. (1984). Identification of the breakage-reunion subunit of T4 DNA topoisomerase. *J. Biol. Chem.* **259**, 9177–81.

Saito, H., Yoshikawa, H., Nishimura, Y., Kondo, S., Takeuchi, T. and Umezawa, H. (1986a). Studies on Lignan Lactone anti-tumor agents I. Synthesis of aminoglycosidic lignan variants related to podophyllotoxin. *Chem. Pharm. Bull.* **34**, 3733–40.

Saito, H., Yoshikawa, H., Nishimura, Y., Kondo, S., Takeuchi, T. and Umezawa, H. (1986b). Studies on Lignan Lactone anti-tumor agents II. Synthesis of N-alkylamino- and 2,6-dideoxy-2-aminoglycosidic lignan variants related to podophyllotoxin. *Chem. Pharm. Bull.* **34**, 3741–6.

Seeber, S., Osieka, R., Gottfried, Sc., Achterrath, W. and Crooke, S.T. (1982). *In vivo* resistance towards antracyclines, etoposide and cis-diammine-dichloroplatinum (II). *Cancer Res.* **42**, 4719–25.

Sinha, B.K., Trush, M.A. and Kalyanaraman, B. (1983). Free radical metabolism of VP-16 and inhibition of anthracycline-induced lipid peroxidation. *Biochem. Pharmacol.* **32**, 3495–8.

Sinha, B.K., Trush, M.A. and Kalyanaraman, B. (1985). Microsomal interactions and inhibition of lipid peroxidation by etoposide (VP-16, 213): implications for mode of action. *Biochem. Pharmacol.* **34**, 2036–40.

Sinha, B.K., Eliot, H.M. and Kalayanaraman, B. (1988). Iron-dependent hydroxyl radical formation and DNA damage from a novel metabolite of the clinically active antitumor drug VP-16. *FEBS Lett.* **227**, 240–4.

Smith, P.J., Anderson, C.O. and Watson, J.V. (1986). Predominant role for DNA damage in etoposide-induced cytotoxicity and cell cycle perturbation in human SV40-transformed fibroblasts. *Cancer Res.* **46**, 5641–5.

Stewart, D.J., Rechard, M., Hugenholtz, H. and Dennery, J. (1983). VP-16 and VM 26 penetration into human brain tumour. *Proc. Amer. Cancer Res.* **24**, 133.

Stähelin, H. and von Wartburg, A. (1989). From podophyllotoxin glucoside to Etoposide. *Progr. in Drug Res.* **33**, 169–266.

Stiff, P.J. and Koester, A.R. (1987). *In vitro* chemoseparation of leukemic cells from murine bone marrow using VP16-213: Importance of stem cell assays. *Exp. Hematol.* **15**, 263–8.

Strife, R.J., Jardin, I. and Covin, M. (1980). Analysis of the anticancer drug VP-

16-213 and VM-26 and their metabolites by high performance liquid chromatography. *J. Chromatog.* **182**, 211–20.

Sugino, A., Peebles, C.L., Kreuzer, K.N. and Cozzarelli, N.R. (1977). Mechanism of action of nalidixic acid: purification of Escherichia coli nal A gene product and its relationship to DNA gyrase and a novel nicking-closing enzyme. *Proc. Natl. Acad. Sci. USA.* **74**, 4767–71.

Sullivan, D.M., Glisson, B.S., Hodges, P.K., Smallwood Kentro, S. and Ross, W.E. (1986). Proliferation dependence of topoisomerase II mediated drug action. *Biochemistry* **25**, 2248–56.

Sullivan, D.M., Latham, M.D. and Ross, W.E. (1987). Proliferation-dependent topoisomerase II content as a determinant of antineoplastic drug action in human, mouse, and Chinese hamster ovary cells. *Cancer Res.* **47**, 3973–9.

Synder, M. and Drlica, K. (1979). DNA gyrase on the bacterial chromosome: DNA cleavage induced by oxolinic acid. *J. Mol. Biol.* **131**, 287–302.

Thurston, L.S., Irie, H., Tani, S., Han, F.S., Liu, Z.C., Cheng, Y.C. and Lee, K.H. (1986). Antitumor agents. Inhibition of human DNA topoisomerase II by podophyllotoxin and α-peltatin analogues. *J. Med. Chem.* **29**, 1547–50.

Wang, J.C. (1987). Recent studies of DNA topoisomerases. *Biochim. Biophys. Acta* **909**, 1–9.

White, J.C., Hines, L.H. and Rathmell, J.P. (1985). Inhibition of 1-β-D-arabinofuranosylcytosine transport and net accumulation by teniposide and etoposide in Ehrlich ascites cells and human leukemic blasts. *Cancer Res.* **45**, 3070–5.

Williams, S.D., Birch, R., Einhorn, L.H., Irwin, L., Greco, F.A. and Loehrer, P.J. (1987). Treatment of disseminated germ-cell tumors with cisplatin, bleomycin, and either vinblastine or etoposide. *N. Engl. J. Med.* **316**, 1435–40.

Wozniak, A.J. and Ross, W.E. (1983). DNA damage as a basis for 4′-demethylepipodophyllotoxin-9-(4,6-O-ethylidene-beta-D-glucopyranoside) (etoposide) cytotoxicity. *Cancer Res.* **43**, 120–4.

Wright, S.E. and White, J.C. (1987). Teniposide-induced changes in the physical properties of phosphatidycholine liposomes. A calorimetric study. *Biochem. Pharmacol.* **35**, 2731–5.

Yalowich, J.C. and Ross, W.E. (1985a). Verapamil-induced augmentation of etoposide accumulation in L1210 cells *in vitro. Cancer Res.* **45**, 1451–6.

Yalowich, J.C., Zucali, J.R., Gross, M. and Ross, W.E. (1985b). Effects of verapamil on etoposide, vincristine, and adriamycin activity in normal human bone marrow granulocyte-macrophage progenitors and in human K562 leukemia cells *in vitro. Cancer Res.* **45**, 4921–4.

Yalowich, J.C. (1987). Effects of microtubule inhibitors on Etoposide accumulation and DNA damage in Human K562 cells *in vitro. Cancer Res.* **47**, 1010–15.

Yarbro, J.W. (ed.) (1986). New developments in the treatment of lung cancer. Proceedings of a meeting. In *Seminars in Oncology* XIII 3.

5

Isolation, purification and initial characterisation

Lignans (Whiting, 1987; Chatterjeee *et al.*, 1984) occur typically in vascular plants (Hearon and MacGregor, 1955) and are found in roots and rhizomes and the woody parts, stems, leaves, seeds and fruits. In primitive species, such as those of the Podophyllaceae, they are the principal organic inclusions. With a few notable exceptions these sources do not provide commercially useful quantities. However, the wound resins of trees are a valuable major source of lignans, which here occur in simple mixtures with other natural products and are readily separated in substantial quantities (Table 5.1).

This review has been subdivided on the basis of parts of plants as sources because this is still the approach most often taken by chemists. It is unfortunate that relatively few studies have been made on the variations of inclusion compounds between the parts of whole plants.

Plant root sources

The roots and rhizomes of *Podophyllum* plants yield up to ten individual aryltetralins (Jackson and Dewick, 1985) which may also be found as their glycosidic variants. The flavones quercetin and rhamnetin are the only other identified products (Hartwell and Schrecker, 1958).

The roots and rhizomes of *Podophyllum hexandrum* afford commercially useful quantities of podophyllotoxin (**2.357**) with yields in the range of 1.5–4.0% of the dry weight (Hartwell and Schrecker, 1958). This will depend on the age of the plant but those which have come to maturity produce the lignan more economically than any existing laboratory synthesis. The American May apple, *Podophyllum peltatum* produces commercial quantities of podophyllotoxin and also of α-peltatin (**2.354**) and β-peltatin (**2.355**). The combined yield of the three compounds is about 2% but both the combined and relative amounts vary with the season (Ayres, 1969). In

Table 5.1. *Resin and heartwood sources*

Lignan	Source	Yield	Reference
Guaiaretic acid Dihydroguaiaretic acid	*Guaiacum officinale*	12%	A
Hinokinin	*Chameacyparis obtusa*	30	B
Matairesinol	*Podocarpus spicatus*	50	C,D
Galbacin	*Himantandra baccata*	*	E
Galgravin	*H. belgraviana*	0.6	E
Lariciresinol	*Larix decidua*	11	F,G
Olivil	*Olea Europa*	45	H
Eudesmin	*Eucalyptus hemiphloia*	10	I
Gmelinol	*Gmelina leichardtii*	2.3	J,K
Pinoresinol	*Pinus, Picea* sp.	35	L
Symplocosignol	*Symplocos lucida*	0.3	M
Conidendrin	*Tsuga, Picea* sp.	**	N
Cyclolariciresionol	*Fitzroyia cuppressoides*	0.3	O
Cyclotaxiresinol	*Taxus baccata*	1.1	P
Cyclo-olivil	*Olea cunninghamii*	*	Q

* Details lacking but appreciable amounts were used for experimentation.
**Commercial quantities.

(A) Haworth, Mavin and Sheldrick (1934)
(B) Yoshiki (1933)
(C) Haworth and Richardson (1935)
(D) Easterfield and Bee (1910)
(E) Hughes and Ritchie (1954)
(F) Bamberger (1897)
(G) Haworth and Kelly (1937)
(H) Korner and Vanzetti (1903)
(I) Maiden and Smith (1896)
(J) Smith (1913)
(K) Birch and Lyons (1938)
(L) Bamberger (1894)
(M) Nishida *et al.* (1951)
(N) Pearl (1945)
(O) Erdtman and Tsuno (1969)
(P) King *et al.* (1952)
(Q) Briggs and Frieberg (1937)

Podophyllum hexandrum the yield is maximised in July/August in England (Ayres, 1969). It must be stressed that such seasonal variations are common (Kamil and Dewick, 1986) and that restricting sampling simply to one point in time can lead to an inclusion compound being overlooked. In the Taiwanese plant *Podophyllum pleianthum* the yields of aryltetralins are variable (Jackson and Dewick, 1985).

Other root sources are those of *Carissa edulis* from which five lignans were obtained (Achenbach *et al.*, 1983) in a combined yield of 5%. The major components were (−)-nortrachelogenin (**2.086**) and carinol (**2.018**) with a minor component of the same class (−) secoisolariciresinol (**2.035**). Two furanoid lignans – olivil (**2.129**) and (+)-lariciresinol (**2.119**) – were also isolated together with the sesquiterpene carisson, a flavone, acetophenones and 4-hydroxy-3-methoxy-cinnamaldehyde.

The roots of seven closely related species of *Artemisia absinthum* were combined and gave rise (Greger and Hofer, 1980) to fourteen furofurans in low yields, of which that of sesamin (**2.223**, 0.04%) was typical. Fargesin (**2.203**) was obtained (Fagbule and Olatunji, 1984) in a similar amount (0.023%) from *Aristolochia albida*. Separation of these extracts may be complicated not only by the range of individual lignans present but also by other co-occurring natural products; for example, roots of *Flaveria chloraefolia* afforded (Bohlmann *et al.*, 1978) the di-isovaleryl ester of 8,8-*bis*-dihydroconiferyl alcohols (**5.1**, 0.02% yield) together with six thiophen derivatives (**5.2** and others).

Leaf and stem sources

The quantities obtained are normally low; thus from the leaves of *Magnolia kobus* five furofuranoid lignans were isolated (Ida *et al.*, 1982) with yields in the range of 0.003–0.06%. The peroxysesquiterpenes were also extracted from this source.

The furan neo-olivil (**2.166**, 0.033%) was yielded (Hernandez *et al.*, 1981) by the whole plant of *Thymus longiflorus* together with monoterpenes and oleanolic acid. Three other lignans of this class were found (Takaoka *et al.*, 1976) in the leaves of *Machilus japonica*, namely, (+)-galbacin (**2.153**, 0.3%), (−)-machilusin (**2.164**, 0.01%), and (+)-galbelgin (**2.155**, 0.005%). One must however be wary of generalisations for the stem and leaves of the creosote bush, *Larrea divaricata*, provide (Merck, 1983) a rich source of nordihydroguaiaretic acid (**2.011**, 12% yield).

Seeds as sources

The seed of *Ovigera hernandia* L. has been reported (Yamaguchi and Nakajima, 1984) to give a 'good yield' of desoxypodophyllotoxin (**2.360**). This lignan has been recorded (Cairnes *et al.*, 1981) as occurring in nine plant species; it is found (0.25% yield) with *β*-peltatin (**2.355**, 0.75%) in *Thujopsis dolabrata* (Hasegawa and Hirose, 1980).

The seed fat of *Virola sebifera* is valued for the treatment of rheumatism but the activity cannot arise from the lipid components; further ana-

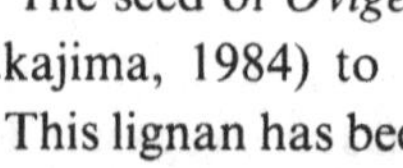

5.1

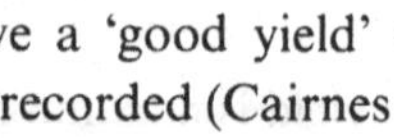

5.2

5.3

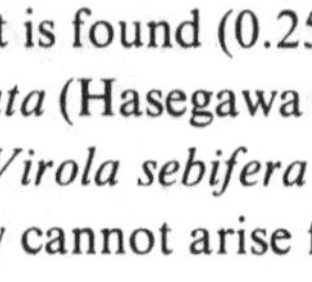

lysis (Lopes *et al.*, 1982) led to the isolation of a polyketide (0.18%) and a group of oxo-otobains (**2.341–2.343**) in yields ranging from 0.05 to 0.93%.

High yields of diarylcyclo-octadiene lignans occur (Song, 1981) in the seeds of *Schizandra* species and lie in the range 7–19%. The stems of the same plants afford 1–10% of compounds of this same class, which are active in lowering the levels of glutamic–pyruvic transaminase in serum. These yields also vary with the season and are maximal in May–July.

The interesting compound (**5.3**) which falls in both the cyclo-octadiene and furan classes was isolated (Spencer and Flippen-Anderson, 1981) in 5% yield from the seeds of *Clerodendron inerme*, together with two isomers in (9% yield) and also acyl-glycerols.

Fruits as sources

The fruits of *Piper cubeba* have been under investigation since the first half of the 19th century (Hearon and MacGregor, 1955); they give the lactol cubebin in 1–5% yield.

After the first report (Kochetkov *et al.*) in 1961 of the extraction of the cyclo-octadienyl lignans schizandrin, schizandrol, and γ-schizandrin (**2.457**) from the seed oil of *Schizandra chinensis* Baill., a number of sesquiterpenes were obtained (Ohta and Hirose, 1968). Research in depth subsequently revealed a large group of cyclo-octadienyl lignans in the fruits of this plant and of related species of the *Schizandraceae* family. Thus the occurrence of the gomisins A, B and C (**2.426–2.428**) in *S. chinensis* was reported in 1975 by Taguchi and Ikeya. Another study (Chen, Y-Y. *et al.*, 1976) was prompted by the activity of these fruits against hepatitis and led to the isolation of five lignans of this class. They were named as the 'wuweizisu' group, but the allocated structures, before or after revision, correspond to those described as schizandrins or schizantherins by other workers. The latter group, the schizantherins A–E (**2.427–2.428**; **2.462–2.464**), were obtained in 1978 by C-S. Liu and colleagues during work on the fruits of *S. sphenanthera*. Schizantherin A was correlated with gomisin C and Schizantherin B with gomisin B, whilst schizantherin C is the angeloyl ester of gomisin B. Schizantherin D (**2.463**) and schizantherin E (**2.464**) have not been found in *S. chinensis*. These fruits were used in traditional oriental medicine as tonics and anti-tussives.

Significant yields of schizandrin (0.26%) and gomisin A (0.19%) were reported (Ikeya, 1979) as a result of further work on *S. chinensis*, together with gomisins B, C, F, and G (yields of 0.022, 0.0035, 0.0035, and 0.016% respectively); by 1980 this group was extended (Ikeya *et al.*, 1980) to include eighteen compounds. Gomisin H (**2.433**) occurred in substantial yield (0.31%) with other gomisins in the range of 0.023–0.001%.

Once kadsurin (**2.452**) and kadsurarin (**2.451**) had been extracted (Chen *et al.*, 1973) from the stems of *Kadsura japonica* Dunal., Ookawa *et al.* (1981) obtained the acetyl-, angeloyl-, and caproyl-esters of binankadsurin A (**2.421**) from the fruits in yields of 0.19%, 0.19%, and 0.02% respectively. Thirty-three cyclo-octadienyl lignans, including various esters, had been obtained from *Schizandra* sp. by the year 1982.

The possibility of winning other types of new lignans in significant numbers from *Schizandra* sp. does exist because the aryltetralin, enshicine (**2.321**, 0.07%), has been obtained (Liu *et al.*, 1984) from *S. henryi* Clarke. The occurrence of a furanoid lignan, chicanine (**2.142**), in the fruit of a related species should be noted (Liu *et al.*, 1981).

The large fruits of *Iryanthera grandis* Ducke contain (Viera *et al.*, 1983) triglycerides, a number of tocotrienols and the interesting aryltetralin (**2.306**, 0.05% yield) with monooxygenated aryl groups. Additional lignans of this class have been obtained (Lopes *et al.*, 1984) from fruits of *Virola sebifera*. The fruits of *Forsythia* sp. also include (Nishibe *et al.*, 1984) matairesinol (**2.062**) and its methyl ether.

The toxicity of the squirting cucumber, *Ecballium elaterium* L., has been known since Biblical times (Elisha). In common with other plants of the *Cucurbitaceae* family it contains bitter principles which are tetracyclic triterpenes; these have been widely studied on account of their neoplastic properties (Lavie and Glotter, 1971). A resin (125 g) obtained (Rao and Lavie, 1974) from a large volume of the juice gave a number of phenols, including 2-nitroquinol, in very low yield together with the furofuranoid lignan – ligballinol (**2.207**, 0.06%). The furofurans (+)-pinoresinol, its glucoside and phillyrin (**2.273**) occur (Tsukamoto *et al.*, 1983) in the fruits of *Forsythia suspensa* and *F. koreana*.

Resin and heartwood sources

The resins are of particular interest from the point of view of commercial interest in lignans as precursors for pharmaceuticals. To some extent this is also true for heartwood sources. In the past there have been no significant developments of those lignans which are available in commercial quantities for example, conidendrin. With the extension and refinement of synthetic methods over the last decade this is an area of research with promise for the future. Table 5.1 (p. 139) summarises those sources of greatest interest.

Factors which favour commercial development of resin sources are:

(a) The yields of included lignans are much higher than is usual in plants.

(b) Resins are usually simple mixtures of major components. For example, olivil (Korner, 1903) is readily obtained in a pure form from *Olea europa*, whilst only the sesquiterpene aromadendrene co-occurs with eudesmin (Maiden and Smith, 1896). Lariciresinol (**2.119**) was also separated (Bamberger and Landsied, 1987) in an essentially pure condition by ethanolic extraction of the resin and precipitation of its potassium salt.

(c) Lignans are typically chiral and are potentially useful as precursors in stereoselective syntheses.

(d) Methods for the interconversion of various classes of lignans are well established.

As seen in the sections above, commercially useful quantities can also be won from sources such as the fruits of *Schizandra* sp. and from the roots and rhizomes of *Podophyllum* sp., to name but two.

Extraction of the wood may provide detectable amounts of compounds not found in other parts – *Guazuma tomentosa* Kunth. was the first of the Sterculiaceae family to yield (Anjaneyulu and Murty, 1981a) a lignan: (−)-syringaresinol (**2.229**, 0.015%), together with two coumarins, three triterpenes, hexacosanol ($C_{26}H_{53}OH$), β-sitosterol and its glucoside; all in yields below 0.04%.

The range of lignans which may occur is illustrated by work on *Virola elongata* which afforded (Macrae and Towers, 1984) thirteen furofurans and also alkaloids, whilst knots of *Araucaria angustifolia* gave (Fonseca *et al.*, 1978) the lignans dimethylpinoresinol (**2.217**), secoisolariciresinol (**2.035**), lariciresinol (**2.119**), cyclolariciresinol (**2.313**) and its 4′-methyl ether with yields in the range of 0.004 – 0.11%.

Bark of *Knema attenuata* on extractions (Joshi *et al.*, 1979) with a large volume of hexane gave the aryltetralin attenuol (**2.290**, 0.006%), in which the pendent ring bears only a single 4′-hydroxyl group. These authors also isolated *meso*-inositol (0.02%) by methanolic extraction. Petroleum was sufficient for the extraction (Pelter *et al.*, 1976) of the three fully alkylated furofurans methyl piperitol (0.025% yield), (+)-epieudesmin (0.010%) and (+)-episesamin (**2.194**, 0.012%) from stem bark of *Zanthoxylum acanthopodium* D.C. A point to note was that two lignans of this class: (−)-asarinin and (−)-sesamin were both obtained (Vaquette *et al.*, 1978) in similar low yields (0.004 and 0.014% respectively) from the leaves and bark of *Zanthoxylum* sp. The leaves themselves afforded in addition the dimethyl ether of piperitol (**2.219**, 0.006%). In concluding this summary of lignan sources attention is again drawn to the range of possible co-occurring compounds; for example, the bark of *Virola elongata* yielded

(Macrae and Towers, 1985) the furans dihydrosesartemin (cf. **2.226**, 0.0015%) and β-dihydroyangambin (**2.134**, 0.001%), together with four furofurans: episesartemin (**2.197**, 0.017%), sesartemin (**2.226**, 0.002%), epiyangambin (**2.199**, 0.01%), and yangambin (**2.232**, 0.009%). In addition *cis*- and *trans*-stilbenes, sitosterol, and two neolignans were also obtained from this same source. Neolignans have been found predominantly in plants of the orders Magnoliales and Liliales (Gottlieb, 1978). The wood of *Schizandra nigra* Max. yielded (Takani *et al.*, 1979) a glucoside of 4-hydroxy-3-methoxyacetophenone, a sesquiterpene alcohol, a triterpenoid ketoacid, a glucoside of (+)-catechin and an aryltetralin lignan – schizandriside (cf. **2.313**). This last (0.0016%) is a β-D-xyloside of (+)-isolariciresinol. An isomeric β-D-glucoside and an α-L-arabofuranoside of this lignan are also known (Popoff and Theander, 1977). It should also be recalled that a successful search for useful compounds may be dependent on the season of choice; for example, occurrence (Inoue *et al.*, 1981) of kigeliol (**2.259A**) in the wood of *Kigelia pinnata* rises through the months of January to April and thereafter declines to a low level in October. The yield of β-sitosterol in this wood remained steady through the year.

Lignan glycosides

Phillyrin (**2.273**) was the first glycoside to be isolated by Carboncini in 1837. It can be obtained (Kramer, 1933) in yields up to 1.6% from the bark of *Phillyrea latifolia* L.

Arctin (18% yield) was isolated (Shinoda and Kawagoye, 1929) with its aglucone arctigenin (**2.045**, 15%) from the seeds of *Arctium lappa*. Symplocosin (**2.277**; Nishida *et al.*, 1951) and glucosides of the podophyllotoxin group (Stoll) had also become available in useful yield by 1954; however the aglycones have always preponderated. There are two main reasons for this. Firstly, moisture and acids are commonly present during the extraction of plant material and hence glycosides are cleaved by hydrolysis. Secondly, the procedures required for the separation and characterisation of glycosides are more exacting than those needed for the aglycones themselves (see Scheme 5.1); hence relatively small amounts may be overlooked.

With the increasing efficiency of separation procedures and analytical methods one expects that the number and variety of known lignans and their glycosides will increase considerably in the future. Of particular interest is the isolation of the first C-glucoside (**2.087**) by Abe and Yamauchi (1986), where the sugar is inserted as a 3-substituent in matairesinol.

Isolation procedures

A wide range of methods is available and the following review is necessarily selective. It is based on the writer's experience and is largely taken from the more recent literature.

(i) *Solvent extraction*

Lipid substances occur in all plants, especially in seeds, but also as the coating and wax of fruits and up to 7% of the dry weight of leaves. Since they are in general soluble in weak organic solvents it is convenient to separate them at an early stage by extraction with hexane or petroleum ether. Typical of the lipid substances which may be removed in this way are fatty acids, glyceryl esters, steroids such as β-sitosterol and terpenoids including waxy triterpenes such as oleanolic acid.

Phenolic natural products have only limited solubility in hexane, especially when a free hydroxyl group is present; nevertheless in a search for lignans it is necessary to monitor a concentrated extract by thin layer chromatography (TLC, see p. 149) and if possible by high performance liquid chromatography (HPLC, see p. 153). The need for this precaution is illustrated by the observation (Pelter *et al.*, 1976) that fully alkylated furofuranoid lignans are appreciably soluble in hexane, although they may be separated from the less soluble sitosterol by recrystallisation from this solvent. Amongst the more polar lignans, the arylnaphthalenes Justicidin A and B were partly extracted (Sheriha and Amer, 1984) by petrol together with sitosterol, long-chain alkanols and ketones and quinoline alkaloids. When the bark of *Knema attenuata* (2.4 kg) was treated (Joshi *et al.*, 1979) with a large volume of hexane (2 × 20 l) even the free phenolic lignan attenuol (**2.290**, 150 mg) was extracted. The tendency to extract lignans, including free phenols, is accentuated if a mixture of hexane–diethyl ether is used (Greger and Hofer, 1980) to remove lipids. The superiority of diethyl ether as compared to hexane as a solvent for lignans is shown by the relative solubilities of galbacin, veraguensin (**2.168**) and the neolignan alcohol surinamensin in these two solvents (Barata *et al.*, 1978).

A distinction between benzene and hexane extraction has been drawn by Fonseca *et al.* (1978). They removed lipids from the evaporated benzene extract with hexane and obtained a residue containing dimethylpinoresinol (cf. **2.217**) as well as the phenols lariciresinol (**2.119**), secoisolariciresinol (**2.035**), cyclolariciresinol (**2.313**) and cyclolariciresinol methyl ether. Benzene extraction (Dias *et al.*, 1982) of the wood of *Urbandendron verrucosum* yielded terpenoids, neolignans and furanoid lignans including free phenols.

Another approach is to attempt the complete removal of soluble organic compounds from the raw material with a hot polar solvent such as acetone, ethanol or methanol and then to extract lipids by treatment of the initial evaporate with hexane. Steam distillation has been used to remove volatile compounds from non-volatile lignans; for example, in thc separation (Takaoka *et al.*, 1976) of a group of furanoid lignans from the leaves of *Machilus japonica.*

(ii) *Fractional extraction*

An isomer (**2.255A**) of 9-hydroxypinoresinol (**2.255**) was obtained (Khan and Shoeb, 1985) by the initial extraction of the aerial parts (2.4 kg) of *Lonicera hypoleuca* with ethanol:water (95:5) and fractionally extracting the residue (155 g) with hexane (2.5 g obtained). In this instance lipid removal was only completed on treatment with chloroform (14 g obtained) and this fraction was separated by silica gel chromatography (p. 151) into nonacosanol, β-sitosterol, scopoletin, syringic acid and the lignan. A similar procedure was followed by Agrawal and Rastogi (1982) after a preliminary extraction of the wood (36 kg) of *Cedrus deodara* by ethanol (300 ml), followed by successive treatment of this material (1.96 kg) with hexane (280 g obtained), chloroform (245 g), *n*-butanol (820 g) and finally with water (180 g obtained).

In a detailed study by Cambie *et al.* (1979) the wood of *Dacridium intermedium* was treated with methanol and the concentrate mixed with Celite, when the dry friable material was successively extracted with petroleum, diethyl ether, ethyl acetate and methanol. The petroleum fraction yielded waxes, esters and the terpene manool. The ethereal fraction included three aryltetralin lignans, but they were isolated mostly from the methanol fraction. This final separation required HPLC of the derived acetates on a reversed phase column. In a procedure (Hernandez *et al.*, 1981) dependent on the initial treatment of the whole plant of *Thymus longiflorus*, but with no other fractionation, difficulties were experienced in purifying the extracted neo-olivil (**2.166**), which was however characterised as its tetra-acetate.

It is to be expected that hot alcoholic solvents, especially butanols, will remove all lignans and their glycosides from plant material. Glycosides may be concentrated by partitioning them into the aqueous phase of a two-phase system. An example (Deyama, 1983) of this procedure is the isolation of (+)-medioresinol by extraction of the bark of *Eucommia ulmoides* with methanol followed by partitioning the evaporate between water and chloroform:methanol:water (70:30:5). (+)-Acetoxypinoresinol-4β-D-glucoside was obtained in a similar way (Chiba *et al.*, 1979) from the stems of *Olea europaea*; the aqueous solution of the extract was freed from non-glucosidic components by treatment firstly with diethyl ether and secondly with chloroform. An initial extract of the roots of *Carissa edulis* was partitioned in methanol:water (1:1) and cleared with petrol, carbon tetrachloride to remove terpenes, and finally with diethyl ether. This yielded (Achenbach *et al.*, 1983) six lignans: nortrachelogenin (**2.085**), lariciresinol (**2.119**), carissanol (**2.019**), carinol (**2.018**), secoisolariciresinol (**2.035**) and (−)-olivil (**2.129**). The following Scheme 5.1 gives the details of a procedure used (Kudo *et al.*, 1980) for the separation of glycosides from the bark of *Ligstrum japonicum*.

Another method of extracting glycosides is direct treatment with water with the removal of less water-soluble compounds by back-extraction with

Scheme 5.1

Procedure for the separation of lignan glucosides from *Ligstrum japonicum* (KUDO, 1980)

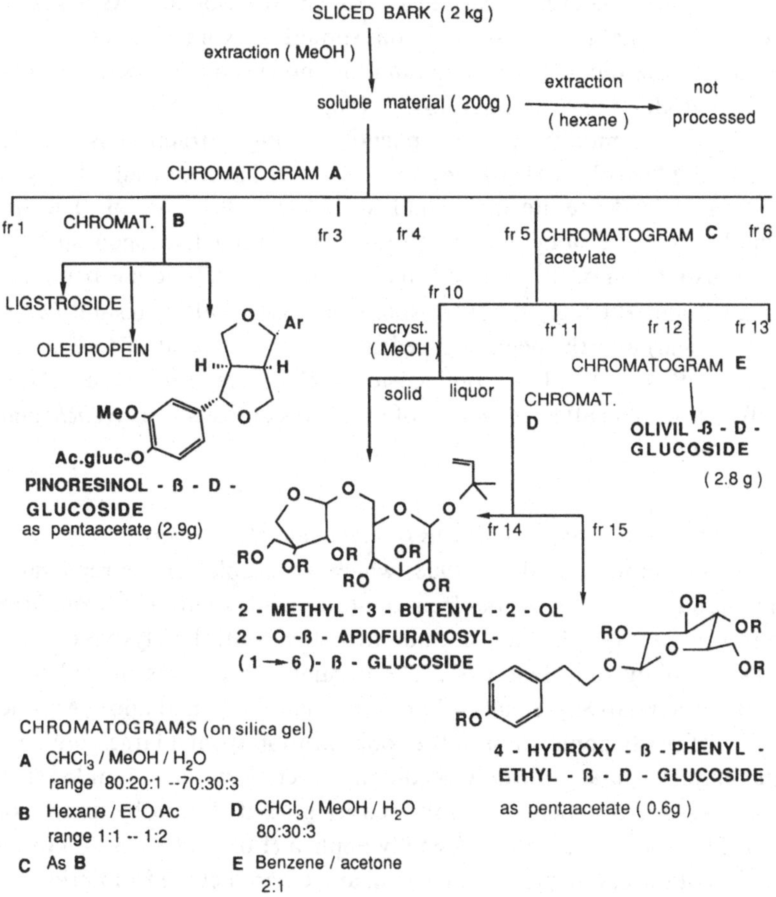

SLICED BARK (2 kg)
extraction (MeOH)
soluble material (200g)
extraction (hexane)
not processed
CHROMATOGRAM **A**
fr 1 CHROMAT. **B**
fr 3
fr 4
fr 5 CHROMATOGRAM **C** acetylate
fr 6
LIGSTROSIDE
OLEUROPEIN
O Ar
H H
MeO
O
Ac.gluc-O
PINORESINOL - ß - D - GLUCOSIDE
as pentaacetate (2.9g)
fr 10
recryst. (MeOH)
solid liquor
fr 11 fr 12 fr 13
CHROMATOGRAM **E**
CHROMAT. **D**
OLIVIL -ß - D - GLUCOSIDE
(2.8 g)
fr 14 fr 15
RO OR OR OR RO OR
2 - METHYL - 3 - BUTENYL - 2 - OL
2 - O -ß - APIOFURANOSYL-
(1→6)- ß - GLUCOSIDE
RO OR OR OR O
RO
4 - HYDROXY - ß - PHENYL - ETHYL - ß - D - GLUCOSIDE
as pentaacetate (0.6g)

CHROMATOGRAMS (on silica gel)

A $CHCl_3$ / MeOH / H_2O
range 80:20:1 --70:30:3

B Hexane / Et O Ac
range 1:1 -- 1:2

C As **B**

D $CHCl_3$ / MeOH / H_2O
80:30:3

E Benzene / acetone
2:1

hexane, diethyl ether and/or chloroform. This ensures the recovery from the aqueous phase of polar lignans such as cyclotaxiresinol (**2.319**). This pentahydroxy-aryltetralin was obtained (King *et al.*, 1952) from the heartwood of yew, *Taxus baccata*, in this way.

(iii) *Separation by precipitation*

Those lignans with free phenolic groups may be obtained from alcoholic extracts by the addition of concentrated potassium hydroxide solution to precipitate their potassium salts; this procedure was followed in the original separation (Bamberger, 1894) of pinoresinol. The more-soluble sodium salts can sometimes be leached directly from the plant material into aqueous sodium hydroxide, followed by acidification and extraction into an organic solvent. When this method was adopted by Gisvold (1948), bisulphite was added to prevent oxidation of the anion of the nordihydroguaiaretic acid (**2.011**) which was isolated. As a general rule strongly alkaline or acid conditions should be avoided, unless they are employed to extract known compounds of known stability, because of the possible production of artefacts (see p. 156).

The risk of modifying these phenols during extraction is avoided (Waller and Gisvold, 1945) if they are precipitated as their lead salts by the addition of lead acetate in ethanol to an alcoholic solution. The free phenols are then obtained in solution by passing hydrogen sulphide into a suspension of the lead salt in alcohol. This method has been used (Agrawal and Rastogi, 1982) to isolate *meso*-secoisolariciresinol (**2.037**, 0.001% yield) and the neolignan cedrusin from the wood of *Cedrus deodara*. (+)-Piperitol and its β-D-glucoside (**2.275**) were also isolated (Kisiel, 1980) as lead salts after extraction of the leaves and stems of *Helichrysum bracteatum*.

(iv) *Extraction of polar lignans from biological materials*

A method (Bradlaw, 1968) which is suitable for the removal of lignans and/or their glycosides from dilute aqueous solution or from urine samples is passage through a column of the neutral polystyrene resin XAD-2 (Rohm and Haas Co.). The organic compounds may then be eluted by a relatively small volume of ethanol or methanol. Another method of ready application to the concentration from a large volume of water employs passage through a column of octadecylsilane-bonded silica (Water Assoc. Inc., Sep-Pak), followed by graded elution by mixed solvents. This technique was successfully applied (Fotsis *et al.*, 1982) to the isolation of the conjugates of enterodiol and enterolactone from urine.

Chromatographic methods

A good general source of information is the series of reviews (Giddings *et al.*) published in *Advances in Chromatography*. Regular annual reviews are published by the American Chemical Society in *Analytical Chemistry* and also by the Royal Society of Chemistry in *Analytical Proceedings*.

Thin layer chromatography (TLC)

A definitive work is that by Kirchner (1978) and other valuable sources of references (Touchstone and Sherma, 1985; Wagner *et al.*, 1984a) have appeared recently.

The technique was first applied to the separation of the lignans of podophyllum on Kieselgel G by Stahl and Kaltenbach (1961) who resolved three glucosides and five aglycones by a two-step procedure. In the first step the glucosides were eluted with chloroform:methanol (9:1) and in the second the dried plate was treated with chloroform:acetone (2:1) to elute the lignans. The latter were the subject of detailed analysis (Jackson and Dewick, 1984) in extracts of *Podophyllum hexandrum* and *P. peltatum*. TLC has now come into routine use for the initial examination of plant material and for monitoring the composition of extracts and fractions obtained from chromatography columns at the various stages of purification. Kieselgel G or alumina plates are generally satisfactory.

Wholly aromatic lignans, such as those of the justicidin group (**2.399–2.405**) are strongly and beautifully fluorescent when irradiated with light in the wavelength range of 254 or 365 nm. Those lignans with a double bond in conjugation, as in podophyllotoxone (**2.364**) or α-apopicropodophyllin (**2.371**), may also be detected in this way.

Since all lignans absorb UV light in the region of 254 nm they are clearly shown on dye-marked plates (GF_{254}) as dark spots on a green fluorescent background; exposure to iodine vapour may also be used. Another method of detection is to spray with a 5% solution of concentrated sulphuric acid in ethanol or acetic anhydride, followed by heating for a few minutes at about 100°C. This gives satisfactory colour development for all lignans with a sensitivity of the order of 0.1 μg. Similar heating after a spray with phosphotungstic acid (5% in ethanol) is a good method and in common with sulphuric acid can be used to differentiate between types of lignan. For example, the acetal sesamolin (**2.283**) gives the typical blue coloration before heating to 100°C and a more rapid response is to be expected for other easily oxidised groups such as free phenols. The formation of intensely red *o*-quinones on treatment of syringyl compounds with nitric acid has been used (Jackson and Dewick, 1984) to detect demethyl-podophyllotoxins and to distinguish them from the trialkyloxyanalogues. The latter also give a positive response, although more slowly, whilst mono- and dioxy compounds are unresponsive. It is probable that this method could be employed more widely and at least to the other classes of lignans.

General surveys of reagents suitable for colour development on TLC plates have been published (Kirchner, 1978; Stahl and Kaltenbach, 1969) and the latter work includes a short but informative specific review of methods of detection and choice of eluents for the TLC of lignans.

An example of the effectiveness of TLC applied to the analysis of plant material is provided by work (Ghosal *et al.*, 1980) on the whole plant (500 g) of *Justicia simplex*, which was extracted separately with petrol (b.p. 60–80°C) and with ethanol. The petrol fraction was run on silica gel using benzene-chloroform as eluent and *four* substances responded to a spray containing 2,4-dinitrophenylhydrazine. However, detection by sulphuric acid revealed *twelve* spots on the same chromatogram. Subsequent large-scale chromatography on silica led to the isolation of sesamin (**2.223**, 70 mg, 0.014%), asarinin (**2.194**, 55 mg, 0.011%), and sitosterol (115 mg, 0.023%) by elution with benzene. A more polar eluent (benzene:chloroform as 9:1) yielded a new phenolic lignan justicolin (**2.205**, 45 mg, 0.009%). The ethanol extract from this wood (*J. simplex*) was concentrated and diluted with water but, probably owing to the presence of residual ethanol, TLC showed all the above three lignans to be present in both petrol and chloroform extracts of water-soluble material. Final extraction with ethyl acetate removed the glucoside, simplexoside (**2.275**) from the aqueous solution. Other recent examples of TLC applied to monitoring of extracts include: the separation (Greger and Hofer, 1980) of thirteen lignans from *Artemisia absinthum*; isolation (Lopes *et al.*, 1982) of aryltetralones from *Virola sebifera*; extraction (Badawi *et al.*, 1983) of furofurans from *Dirca occidentalis*; and component arylnaphthalenes in the leaves of *Justicia flava* (Olaniyi and Powell, 1980).

Preparative layer chromatography (PLC) has been used *inter alia* in separating (Powell and Plattner, 1976), a lignan alcohol from carboxylic acids, for resolving (Cairnes, Ekundayo and Kingston, 1980) podophyllotoxins from the leaves of *Juniperus phoenicea*, and a group of furan and furofuran lignans from the bark resin of *Virola elongata* (Macrae and Towers, 1985). This last paper describes colour development in detail.

Owing to their propensity for complexation some lignans may be subject to co-chromatography and elute as a mixture. A recent example is that of the tetralones aristoligone (**2.343B**) and aristosynone (**2.343C**), which formed a stable 2:1 complex (Urzua and Shamma, 1988).

A procedure (Stahl and Kaltenbach, 1961) for the separation of glycosides has been mentioned above. Other methods are described by Sheriha and Amer (1984) and a separation of the diglycoside, eleutheroside E (**2.266**), from *Eleutherococcus senticosus* and mistletoe preparations was achieved by Wagner *et al.* (1982, 1984b).

(ii) *Column chromatography*

The choice of conditions is best made by preliminary trials with TLC and the most useful media are alumina and silica gel. Suitable solvents for column chromatography with each of these materials will be of the same type as for TLC, but they are likely to be slightly less polar. It is inadvisable to use solid phases of high activity because of the risk of reaction on-column (see below: Formation of artefacts).

Lignans of the podophyllotoxin group are efficiently separated (Hartwell and Schrecker, 1958) on alumina columns on elution with ethanol: benzene mixtures. Lower polarity lignans of the furan class, isolated (Takaoka *et al.*, 1976) from *Machilus japonica*, were purified by elution with hexane:ethyl acetate (from 95:5 to 1:1). When re-chromatography is necessary there may be advantage in using alumina and silica gel in tandem. The polar lignan lyoniresinol (**2.328**) was partially separated by Vecchietti *et al.* (1979) from alkaloids and cinnamyl amides of tryptamine on alumina using methylene chloride:methanol; the resolution was completed on silica with cyclohexane:ethyl acetate as the eluent. The same workers described the chromatography of the xylopyranoside of 5′-methoxycyclolariciresinol (**2.315**), firstly on silica with methanol in ethyl acetate and secondly by elution from alumina with methanol in methylene chloride; a TLC procedure for the identification of D-xylose was also given.

Silica gel has been applied more often than any other material and lignans of wide-ranging polarity have been successfully chromatographed in this way. Low polarity furans from *Urbanodendron verrucosum* (Dias *et al.*, 1982) were eluted with methylene chloride in benzene; those with a phenolic hydroxyl group required (Liu *et al.*, 1981) ethyl acetate in benzene. Work (LeQuesne *et al.*, 1980) on the leaves and stems of *Nectandra rigida* describes the separation of galgravin (**2.156**) from the mono- and dihydric phenols nectandrin A and B (**2.165**) using mixtures of toluene and ethyl acetate: this study also gives details of the removal of alkaloids by extraction with hydrochloric acid (2%) or tartaric acid (5%).

Furofurans including free phenols can be separated on silica by elution with benzene and benzene:chloroform (Badawi *et al.*, 1983); glycosides of this class may also be run on the same column and eluted with chloroform:methanol:water (95:4:1). A large group of furofurans extracted from the roots of *Artemisia absinthum* were resolved (Sheriha and Amer, 1984) in a similar manner on silica using solvents ranging from a low polarity mixture of petrol:diethyl ether to methanol (10%) in diethyl ether. The acetal simplexolin (**2.284A**) was eluted (Ghosal *et al.*, 1979) with benzene:chloroform.

A procedure (Kudo *et al.*, 1980) for the separation of a group of furan and furofuran glucosides was outlined above (Scheme 5.1). In this work the peracetate of pinoresinol glucoside (**2.217**) was eluted from silica with hexane:ethyl acetate, whereas for the more polar *bis*-glucoside of syringaresinol (**2.229**) and the monoglucoside of olivil (**2.129**) a mixture of chloroform:methanol:water (8:2:1) was needed. Other papers (Kisiel, 1980; Jolad *et al.*, 1980; Hernandez *et al.*, 1981) report the use of chloroform:methanol mixtures for the chromatography of glycosides but, depending on the activity of the silica, it may be necessary to add some methanol to chloroform in order to elute furofuran aglycones.

Pygeoside, a xyloside (**2.335**; Chandel and Rastogi; 1980) of the tetralin diol has been chromatographed on silica with wet chloroform:methanol. A combination of chloroform and petroleum was sufficient (Olaniyi and Powell, 1980) for the chromatography of lactones of the naphthalene class extracted from *Justicia flava*, but addition of methanol was needed to recover cyclo-lariciresinol (**2.313**). A related group of diols isolated (Fonseca *et al.*, 1978) from *Araucaria angustifolia* was separated using acetone to elute the more strongly retained compounds.

An interesting paper (Anjaneyulu and Murty, 1981a) describes the separation of a range of lignans of different classes including the naphthol lactone diphyllin (**2.394**) and its glycosides.

Other media have occasionally been used for the chromatography of lignans. For example, mono- and diglucosides of pinoresinol, the diglucoside of medioresinol (**2.214**) and furofuran aglycones were separated (Deyama, 1983) on charcoal. Glucuronides have been run (Fotsis *et al.*, 1982) on the anion exchange resin Sephadex A-25. Celite 535 was suitable for the separation of the esters of the cyclo-octadienyl lignans from *Kadsura japonica* (Ookawa *et al.*, 1981), whilst Hyflo-supercel was employed (Tandon and Rastogi, 1976) for the chromatography of pinoresinol and three hydroxylated butyrolactones.

Traditional column chromatography, monitored by TLC, is likely to remain in long-term use for the separation of material of gramme quantities. PLC is a less attractive technique because lignan recovery may fall as low as 50%. With both of these methods there is a risk of modification of material whenever it is in prolonged contact with a stationary phase, especially when this is composed of one of the more active forms of alumina or silica (see below: Formation of artefacts).

(iii) *High-performance liquid chromatography* (*HPLC*)

HPLC is superior to the methods discussed above in terms of resolution, recovery, detection and speed of operation. It may be applied analytically, semi-preparatively (100 mg scale) or on a larger scale.

There is a growing awareness of the importance of this technique for the purification of organic compounds, although its full potentialities have yet to be generally appreciated by those active in lignan chemistry. Useful reviews of the method (Bristow, 1976) and scope (Horvath, 1986) of HPLC have been published; its advantages over column chromatography are illustrated by the procedure for separation (Chaudhuri and Sticher, 1981) of the antileukaemic diglucoside liriodendrin (**2.271**) from the iridoid glucosides and syringin isolated from *Globularia alypum* L. Here, after a three-stage procedure involving alumina (once) and silica (twice), resolution was completed using an HPLC column of silylated silica (reverse phase) and elution with methanol:water (3:7). In related work (Wagner *et al.*, 1984b) gradient elution with acetonitrile:water was applied to the purification of the diglucoside eleutheroside E (**2.266**). Furofuran aglycones have been chromatographed on reverse phase silica where elution with methanol:water (4:1) resolved (Tatematsu *et al.*, 1984) the phenolic lirioresinol B (**2.209**) from pinoresinol and matairesinol. The phytoalexin (**2.233**) was purified by HPLC on silica in the direct mode by elution with hexane:ethyl acetate (range 1:3–1:1). This paper (Kobayashi and Ohta, 1983) also described the identification of D-glucose by co-HPLC on elution with water at 65°C. Eleven lignans extracted from spruce and fir wood were chromatographed (Gorokhova *et al.*, 1979) directly on silica with iso-octane:chloroform (93:7) and also by reverse phase on elution by combinations of methanol:acetic acid:water.

Ikeya *et al.* (1982) reported the separation of the dibenzocyclooctadiene lignans γ-schizandrin (**2.457**) and its isomer gomisin N (**2.442**) on reverse phase silica using acetonitrile:methanol:water (1:1:1). The same group published (Ikeya *et al.*, 1982) a procedure for characterisation of the angeloyl esters of gomisin O and isogomisin O.

The efficiency of HPLC has been demonstrated (Momose *et al.*, 1978) by the separation on a semi-preparative scale (50 mg) of the isomers (**5.4**) and (**5.5**) formed by photolysis of taiwanin A (**2.108**).

The acetates of aryltetralin acetals such as (**2.311**) were separated (Hernandez *et al.*, 1981) by preparative HPLC on reverse phase with methanol:water (52.5:47.5) as eluent. The free phenols were themselves separated directly on silica by elution with benzene:acetone (97.5:2.5).

Otobain (**2.304**) was isolated from the fruits of *Myristica otoba* by classical chromatography (Gilchrist *et al.*, 1962), but seven other tetralin lignans (e.g. **2.307–2.309**) from this same source were separated by later workers (Nemethy *et al.*, 1986) using HPLC. The technique was also used (Cairnes *et al.*, 1980), to monitor the isolation of deoxypodophyllotoxin (**2.360**) and of methyl β-peltatin A from *Juniperus phoenicea* after their separation by PLC. The same group reported (Cairnes *et al.*, 1981) the comparative chromatography of four members of this class with burseran (**2.111**), sesamin (**2.223**) and the butyrolactone yatein (**2.088**). C.K. Lim and Ayres (1983) studied the factors governing the separation of the seven known diastereoisomers of podophyllotoxin and the necessary choice of eluents for satisfactory resolution of functional derivatives (Scheme 5.2).

The power of the method has been further demonstrated (Lim, 1984) by the determination of the anticancer agents Etoposide and Teniposide (see Chapter 4) in serum and by the resolution of these drugs from their picro-analogues (see also: Formation of artefacts).

Scheme 5.2 HPLC separation of some podophyllum lignans

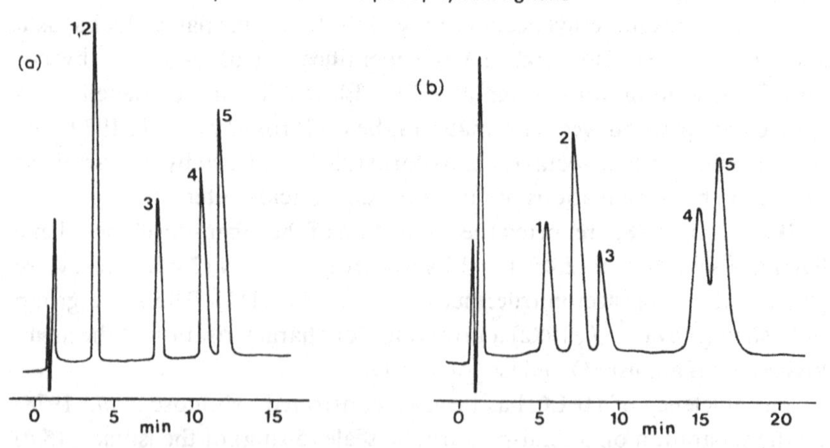

Separation of picropodophyllin (1), podophyllotoxin (2), podophyllotoxone (3), deoxypodophyllotoxin (4) and acetylpodophyllotoxin (5). Column, ODS-Hypersil (10 cm × 5 mm); eluent, (a) acetonitrile–water (40:60) and (b) methanol–water (50::50); flow-rate, 1 ml/min.

Gas–liquid chromatography (GLC)

Suitable conditions for the gas chromatography of the main classes of lignans were determined by Ayres and Chater in 1969. It was shown that, if allowance is made for functionality, lignans of different classes have characteristic retention times relative to that of cholestane. Liovil (**2.113**) and pinoresinol (**2.217**) were identified (Lutskii *et al.*, 1978) in *Picea obovata* and other spruce species by GLC of their silylated derivatives. The method was also used by Rainer (1979) to establish the distribution pattern of lignans in the wood of *Picea abies* – the level in sapwood was negligible, but in heartwood it ranged up to 0.5% and was maximal in the lower part of the trunk.

GLC has found clinical applications (Fotsis *et al.*, 1982) in the identification of excreted enterolactone and enterodiol (**2.028**) where silicone ester SE-30 and OV-210 capilliary columns were used with mass spectrometric detection. The same workers (Boriello *et al.*, 1985) characterised these lignans as the products of biological reduction of secoisolariciresinol (**2.035**) and matairesinol (**2.062**) in semi-synthetic diets by GC/MS. Gas chromatography has advantages for work of this type because of the high sensitivity of detection. However, HPLC is otherwise to be preferred for the analysis of lignans as a wide range of possible eluting solvents is available and it may be directly applied to polar substances.

GLC is of value for the determination of conjugated compounds which are UV transparent. For example, 12-L-methyltetradecanoic acid was confirmed (Powell and Plattner, 1976) by comparative GLC to be one of the residues and ferulic acid the other in a naturally occurring diester (**5.6**) of secoisolariciresinol. This large molecule ($C_{45}H_{62}O_{10}$, M = 762) was successfully chromatographed on 3% OV–1 in the temperature range of 200–400°C. GLC has also been used (Ikeya *et al.*, 1982) to identify angelic and tiglic acids released on hydrolysis of constituents of *Schizandra chinensis*.

Comparative paper chromatography has been applied (Bradlaw, 1968; Chandel and Rastogi, 1980) to the identification of sugars conjugated with lignans, but for this purpose GLC of silyl derivatives is clearly advantageous. Deyama (1983) confirmed the conjugation of glucose with medioresinol, pinoresinol and syringaresinol by hydrolysis in sulphuric acid (10%), ether extraction and passage of the aqueous hydrolysate

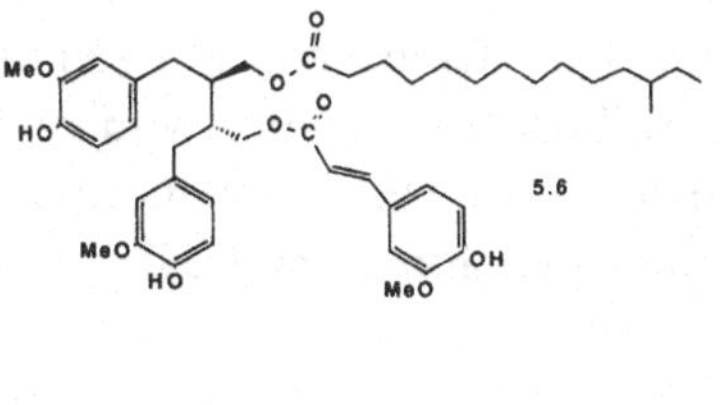

5.6

through Amberlite IRA-45 resin. The concentrate from this was then reduced with borohydride and the peracetylated product chromatographed on a stationary phase of OV-17 at 200°C. In later work (Deyama *et al.*, 1985) this group hydrolysed the diglucosides of (−)-olivil (**2.129**) and 8-hydroxypinoresinol (**2.252**) enzymically, before characterising the sugar by freeze-drying and GLC of the silyl compound. Anjaneyulu *et al.* (1981b) showed that cleistanthin D was the D-xyloside of diphyllin (**2.394**) by gas chromatography of the methylated sugar on diethyleneglycol succinate at 180°C.

Droplet counter-current chromatography (DCCC) is applicable to the isolation of polar lignans (Hostettman, 1980) and was employed (Kobayashi and Ohta, 1983) in the separation of furofuran metabolites.

Formation of artefacts

It is essential that heating of natural products, including lignans, is kept to the necessary minimum. Solvents must be removed under vacuum and aqueous solutions of polar lignans and glycosides should be concentrated either by adsorption, as in the Sep-Pak methods (Fotsis *et al.*, 1982) or by freeze-drying (lyophilisation). Examples of the latter method are to be found in the work of Jolad *et al.* (1980) and that of Chaudhuri and Sticher (1981) on the extraction of the diglucoside liriodendrin (**2.271**).

Whenever material is milled prior to extraction frictional heating should be avoided by the use of a cooled mill. Material should be stored in sealed refrigerated containers in order to limit the risk of attack by organisms such as moulds and rot fungi; some examples of phytoalexin formation in this way were given in the Introduction (pp. 104–5).

Precautions taken during extraction may be vitiated by inefficiencies at a later stage. Attention has been drawn (Lederer and Lederer, 1957) to the range of reactions which may occur on active alumina, although chromatography on alumina partially inactivated by the addition of water is less likely to produce artefacts than is alumina of activity one (Brockman and Schodder, 1941). The topic has also been reviewed by Posner (1978) who gave examples of aldolisation, hydration, elimination of −OH and alkoxy groups, hydride transfer and a number of skeletal rearrangements.

(i) *Changes induced by bases*

The enolisation and consequent stereochemical equilibration of butyrolactones and other lactones will occur on heating with strong bases such as alkali hydroxides and alkoxides. More facile and more easily overlooked is the epimerisation of strained enolisable molecules such as lactones and ketolactones of the podophyllotoxin group (**2.357** and **5.7–5.9** below). It has long been known (Hartwell and Schrecker, 1958) that even weak bases like ammonia and sodium acetate rapidly initiate epimerisation at C-8′, with formation of mixtures containing as little as 1% of the original physiologically active compound. These 'picro' isomers are much less soluble than the active precursors and this change may well be the cause of precipitation which is sometimes observed in saline preparations of the derived drugs Etoposide and Teniposide. Indeed assay of these substances and of their metabolites may be invalidated by contact with bases and failure to extract the artefacts. It is also conceivable that loss of activity may result *in vivo* from this cause if these drugs are administered orally when the small intestine is in an alkaline condition.

In the ketolactone (**5.7**) it is of interest that bases lead to strain-relief (Jackson and Dewick, 1985) through enolisation of the lactone, presumably due to the geometry favouring interaction between a carbanion developing at C-8′ and the lactone carbonyl. However, on heating the 160°C equilibration takes place at C-8 with formation of isopicropodophyllone (**5.9**), the product of inversion at this position. It has been noted (Dewick and Jackson, 1981) that commercial samples of *Podophyllum hexandrum* and *P. peltatum* contain artefacts of this type formed during extraction and/or drying. Related changes probably led to the isolation (Murphy *et al.*, 1975) of austrobailignan-4 (**5.10**).

Another type of base-induced change (Robertson and Waters, 1933) in this group of lignans is the shift of the strained α-double bond in apo-compounds to the essentially strain-free β-position. This also occurs on contact with weak bases or on warming.

(ii) *Changes induced by acids*

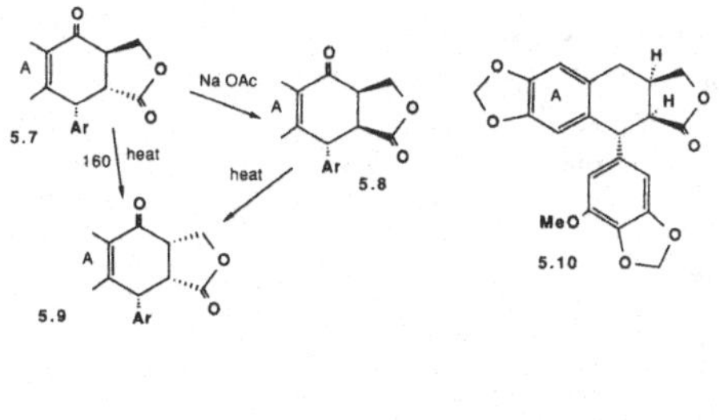

Instances have been recorded of the formation of conjugated alkenes during chromatography of tertiary alcohols of the aryltetralin class. One example is that of hydroxyotobain (**5.11**) which, when isolated (Maclean and Stevenson, 1966) from a solution in petrol, was mixed with otobaene (**5.12**) formed by contact with neutral alumina at room temperature. Otobaene was also formed as an artefact on TLC or silica column chromatography of the extracts of *Virola cuspidata* (Blair *et al.*, 1969) and Martinez *et al.* showed (1985) that the same reaction occurred when a chloroform solution of methoxyhydroxyotobain was wetted with dilute hydrochloric acid. The same authors (Lopes *et al.*, 1984) have discussed the possibility of artefact formation during the treatment of extracts from other *Virola* sp. This linked the formation of an aryltetralin with the *in vitro* Friedel-Craft cyclisation of the co-occurring hydroxydibenzylbutane (**5.13**). Enol-dehydration of the diketone (**5.14**) was considered to be the *in vitro* route to a co-occurring arylfuran (**5.15**), but neither mechanism was thought to be of biogenetic significance (for further discussion see Chapter 7, pp. 292–3).

There are many examples of the isomerisation of furan and furofuran lignans at the α-positions. This takes place because protonation of an ether-oxygen atom leads to reversible ring-opening through a mesomerically stabilised carbocation.

In furofurans equilibrium is established between equatorial (**5.16**), axial-equatorial (**5.17**), and diaxial (**5.18**) types. Briggs *et al.* (1968) demonstrated that lirioresinol-C (**5.18**; Ar = 3,4,5-trimethoxyphenyl) was isomerised on alumina to the diequatorial isomer (**5.16**). It has also been recorded (Dickey, 1958) that changes of this kind can take place during the hydrolysis of furofuran lignans. In view of this susceptibility, caution is needed whenever epimers of these compounds are isolated from a common source as for example, sesartemin and episesartemin; yangambin and epiyangambin from *Virola elongata* (Macrae and Towers, 1985). This is most important whenever the natural source includes those most at risk

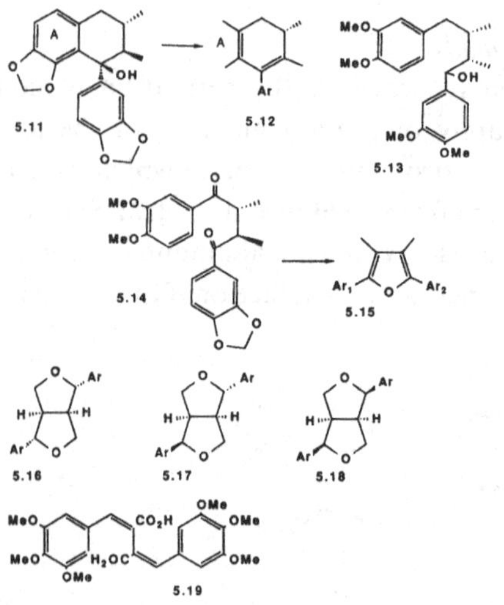

5.11 5.12 5.13

5.14 5.15

5.16 5.17 5.18

5.19

namely diaxial types like diayangambin (**2.186** or lirioresinol-C) and diasesartemin (**2.185**). One method of confirmation would be to analyse extracted material by GLC on silicon ester or other non-polar phase, which will resolve (Ayres, 1969) these isomers, before exposure to silica or alumina.

Lignan carboxylic acids are rare but the dienedioic acid (**5.19**) occurs in *Phebalium nudum* although it was isolated (Brown *et al.*, 1975) as its diethyl ester after extraction of the bark with ethanol. Ethyl and methyl acetal groups have been said to co-occur in furofuran (Anjaneyulu *et al.*, 1975) and aryltetralin (Cambie *et al.*, 1979) lignans, but a real possibility of ethoxyl exchange for methoxyl exists because the natural materials were surface adsorbed on to silica or alumina and then treated with solvent (diethyl ether, ethyl acetate) which could contain ethanol. This possibility is strengthened by the observations:

(1) that ethyl groups have only been reported at positions which are liable to acid-catalysed exchange;
(2) lignans solvate preferentially with alcohols even when dissolved in a large volume of aprotic solvent.

This matter is of considerable biogenetic significance (Anjaneyulu *et al.*, 1975; also see Chapter 7) and must be regarded as an open question, which could be resolved by mass spectrometry of extracted material before contact with potential catalysts.

References

Abe, F. and Yamauchi, T. (1986). Lignans & lignan glucosides from *Trachelospermum asiaticum*. *Chem. Pharm. Bull.* **34**, 4340–5.

Achenbach, H., Waibel, R and Mensah, I.A. (1983). Lignans & other constituents from *Carissa edulis*. *Phytochemistry* **22**, 749–53.

Agrawal, P.K. and Rastogi, R.P. (1982). Two lignans from *Cedrus deodara*. *Phytochemistry* **21**, 1459–61.

Anjaneyulu, A.S.R., Rao, K.J., Rao, K.V., Row, L.R and Subramanyam, C. (1975). The stuctures of lignans from *Gmelina arborea* Linn. *Tetrahedron* **31**, 1277–85.

Anjaneyulu, A.S.R. and Murty, V.S. (1981a). Chemical examination of *Guazuma tomentosa* Kunth. *Indian J. Chem.* **20B**, 85–7.

Anjaneyulu, A.S.R., Ang, A.S.P., Ramiah, P.A., Row, L.R. and Venkateswarlu, V. (1981b). New lignans from the heartwood of *Cleistanthus collinus*. *Tetrahedron* **37**, 3641–52.

Ayres, D.C. (1969). Incorporation of L-[U-^{14}C]-β-phenylalanine into the lignan podophyllotoxin. *Tetrahedron Lett.* 883–6.

Ayres, D.C. and Chater, R.B. (1969). Lignans & related phenols. Part 10. Classification of the principal groups of lignans by gas chromatography. *Tetrahedron* **25**, 4093–8.

Badawi, M.M., Handa, S.S., Kinghorn, A.D., Cordell, G.A. and Farnsworth, N.R. (1983). Plant anti-cancer agents XXVII. Anti-leukemic & cytotoxic constituents of *Dirca occidentalis*. *J. Pharm. Sci.* **72**, 1285–7.

Bamberger, M. (1894). The study of resin extractives. *Monatsch.* **15**, 505–18.

Bamberger, M. and Landsied, A. (1897). The extractives of larch. *Monatsch.* **18**, 500–9.

Barata, L.E.S., Baker, P.M., Gottlieb, O.R. and Ruveda, E.A. (1978). Neolignans of *Virola surinamensis*. *Phytochemistry* **17**, 783–6.

Birch, A.J. and Lyons, F. (1938). The constitution of gmelinol. *J. Proc. Roy. Soc. N.S. Wales* **71**, 391–405.

Blair, G.E., Cassady, G.M., Robbers, J.E., Tyler, V.E. and Raffauf, R.F. (1969). Isolation of 3,4′,5-trimethoxy-*trans*-stilbene, otobaene, and hydroxyotobain from *Virola cuspidata*. *Phytochemistry* **8**, 497–500.

Bohlmann, F., Lonitz, M. and Knoll, K.H. (1978). New lignan derivatives from the *Heliantheae* family. *Phytochemistry* **17**, 330–1.

Boriello, S.P., Setchell, K.D.R, Axelson, M. and Lawson, A.M. (1985). Production & metabolism of lignans by human faecal flora. *J. Appl. Bacteriol.* **58**, 37–43.

Bradlaw, H.L. (1968). Extraction of steroid conjugates with a neutral resin. *Steroids* **11**, 265–72.

Briggs, L.H. and Frieberg, A.G. (1937). The resinol of *Olea Cunninghamii* (Maire). *J. Chem. Soc.* 271–3.

Briggs, L.H., Cambie, R.C. and Couch, R.A.F. (1968). Lirioresinol C dimethyl ether, a diaxially substituted 3,7-dioxabicyclo-[3,3,0]octane lignan from *Macropiper excelsum*. *J. Chem. Soc. C* 3042–5.

Bristow, P.A. (1976). *Liquid Chromatography in Practice*. Wilmslow.

Brockman, H. and Schodder, H. (1941). Aluminium oxide with buffered adsorptive properties for chromatography. *Chem. Ber.* **74**, 73–8.

Brown, K.L., Burfitt, A.I.R., Cambie, R.C., Hall, D. and Mathai, K.P. (1975). The constituents of *Phebalium nudum* III. The structure of phebolin & phebalarin. *Aust. J. Chem.* **28**, 1327–37.

Cairnes, D.A., Ekundayo, O. and Kingston, D.G.I. (1980). Plant anti-cancer agents X. Lignans from *Juniperus phoenicea*. *J. Nat. Prod.* **43**, 495–7.

Cairnes, D.A., Kingston, D.G.I. and Rao, M.M. (1981). HPLC of podophyllotoxins & related lignans. *J. Nat. Prod.* **44**, 34–7.

Cambie, R.C., Pang, G.T.P., Parnell, J.C., Rodrigo, R. and Weston, R.J. (1979). Chemistry of the *Podocarpaceae* LIV. Lignans from the wood of *Dacrydium intermedium*. *Aust. J. Chem.* **32**, 2741–51.

Carboncini (1837). Phillyrin, a plant substance from *Phillyrea latifolia*. *Liebigs Ann. Chem.* **24**, 242–5.

Chandel, R.S. and Rastogi, R.P (1980). Pygeoside, a new lignan xyloside from *Pygeum acuminatum*. *Indian J. Chem.* **19B**, 279–82.

Chatterjee, A., Banerji, A., Banerji, J., Pal, S.C. and Ghosal, T. (1984). Recent advances in the chemistry of lignans. *Proc. Indian Acad. Sci.* **93**, 1031–57.

Chaudhuri, R.K. and Sticher, O. (1981). New iridoid glucosides & a lignan diglucoside from *Globularia alypum* L. *Helv. Chim. Acta* **64**, 3–15.

Chen, Y., Lin, R., Hsu, H., Yamamura, S., Shizuri, Y. and Hirata, Y. (1973). The structures & conformations of two new lignans, kadsurin & kadsurarin. *Tetrahedron Lett.* 4257–60.

Chen, Y-Y., Shu, Z-B. and Li, L-N. (1976). Studies on fructus *Schizandrae*. Part 4. Isolation & determination of the active compounds of *Schizandra chinensis* Baill. *Sci. Sin.* **19**, 276–90.

Chiba, M., Okabe, K., Hisada, S., Shima, K., Takemoto, T. and Nishibe, S.

(1979). Elucidation of the structure of a new glucoside from *Olea Europaea* by CMR spectroscopy. *Chem. Pharm. Bull.* **27**, 2868–73.

Dewick, P.M. and Jackson, D.E. (1981). Cytotoxic lignans from podophyllum & the nomenclature of aryltetralin lignans. *Phytochemistry* **20**, 2277–80.

Deyama, T. (1983). The constituents of *Eucommia ulmoides* Oliv. I. Isolation of (+)-medioresinol di-O-β-D-glucopyranoside. *Chem. Pharm. Bull.* **31**, 2993–7.

Deyama, T., Ikawa, T. and Nishibe, S. (1985). The constituents of *Eucommia ulmoides* Oliv. Part 2. Isolation & structure of three new lignan glycosides. *Chem. Pharm. Bull.* **33**, 3651–7.

Dias, A. de F., Giesbrecht, A.M. and Gottlieb, O.R. (1982). Neolignans from *Urbanodendron verrucosum. Phytochemistry* **21**, 1137–9.

Dickey, E.E. (1958). Liriodendrin, a new lignan diglucoside from the inner bark of yellow poplar (*Liriodendron tulipfera* L.). *J. Org. Chem.* **23**, 179–84.

Easterfield, T.H. and Bee, J. (1910). The resin acids of the *conifereae*. Part II: Matairesinol. *J. Chem. Soc.* **97**, 1028–32.

Elisha, II Book of Kings, Chap. 4, vv. 38–41.

Erdtman, H. and Tsuno, K. (1969). The chemistry of the order *Cupressales. Acta Chem. Scand.* **23**, 2021–4.

Fagbule, M.O. and Olatunji, G.A. (1984). Isolation and characterisation of the furofuran lignan fargesin. *Cellulose Chem. and Technol.* **18**, 293–6.

Fonseca, S.F., Campello, J. de P., Barata, L.E.S. and Ruveda, E.A. (1978). CMR spectral analysis of lignans from *Araucaria angustifolia. Phytochemistry* **17**, 499–502.

Fotsis, T., Heikkinen, R., Adlercreutz, H., Axelson, M. and Setchell, K.D.R. (1982). Capilliary GC method for the analysis of lignans in human urine. *Clin. Chim Acta* **121**, 361–71.

Ghosal, S., Banerjee, S. and Srivastava, R.S. (1979). Simplexolin a new lignan from *Justicia simplex. Phytochemistry* **18**, 503–5.

Ghosal, S., Banerjee, S. and Jaiswal, D.K. (1980). New furofurano lignans from *Justicia simplex. Phytochemistry* **19**, 332–4.

Giddings, J.C., ed., *Advances in Chromatography* (to 1987), vols. 1–27. Dekker: New York.

Gilchrist, T., Hodges, R. and Porte, A.L. (1962). The structure of otobain. *J. Chem. Soc.* 1780–6.

Gisvold, O. (1948). A preliminary survey of the occurrence of NDGA in *Larrea divaricata. J. Amer. Pharm Assoc.* **37**, 194–6.

Gorokhova, V.G., Gorokhova, K., Tyukavkina, N.A., Leonteva, V.G., Babkin, V.A. and Modonova, L.D. (1979). Liquid chromatography of plant phenolic compounds Part 6. Adsorption & reverse phase chromatography of lignans. *Khim. Drev.* 103–6. *Chem. Abs.* **92**, 54376.

Gottlieb, O.R. (1978), Neolignans. *Progr. Chem. Org. Nat. Prod.* **35**, 1–72.

Greger, H. and Hofer, O. (1980). New unsymmetrically substituted tetrahydrofurofuran lignans from *Artemisia absinthum. Tetrahedron* **36**, 3551–8.

Hartwell, J.L. and Schrecker, A.W. (1958). Lignans of podophyllum. In Zechmeister, L., *Progr. Chem. Org. Nat. Prod.* **15**, 83–166.

Hasegawa, S. and Hirose, Y. (1980). A diterpene glycoside and lignans from seed of *Thujopsis dolabrata. Phytochemistry* **19**, 2479–81.

Haworth, R.D., Mavin, C.R. and Sheldrick, G. (1934). The constituents of *guaiacum* resin. *J. Chem. Soc.* 1423–9.

Haworth, R.D. and Richardson, T. (1935). The constituents of natural phenolic resins Part I: Matairesinol. *J. Chem. Soc.* 633–6.

Haworth, R.D. and Kelly, W. (1937). The constituents of natural phenolic resins.

VIII: Lariciresinol, cubebin and some stereochemical relationships. *J. Chem. Soc.* 384–91.

Hearon, W.M. and MacGregor, W.S. (1955). The naturally occurring lignans. *Chem. Rev.* **55**, 957–1068.

Hernandez, A., Pascual, C. and Valverde, S. (1981). Neo-olivil, a new lignan from *Thymus longiflorus. Phytochemistry* **20**, 181–2.

Horvath, C., ed. (1986). *High Performance Liquid Chromatography, Advances & Perspectives*, vols. 1–4.

Hostettman, K. (1980). Droplet counter-current chromatography & its application to the preparative scale separation of natural products. *Planta Med.* **39**, 1–18.

Hughes, G.K. and Ritchie, E. (1954). Chemical constituents of *Himantandra* species. *Aust. J. Chem.* **7**, 104–12.

Ida, T., Nakano, M. and Ito, K. (1982). Hydroperoxysesquiterpene and lignan constituents of *Magnolia kobus. Phytochemistry* **21**, 673–5.

Ikeya, Y., Taguchi, H., Yosioka, I. and Kobayashi, H. (1979). The constituents of *Schizandra chinensis* Baill. VI. Isolation and structure determination of five new lignans, Gomisin A, B, C, F, and G, and the absolute structure of schizandrin. *Chem. Pharm. Bull.* **27**, 1383–94.

Ikeya, Y., Taguchi, H., Sasaki, H., Nakajima, K. and Yosioka, I. (1980). The constituents of *Schizendra chinensis* Baill. VI. *Chem. Pharm. Bull.* **28**, 2414–21.

Ikeya, Y., Ookawa, N., Taguchi, H. and Yosioka, I. (1982). The constituents of *Schizandra chinensis* Baill XI. The structure of three new lignans, angeloyl gomisin O, angeloyl- and benzoylisogomisin O. *Chem. Pharm. Bull.* **30**, 3202–6.

Ikeya, Y., Taguchi, H. and Yosioka, I. (1982). The constituents of *Schizandra chinensis* Baill X. The structures of O-schizandrin and four new lignans, (−)-gomisin L_1 & L_2, (±)-gomisin M_1, and (+)-gomisin M_2. *Chem. Pharm. Bull.* **30**, 132–9.

Inoue, K., Inouye, H. and Chen, C-C. (1981). A naphthoquinone and a lignan from the wood of *Kigelia pinnata. Phytochemnistry* **20**, 2271–6.

Jackson, D.E. and Dewick, P.M. (1984). Aryltetralin lignans from *Podophyllum hexandrum & P. peltatum. Phytochemistry* **23**, 1147–52.

Jackson, D.E. and Dewick, P.M. (1985). Tumour-inhibiting aryltetiralin lignans from *Podophyllum pleianthum. Phytochemistry* **24**, 2407–9.

Jolad, S.D., Hoffmann, J.J., Cole, J.R., Tempesta, M.S. and Bates, R.B. (1980). Cytotoxic agent from *Penstemon deustus* (*Scrophuliaceae*). Isolation and stereochemistry of liriodendrin, a symmetrically substituted furofuranoid lignan diglucoside. *J. Org. Chem.* **45**, 1327–9.

Joshi, B.S., Viswanathan, N., Balakrishnan, V., Gawad, D.H. and Ravindranath, K.R. (1979). Attenuol, structure, stereochemistry and synthesis. *Tetrahedron* **35**, 1665–71.

Kamil, W.D. and Dewick, P.M. (1986). Biosynthesis of the lignans α- and β-peltatin. *Phytochemistry* **25**, 2089–92.

Kelleher, J.K. (1977). Tubulin binding affinities of podophyllotoxin and colchicine analogues. *Mol. Pharmacol.* **13**, 232–41.

Khan, K.A. and Shoeb, A. (1985). A lignan from *Lonicera hypoleuca. Phytochemistry* **24**, 628–32.

King, F.E., Jurd, L. and King, T.J. (1952). Isotaxiresinol, a new lignan extracted from the heartwood of *Taxus baccata. J. Chem. Soc.* 17–24.

Kirchner, J.G. (1978). *Thin Layer Chromatography*, 2nd edn, ed. E.S. Perry. Wiley: New York.

Kisiel, W. (1980). Lignans from *Helichrysum bracteatum. Planta Med.* **38**, 285–7.

Kobayashi, M. and Ohta, Y. (1983). Induction of stress metabolite formation in suspension cultures of *Vigna angularis*. *Phytochemistry* **22**, 1257–61.

Kochetkov, N.K., Khorlin, A., Chizor, S. and Schienchenko, V.I. (1961). Schizandrin lignan of unusual structure. *Tetrahedron Lett*. 730–4.

Korner, G. and Vanzetti, L. (1903). Olivil, its composition and constitution. *Atti R. Accad. Lincei* **12**, 122–5; *J. Chem. Soc. Abs*. **1903**, 430.

Kramer, A. (1933). Phillyrin and its hydrolysis by emulsion. *Bull. Soc. Chim. Biol.* **15**, 665–84.

Kudo, K., Nohara, T., Komori, T., Kawasaki, T. and Schulten, H-R. (1980). Field desorption mass spectroscopy of natural products VII. Lignan glucosides from the bark of *Ligstrum japonicum*. *Planta Med*. **40**, 250–61.

Lavie, D. and Glotter, E. (1971). The Cucurbitanes a group of tetracyclic triterpenes. *Prog. Chem. Org. Nat. Prod*. **29**, 307–62.

Lederer, E. and Lederer, M. (1957). Chromatography – A review of principles & applications, pp. 61–7. Elsevier: Amsterdam.

LeQuesne, J.P.W., Larrahondo, J.E. and Raffauf, R.E. (1980). Anti-tumour plants X. Constituents of *Nectandra rigida*. *J. Nat. Prod*. **43**, 353–9.

Lim, C.K. and Ayres, D.C. (1983). HPLC of aryltetrahydronaphthalene lignans, *J. Chromatog*. **255**, 247–54.

Liu, C-S., Fang, S-D., Huang, M-F., Kao, Y.L. and Hsu, J-S. (1978). Studies on the active principles of *Schizandra spenanthera*. *Sci. Sin*. **21**, 483–502.

Liu, J-S., Huang, M-F., Gao, Y-L. and Finlay, J.A. (1981). The structure of chicanine, a new lignan from *Schizandra* sp. *Can. J. Chem*. **59**, 1680–4.

Liu, J-S., Huang, M-F., Ayer, W.A. and Nakashima, T.T. (1984). Structure of enshicine from *Schizandra henryi*. *Phytochemistry* **23**, 1143–5.

Lopes, L.M.X., Yoshida, M. and Gottlieb, O.R. (1982). 1,11-Diarylundecane-1-one and 4-aryltetralone lignans from *Virola sebifera*. *Phytochemistry* **21**, 751–5.

Lopes, L.M.X., Yoshida, M. and Gottlieb, O.R. (1984). Aryltetralone and arylindanone neolignans from *Virola sebifera*. *Phytochemistry* **23**, 2021–4.

Lopes, L.M.X., Yoshida, M. and Gottlieb, O.R. (1984). Further lignoids from *Virola sebifera*. *Phytochemistry* **23**, 2647–52.

Lutskii, V.I., Leont'eva, V.G., Modonova, L.D. and Tyukavkina, N.A. (1978)). Use of GLC during the study of guaialignans. *Khim. Drev*. 87–90. *Chem. Abs*. **89**, 45221.

Maclean, I. and Stevenson, R. (1966). The constitution of otobaphenol. *J. Chem. Soc*. 1775–80.

Macrae, W.D. and Towers, G.H.N (1984). An ethnopharmacological examination of *Virola elongata* bark: a South American arrow poison. *J. Ethnopharmacol* **12**, 75–92.

Macrae, W.D. and Towers, G.H.N. (1985). Non-alkaloidal constituents of *Virola elongata* bark. *Phytochemistry* **24**, 561–6.

Maiden, J.H. and Smith, H.G. (1896). On aromadendrin or aromadendric acid from the turbid group of eucalyptus kinos. *Amer. J. Pharm*. 679–87.

Martinez, J.C., Cuca, L.E., Yoshida, M. and Gottlieb, O.R. (1985). Neolignans from *Virola calophylloidea*. *Phytochemnistry* **24**, 1867–8.

The Merck Index (1983), 10th edn, entry 6534. Merck and Co.: Rahway, N. Jersey.

Momose, T., Kanai, K. and Nakamura, T. (1978). Synthetic studies on lignans & related compounds VI. Photochemical rearrangement of β-apolignans. *Chem. Pharm. Bull*. **26**, 1592–7.

Murphy, S.T., Ritchie, E. and Taylor, W.C. (1975). Some constituents of *Austrobaileya scandens*: structures of seven new lignans. *Aust. J. Chem*. **28**, 81–90.

Nemethy, E.K., Lago, R., Hawkins, D. and Calvin, M. (1986). Lignans of *Myristica otoba*. *Phytochemistry* **25**, 959–60.

Nishibe, S., Tsukamoto, H. and Hisada, S. (1984). Effects of *O*-methylation and *O*-glycosylation on CMR chemical shifts of matairesinol, (+)-pinoresinol, and (+)-epipinoresinol. *Chem. Pharm. Bull.* **32**, 4653–7.

Nishida, K., Sumimoto, M. and Kondo, T. (1951). A constituent of *Symplocos lucida*. *J. Japan Forest Soc.* **33**, 235–9.

Nishida, K., Sumimoto, M. and Kondo, T. (1951). *J. Soc. Forestry Japan* **33**, 269–72.

Ohta, Y. and Hirose, H (1968). New sesquiterpenes from *Schizandra chinensis*. *Tetrahedron Lett.* 2483–5.

Olaniyi, A.A. and Powell, J.W. (1980). Lignans from *Jucticia flava*. *J. Nat. Prod.* **43**, 482–6.

Ookawa, N., Ikeya, Y., Taguchi, H. and Yosioka, I. (1981). The constituents of *Kadsura japonica* Dunal. I. The structure of three new lignans, acetyl-, angeloyl- and caproyl-binankadsurin A. *Chem. Pharm. Bull.* **29**, 123–7.

Pearl, I.A. (1945). Conidendrin from Western hemlock sulphite waste liquor. *J. Org. Chem.* **10**, 219–21.

Pelter, A., Ward, R.S., Rao, E.V. and Sastry, K.V. (1976). Revised structures for pluviatilol, methyl pluviatilol, and xanthoxylol. *Tetrahedron* **32**, 2783–8.

Popoff, T. and Theander, O. (1977). The constituents of conifer needles. VI: Phenolic glycosides from *Pinus sylvestris*. *Acta Chem. Scand.* **B31**, 329–37.

Posner, G.H. (1978). Organic reactions at alumina surfaces. *Angew. Chem. Int. Edit.* **17**, 487–96.

Powell, R.G. and Plattner, R.D. (1976). Structure of a secoisolariciresinol diester from *Salvia plebeia* seed. *Phytochemistry* **15**, 1963–5.

Rainer, E. (1979). Distribution of lignans in Norway spruce. *Acta Acad. Abu. Ser. B* **39**. *Chem. Abs.* **91**, 212771.

Rao, M.M. and Lavie, D. (1974). The constituents of *Ecballium elaterium* L.XXII. Phenolics as minor components. *Tetrahedron* **30**, 3309–13.

Rideout, J.M., Ayres, D.C., Lim, C.K. and Peters, T.J. (1984). Determination of etoposide (VP16-213) and teniposide (VM-26) in serum by HPLC with electro-chemical detection. *J. Pharm. & Biomed. Anal.* **2**, 124–8.

Robertson, A. and Waters, R.B. (1933). Podophyllotoxin. *J. Chem. Soc.* 83–6.

Sheriha, G.M. and Amer, K.M.A. (1984). Lignans of *Haplophyllum tuberculatum*. *Phytochemistry* **23**, 151–3.

Shinoda, J. and Kawagoye, M. (1929). Constituents of the seeds of *Arctium lappa* L. *J. Pharm. Soc. Japan* **49**, 565–75.

Smith, H.G. (1913). Crystalline deposit occurring in timber of the Colonial Beech. *Chem. News* **108**, 169–72, *Chem. Abs.* **7**, 4065.

Song, W. and Tong, Y. (1981). Occurrence of some important lignans in Wu Wei Zi (*Schizandra chinensis*) and its allied species. *Chem. Abs.* **99**, 19670.

Spencer, G.F. and Flippen-Anderson, J.L. (1981). Isolation and X-ray structure determination of a lignan from *Clerodendron inerme* seeds. *Phytochemistry* **20**, 2757–9.

Stahl, E. and Kaltenbach, A. (1961). Thin-layer chromatography, Part IX. Rapid separation of digitalis & podophyllum–glycoside mixtures. *J. Chromatog.* **5**, 458–60.

Stahl, E. (1969), ed. *Thin-layer Chromatography*, 2nd edn. Springer; Berlin.

Taguchi, H. and Ikeya Y. (1975). The constituents of *Schizandra chinensis* Baill. I: The structures of Gomisin A, B, and C. *Chem. Pharm. Bull.* **23**, 3296–8.

Takani, M., Ohya, K. and Takahashi, K. (1979). Studies on the constituents of medical plants. XXII: Constituents of *Schizandra nigra* Max. *Chem. Pharm. Bull.* **27**, 1422–5.

Takaoka, D., Watanabe, K. and Hirui, M. (1976). Studies on lignoids in *Lauraceae* II. Studies on lignans in the leaves of *Machilus japonica* Sieb. et Zucc. *Bull. Chem. Soc. Japan* **49**, 3564–6.

Tandon, S. and Rastogi, R.P. (1976). Wikstromol a new lignan from *Wikstroemia viridiflora. Phytochemistry* **15**, 1789–91.

Tatematsu, H., Kurokawa, M., Niwa, M. and Hirata, Y. (1984). Piscicidal constituents of *Stellera chamaejasme* L. *Chem. Pharm. Bull.* **32**, 1612–18.

Touchstone, J.C. and Sherma, J. (1985). Technique & applications of thin-layer chromatography. New York: Wiley.

Tsukamoto, M., Hisada, S. and Nishibe, S. (1983). Studies on the lignans from *Olaceae* plants. *Chem. Abs.* **100**, 117805.

Urzua, A. and Shamma, M. (1988). The 4-aryltetralones of *Aristolochia chilensis. J. Nat. Prod.* **51**, 117–21.

Vaquette, J., Cave, A. and Waterman, P.G. (1978). Alkaloids, triterpenes & lignans from *Zanthoxylum* sp. *Sevenet* – a species indigenous to New Caledonia. *Planta Med.* **35**, 42–7.

Vecchietti, V., Ferrari, G., Orsini, F. and Pelizzoni, F. (1979). Alkaloid & lignan constituents of *Cinnamosma Madagacariensis. Phytochemistry* **18**, 1847–9.

Viera, P.C., Gottlieb, O.R. and Gottlieb, H.E. (1983). Tocotrienols from *Iryanthera grandis. Phytochemistry* **22**, 2281–6.

Wagner, H., Heur, Y. H., Obermeier, A., Tittel, G. and Bladt, S. (1982). DC & HPLC analysis of *Eleutherococcus. Planta Med.* **44**, 193–8.

Wagner, H., Bladt, S. and Zgainski, E.M. (1984a). *Plant Drug Analysis: a TLC Atlas.* Springer: Berlin.

Wagner, H., Feil, B. and Bladt, S. (1984b). *Viscum album*–mistletoe. Analysis & standardisation of medicinal drugs & plant preparations by HPLC & other chromatographic methods. *Dtsch. Apoth. Ztg.* **124**, 1429–32.

Waller, C.W. and Gisvold, O. (1945). A phytochemical investigation of *Larrea divaricata. J. Amer. Pharm. Assoc.* **34**, 78–8.

Whiting, D.A. (1987). Lignans, neolignans & related compounds. *Nat. Prod. Rep.* 499–525.

Yamaguchi, H. and Nakajima, S. (1984). Studies on the constituents of the seeds of *Hernandia ovigera* L. Part 4. Synthesis of β-peltatin A & B methylene ethers from podophyllotoxin. *Chem. Pharm. Bull.* **32**, 1754–60.

Yoshiki, Y. and Ishiguro, T. (1933). Crystalline constituents of hinoki oil. *J. Pharm. Soc. Japan* **53**, 73–151.

6

Determination of structure

This chapter will in the main concentrate attention on spectroscopic methods for establishing lignan structures, but complementary chemical methods will be mentioned as appropriate.

Ultraviolet absorption spectra

All lignans show a basic UV absorption pattern typical of aromatic compounds with three bands in the regions of 210, 230, and 280 nm which correspond to the singlet excited states defined by Platt (1949; see Murrell, 1971) as $^1B_{a,b}$; ^1L_a and ^1L_b respectively. As the greater majority of lignans are optically active these maxima may usually be correlated with those in circular dichroism (CD) plots (p. 000). Clearly this basic absorption pattern will be modified in compounds with additional conjugation including those lignans which are fully aromatic.

The most intense absorption maximum arising from a component of the ^1B excitation in the region of 210 nm has a molecular extinction (ε) of about 5×10^4 but this is rarely recorded. Peaks in the range 220–240 nm ($\varepsilon = 10$ to 20×10^3) and at about 280 nm ($\varepsilon = 2$ to 10×10^3) are of diagnostic value since the absorption is of sufficient intensity to allow monitoring at all stages of the isolation procedure (p. 000).

In lignans where there is no conjugation to or between the aryl groups the absorption at the longest wavelength is sensibly the sum of their molar extinction coefficients. This may be verified by summing two aromatic components (from entries 1–4) and making the comparison with the extinction of a diarylbutane (e.g. entry 5) and of a butyrolactone (entry 6). Even in the aryltetralins (entries 7–12), where these two chromophores interact through a common benzylic carbon atom, a useful guideline remains. It is likely that in α-conidendrin, dimethylcyclo-olivil, and polygamain (entries 10–12 respectively) both rings are dioxygenated, since

Table 6.1 *Ultraviolet absorption of typical lignans*

Compound	Substituents	(nm) (ε, 10^{-3})
1 Guaiacol	1-OH, 2-OMe	277(2.73)289
2 Dimethoxybenzene	1-OMe, 2-OMe	274(3.92)
3 Trimethoxybenzene	1-OMe, 2-OMe, 3-OMe	267(0.69)
4 Methylenedioxybenzene	1,2-O-CH_2-O	282(3.24)
5 Austrobailignan-5	C-3,4; C-3′,4′:O-CH_2-O	236(8.20)288(7.60)
6 Di-*O*-methyldihydroxy thujaplicatin methyl ether	C-3,4,4,5,5′:OMe C-8,8′:OH	240(18.6)280(5.63)
7 Podophyllotoxin	C-4,C-5:O-CH2-O; C-3′,4′,5′:OMe	236(13.6)292(4.88)
8 β-Peltatin	C-4,C-5:O-CH_2-O; C-6:OH C-3′,4′,5′:OMe	— 276(1.77)
9 Dimethyl lyoniresinol	C-3,4,5:OMe C-3′,4′,5′:OMe	234(17.2)273(1.98)
10 α-Conidendrin	C-3′,C-5:OMe C-4′,C-4:OH	227(15.9)283(7.07)
11 Dimethylcyclo-olivil	C-4,C-5:OMe C-3′,C-4′:OMe	230(17.6)283(7.10)
12 Polygamain	C-3′, C-4′:O-CH_2-O C-4,C-5:O-CH_2-O	288(3.09)
13 Veraguensin	C-3,3′,4,4′:OMe	232(18.8)273(6.20)
14 Calopiptin	C-3,C-4:OMe C-3′,C-4′:O-CH_2-O	234(13.2)273(6.70)
15 Yangambin	C-3,4,5:OMe C-3′,4′,5:OMe	232(10.8)270(1.20)
16 Epiyangambin		232(16.6)270(1.90)
17 Diayangambin		231(14.5)270(1.40)
18 Sesamin	C-3,C-4:O-CH_2-O C-3′,C-4′:O-CH_2-O	238(12.7)287(10.1)

they all have ε values greater than that of podophyllotoxin (entry 7). Comparison with 1,2,3-trimethoxybenzene (entry 3) would lead one to expect a significant decrease in the absorption of podophyllotoxin; this arises from cancellation of spectroscopic moments in these trioxybenzenes (Platt, 1949). The effect is most clearly seen in the spectra of dimethyl lyoniresinol (entry 9) and in the three isomeric yangambins (entries 15–17). In the latter group of furofurans one sees some evidence of dependence of the extinction upon configuration, but the spectra are too similar overall for any reliable correlation with their geometry to be made. One again notes that here (entry 18) and also in the absorption of the two furans (entries 13 and 14) that, although ε values are not strictly additive, they still reveal the pattern of oxygenation.

The UV spectra of free phenolic lignans differ very little from those of fully etherified compounds, but the phenols may be distinguished by a shift to longer wavelength absorption in their anions. For example (Evcim, 1986), when a trace of base (1 drop of a 2% solution of NaOH or NaOMe) is added to a solution of haplomyrfolin (**2.047A**) the maxima at 230 ($\varepsilon = 15.1 \times 10^3$) and 284 nm ($\varepsilon = 10 \times 10^3$) shift to 242 nm ($\varepsilon = 18.2 \times 10^3$) and 290 nm ($\varepsilon = 11.7 \supset 10^3$). Similar bathochromic shifts have been reported for the tetralin glucoside pygeoside (**2.335**; Chandel and Rastogi, 1980) and for the furan nectandrin A (**2.163**; Le Quesne *et al.*, 1980). Of particular interest are keto-lignans such as lirionol (**6.1**; Chen and Chang, 1978) and magnolenin C (**6.2**; Rao and Wu, 1978). In the former, conjugation between one aryl group and the carbonyl group results in shifts to longer wavelength maxima at 242 ($\varepsilon = 21.4 \times 10^3$) and 291 nm ($\varepsilon = 11.0 \times 10^3$); these are markedly extended to 262 ($\varepsilon = 12.0 \times 10^3$) and 371 nm ($\varepsilon = 21.7 \times 10^3$) by the addition of sodium methoxide. Through its dissociation magnolenin C has a spectrum akin to that of syringaldehyde and on basification the longest wavelength maximum at 305 nm shifts to 365 nm. This behaviour is sufficient to show that glucosylation has occurred on the remote syringyl group.

Conjugation with a C═C, as in guairetic acid (**2.010**) shifts the absorption maximum towards the visible. When the system is extended so as to include the carbonyl group in the butyrolactones a further bathochromic shift occurs (Scheme 6.1):

> Chaerophyllin 237(15.8×10^3); 295(13.2×10^3); 331(18.6×10^3) (**2.097**, Mikaya, 1981).
> Hibalactone 294(1×10^3); 334(10×10^3), see Scheme 6.1, (**2.099**, Yamaguchi, 1970).

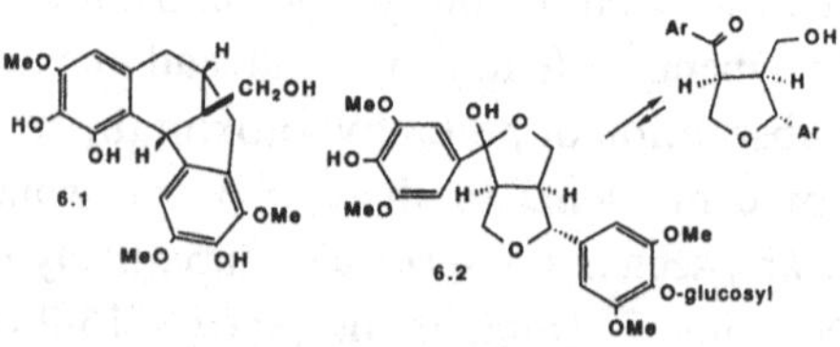

MeO HO OH H H CH₂OH 6.1 OMe MeO OH

MeO HO MeO OH O H H O 6.2 OMe O-glucosyl OMe

Ar O OH H H O Ar

The UV spectrum is sensitive to geometrical isomerism about the C=C double bond although these geometrical isomers are best distinguished by differences in their PMR spectra (p. 199).

When apotetralins are formed by elimination of a hydroxyl group in ring B evidence of conjugation is found (Hartwell, 1952) although in α-apopicropodophyllin (**6.3** and Scheme 6.1) this is not so marked as in hibalactone owing to the geometrical constraint imposed by this strained system. α-Apopicropodophyllin is very easily isomerised through its conjugate base to the β-isomer (**6.4**) and here the absorption (Scheme 6.1) is very like that of podophyllotoxin, save that the minimum at 265 nm is less marked owing to the addition of a component from the butenolide. With further movement of the double bond into the cross-conjugated 7′- position the ^1L_b band shifts to 350 nm. It should be noted also that the integrated absorption is even greater here than it is for the other unsaturated lactones.

Scheme 6.1 The ultraviolet absorption spectra of some lignans (Schrecker and Hartwell, 1958).

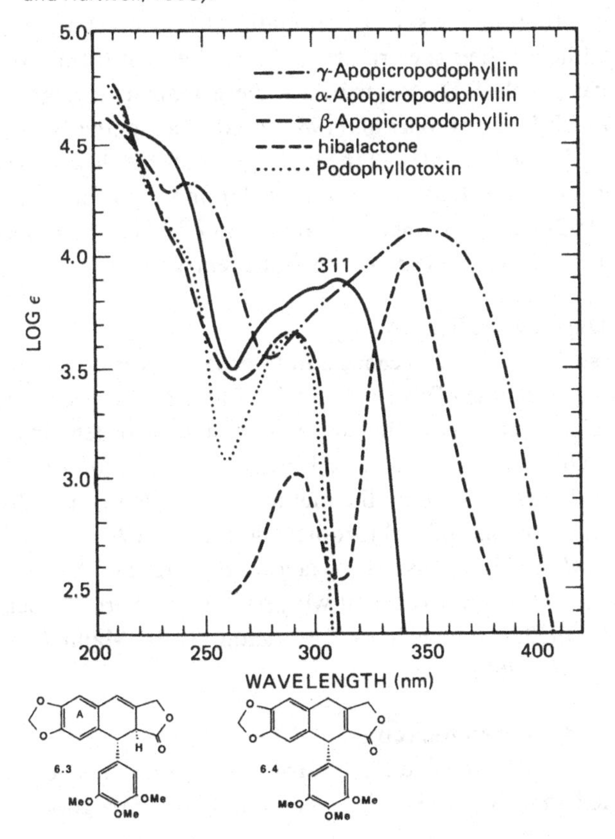

Table 6.2. *UV Absorption of biaryl and arylnaphthalene lignans* (λ nm, $\varepsilon \times 10^{-3}$)

1 Biphenyl	246(19.3)
2 4,4′-dihydroxybiphenyl	287(21.7)
3 γ-Schizandrin (**2.457**)	220(47.0) 253(12.6) 279(4.0)
4 Gomisin H (**2.433**)	219(64.6) 248(20.9) 276(5.75) 285(4.5)
5 Gomisin L_1 (**2.438**)	222(38.9) 252(10.0) 282–288 (2.75)
6 Steganone (**2.468**)	238(27.6) 276(9.2) 317(5.7)
7 Dehydroanhydropodophyllotoxin	258(55.0) 310(10.5) 350(5.7)
8 Dehydroguaiaretic acid (**2.391**)	237(102) 284(13.8) 314(4.16) 329(5.5)

Dibenzocyclo-octadiene lignans

When the aromatic rings of biaryls can interact in a common plane the UV absorption (Sadtler, 1979) is strongly enhanced by conjugation (Table 6.2, entries 1 and 2). However, in lignans of this class rotation is restricted by substitution of all four *ortho*-positions and also by the condensed alicyclic ring. As a result the aromatic rings interact at an angle of about 60° and hence their spectra are not very different from those summarised above (Table 6.1). Nevertheless there remains a degree of enhancement which Kochetkov *et al.* (1961) used in assigning the structure of schizandrin (Table 6.2, entry 3) the first lignan of this class to be identified. Gomisin H (entry 4; Ikeya *et al.*, 1979c) and gomisin L_1 (entry 5; Ikeya *et al.*, 1982a) are two other typical lignans of this type. The effect of further conjugation may be seen in steganone (entry 6).

Arylnaphthalene lignans

The strong UV fluorescence and intense absorption of these substances are unmistakable (Table 6.2, entries 7 and 8). Their spectra serve not only to characterise naphthalenic lignans but also dehydrogenated aryltetralins, which can include those formed by cyclisation of other lignans (pp. 330–3, 357–9). Among the methods available for dehydrogenation may be cited the use of dichlorodicyanoquinone (Ward *et al.*, 1981; Yamaguchi *et al.*, 1982), the use of oxygen/base (Ayres and Mundy, 1968; Momose *et al.*, 1978), and treatment with palladium/charcoal (Schrecker and Hartwell, 1953; Noguchi and Kawanami, 1940; Majumder *et al.*, 1972; Kohen *et al.*, 1966).

Infrared absorption spectra

To ensure that reliable IR spectra are recorded it is advisable to dry crystalline lignans for several hours at about 40°C in a good vacuum, which may well be enhanced by connecting a thimble of a molecular sieve chilled in liquid nitrogen to the evacuated drying pistol. These measures are necessary because of the tendency of lignans to solvate small molecules and these precautions should also be taken whenever a sample is prepared for elemental analysis.

(i) *Lignan solvates*

All lignans tend to become solvated by small guest molecules whose spectral responses may lead to false conclusions about the structure of the host. As an example one may cite podophyllotoxin (**6.5**), which is polymorphic and also forms five distinct solvates with water and/or benzene (Hartwell and Schrecker, 1958) as well as inclusions of guests such as ethanol and chloroform. It is indeed probable that the spaces in crystalline lignans are chiral and capable of resolving small asymmetric molecules. In forming inclusion compounds of this type lignans of the podophyllotoxin class may be compared to intensively researched host molecules, which include Dianin's compound (**6.6**, X = O) and its thioanalogue (**6.6**, X = S). Evidence that lignans attract guest molecules in solution is available from NMR studies (Ayres *et al.*, 1972a).

(ii) *Preparation of the sample*

Nujol and hexachlorobutadiene mulls are satisfactory, but solution spectra are to be preferred. Carbon tetrachloride and carbon disulphide have large 'windows', but may fail to take up sufficient material and the use of alcohol-free chloroform is recommended.

Potassium bromide discs give excellent spectra provided that this material (200 mg to include 1.5 mg of sample) is thoroughly dried in order to minimise the water peaks at 3400 and 1640 cm^{-1}. It is also advisable to run a blank disc, when desolvated lignan alcohols and phenols may be characterised more reliably and with greater sensitivity than is possible with NMR spectra (pp. 196–7). The latter technique is best when applied to *O*-acetyl derivatives.

(iii) *Functional group responses*

Many good summaries of the relevant data have been published (e.g. Colthup *et al.*, 1975; Yamaguchi *et al.*, 1970; Toda *et al.*, 1981) and

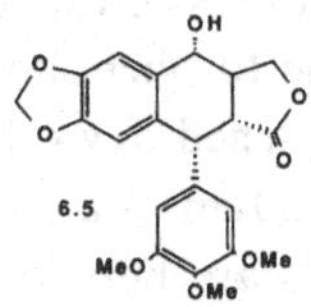

OH O O O O 6.5 MeO OMe OMe

X 6.6 OH

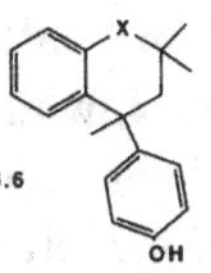

therefore an extensive discussion in this book is unnecessary. However, one can illustrate the use of IR by reference to Scheme 6.2; in all five of these lignan spectra their aromatic character is shown by:

(a) The Ar-H stretch in the region of 3000–3100 cm^{-1} at a slightly higher frequency than the absorption typical of the saturated portion in the range of 2800–3000 cm^{-1}.
(b) Ar-H wagging bands which fall in the range of 800–900 cm^{-1} for oxygenated rings.
(c) A strong band in the region of 1585–1620 cm^{-1} which arises from quadrant stretching of the ring (Colthup *et al.*, 1975) and which is insensitive to the substitution pattern. There is also another band at lower frequency (*ca.* 1500 cm^{-1}) which results from semicircular stretching of the ring, but which is more variable owing to mixing with the C—H in-plane bending mode.

A strong band in this region can be seen at 1510 cm^{-1} in the 3,4-dioxy compounds (**1,3,4**) but in trioxyethers such as podophyllotoxin (**2**) and wuweisisu C (**5**) a lower frequency peak appears at 1480 cm^{-1}.

Alkenes which are unsymmetrical show additional absorption in this region. In the conjugated lignan calocedrin this appears at 1640 cm^{-1} (Fang *et al.*, 1985a).

Carbonyl stretching clearly characterises lactones such as matairesinol (**1**, $\nu = 1755$ cm^{-1}) but when the five-membered ring is further strained as in podophyllotoxin (**2**) this frequency rises to 1775 cm^{-1}. In artefacts of the C-2 epimeric 'picro' series the peak falls to 1755 cm^{-1} and this important distinction may be reliably based upon the IR spectrum provided that the wavelength scale is aligned by reference to the absorption of a polystyrene film. In conjugated ketones the carbonyl absorption falls typically near to 1670 cm^{-1}.

Ether absorption arising from asymmetric stretching of the tetrahydrofuran ring (**3,4**) gives rise to a strong band near 1070 cm^{-1} with another due to the symmetric stretch or breathing frequency near 910 cm^{-1}. This region is crowded, however, as absorption from the Ar-OMe groups falls in the range of 1200–1300 cm^{-1}, with the Ar-OMe stretch coincident with the longer wavelength peaks of tetrahydrofurans and the appearance here of additional C—O bands will further complicate the spectra of lignan alcohols. An isolated peak at 2840 cm^{-1} attributable to phenolic methyl ethers is to be seen in spectra **1** and **4**, whilst methylenedioxy ethers are indicated (Briggs *et al.*, 1957) by absorption near 930 cm^{-1} as in podophyllotoxin (**2**) and wuweisisu C (**5**). Methyl acetals are best identified,

Scheme 6.2 Infrared spectra of lignans of the major classes.

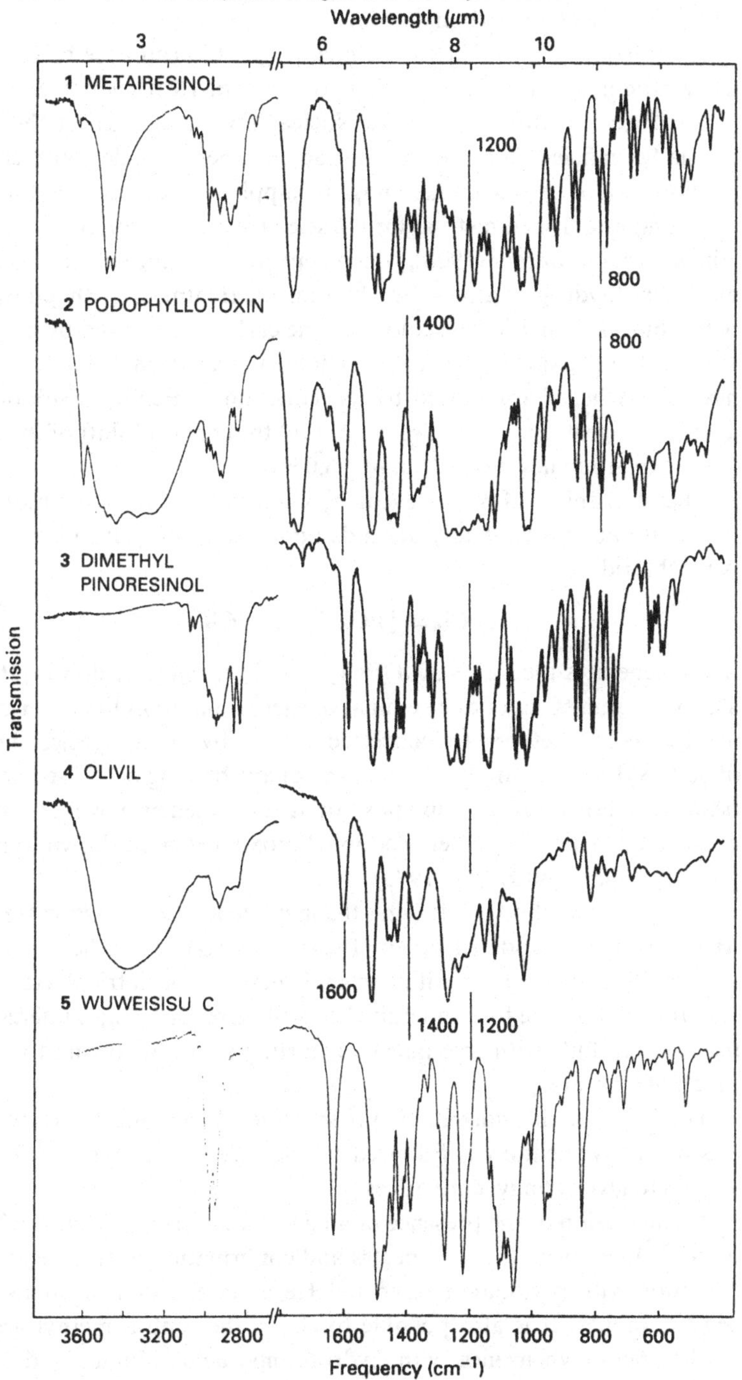

after acid hydrolysis, by the loss of the alkyl characteristics and the emergence of those of the carbonyl group.

It is fortunate that PMR experiments (p. 195) can now be conducted using little more material than that required for IR measurements. These results together with those of the Nuclear Overhauser Effect (NOE) are normally sufficient for the orientation of ethers. Single-stage chemical methods can be used with advantage to supplement the recorded spectra.

A long-established method for the selective cleavage of methylenedioxy ethers is that of Barger (1908), which depends upon the substitution of the methylene hydrogen atoms by chlorine on treatment with phosphorus pentachloride in methylene chloride. The carbonate derived by hydrolysis loses carbon dioxide in the acidic medium and a catechol is revealed. This may be further characterised by its reduction of Fehling's solution and Tollen's reagent. In NMR spectra (p. 196) the chemical shift of the methylenedioxy group lies typically at $\delta = 6.0$ Hz.

The mechanism of methylenedioxy chlorination has never been determined although it probably depends upon the disproportionation of the pentachloride:

$$2PCl_5 \rightleftharpoons [PCl_4]^+ + [PCl_6]^-$$

which generates the Lewis acid $[PCl_4]^+$ and the chloride donor $[PCl_6]^-$. The bolstering effect of one geminal oxygen atom upon the other exposes methylenedioxy ethers to concerted attack by these species, but 1,2-dimethoxybenzenes are unaffected. In lignans bearing a 1,2,3-trimethoxybenzene, substitution at a ring position takes precedence over the reaction of any methylenedioxy ether. Podophyllotoxin has been shown to react in this way (Ayres and Harris, 1973).

A superior method of cleaving the methylenedioxy group is treatment with boron trichloride (Schreier, 1964). This reaction is best conducted (Evans, 1986) by titration with one equivalent of boron trichloride in hexane. In alcohols one further equivalent will be required, and interestingly, the reaction fails with free phenolic starting materials owing to precipitation of a borate.

The boron trichloride reaction is selective, as in a one-pot reaction the loss of methylenedioxy is followed by the slower cleavage of trimethoxyethers to give syringyl derivatives.

An indication of the presence of guaiacyl and syringyl groups will have been obtained from the UV spectra and confirmation is obtained by their oxidation with periodate (Adler and Hernestam, 1955) to intensely red *ortho*-quinones. It is also possible to cleave the central methyl group in 1,2,3-trimethoxybenzenes with hydrobromic acid (Hunter and Levine,

1926) or with nitric/acetic acids. In the latter medium oxidation follows demethylation to afford the same quinone as that obtained from 4′-demethylpodophyllotoxin by the action of periodate (Ayres and Harris, 1973). This reaction has been used to detect podophyllotoxins on chromatography plates (Jackson and Dewick, 1985) and may well prove to be of general application to other natural products.

Mass spectra (MS)

Lignan glycosides are too involatile for their mass spectra to be obtained by direct vapour injection. The method of chemical ionisation (Milne *et al.*, 1971) is helpful for compounds of low volatility depending as it does on bombardment of the sample with ions of the type produced from methane by the ion–molecule reaction:

$$CH_4^{\cdot +} + CH_4 \rightarrow CH_5^+ + CH_3^{\cdot}$$

whence $$\text{Sample (M)} + CH_5^+ \rightarrow \text{Sample } (M + H)^+ + CH_4$$

The sample must still be heated to produce the ion $(M + H)^+$ but the passage of methane over the sample is an aid to evaporation.

The technique of field desorption is valuable for samples of low volatility (Rose and Johnstone, 1982). Here positive ions are produced on, and then repelled from, a positive electrode of high curvature when the sample comes into contact with the high potential gradient. Lignans with their high oxygen numbers give relatively stable even-electron low energy (*ca.* 10 ev) ions of this kind, which give strong molecular ions and few fragment ions. The simplicity of the individual spectra makes it possible to apply this technique to the initial examination of mixtures from plant sources.

A procedure of increasing importance for the analysis of involatile samples is that of fast atom bombardment (FAB; see Rose and Johnstone, 1982). Here xenon or argon radical ions exchange their high energy with gas molecules at thermal energy levels:

$$A^{\cdot +}(\text{fast}) + A(\text{thermal}) \rightarrow A(\text{fast}) + A^{\dot{+}}(\text{thermal})$$

The lower energy exchanged ions are deflected and the high energy argon atoms are directed onto the sample, which is dispersed at ambient temperature in a thin mobile film of glycerol so as to ensure a continuous supply of ions $(M + H)^+$. The molecular ion (+H) and a limited number of daughter ions can be detected in these experiments against a background of ions arising from the solvent; these include $\{(C_3H_8O_3)_n . H\}^+$ and their fragments.

It is to be expected that almost all lignan aglycones will give satisfactory mass spectra by the normal direct injection technique. In many instances these highly energised (*ca.* 70 ev) flexible molecules undergo extensive fragmentation and although their spectra show few significant daughter ions, these do serve to characterise the aromatic rings (as **6.7**) and the basic dimeric structure (as **6.8**). The accurate mass of the molecular ion may usually be determined and this knowledge limits the possibilities which need to be considered for the structure of any isolated phenolic fraction. Thus the lowest mass number likely to be encountered corresponds to the hypothetical structure **6.9**, whilst the upper limit for a lignan must be close to that of the structure **6.10**.

At this point in any investigation the mass spectrum will be of value in distinguishing a lignan from other closely related natural isolates. The flavonoids, whose acetate-biogenesis also involves arylpropane units, pro-

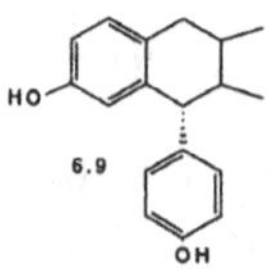

$ArCH_2^+$ **6.7**

$Ar{-}CH{=}CH{=}CH_2^+$ **6.8**

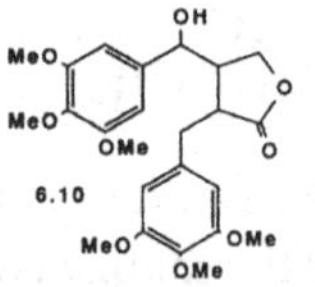

6.9 6.10

Scheme 6.3 Mass spectra of flavones & flavanones

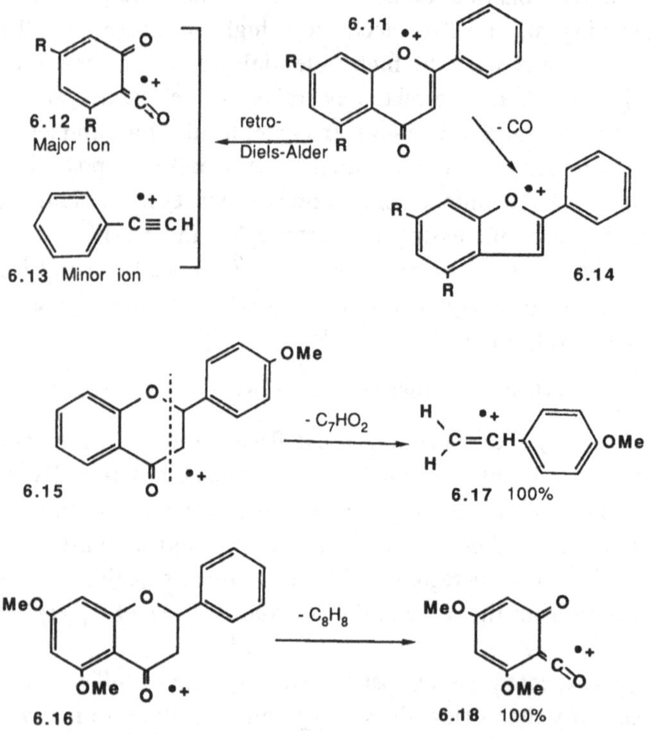

6.11 retro-Diels-Alder → 6.12 Major ion; 6.13 Minor ion

6.11 − CO → 6.14

6.15 − C_7HO_2 → 6.17 100%

6.16 − C_8H_8 → 6.18 100%

vide a relevant illustration and details of their mass spectra have been discussed (Mabry and Markham, 1975). These less flexible conjugated structures (e.g. **6.11**) give rise to intense molecular ions and so differ from most lignans. Fragments (**6.12, 6.13**) are formed by the retro Diels-Alder reaction and the carbonyl group is also lost as carbon monoxide to give **6.14**. The flavanone (**6.15**) fragments to give major C6-C2 daughter ions (e.g. **6.17**) and also (**6.12**), whilst the compound (**6.16**) gives (**6.18**) and (**6.13**) as major ions, in contrast to the behaviour of lignans. A noteworthy feature of these spectra is the preferred survival of the more highly oxygenated structural elements (**6.12, 6.17, 6.18**).

Diarylbutanes and diarylbutyrolactones

There have been relatively few studies of the former group of lignans where initial cleavage of the molecular ion occurs at the congested 8,8′-linkage. For example, austrobailignan—5 (**2.002**; Murphy *et al.*, 1975) gives a molecular ion of 326 m/z (40%) with a fragment of 163 m/z (5%). In this spectrum the base peak (135 m/z, 100%) arises from the methylenedioxybenzyl ion, but in the unsymmetrical lignan (**6.19**) this is less intense than the dimethoxybenzyl fragment (m/z = 151) which becomes the base peak. Other studies of the mass spectra of diarylbutanes have been reported for conocarpol (**2.005**) by Hayashi and Thomson (1975) and for norisoguaiacin (**2.300**) by Gisvold and Thaker (1974).

meso-Secoisolariciresinol (**6.20**, but *cis*-8,8′, $R_1 = R_2 = H$; Agrawal and Rastogi, 1982) gives a molecular ion (m/z = 362, 15%) with a minor peak (m/z = 181, 4%) arising from symmetrical cleavage, but the base peak arises from fragmentation of the 4-hydroxy-3-methoxybenzyl group (m/z = 137, $C_8H_9O_2$). In this diol the pattern diverges (Powell and Plattner, 1976) from that of the butanes as the loss of two water molecules precedes benzylic cleavage, to give a significant ion (**6.21**, m/z = 189, 11%).

In all these spectra the relative abundance of ions will vary with the operating conditions. It is worthwhile to compare the spectra (Table 6.3) obtained for (−)-seco-isolariciresinol (**6.20**; *trans*-8,8′, $R_1 = R_2 = H$) and the *meso*-isomer when care was taken (Andersson *et al.*, 1975) to achieve comparability. Mass losses due to CO (26) and to CH_3 can be dis-

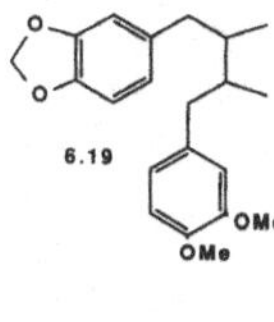

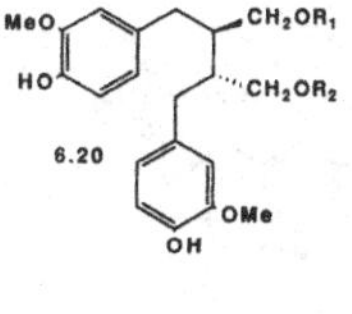

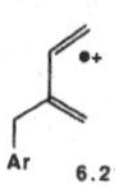

Table 6.3. *Principal MS fragments m/z (% abundance)*

meso-secoisolariciresinol		(−)-secoisolariciresinol	
M+ 362(12%)	137(100%)	M+ 362(22)	137(100%)
194(4)	122(10)	194(6)	122(16)
189(5)	94(8)	189(10)	94(12)
163(4)	91(4)	163(7)	94(12)
138)21)		138(38)	

tinguished in these spectra, but under these conditions no peak (m/z = 181) arises from initial symmetrical cleavage.

In the mixed diester of secoisolariciresinol (Powell and Plattner, 1976; **6.20**, *trans*-8,8′; R_1 = (+)-12-L-methyltetradecanoyl, R_2 = ferulyl) the molecular ion appears ($C_{45}H_{62}O_{10}$, m/z = 762, 7%) with prominent peaks due to the initial loss of the ferulyl group, to give an ion of m/z = 568 ($C_{35}H_{52}O_6$, 17%), followed by loss of the methyl-decanoyl group to give a fragment $C_{20}H_{22}O_4$ (m/z = 326, 25%). The latter is equivalent to the loss of two water molecules from the parent diol (m/z = 362 less 36). In these compounds benzylic cleavage is again preferred and gives rise to the base peak ($C_8H_9O_2$, m/z = 137).

A detailed mass spectroscopic analysis of the human metabolites enterodiol (**2.028**) and enterolactone (**2.049**) has been made (Setchell *et al.*, 1981).

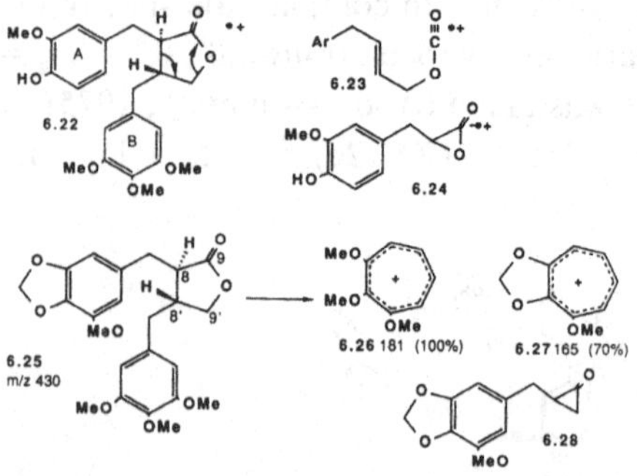

In diarylbutyrolactones benzylic cleavage characterises the aromatic residues and where these differ their relation to the carbonyl group may be established. Thus the lactone (**6.22**) isolated by Nishibe *et al.* (1981) from *Trachelospermum asiaticum* gave benzylic fragments of m/z = 137 and 181 from ring A and ring B respectively together with the alternative pair of cleavage products of structure (**6.23**). Formation of the ion (**6.24**) by the indicated bonding changes showed that the 4-hydroxy-3-methoxybenzyl radical was located at C-8. Similar criteria were used by Lopes *et al.* (1983) to orientate the substituents in four other butyrolactones isolated from fruits of *Virola sebifera.*

Badheka *et al.* (1986) have examined a group of six butyrolactones including unsymmetrically substituted compounds such as **6.25** and they identified the fragment ions **6.26** and **6.27**. The relationship of the methoxy-methylenedioxy ring to the lactone was confirmed by formation of the ion **6.28**. A similar family of fragment ions was obtained from the lignan isolated (Suzuki *et al.*, 1982) from *Wikstroemia indica* and comparison with earlier work (Nishibe *et al.*, 1981) led to assignment of a structure enantiomeric with (**6.29**; X = H; R_1, R_2 = Me). A detailed study of these fragmentation patterns has been made by Harmatha *et al.* (1982) and this covered the determination of the accurate masses of major ions including those with deuterium labels. Yatein (**6.29**; X = OMe, R_1, R_2 = —CH_2—) was selectively deuterated at C-8 by quenching the carbanion with deuterium oxide; fragmentation then gave the ion **6.30** (m/z = 265). The associated fragment of the structure which bore the methylenedioxy group was characterised by the unlabelled ion **6.31**. As a further refinement this work was extended (cf. **6.20**) to include the mass spectra of the diols obtained by reduction of yatein with lithium aluminium hydride and lithium aluminium deuteride. A symmetrical cleavage ion $(MeO)_3C_6H_2$—CH_2—CH—CX_2—OH (X = H, m/z = 225, 2%) was detected in one experiment and the *bis*-deuterated ion (X = D, m/z = 227) in the other. The methenylated fragment from this cleavage was not detected and the spectrum was richer in trimethoxylated ions in keeping with the formation of ions of higher oxygen number from flavones (p. 177).

In compounds like pluviatolide (**2.070**) the free phenolic —OH group can be labelled by measuring the spectrum in the presence of deuterium oxide (Corrie *et al.*, 1970). Labelling by this method (compare **6.29**), or otherwise through the conjugate base, should be undertaken whenever mass spectra are used to assign unsymmetrical structures.

In unsaturated lactones the nature of the benzyl cleavage is unequivocal; thus for chaerophyllin (**6.32**) the accurate mass of all major ions was measured (Mikaya *et al.*, 1981). Here the base peak ($C_{13}H_{13}O_4$) arose

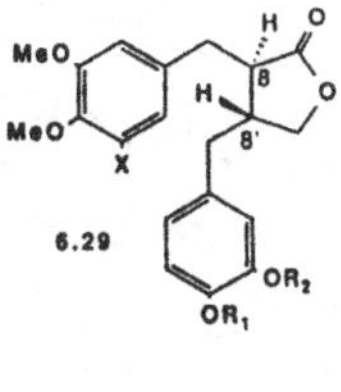

6.29 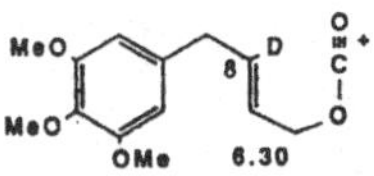 6.30 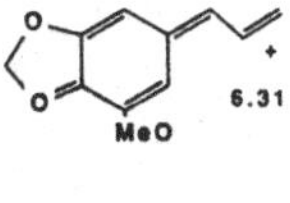 6.31

from cleavage (a) and the associated ion ($C_8H_7O_2$) was also intense, whereas the alternative benzylic fragment ($C_9H_{11}O_2$, m/z = 151) was very weak. These observations correlate with those of Fang *et al.* (1985a) on hibalactone (**2.099**) and calocedrin (**2.096**) and also with those of Turabelidze *et al.* (1982) on hibalactone and nemerosin (**2.102**). In this latter work accurate masses were reported for all major ions, including further evidence for the typical loss of formaldehyde from the methylenedioxybenzyl ion. The mass spectra of C-8-oxygenated compounds show a comparable pattern following elimination of this substituent. An example is provided by the work of Crombie's group on the angeloyl lignan helianthoidin (**6.33**; Burden *et al.*, 1969). As shown above (cf. **6.26** and **6.27**) unsymmetrical saturated lactones give benzyl cleavage ions in similar yield. However a clear distinction between the aryl groups in (**6.33**) was made by preparation of the deuterated analogue (**6.34**; X = D) from the derived alkene; this afforded the deuterated ion (**6.35**) and a deuterium-free ion ($C_9H_{11}O_2$).

Tetrahydrofurans (epoxylignans)

Lignans of type A (**6.36**) have been examined by several groups of workers; they undergo cleavage by two principal routes (Pelter, 1967)

Scheme 6.4 Mass spectra of unsaturated lactones & their precursors

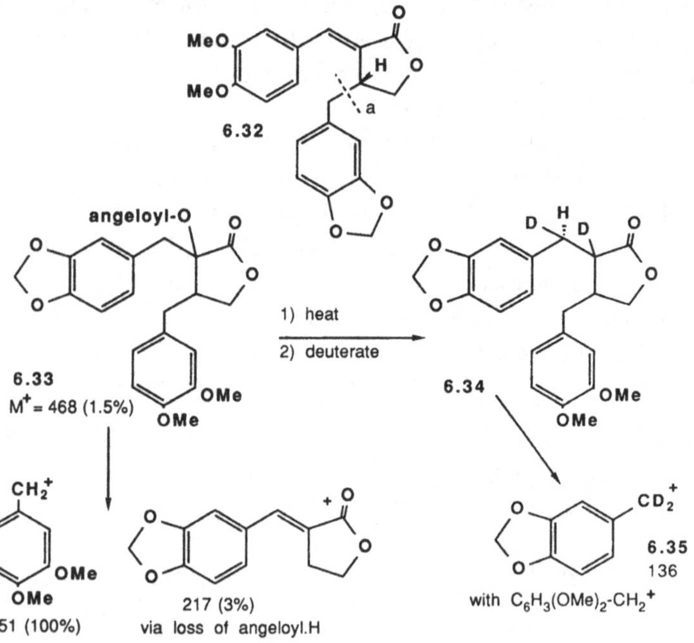

MeO
MeO
H
O
O
a
6.32
angeloyl-O
O
O
O
O
OMe
OMe
6.33
M^+ = 468 (1.5%)
1) heat
2) deuterate
D H D
6.34
CH_2^+
OMe
OMe
151 (100%)
217 (3%)
via loss of angeloyl.H
CD_2^+
6.35
136
with $C_6H_3(OMe)_2$-CH_2^+

where isomers such as galbelgin and galgravin (**6.36**; $Ar_1 = Ar_2 =$ 3,4-dimethoxyphenyl) exhibit similar spectra, illustrated for galgravin (Scheme 6.5). The molecular ion (M = 373) is weak (2%) with a stronger peak (M − H, 10%) and a base peak of m/z = 206, which is produced by the major cleavage. Subsequent fragmentation gave ions which typified the aromatic residues and the pathway to the dimethoxybenzyl ion (**6.39**) was established by the identification of a metastable peak. In these compounds and in galbacin (**6.40**) cleavage of the molecular ion also followed an alternative minor path.

These two types of cleavage were also found (Holloway and Schienman, 1974) for grandisin (**2.157**), but the molecular ion here was (**6.41**; Ar = 3,4,5-trimethoxyphenyl-) and it is possible that the balance between the cleavage modes was energetically biased by a preference for this trioxygenated ion.

No significant differences are to be found in the mass spectra of geometrical isomers. However it was argued by Takaoka *et al.* (1976) that (**6.42**) was the correct structure for machilusin, on the grounds that the base peak (**6.44**) was formed by preferential cleavage of its *cis*-2,3- bond rather than from cleavage of a less strained *trans*-linkage in the possible alternative structure **6.43**. In related work (McAlpine *et al.*, 1968) showed that calopiptin had the structure (**6.45**; X = H); here the base peak (**6.44**) arose by cleavage of the *trans*-substituted C8-C9 bond, whereas the *cis*-related

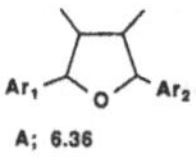

A; 6.36 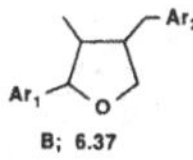 B; 6.37 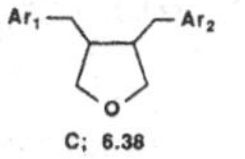 C; 6.38

dimethoxyphenyl group was included in a less abundant ion (**6.46**). This weakens Takaoka's argument concerning machilusin and one notes that McAlpine showed that the dimethoxyphenyl group bore a *cis*-relationship to methyl by comparison of the PMR spectra before and after selective nitration. In this work preferential monosubstitution of calopiptin was proved by the mass spectrum of the product (**6.45**; X = NO_2), which contained the unmodified ion (**6.44**, 85%) and another of m/z = 206. It was shown that the latter ion differed from the unmodified α-cleavage fragment (**6.46**), and that nitration had occurred, by determination of its accurate mass (206.0826, $C_{11}H_{12}NO_3$ requires 206.0817). The same group also reported the mass spectrum of veraguensin (**2.168**).

In tetrahydrofurans of type B the fragmentation pattern has been shown (Pelter, 1967) to change markedly in derived alcohols. Here oxy-substitution facilitates the elimination of an adjacent benzyl radical or ion to give fragments of the type found for di-*O*-methylolivil (**6.47** equivalent

Scheme 6.5 Cleavage of 7,7'-epoxylignans

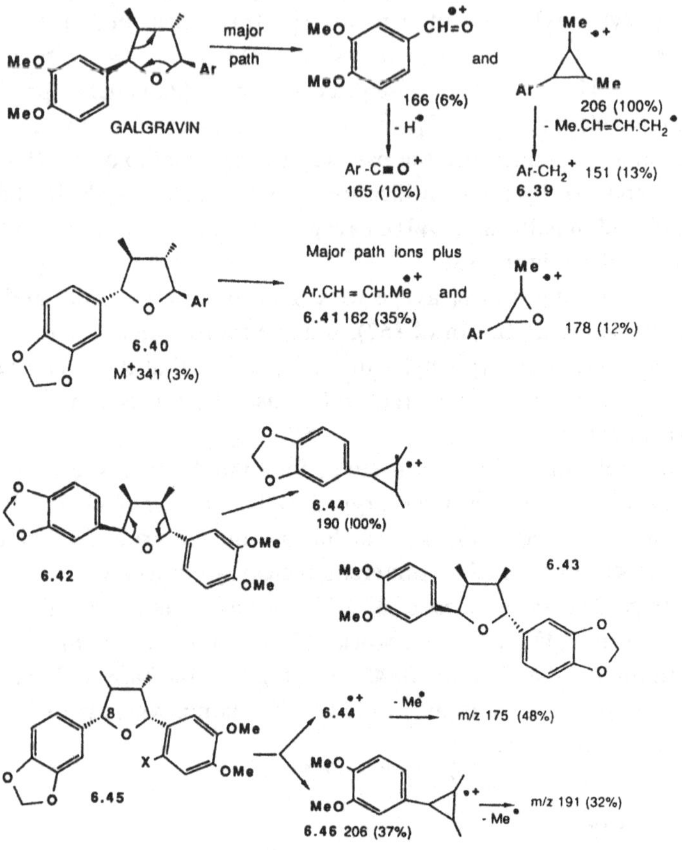

to dihydrogmelinol). In the primary alcohol taxiresinol (**6.48**; Majumder *et al.*, 1972) a minor peak arises from loss of a water molecule, but the hydroxymethyl group is characterised by a major ion ($C_{18}H_{20}O_5$; m/z = 316, 21%) formed by the loss of formaldehyde.

The mass spectrum of the type C lignan burseran (**6.49**) has been reported by Cole *et al.* (1969). This included the molecular ion and others typical of both oxybenzyl groups, together with one of mass 69 (C_4H_5O) which arose from the ring residue.

The structure (**6.50**) for dihydroneogmelinol was originally proposed (Birch *et al.*, 1964) on the assumption that neogmelinol had structure (**6.51**). It was subsequently shown (Birch *et al.*, 1967) that dihydroneogmelinol and dihydrogmelinol had identical mass spectra and that loss of a benzyl group (M − 151) occurred rather than loss of a α-hydroxybenzyl group (M − 167). It therefore followed that (**6.52**) was correct for

Scheme 6.6 Fragmentation of hydroxylated lignans

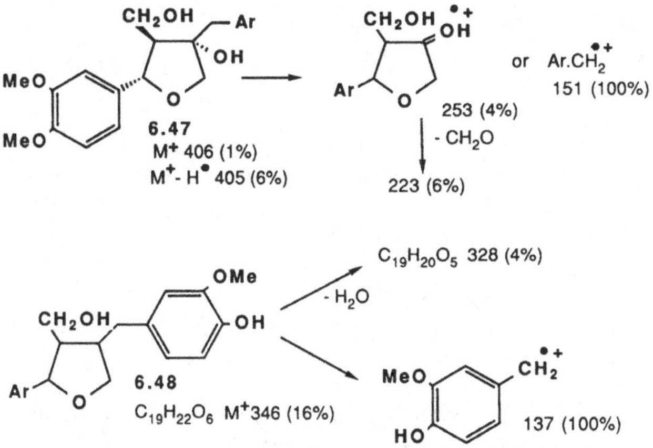

dihydroneogmelinol. Hence neogmelinol was not a position isomer (as **6.51**) of gmelinol (**6.53**; X = OH), but rather a geometrical isomer (**6.54**; X = OH; see discussion of the NMR, pp. 215–16).

Furofurans (bisepoxylignans)

A detailed study of this group (**6.53**, **6.54**; X = H, Ar = 3,4-dimethoxyphenyl) has been made by Pelter (1967). There have been difficulties in distinguishing between the typical 7,7′-substitution pattern and that of 7,9′-substitution as illustrated above for neogmelinol. At the time of writing no naturally occurring 7,9′-diarylfurofuran has been identified, although claims have been made. The confusion arose because of difficulty in distinguishing between the MS and PMR spectra of these alternatives – CMR spectra give reliable guidance (see pp. 211–12).

The molecular ions in these spectra are strong, possibly due to the rigidity of the bicyclic structure, and study of metastable peaks (Pelter, 1967) has shown that they fragment directly to daughter ions of types (**6.55**) and (**6.56**). The ions **6.55** and **6.57** are formed by 'vertical' cleavage and typify the basic C6-C3 unit as seen in the MS of the diarylbutanes. The peaks

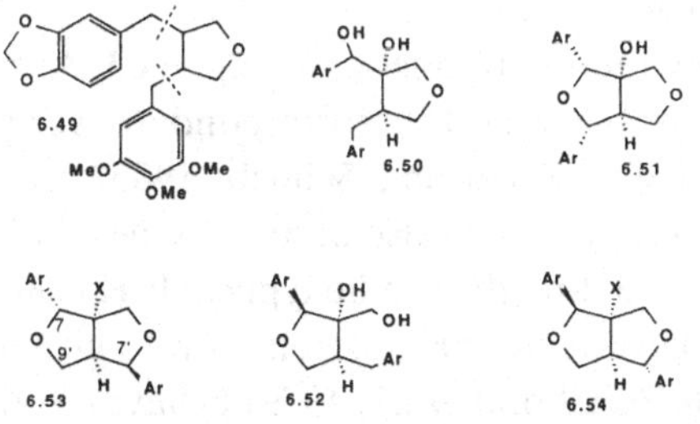

Scheme 6.7 Typical fragment ions in bisepoxylignan mass spectra

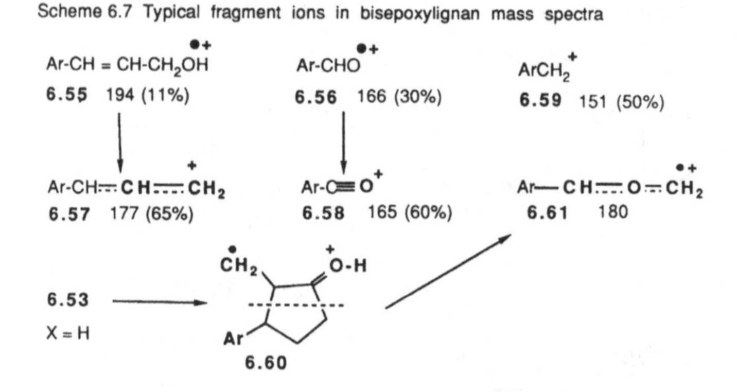

6.56 and **6.58** arise from α-cleavage and may be compared to similar fragments in furan spectra (e.g. galgravin, p. 181). Here also these ions characterise the aryl substituents and locate them next to ring oxygen atoms. These substituents may also be identified by the abundant benzyl ions (**6.59**).

The mass spectrum was important in assigning (Lavie *et al*., 1974; Rao and Lavie, 1974) the unusual mono-oxyarylated structure of ligballinol (**2.207**).

A minor peak (**6.61**) in these spectra arises from 'horizontal' cleavage. This type of fragment (**6.61**) is enriched in hydroxylated furofurans such as gmelinol (**6.53**; X = OH) where, as seen for dimethylolivil (**6.47**), stabilisation of a radical adjacent to the —OH group favours the precursor **6.60**. This also shows that the aryl substituents are located on different rings and excludes the possibility of 7,9′-arylation. The latter structure would be expected to generate a significant ion of the type Ar—CH—O—CH—Ar $^+$ (m/z = 316) which is not seen.

The mass spectra of some other hydroxylated furofurans have been determined recently (Tsukamoto *et al*., 1984a,b). In the spectrum of (+)-fraxiresinol (**6.62**; Ar_1 = syringyl, Ar_2 = guaiacyl) the molecular ion is most abundant and major fragments of both types **6.57** and **6.59** appear. The aromatic substituents are also characterised by peaks of type **6.56** and by horizontal cleavage (cf. **6.60**).

In furofuran acetals related to hydroxysesamin (e.g. **6.63**; Ar = guaiacyl) fragment ions of m/z = 163 and 166 correspond to vertical and horizontal cleavage respectively (Khan and Schoeb, 1985) and loss of water from the molecular ion (m/z = 374) also occurs. In the related structure of sesamolin (**6.64**; Ar = 3,4-methylenedioxyphenyl; Haslam, 1970) the aryloxy residue is eliminated from the molecular ion to give **6.65** and the *bis*-acetal simplexolin (**6.66**; Ghosal *et al*., 1979) behaves in the same way. The mass spectrum of sesamolin includes an ion (m/z = 203) formed by the loss of formaldehyde from **6.65** and the analogous aryloxyion (m/z = 219) occurs in the spectrum of simplexolin. In both of these spectra the eliminated aryloxy group is characterised by the sesamol ion (**6.67**). A further discussion of hydroxyl substitution effects on the mass spectra of furofurans has been published (Pelter and Ward, 1978).

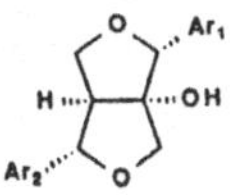

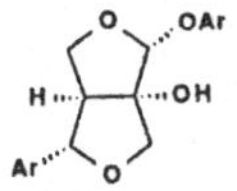

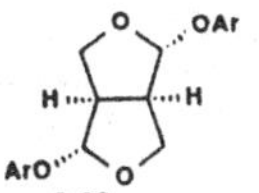

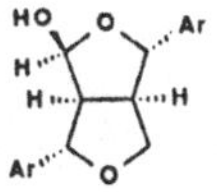

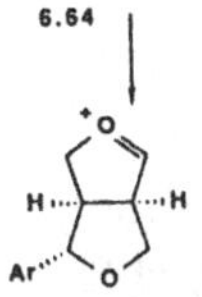

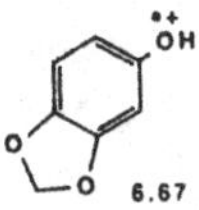

Oxofurofurans

A small group of these lactones exists and structural assignments were initially confused by reports (Lavie *et al.*, 1974; Brieskorn and Huber, 1976) of a basic 7,7′-diaryl structure. Both these reports have now been revised as a result of reinterpretation of their mass spectra together with other spectral analyses.

The germination inhibitor isolated (Lavie *et al.*, 1974) from *Aegilops ovata* L. was assigned structure **6.68** and both (+)-aptosimon (**6.69**; X, Y = O) and (+)-aptosimol (**6.69**; X, Y = H, OH) from *Aptosimon spinescens* were also regarded (Brieskorn and Huber, 1976) as being 7,7′-diarylated. In both instances the ions [M − 84(85)] were taken to characterise ring B and to arise from the molecular ions by cleavage of the lactone ring. Lavie later (see Cooper *et al.*, 1979) revised his earlier assignment to **6.70** following synthesis of the model compound **6.71**, which gave a daughter ion ($C_{17}H_{16}O$) by cleavage route (a) and another (C_7H_5O) by route (b). This model also gave rise to the C_9H_9 ion by vertical cleavage. Comparable behaviour was shown by the synthetic lactone **6.72** and here the differently substituted aryl groups were distinguished by production of the ion **6.73** by vertical cleavage. Loss of carbon dioxide (route (a) above) and α-cleavage (route (b), **6.71**) were also significant but in neither model was loss of the lactone ring observed. NMR experiments confirmed this structural revision and synthetic work by Stevens and Whiting (1986) also bears on this point and on the revision of structures **6.69** for aptosimon and aptosimol.

Work by Ululeben *et al.* (1978) has shown that the anti-tumour agent styraxin (**6.74**) also possesses the indicated diarylation pattern since the relative positions of these substituents were placed by identification of a vertical cleavage fragment $C_{10}H_9O_2$ (m/z = 161), with daughter ions derived from this by the loss of formaldehyde and carbon monoxide. The relationship of the guaiacyl residues to the heterocyclic oxygen atoms was indicated by the ion $C_7H_7O_2 . C{\equiv}O^+$ and the whole assignment was confirmed by an X-ray determination.

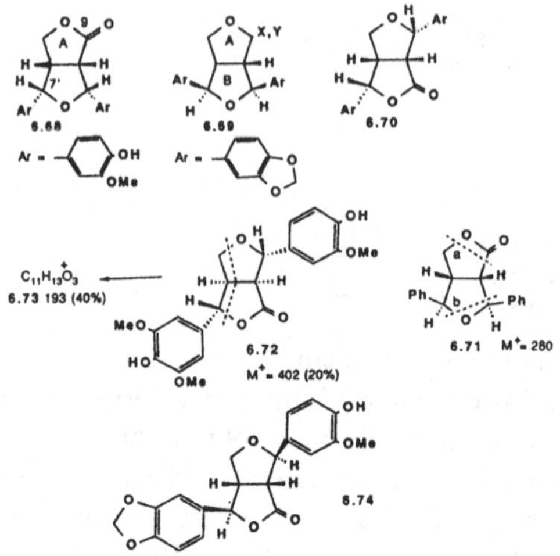

Aryltetralins

An early study was made by Pelter (1968) who showed that the retro-Diels-Alder reaction was a feature of spectra of compounds such as galbulin (**6.75**) and galcatin (**6.76**; R_1, R_2 = —CH_2—; R_3, R_4 = Me). Both these spectra included strong M^+ ions (67 and 55% respectively), with peaks arising from the loss of C_4H_8 (or C_4H_8 + H), but where the base peak (M − 87) was derived by the loss of C_4H_8 together with 31 m.u. The spectrum of isogalcatin (**6.76**; R_1, R_2 = Me; R_3, R_4 = —CH_2—) differed from that of galcatin in that the molecular ion now formed the base peak and the loss of C_4H_8 (56 m.u.) or C_4H_8.H (57 m.u.) was more pronounced: significant ions were also formed by the combined loss of 56 + 31 m.u. (OMe) and 56 + 30 m.u. (CH_2 = O). This showed that both rings A and C were involved in these early fragmentation steps.

It has been proposed (Scheme 6.8; Pelter, 1968) that a cyclo-addition follows the retro-Diels-Alder reaction to give **6.77**, which achieves the more stable even-ionic structure **6.78** by migration of hydrogen followed by ejection of the methoxyl radical.

In lignan primary alcohols of this class, such as dimethyl cyclolariciresinol (**2.317**) and pygeoresinol (**2.334**; Chandel and Rastogi, 1980) the

Scheme 6.8 Fragmentation of aryltetralins

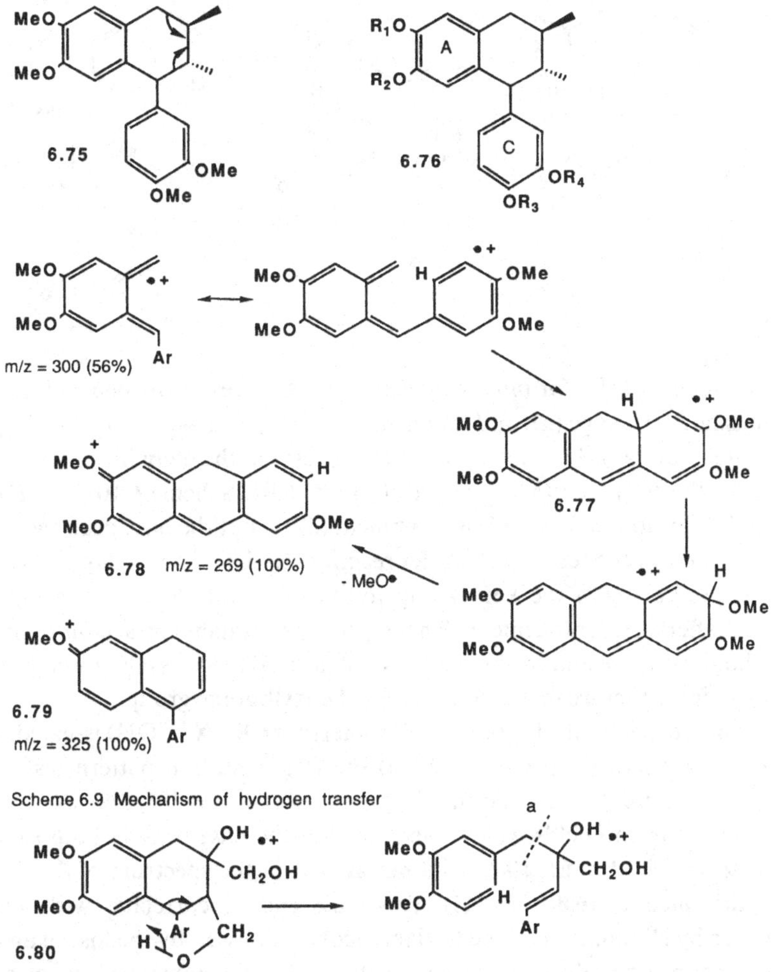

6.75

6.76

m/z = 300 (56%)

6.77

6.78 m/z = 269 (100%)

- MeO•

6.79

m/z = 325 (100%)

Scheme 6.9 Mechanism of hydrogen transfer

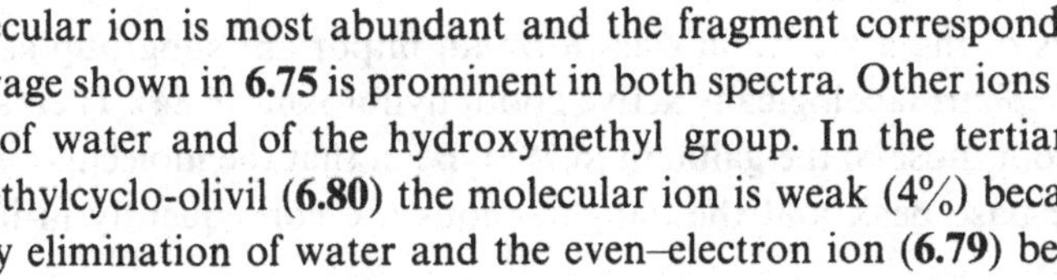

6.80

molecular ion is most abundant and the fragment corresponding to the cleavage shown in **6.75** is prominent in both spectra. Other ions arise from loss of water and of the hydroxymethyl group. In the tertiary alcohol dimethylcyclo-olivil (**6.80**) the molecular ion is weak (4%) because of the ready elimination of water and the even–electron ion (**6.79**) becomes the base peak after the loss of both hydroxymethyl groups (cf. p. 177). Fragments corresponding to **6.79** give rise to weak peaks in the spectra of dimethylcyclolariciresinol and pygeoresinol.

Other features of lignan alcohol spectra are those which follow hydrogen transfer from a hydroxyl group to an aromatic ring; Scheme 6.9 shows

Scheme 6.10 McLafferty rearrangement

6.81

6.82
M^+ 356 (7%)

m/z = 194 (100%)

and

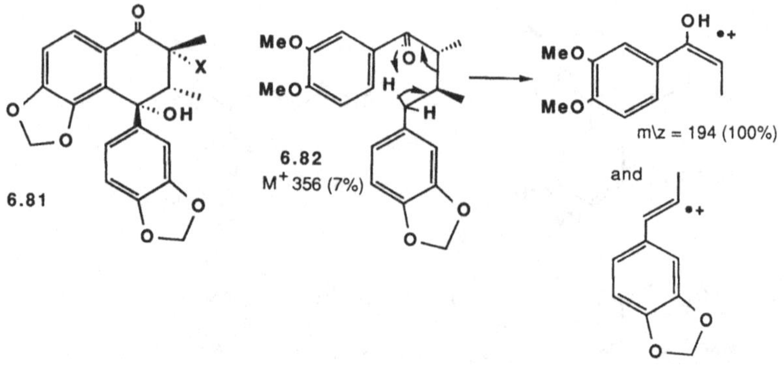

one such transfer for dimethyl cycloolivil. As a result the benzyl ion is intensified (cleavage (a)) and fragments of the type Ar—CH—CH—CH_2 are more abundant than those seen in the spectra of the galbulin subgroup.

In the tetralone (Martinez *et al.*, 1985; **6.81**, Scheme 6.10; X = H) the molecular ion (m/z = 386) is prominent (37%) but the base peak (M − 56) is formed by the retro-Diels-Alder reaction. In the co-occurring open chain compound (**6.82**) the keto-group initiates central cleavage through the McLafferty rearrangement. These principal fragments account for the whole structure and the mass spectra of derived ketones could well be used to assign structure to alcohols of the dibenzylbutane group.

In a related pair of geometrical isomers (of **6.81**; X = OH) isolated from *Virola sebifera* (Lopes *et al.*, 1984) the fragmentation pattern was completely different. The molecular ion was weak (m/z = 386; 8%), as was that resulting from Dields-Alder reversion, and the base peak in both isomers (m/z = 325; M − 61, 100%) did not appear in the spectrum of the monohydric alcohol (**6.80**; X = H). This mass change correlates with loss of water by elimination of one tertiary alcohol followed by the loss of an acetyl group; a possible mechanism for this which generates the even-electron ion (**6.83**) is shown below.

Lactones of the aryltetralin class form an important subgroup as this includes the pharmacologically active podophyllotoxins (**6.84**). Their spectra differ from those of the galbulin (**6.75**) type in that the molecular ion is usually the base peak and the daughter ions are consequently in lower

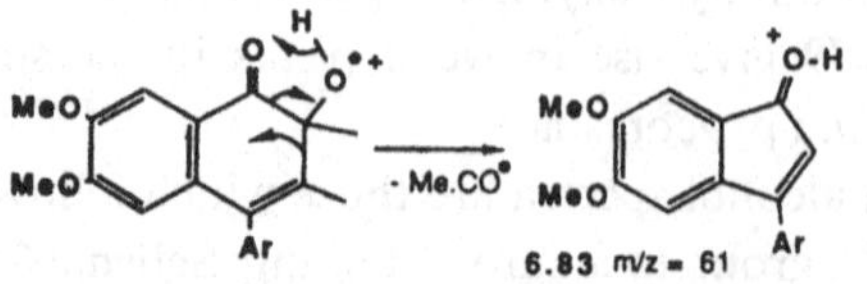

- Me.CO•

6.83 m/z = 61

Table 6.4. *Principal ions (%) in the mass spectra of some analogues of podophyllotoxin*

6.84 variation	M + 1	M	M − 18	M − 115	Ring C.H.
R1 = H; R2 = H; 4β—OH	18	59	77	—	44
R1 = H; R2 = H; 4α—OH	31	100	—	—	14
R1 = H; R2 = Me; 4α—OH	23	74	28	10	26
R1 = Me; R2 = Me; 4α—OH	24	100	10	—	14

H,OH 9 7 7' O O O O C
6.84 MeO OR₂ OR₁

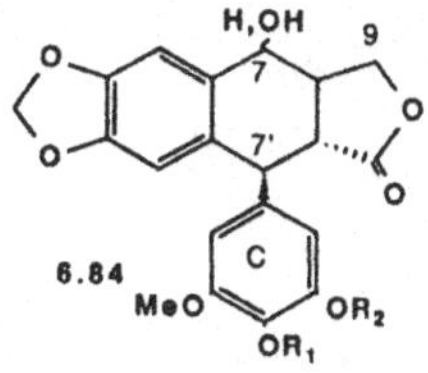

abundance. In 7-deoxypodophyllotoxin and the peltatins (**2.354**, **2.355**) low-abundance ions [M − (84 + H)] and [M − (84 + OMe)] arise from the retro-Diels-Alder reaction. In methyl conidendrin an abundant ion [M − (84 + OMe), 44%] is formed by the retro-reaction; in conidendrin (**2.348**) itself the corresponding ion (m/z = 241) has an abundance of 18% (Pelter, 1968; Duffield, 1967).

Benzyl ions characteristic of both aromatic rings occur in all these spectra, although it is not immediately obvious how they arise since both aryl groups are doubly-linked into the carbocyclic skeleton. It is possible that this type of cleavage is initiated by hydrogen atoms associated with the molecular ion. These ions (M + 1) and (M + 2), commonly occur at greater abundance than those calculated for 13C nuclides (20% and 2% for a 20-carbon compound including one or two carbon nuclides respectively).

In podophyllotoxins elimination of the benzylic —OH group often generates an (M − 18) ion and this alternative pathway reduced the intensity of the molecular ion. The extent of this component is variable as is shown by reference to the data (**6.84** and Table 6.4; Ritchie, 1986) for some analogues. One notes especially the differences in the records for the epimeric pair which shows that these spectra can be misleading when detailed structural assignments are sought.

Electrophilic substitution in podophyllotoxins occurs in ring C at C-2′ and the mass spectra of these hindered derivatives may vary considerably from those of the natural precursors. This is apparent in the fragmentation of 2′-bromoepipodophyllotoxin (Table 6.5 and structure **6.85**; Ayres and Lim, 1972b), which still has the molecular ion as the base peak

Table 6.5. *Principal ions in the fragmentation of 2′-bromoderivates*

6.85 var	M	M − ROH	M − Br	M − Br − ROH
R═H	494(100%) 492(100%)	476(2) 474(2)	413(24)	395(11)
R═Ac	536(92%) 534(95%)	476(50) 474(40)	455(5)	395(100)

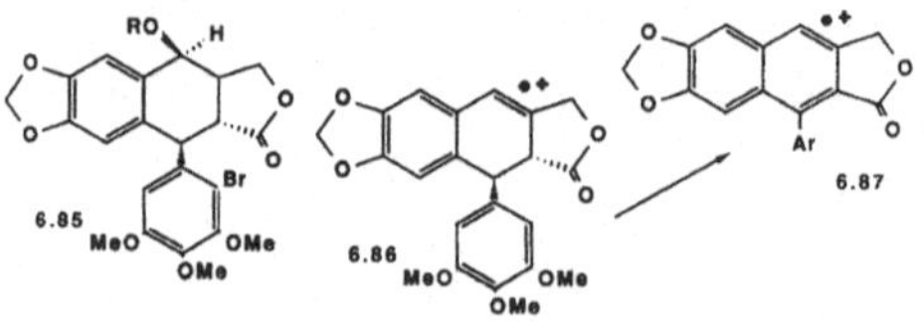

6.85 6.86 6.87

but the loss of water is now only a minor pathway. However, when C-4 elimination is enhanced by acetylation it becomes dominant. The ion (**6.86**; m/z = 395) formed by the elimination of acetic acid and the loss of bromine provides a link to the arylnaphthalenes, in that further loss of two H atoms gives an ion (**6.87**) common to that obtained by the fragmentation of these fully aromatic lignans.

Arylnaphthalenes

An early study (Burden *et al.*, 1969) of helioxanthin (**6.88**; X = H) showed that fragmentation was initiated by loss of one methylenedioxy group with the apparent transfer of a hydrogen atom (M − 29, 16%). A sequence was then followed in which oxygen atoms in association with single carbon atoms were lost in turn – [(M − 29 − 28, 4%; M − 57 − 58, 6%) etc]. The ions and associated transitions were characterised by accurate mass measurements and detection of metastable peaks; a similar result has been recorded (Ghosal *et al.*, 1979) for prostalidin C (**6.88**: X = OH). The sequence also resembles that of the ion (**6.86**) found in the fragmentation of podophyllotoxin derivatives.

A distinctive feature of these mass spectra is the stability of the molecular ion, which is normally the base peak, leading to a distribution of daughter ions of relatively low abundance. The presence of doubly-charged ions is another feature which may extend to fragments. For example, Kuo *et al.* (1985) observed that the spectrum of taiwanin H (**6.89**; X = H), like that of plicatinaphthol (**6.89**; X = OH; MacLean and MacDonald, 1969), included such ions in 4–6% abundance in addition to M^+.

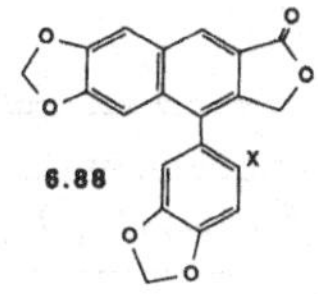

6.88

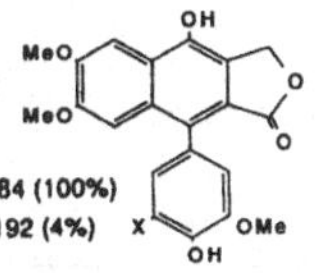

6.89
$M^{+} = 384$ (100%)
$M^{++} = 192$ (4%)

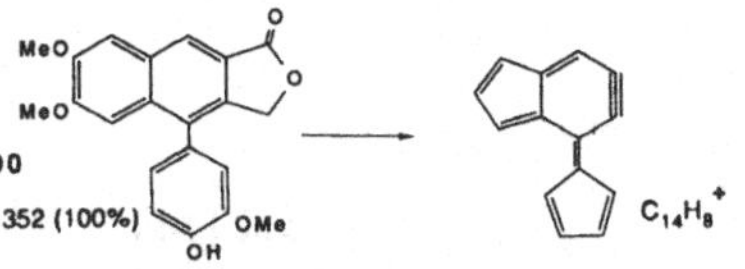

6.90
$M^{+\bullet}$ 352 (100%)

$C_{14}H_8^+$

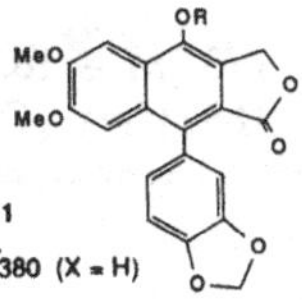

6.91
$M^{+\bullet}$ 380 (X = H)

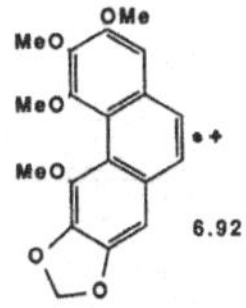

6.92

A significant fragment was that arising from loss of the lactone ring (M − 58) and this is detectable as an early fragment in all these spectra. This ion was prominent in the spectrum of detetrahydroconidendrin (**6.90**) but loss of methylenedioxy groups or of a C4-oxy group may take precedence.

In plicatinaphthol and taiwanin H the hydroxyl group is lost with transfer of a hydrogen atom to give an abundant ion (M − 16, 18%); fragmentation of the lactone is also seen (M − 58, 6%; M − 89, 12%). The naphthyl ether justicidin A (**6.91**; R = Me, Khalid, 1981) fragments with the loss of 43 mass units (C + OMe) and of the lactone ring to give an ion (m/z = 293, 18%) also common to the spectrum of diphyllin (**6.91**; R = H).

Dibenzocyclo-octadienes

There has been no specialised study of the spectra of this group and the method has been used principally to characterise individual compounds by accurate mass measurement. Typical examples are those of wuweizisu C (**2.473**) and gomisin R (**2.447**; Ikeya *et al.*, 1982b), which show the molecular ion as the base peak. The many minor fragment ions include those at (M − 56) which probably arise from retro- Diels-Alder cleavage. Similar results have been reported recently (L-Niang *et al.*, 1985) for five lignans from *Kadsura coccinea* including gomisin E (**2.430**). The most detailed available data was published by Ikeya and colleagues (1979*d*). This included mass spectral characterisation of the side-chain conjugate of gomisin D and a scheme for fragmentation of the gomisins B, C, and F to the intense phenanthryl radical ion (**6.92**).

Table 6.6. *Field desorption mass spectra*

Peracetyl pinoresinol glucoside (M = 730, $C_{36}H_{42}O_{16}$)			Pinoresinol glucoside (M = 568, $C_{26}H_{32}O_{11}$)
753 [M + Na]$^+$	70%		
730 [M]$^+$	100		
711 [M + Na − 42]$^+$	50		
688 [M − 42]	46		
		8%	520 [M]$^+$
399 [M − 331]$^+$	38		
331 [$C_6H_{11}O_5$.4Ac]$^+$	44%		
		23	179 [M − 163 + H]$^{2+}$
		9	163 [$C_6H_{11}O_5$]$^+$

Lignan conjugates

The involatility of polar compounds like the lignan glycosides prevents the direct application of electron-impact spectroscopy, but some use has been made of low-energy methods of excitation (p. 175). Another approach is to block hydrogen-bonding groups by acetylation or trimethylsilylation. These procedures not only increase the volatility of the sample, but tend also to enhance the stability and hence the yield of the molecular ion.

Field desorption was used (Kudo *et al.*, 1980) for the first time in lignan chemistry to characterise glucosides of pinoresinol (**2.217**), olivil (**2.129**), and the *bis*-glucoside of syringaresinol (**2.229**) obtained from the fruits of *Ligstrum japonicum*, which are known as 'Jotei' in Chinese pharmacy. It is of interest to compare these results for pinoresinol glucoside with those obtained for the peracetylated derivative (Table 6.6).

It can be seen that the spectrum of the acetyl derivative is richer in higher-mass ions including those which carry Na^+ ions when desorbed from the matrix. Acetyl groups are cleaved with loss of ketene ($CH_2{=}C{=}O$, M = 42) and transfer of a hydrogen atom. The glucosyl residue is characterised by the ion (**6.93**; R = Ac, m/z = 331) and the complementary cleavage fragment (399, M − 331) is also present. Some information was also obtained from the spectrum of the unmodified pinoresinol glucoside, where a prominent feature was the abundant doubly-charged ion of the aglycone [m/z = 179, $(M - 162)^{2+}$]; the glucosyl fragment gave the ion (**6.93**; R = H, m/z = 163).

FAB mass spectra have been obtained (Ayres and Evans, 1987) for a series of variants of Etoposide (p. 113). For example epipodophyllotoxin cellobioside (**6.94**; β-$OR_1 = C_{12}H_{21}O_{11}$, R_2 = Me; $C_{34}H_{42}O_{18}$, M = 738) gives a molecular ion (23%) with $[M + 1]^+$ of 29% abundance, whilst the base peaks corresponds to the aglycone (M − ROH). This spectrum also included ions of mass (397 + H; 49%) and (397 + 2H; 26%) and others corresponding to that of the free lignan at M − 324. The hepta-acetyl derivative of epipodophyllotoxin cellobioside ($C_{48}H_{56}O_{25}$; M = 1032) gave a more intense molecular ion (74%) with (M + 1) as the base peak. Loss of the carbohydrate fragment from the latter accounted for a peak (m/z = 414) characteristic of the lignan.

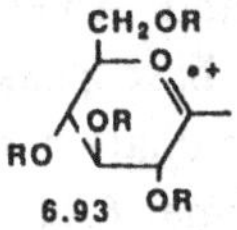

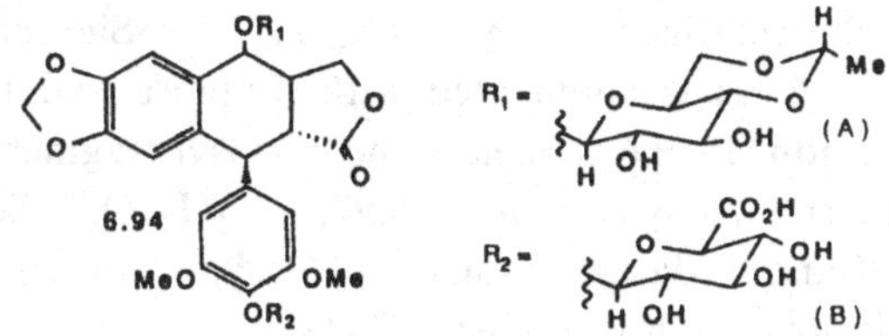

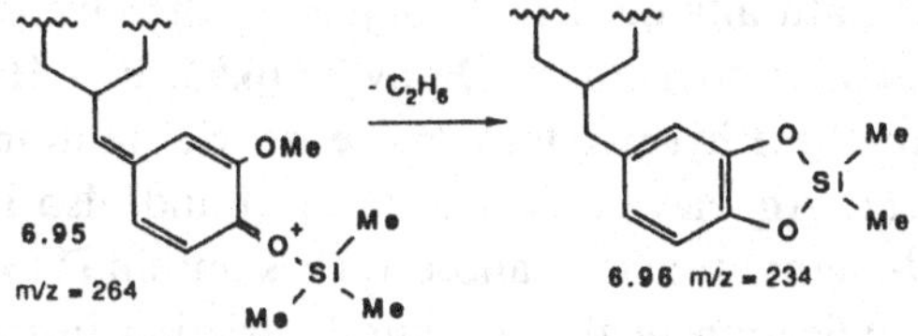

FAB spectra have also been obtained (Holthuis, 1985) for Etoposide itself and also for Teniposide (p. 113). In this work the glucuronide (**6.94**; R_1 = A, R_2 = B) was identified in urine as a metabolite of Etoposide, which amply demonstrates the potentialities of the FAB technique in this context.

The electron-impact spectrum of Etoposide was obtained by Colombo *et al.* (1985) after silylation during a study of the metabolic pathway in perfused liver. Here the formation of a fragment with m/z = 455 (Lignan + R2 = Me_3Si—) pointed to modification of the glycosyl residue, with the phenolic —OH group free to accept the silyl group. Silylation in addition to marking free hydroxyl groups can also provide information about their orientation. Thus Le Quesne *et al.* (1980) showed that the daughter ion (**6.95**) from silylated nectandrin (**2.165**) lost ethane through the interaction of neighbouring methyl groups to give (**6.96**).

Acetylation has been widely used to improve the yield of ions from electron-impacted molecules. One example which demonstrates its value for highly polar substances is given by Deyama (1983) who derivatised medioresinol di-*O*-β-D-glucoside (cf. **2.269**) as its octa-acetate.

In tuberculatin (M = 512, $C_{26}H_{24}O_{11}$; Sheriha and Amer, 1984) diphyllin (**6.91**) is conjugated with D-apiose and the base peak is the diphyllin ion. In the triacetate the glycosyl fragment is characterised by the triacetylapiosylium ion (**6.97**, $C_{11}H_{16}O_7$; Scheme 6.11, m/z = 259.081) and its daughter ion (**6.98**; $C_7H_7O_3$, m/z = 139.038); the diphyllin ion is now a minor fragment (34%).

In some naturally occurring conjugates the glycosyl hydroxyl groups are blocked as methyl ethers. Diphyllin (**6.91**; R = H) occurs conjugated with 3,4-di-*O*-methyl-D-xylopyranose as cleistanthin A (**6.91**; R = A, Scheme 6.11; Anjaneyulu *et al.*, 1981a,b) and also in conjugation with 2,3,5-tri-*O*-methyl-D-xylofuranose as cleistanthin D (**6.91**; R = B, Scheme 6.11). An indication of the structural variation in these two conjugates (structures A and B) came when the electron-impact spectrum of cleistanthin D was compared with that of fully methylated cleistanthin A. The molecular ion of cleistanthin D has 8% abundance compared to only 1% in methyl cleistanthin A, and the aglycone fragment (m/z = 380.81%) is no longer the base peak in cleistanthin D. Both spectra include the ion (m/z = 293). An abundant ion in the spectrum of cleistanthin D has m/z = 175 (69%) which arises from the trimethylxylosyl fragment ($C_8H_{15}0_4$) and this loses methanol to generate the base peak (m/z = 143) – both these ions are prominent in the spectrum of methyl cleistanthin A.

Further correlation was forthcoming from related work by Sastry and Rao (1983). They showed that the mass spectrum of the tetraacetate of

Scheme 6.11 Fragmentation of diphyllin conjugates

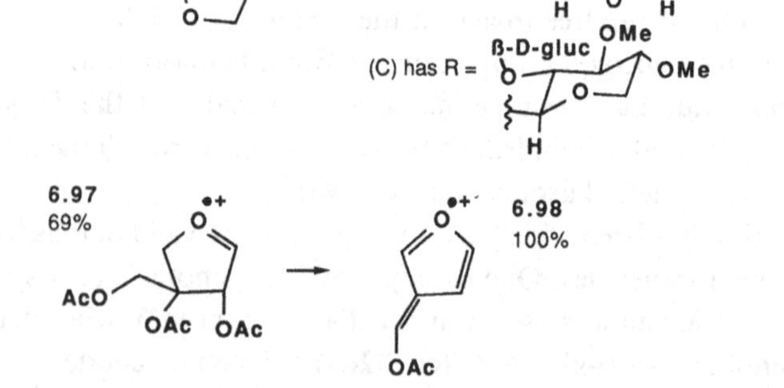

cleistanthanoside A (**6.91**; R = β-D-glucosyl-β-3,4-di-*O*-methylxylopyranosyl $C_{34}H_{38}O_{16}$, C in Scheme 6.11) included similar ions – the aglycone fragment (m/z = 380) had an abundance of 61% and the base peak (m/z = 169) was again derived from the glycosyl group.

It is apparent that mass spectra give valuable information which typifies lignans and gives insights into their structure. However in determining the geometry and the finer details NMR experiments are of first importance.

Nuclear magnetic resonance spectra

As these spectra are particularly sensitive to structural variations (Kemp, 1986) substantial differences exist between the different classes of lignans. For this reason and also to correlate best the volume of data the discussion is based upon class subdivision.

(i) *Introductory remarks*

Apart from the long-established technique of saturation decoupling some use has been made in lignan chemistry of the INDOR method (Kowaleski, 1969). This is well suited to the problem which arises when critical signals from aliphatic protons are obscured by the strong peaks which arise from the ether groups. There has to date been only limited publication of two-dimensional lignan spectra (Morris, 1986) but these are on the increase and examples are included (pp. 211, 226).

The nuclear Overhauser effect (Noggle and Schirmer, 1971) which depends on the enhancement of absorption of energy from an irradiated proton by others in close proximity was little used in structural evaluation until recently. This arose from the rapid attenuation on separation as the effect is inversely proportional to the sixth power of the internuclear distance; hence the neighbouring signals are often increased by only a small proportion of their original intensity. With the advent of difference spectroscopy (Sanders and Hunter, 1987) this turns out to be a blessing in disguise. In this adaption computerised control of data aquisition makes it possible to collect the two spectra under identical conditions. If that obtained on irradiation is subtracted from the original, small enhancements in the range of a few per cent can be assigned with certainty and the technique is now widely used.

(ii) *Aromatic substitution patterns*

3,4-Dioxysubstitution is the most common (**6.99**) and the aromatic peaks then fall within the typical range (Table 6.7) as a multiplet arising from an ABX system (see Kemp, 1986). Integrated intensities

Table 6.7. *Typical chemical shifts (δ values*) of protons located in lignans*

Proton type	Chemical shift
Ar—H	6.5–7.2
=C—H	6.6–7.5
O—CH_2—O	6.0
Ar—C(OAc)—H	5.7–6.0
Ar—C(OH)—H	4.4–4.9
Ar—O—H	3.5–4.5**
Ar—C(Ar)—H	3.8–4.6
Ar—O—CH_3	3.8–4.1
Ar—CH_2—OH	3.4–3.6
Ar—CH_2—OC=O	3.8–4.1
Ar—CH_2—R	2.2–2.7
—CH—CO_2R	2.6–3.2
R—C(H)	1.7–2.1
R—CH_3	0.9

* In $CDCl_3$ (for $CHCl_3$ δ = 7.25).
** This will vary with concentration and temperature and when exchange is rapid the signal will be lost.

define the number of methyl ether groups whilst a methylenedioxy group falls at distinctly lower field (Table 6.7) as a broad singlet or occasionally as a magnetically non-equivalent AB pair. On substitution of veratryl types, for example by nitration (McAlpine *et al.*, 1968), the spectrum is simplified as two *para*-coupled (*ca.* 0.5 Hz) deshielded peaks with that *ortho*- to the nitro-group at lowest field (δ = *ca.* 7.6). The 3,4,5-trimethoxyphenyl group is characterised as a singlet (2H) arising from equivalent protons; when the adjacent proton (7H) is irradiated these protons at 2,6-H can be identified by decoupling or by NOE (p. 199).

A free phenolic —OH group may give a broad peak in deuterochloroform at normal temperature and in dilute solution (<5% wt/vol) in the range δ = 3.5–4.5. This is sensitive to H-bonding (Vieira *et al.*, 1983) and will shift on the addition of small amounts of a polar solvent such as acetone and will be eliminated by exchange with D_2O. The number of protons *ortho* to a phenolic —OH group may be determined by heating with D_2O, when over several hours these C—H protons will exchange and their signals will also be lost. This procedure was used (Hayashi and Thomson, 1975) to show that conocarpol (**2.005**) was a *bis*-phenol with four exchangeable protons. Acetylation of phenolic —OH groups introduces a marking methyl group ($\delta = 2.0$) and any *ortho*-hydrogen atom is deshielded. In alcohols the —OH proton resonance appears near $\delta = 9.0$ but it is also liable to be lost through intermolecular exchange.

(iii) *Dibenzylbutanes and dibenzylbutyrolactones*

There have been few studies of the NMR spectra of lignans of type (**6.100**, Y = H). Assignment of most of the structural features is straightforward (Table 6.7) with terminal methyl groups appearing at $\delta = 0.8$ (J = 6 Hz) and benzyl protons in the region of $\delta = 2.2$–2.7. Some of these compounds, such as conocarpol, 2′-methoxyconocarpol (**2.007**), and dihydroguaiaretic acid (**2.009**) occur as *meso*-forms but the majority are asymmetrical and hence optically active. The chemical shifts of the pair of methine protons occur typically at $\delta = 1.7$–2.1 but isomers of this type cannot be distinguished by inspection of the spectrum. This is partly on practical grounds since in molecules of this size solubility is limited and the details of these broad multiplets cannot be seen. It should also be noted that in diastereoisomeric diols (**6.100**, Y = OH) at 100 MHz the racemic mixture gave a multiplet (8,8′-H) of width 20 Hz centred on $\delta = 1.90$ whereas the corresponding signal in the *meso*-isomer had a width of 30 Hz (Setchell *et al.*, 1980). It is not possible to account for these relative signal widths using the first-order approximation (Emsley *et al.*, 1965), although attempts have been made in the past to assign 'typical' coupling constants on this basis.

As a working rule the first-order character is likely to be distorted whenever the chemical shift difference is less than five times the coupling constant; in these circumstances second-order character develops – the number of peaks exceeds that predicted from the Pascal triangle and intensities also diverge from those of first-order spectra. Other examples of this problem of assigning peaks in these spectra are to be found in the work of Murphy *et al.* (1975) on austrobailignans-5 and 6 (**2.002, 2.003**) and in that of Anjaneyulu *et al.* (1973) on phyllanthin (**2.034**) and niranthin (**2.032**). The structures of this latter pair were written in error as *meso*-forms although both are optically active. Nordihydroguaiaretic acid (**2.011**) provides a reliable reference in this class of lignans in view of the X-ray analysis of its dibromoderivative by McKechnie and Paul (1969),

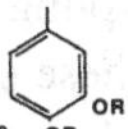

6.99 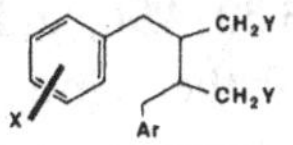 6.100 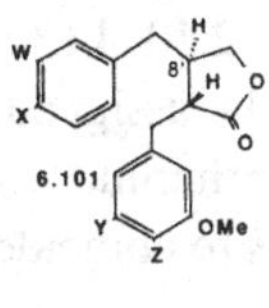 6.101

Table 6.8. *Chemical shifts (δ) of aliphatic protons in lactones*

trans-(Cooley)		*cis*-(Harmatha)
benzyl 7′α	2.500	2.33
7′β	2.610	2.95
benzyl 7α	2.906	2.76
7β	3.000	3.25
C-8′	2.498	2.66
C-8	2.590	3.07
lactone 9′α	3.868	4.04
9′β	4.135	4.06

Scheme 6.12 Preferred conformation of butyrolactones

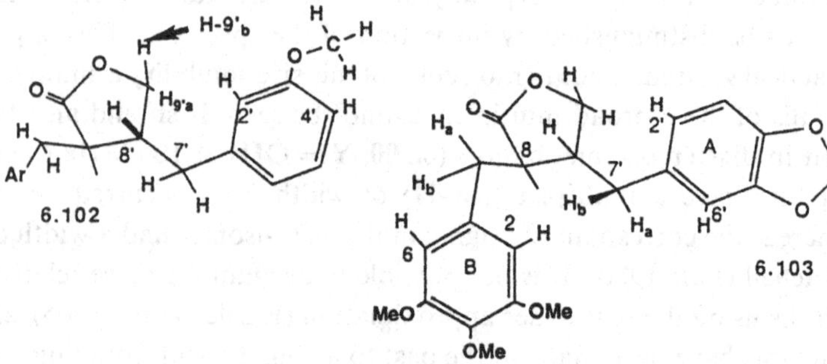

which established that it was a *meso*-compound and not a racemic mixture.

In the related benzylbutyrolactones errors of interpretation have also been made owing to unawareness of second-order character in the NMR spectra. Instances of this occur *inter alia* in work on pluviatolide (**2.070**; Corrie *et al.*, 1970) and prestegane A (**2.075**; Taafrout *et al.*, 1983a,b). The problem was not even avoided in the 250 MHz spectrum of arctigenin (**2.045**; Suzuki *et al.*, 1982), where it was still not possible to escape the confusion which arises from the separation of the benzylic pairs and their part-coincidence with the signals from the 8,8′-protons (Table 6.8).

A comparison has been made (Harmatha *et al.*, 1982) between the spectra of a pair of diastereoisomeric lactones namely, yatein (**6.101**; W, X = O—CH_2—O; Y, Z = OMe; 8′α-H) and its *cis*-isomer (**6.101**; 8′β-H). These workers were only able to resolve the spectra taken at 100 MHz by titration with a lanthanide shift reagent, when the eight aliphatic proton signals shifted downfield from $\delta = 2.50$–4.13 to 3.30–4.96, and coupling constants and chemical shifts could be assigned. Further, since all these shifts were linearly related to the lanthanide concentration the peaks so resolved could be extrapolated so as to coincide with those in a 350 MHz spectrum.

In a related study of enterolactone (**6.101**; W = MeO, X = Y = Z = H; Cooley, 1984) the assignments made by Harmatha *et al.* (1982) were confirmed for a *trans*-lactone using INDOR experiments which are especially useful (Ayres *et al.*, 1972a) when details of the spectrum are lost under strong ether peaks. Complete assignment of proton shifts and a correlation of configuration and conformation in solution with that made by X-ray crystallography was obtained by NOE difference experiments. One of these is shown in Scheme 6.12 where irradiation of $H(9'_b)$ in dimethylenterolactone gave a large positive response in the signal of its geminal partner $H(9'_a)$ and another positive (up to 15%) response in the H-8′ peak. The methoxy peak at $\delta = 3.69$ responded negatively and there were weak positive effects at 2′-H and 4′-H, which are both excited through rotation of the aromatic ring. In a complementary experiment irradiation of H (9′a) did not enhance the 8′-H peak but rather those arising from aromatic protons and the geminal pair at C-7′. It is evident that once the lactone signals are identified NOE experiments will relate the aromatic rings to the lactone function. For example in yatein (**6.101**; W, X, $=O-CH_2-O$; Y, Z = OMe), in the preferred conformation (**6.103**), irradiation of either or both lactone protons (C-9′) will give a response in ring A and conversely. Furthermore irradiation of the clearly defined signal (2H) due to the equivalent ring B pair will enhance the signals arising from protons at C-7,7′ and C-8. NOE difference spectrometry has also been used by Matlin *et al.* (1984) to assign a structure to thujaplicatin methyl ether (**2.079**). The technique can clearly be used to orientate ether groups with respect to protons at unsubstituted aromatic sites.

Benzylidene isomers of the type shown below can be distinguished through their PMR spectra (Jakupovic *et al.*, 1986). In the 7*Z*-isomer (**6.104**) the vinyl proton is shielded and appears as a broad singlet (δ = 6.51) and the aromatic protons 2-H and 6-H lie at $\delta = 6.70$ and 6.67; in the 7*E*-isomer (**6.105**) the latter pair are now shielded ($\delta = 6.41$ and 6.49) and the vinyl proton moves downfield to $\delta = 7.66$.

(iv) *^{13}-C NMR spectroscopy* (CMR; Breitmaier and Voelter, 1987)

This was first employed for assignment of structure to cinnamyl

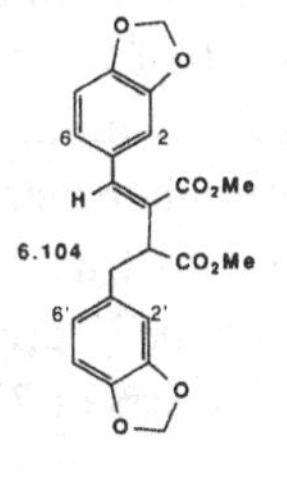

6.104

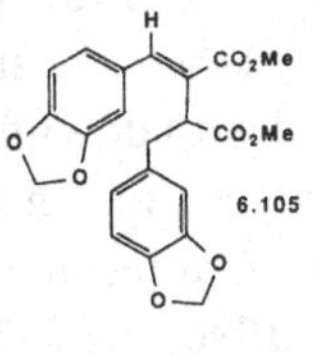

6.105

Table 6.9. ^{13}*C-shifts (ppm) for aryl groups in butyrolactones (Nishibe, 1980, 1981; Ward, 1979)*

Ring carbon	(ring A, R_1 = OH)	(ring B, R_2 = H)	(ring B, R_2 = OMe)
1	130.4	129.5	133.6
2	111.4	111.2	105.5
3	146.6 (OMe)	144.5 (OMe)	153.2 (OMe)
4	147.8 (OH)	146.7 (OMe)	136.7 (OMe)
5	114.1	111.7	153.2 (OMe)
6	122.0	120.5	105.5

Scheme 6.13 CMR assignments for diarylbutanes & diarylbutyrolactones

6.106 6.107 6.108

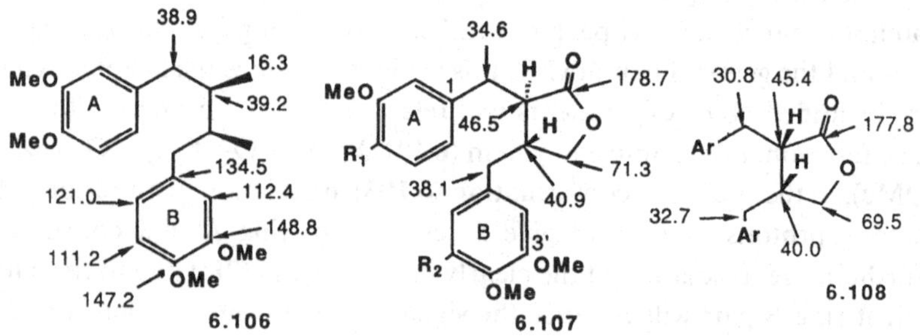

compounds and monolignoid dimers by Ludeman and Nimz (1974) and the topic has since been reviewed in detail by Agrawal and Thakur (1985), who provided a tabulated record of chemical shifts for 277 substances. Scheme 6.13 and Table 6.9 show some of the principal assignments for diarylbutanes and diarylbutyrolactones.

Typical figures for a butane model are given for the symmetrical lignan dimethyl *meso*-dihydroguaiaretic acid (**6.106**). The aromatic ring carbons are typified by a downfield shift at the point of attachment of the side-chain (134 ppm) and this is more pronounced at the point of oxygenation (*ca.* 148 ppm). Table 6.9 shows additional chemical shifts for guaiacyl, veratryl, and tri-*O*-methyl pyrogallyl residues (Ward *et al.*, 1979; Nishibe *et al.*, 1981). Methyl ether carbons have a typical shift near 55 ppm and this moves downfield to *ca.* 60 ppm for the central group of a tri-*O*-methyl pyrogallyl residue; for methylenedioxy ethers the shift is *ca.* 101 ppm.

Aliphatic carbon atoms respond with downfield shifts whenever oxygen is substituted for adjacent hydrogen; for instance in nortrachelogenin (**2.085**; Achenbach *et al.*, 1983) the C-8 resonance lies at *ca.* 77 ppm. In 7-hydroxyarctigenin (Nishibe *et al.*, 1980) the C-7 peak appears at 73.9 ppm and in 9,9′-hydroxylated compounds the carbon shift is at 61 ppm (Achenbach *et al.*, 1983).

The 13-C shifts of *trans*-lactones are recorded in a number of papers and those results summarised in Scheme 6.13 are typical. These quoted CMR shifts were all obtained by broad band proton decoupling so that all the resonances appear as single peaks enhanced by NOE. If the efficiency of decoupling is deliberately reduced, a relatively simple spectrum is obtained in which $J_{C,H}$ contracts to about 30 Hz. In methyl, methylene, and methine groups the directly bonded carbon resonance can now be identified as a quartet, triplet, or doublet whilst singlets are retained for

Scheme 6.14 Proton decoupled (A) and partially decoupled (B) ^{13}C-NMR spectra of 2,2′-dimethoxybenzyl-*trans*-butyrolactone.

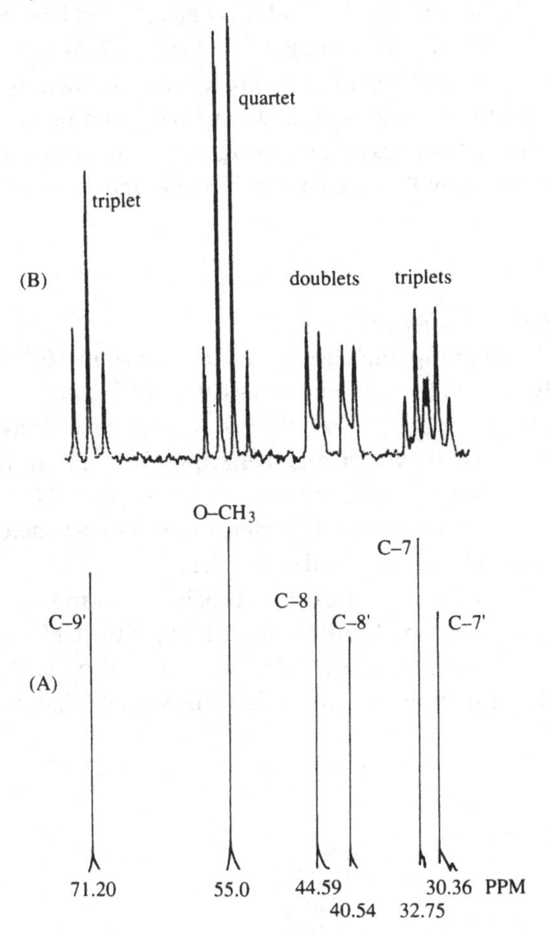

quaternary atoms. This is shown in Scheme 6.14 for 2,2′-dimethoxybenzyl-*trans*-butyrolactone (Kirk, 1987). A superior method is that of 2D correlation (p. 211).

When the individual proton shifts have been determined, as they were for enterolactone (**6.101**; W = OMe, X = Y = Z = H), the carbon resonances may be identified by selective proton decoupling (Cooley *et al.*, 1984).

The CMR spectra themselves do not distinguish certainly between geometrical isomers, although in two recorded instances (Nishibe *et al.*, 1980; Kirk, 1987) the shifts of *all* the aliphatic carbons in *cis*-lactones were at higher field than those in the corresponding *trans*-isomer. It is advisable therefore to assign configuration by comparison of a high-resolution spectrum with that of a known *trans*-isomer whose structure is further substantiated by an X-ray study (Cooley *et al.*, 1984); it is also important to stress the value of chemical methods. The less stable *cis*-lactones on heating with a base are epimerised through the conjugate base (pp. 362–5) to form the more stable *trans*-isomer. Well established procedures are then available (Schrecker and Hartwell, 1955) for the sequential reduction of either lactone to a diol and thence to an alkane, where again one notes that a *meso*-form has been determined by X-ray work (McKechnie and Paul, 1969).

Furans

(i) *9,9′-Epoxylignans*

This small subgroup includes (−)-*trans*-burseran (**6.109**) and (−)-cubebin (**6.110**). Complete assignments of the PMR spectra are not available. In burseran (Cole *et al.*, 1969) the 9,9′-methylene protons give a broad band at $\delta = 3.3$–4.0, whilst the benzylprotons fall at $\delta = 2.5$ resolved from the multiplet arising from the 8,8′-pair ($\delta = 2.0$–2.3). In cubebin (Batterbee *et al.*, 1969) the —OH signal ($\delta = 3.4$) was detectable in deuterochloroform and 9-H appeared at $\delta = 5.22$.

A stereoselective synthesis (Tomioka, 1979a) of both *cis*- and *trans*-burseran led to a useful correlation (Tomioka *et al.*, 1985) of the CMR spectra including those of the butyrolactones (p. 200), since here also the aliphatic carbon shifts (Table 6.10) of the *cis*-isomer lie at higher field than those of the *trans*-isomer.

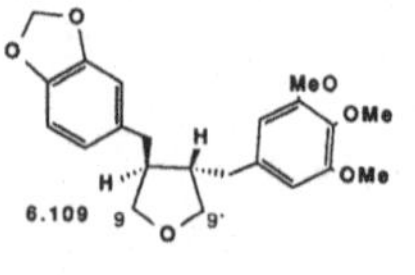

6.109

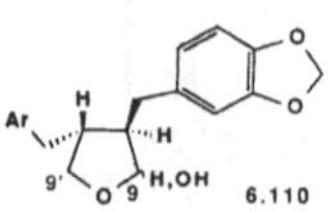

6.110

Table 6.10. *CMR of the burserans (ppm)*

Carbon	*cis-*	*trans-*
9 9′	72.0(t), 71.9(t)	73.2(t), 73.2(t)
8 8′	43.8(d), 43.6(d)	46.6(d), 46.3(d)
7 7′	33.9(t), 33.4(t)	39.9(t), 39.2(t)

The multiplicities (d = doublet, t = triplet) were found by off-resonance decoupling (Scheme 6.14).

6.111 all-*cis* | 6.113 *cis-cis-trans* | 6.115 *cis-trans-trans*

6.112 *trans-cis-trans* | 6.114 *cis-trans-cis* | 6.116 all-*trans*

(ii) *7,7′-Epoxylignans*

Members of this significant subgroup occur in six possible diastereoisomeric forms (**6.111–6.116**). When the 7,7′-aryl substituents are identical two *meso*-forms (**6.111**, **6.112**) are possible together with four others (**6.113–6.116**) which are normally optically active.

The assignment of configuration by application of the Karplus relationship is not reliable but, owing to the constraint imposed by the five-membered ring, shifts can be correlated with position relative to the aryl substituents.

By comparison of the structures with the tabulated chemical shifts (Table 6.11) it is apparent that the two symmetrical forms (**6.111** and **6.112**) and the all-*trans* isomer (**6.116**) have simpler spectra than the rest. In the structures of lower symmetry (**6.113**, **6.115**) a C-9 or C-9′-methyl signal appears at higher field if shielded by a *cis*-aryl group. The 8,8′-H peaks are complex and unresolved in these 60 MHz spectra but doublets which arise from 7,7′-H are resolved and two chemical shifts are distinguishable in the lower symmetry molecules.

It is not possible to relate coupling constants with configuration since puckering of the ring leads to similar J values for *cis*- and *trans*-located protons. For example, in tetrahydrofuroguaiacin-B (all *cis;* **6.111**, **2.167**) $J_{7,8}$ = 6.5 Hz and in galgravin (*trans-cis-trans*, **2.156**) this constant is little

Table 6.11. *Chemical shifts of 7,7′epoxylignans**

C—H	**6.111** all *cis-*	**6.112** *t-c-t*	**6.113** *c-c-t*	**6.115** *c-t-t*	**6.116** *t-t-t* all *trans*
9′-Me	0.55d	1.08d	0.60d	0.67d	1.00d
9-Me			1.02d	1.07d	
8,8′H	2.6m	2.36m	2.50m	1.6–2.5	1.7m
7′-H	4.95d	4.55d	5.46d	5.11d	4.60d
7—H			4.66d	4.43d	

m = multiplet; d = doublet
Naturally occurring examples are:
6.111 tetrahydrofuroguaiacin B
6.112 galgravin
6.113 (+)-chicanine
6.115 (+)-veraguensin
(+)-calopiptin
(+)-austrobailignan-7
6.116 fragransin B_3
(−)-galbelgin
(−)-grandisin

* δ values for $CDCL_3$ save for **6.111** in CS_2.
The figures were compiled from Crossley (1962), King (1964), Sarkanen (1973a), Holloway (1974), and Takaoka (1976).

changed at 6.2 Hz. Other comparisons of J values may be made from published data (Sarkanen and Wallis, 1973a), which were used to estimate conformations in solution. This work is complicated by thermal excitation and pseudorotation in these molecules with consequent averaging of the coupling constants. A CMR study has also been made (Fonseca *et al.*, 1979) and the results well-characterise the geometrical variations (Table 6.12). These shifts are not affected by changes in the aromatic ring substituents (see p. 000). The equivalence of the C7,7′- and C8,8′-atoms is seen in the symmetrical lignans (**6.111, 6.112**). In **6.112** the C7,7′-peaks shift downfield when a *trans*-relation is established and in **6.115** shifts typical of both *cis*- and *trans*-elements appear. In the all *trans*-model (**6.116**) all four of the ring peaks are at low field. The shifts quoted for the fully aromatic lignan di-*O*-methyl furoguaiacin (cf. **2.178**) are unmistakable as also are the PMR and UV spectra of this type.

A study of NOE difference has been made (Urzua, 1987) for these tetrahydrofurans. In (+)-aristolignin, which has the configuration (**6.115**), the benzylic protons at C-7, C-7′ lay at $\delta = 5.14$ and 4.41 respectively. The 2-H signal was assigned on ring A since both it and 5-H were enhanced by irradiation of the methoxyl signals ($\delta = 3.86, 3.88$), whereas in ring B 2′-H alone responded to methoxyl irradiation. The two benzylic protons could

Table 6.12. *CMR shifts (ppm) in 7,7′-epoxylignans*

Position	Isomer				
	6.111	**6.112**	**6.115**	**6.116**	Aromatic
7	82.4	87.1	87.1	88.1	146.4
8	41.2	44.3	45.9	50.9	117.4
8′	41.2	44.3	47.8	50.9	117.4
7′	82.4	87.1	82.8	88.1	146.4

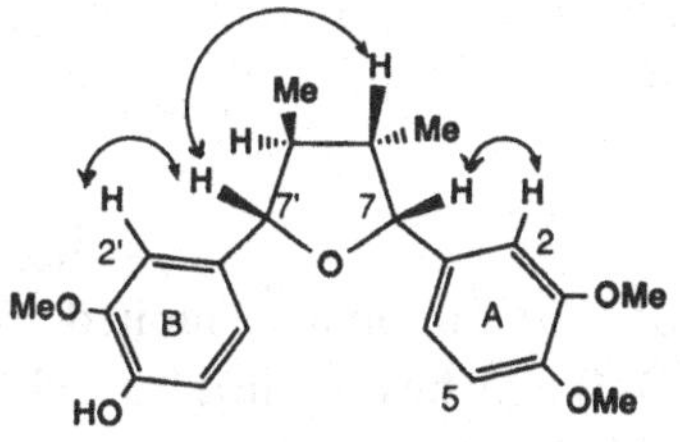

ARISTOLIGNIN

then be located by irradiation of aristolignin and machilin H Ar-H signals, when the lower-field 7′-H peak ($\delta = 4.41$) was enhanced by 10% via 2′-H and 7-H ($\delta = 5.14$) by 29% via 2-H. The larger response indicates that in solution, 7-H resides more in the plane of the adjacent aromatic ring than does 7′-H. The lower field resonance for 7-H bears this out.

Also consistent with the geometry shown for aristolignin is the NOE between 7′-H and 8-H and the shielding of C-9 methyl ($\delta = 0.67$) relative to C-9′ ($\delta = 1.07$). In the related model of machilin H (**6.112A**) a *cis*-8,8′ NOE relationship was established (Shimomura *et al.*, 1988).

A *cis*-configuration (**6.111**) in which the aryl and methyl substituents are all opposed (Blears and Haworth, 1958) makes for an unstable structure which isomerises in acidic solution to give **6.112** (McAlpine *et al.*, 1968). Another example is that of veraguensin (**6.115**) which yields galbelgin (Crossley and Djerassi, 1962) and others are given by Sarkanen and Wallis (1973b) who also used sodium/liquid ammonia reduction to assign configurations at C-8,8′. Here a *cis*-relation in the lignan gives a *meso*-dibenzylbutane wheras a *trans*-configuration yields a chiral compound.

The reductive cleavage (see also p. 214) is also important in that it provides a method for correlating the absolute configurations of natural lignans. Thus (−)-galbelgin (**6.116**; Ar = 4-hydroxy-3-methoxyphenyl) affords (−)-dihydroguaiaretic acid (**6.117**; X = H), which may in turn (Scheme 6.15) be related to the dibenzylbutyrolactones (**6.119**, and p. 10) via the intermediate diols (as **6.118**). More details of chemically established relationships are given by Weinges and Spanig (1967).

Scheme 6.15 Correlation of absolute configurations

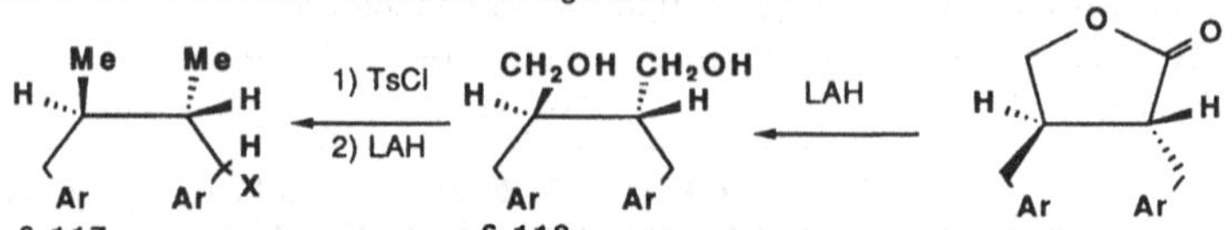

6.117 ← 1) TsCl 2) LAH ← 6.118 (-) - SECOISOLARICIRESINOL ← LAH ← 6.119 (-) - MATAIRESINOL

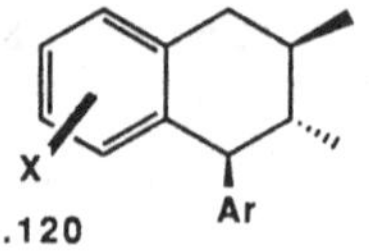

6.120 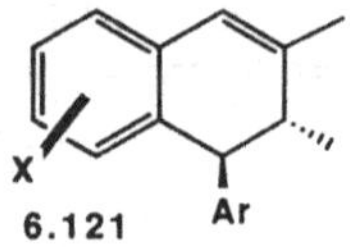 6.121

When the contact time with sodium/ammonia is limited the semi-reduced benzyl alcohol (**6.117**; X = OH) may be obtained from cleavage of the furan ring and this cyclises in acid to yield a tetralin (**6.120**). This structure may similarly be correlated with the natural tetralin lactones and also with apo-compounds of type **6.121**. The latter are also formed when tetrahydrofurans are treated with acid under more forcing conditions (Sarkanen and Wallis, 1973b).

(iii) *7,9′-Epoxylignans*

The principal members of this subgroup are the alcohols depicted in Scheme 6.16. Their structures contrast in an interesting way with those of the 7,7′-epoxylignans, since no alcohol with this structure is known as yet.

The insertion of —OH groups lowers the symmetry still further but at the same time reduces the multiplicity of PMR spectra. There are no methyl peaks in these PMR spectra and in olivil (**6.122**) the doublet due to the benzyl pair lies at $\delta = 4.71$ (Smith, 1963) and clearly differs from the signal arising from this group in lariciresinol (**6.123**; R = Me). By acetylation it is possible to separate the oxymethylene peak from the primary alcohol which coincides in the range of $\delta = 3.57$–4.15 with that due to the furan ring. In 5-methoxylariciresinol (**6.123**; R = Me, X = OMe) the additional methoxyl group was located by comparison of CMR shifts in the two aromatic rings (Duh *et al.*, 1986). The NOE response on irradiation of the 9′-methyl group was used (Iida *et al.*, 1983) to confirm the structure of magnostellin A (**6.124**, Ar = 3,4-dimethoxyphenyl).

(−)-Massoniresinol (**6.125**) is a trihydric alcohol isolated (Shen and Theander, 1985) from *Pinus massoniana*. The additional —OH insertion was indicated by the signal from the oxybenzylic C—H which appeared as

Scheme 6.16 Structures of 7,9'-epoxylignans

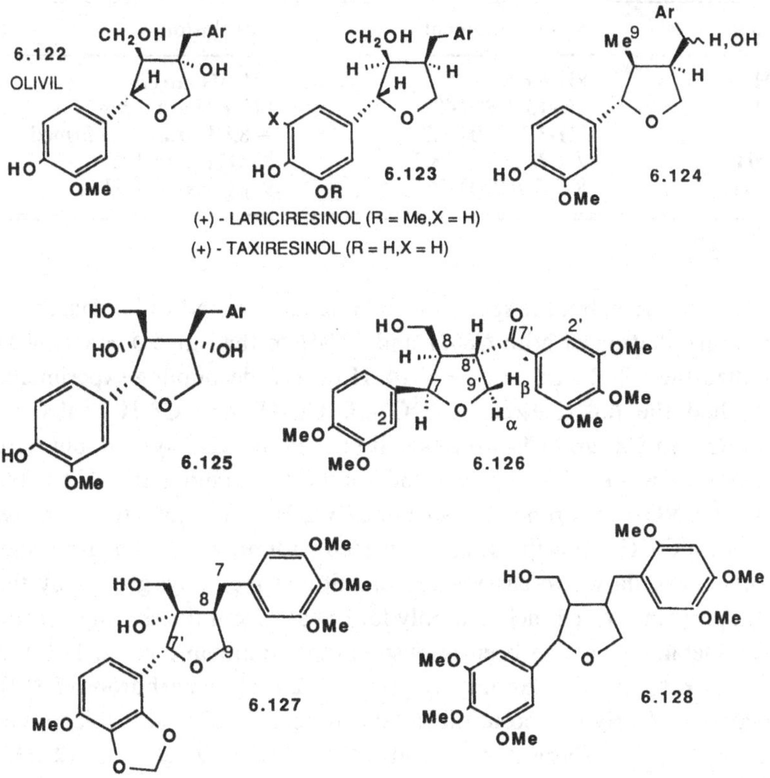

6.122 OLIVIL

6.123

(+) - LARICIRESINOL (R = Me,X = H)

(+) - TAXIRESINOL (R = H,X = H)

6.124

6.125

6.126

6.127

6.128

a *singlet* at δ = 5.01 with a benzylic pair of protons at δ = 2.93 and two oxymethylene resonances in the region of δ = 3.6–4.0. In the triacetate two aryloxyacetates could be distinguished from a primary alkyl acetate, which was resolved at δ = 4.28, 4.45; complete acetylation is difficult, but the use of acetic anhydride/dimethylaminopyridine is effective with similar *bis*-tertiary diols (Inoue *et al.*, 1981). It was possible to complete the orientation of substituents on the furan ring by irradiation, which established that no methylene group was coupled to any other. Massoniresinol includes a geminal diol of the type once wrongly ascribed to olivil; complete alkylation to include the two tertiary —OH groups was achieved with methyl iodide/DMSO/sodium hydride. The structural relation to olivil **(6.122)** was established by CMR and a common absolute configuration in these two lignans was shown by their circular dichroism (p. 252).

Sylvone **(6.126)**, a rare ketonic lignan, was obtained from *Piper sylvaticum* (Banerji *et al.*, 1984). At 200 MHz the monoacetate was resolved as a

Table 6.13. *NOE enhancements in the PMR of sylvone*

Irradiation	% Enhancement	Conclusion
C7-H	8H-6.6%	7H, 8H are *cis*-
C8-H	7H-13.9; 8′H-2.8	8H, 8′H are *trans*-
C8′-H	8H-3.3; 9′βH-9.7	−8,8′H *trans*- confirmed
C9′-αH	7H-3.5; 9′βH-18.3	9′-αH *cis*- to 7-H
C9′-βH	8′H-9.7; 9′αH-26.3	9′-βH *cis*- to 8′-H

singlet at $\delta = 1.76$, but the spectrum still included second-order characteristics as irradiation of both C8′-H and C9′-H in the range $\delta = 4.00$–4.44 simplified the C8-H signal at $\delta = 3.14$. However, decoupling experiments established the link between the C7-H, C8′-H, and C8-H peaks. At 360 MHz and C8- and C9-protons still formed an ABC system, but here the geometry was established by a study of the NOE (summarised in Table 6.13). The NOE experiments also linked C2-H of the dimethoxyphenyl group with C7-H. Thus this ring was placed with respect to the furan ring; the other was shown to carry three substituted methoxyl groups by the CMR spectrum, which included only four peaks from the six ring carbon atoms. Details have also been published (Satyanarayana *et al.*, 1986) of two other keto-lignans, oxodihydrogmelinol (**2.118**) and arborone (**2.117**). The latter is clearly related to the ketal arboreal (**2.237**) with which it co-occurs in *Gmelina arborea*, which is also the source of gmelamone (**2.250**). More details of these CMR spectra will for brevity be given with the discussion of the furofurans (pp. 211–12).

An interesting recent development has been the discovery (MacRae and Towers, 1985) of two naturally occurring 7′,9-epoxylignans, (+)-dihydrosesartemin (**6.127**) and (+)-β-dihydroyangambin (**6.128**), when hitherto dihydro-derivatives of furofurans had only been obtained by reduction of the latter *in vitro*. A number of compounds (**6.138–9**, **6.142**) prepared in this way are shown in Scheme 6.17 and their chemistry will be discussed in the following section.

Furofurans

Useful information about structures of this type (e.g. **6.129–6.131**) may be obtained from PMR experiments, although a number of workers made errors through the invalid application of the Karplus relation to spectra with second-order features. Reliable assignment of structure to these compounds and their dihydro-derivatives may require the use of CMR and a study of NOE is recommended.

Table 6.14. *PMR(δ) and CMR (ppm) shifts of furofurans* (Pelter, 1976)

	6.129	6.130 epi-	6.131 dia-
H-7	4.75(85.77)	4.45(87.53)	4.90(83.96)
H-7′		4.85(81.96)	
H-8	3.15(54.31)	2.90(54.44)	3.15(49.49)
H-8′		3.30(50.11)	
H9′β	3.8–4.0(71.72)	3.35(70.92)	3.3–3.65(68.75)
H9β		3.85(69.64)	
H9′α	4.2–4.4(71.72)	3.85(70.92)	3.65–4.0(68.75)
H9α		4.13(69.64)	

A 400 MHz spectrum of dimethylpinoresinol appears on p. 210 (Figure 6.1).

One again draws attention to the importance of acid-catalysed epimerisation as a test for steric compression. Of the three diastereoisomers shown in Table 6.14, the diaxial model (**6.131**) is destabilised by the non-bonded interaction of both aryl groups on C7- and C7′- with β-H-atoms on C9- and C9′- and consequently forms but a minor component of an equilibrium mixture in which **6.129** predominates. These structural types may be resolved and classified by the use of HPLC. When an aryl group takes up an 'axial' conformation, essentially parallel to the principal axis, the interactive β-H atom falls within its shielding cone. This can be verified by reference to Table 6.14.

A number of points arise from this data:

(1) Structures **6.129** and **6.131** are symmetrical since H7, H7′ and H8, H8′ are equivalent.

(2) In structure **6.130** one of the protons in the 9,9′-pairs is shielded with respect to the others and to those in the spectrum of **6.129**.

(3) For structure **6.131** the shifts of 9,9′-protons indicate that there is shielding and the symmetry establishes that two protons must be affected.

Owing to its symmetry, dimethylpinoresinol (**6.129**; Ar = 3,4-dimethoxyphenyl) has a simple spectrum but 9-βH and 9′β-H are concealed by the strong ether peaks ($\delta = 3.9$). However a 2D COSY spectrum (Figure 6.2) reveals off-diagonal coupling of signals close to $\delta = 3.9$ with those at

FIGURE 6.1. 400 MHz PMR spectra of the aliphatic protons in dimethylpinoresinol and the gmelinols ($CDCl_3$).

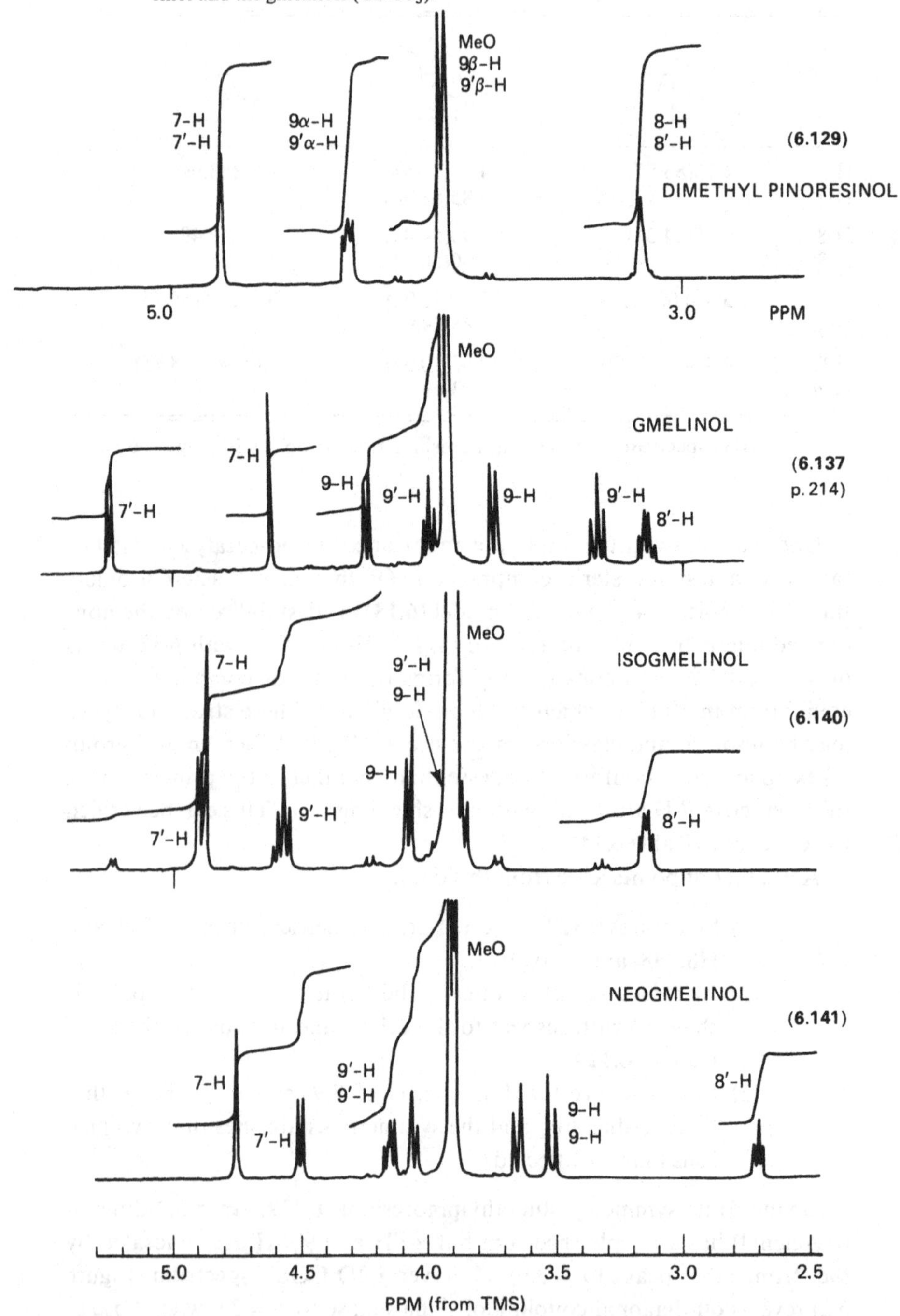

FIGURE 6.2. 2D-COSY spectrum of dimethylpinoresinol at 400 MHz.

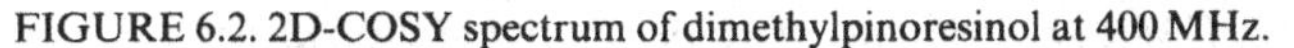

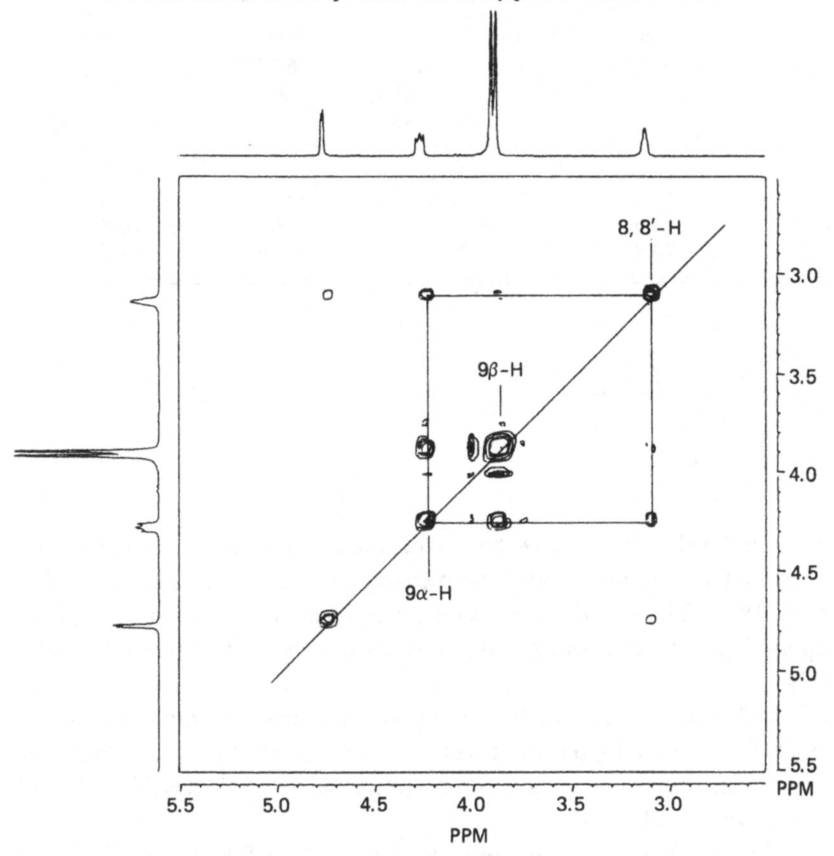

$\delta = 4.25$ (9α-H, 9′α-H) and $\delta = 3.1$(8-H, 8′-H), which confirms that 9β-H, 9′β-H lie beneath the ether peaks.

It can also be seen that the apparent coupling constants do not relate to the geometry since $J_{7,8} = 5$ Hz in both the epilignan (**6.130**) and the dialignan (**6.131**); further there is no sensible difference in the C7,7′-H shifts in the three isomers.

The CMR data (Table 6.14, figures in parentheses) are more relevant to the assignment of stereochemistry. For example, a lack of symmetry in the epi-isomer (**6.130**) is again seen and oxymethylene carbon signals are at highest field when axially substituted (in **6.131**) and lie at lowest field in the diequatorial isomer (**6.129**). In general the CMR aliphatic shifts respond consistently with torsional variations as also do the C1,1′-atoms of the aryl groups. In **6.129** (Ar = 3,4-methylenedioxyphenyl) a single peak appears at 134.04 ppm, in the epi-isomer **6.130** two are seen at 133.51 and 130.81,

Table 6.15. *CMR shifts (ppm) of pinoresinol and derivatives*

Carbon	Pinoresinol	**6.132** (X = OH; Y = H)	**6.132** (X = H; Y = Me)	**6.132** (X = OH; Y = Me)
C-7	85.2	87.1	85.0	86.9
C-8	53.6	91.0	53.7	91.0
C-8′	53.6	60.8	53.7	60.8
C-9	70.9	74.7	71.0	74.7

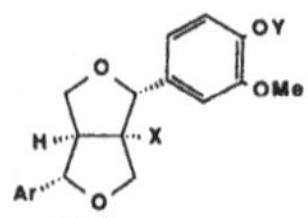

OY
OMe
O
H
X
Ar
O
6.132

whilst in **6.131** only the single higher-field peak appears at 131.38. These C-1 resonances vary only slightly with changes in the oxysubstituents (Pelter *et al.*, 1976). The use of CMR spectra in the placing of glycosides depends upon C-1, C-1′ resonances and is discussed in conjunction with Scheme 6.29.

Another extensive compilation of CMR data has been made by Tsukamoto *et al.* (1984a,b) and the effect of —OH substitution in the aliphatic part of these molecules is recorded for pinoresinol (**6.132**; X = Y = H) and derivatives in Table 6.15.

The C7- and C9-shifts are little changed by C8-hydroxylation. The CMR spectrum of prinsepiol (**6.133**), in which both bridgehead positions are hydroxylated, has been recorded (Kilidhar *et al.*, 1982).

The shifts of aromatic carbon atoms are essentially as listed above (Scheme 6.13 and Table 6.9, p. 200). Further details may be obtained from work on lignans of *Fraxinus mandshurica*, which included furans, furofurans, olivil, and a tetralin type (Tsukamoto *et al.*, 1984b; Nishibe *et al.*, 1984); relevant data are also given in a study of lignans from *Daphne tangutica* (Zhuang, 1984). An example of off-resonance decoupling for the determination of the multiplicity of ^{13}C-signals is recorded for the epi-isomer (+)-sylvatesmin (Banerji and Pal, 1982).

A naturally-occurring 9-hydroxyfurofuran (**6.134**) has been described by Khan and Schoeb (1985). One expects a marked shift downfield at this position which corresponds to an acetal carbon (*ca.* 100 ppm), although the CMR spectrum was not reported the structure was related to that of pinoresinol (cf. **6.129**) by selective reduction with diborane. Oxidative conversion of **6.134** to a type of naturally-occurring lactone is also possible (see p. 100).

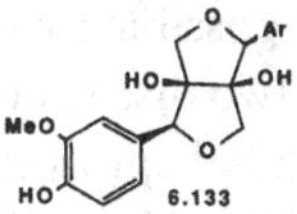

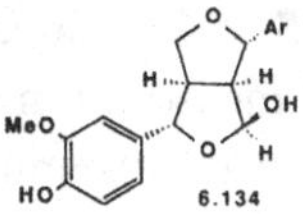

A detailed study of the PMR spectra of lactols and lactones related to furofurans has been made (Vande Velde *et al.*, 1984). Observed coupling constants may be of use in allocating structures for example, when the —OH group occupies an *endo*-position (**6.134**) the dihedral angle between 9-H and 8-H is close to 90° and $J_{8,9}$ is small. These authors also draw attention to second-order characteristics in these spectra, where in the diacetate of pinoresinol (**6.129**; Ar = 4-hydroxy-3-methoxyphenyl) a computer fit was needed to evaluate the coupling constants between aliphatic protons. In this model there was correspondence between coupling constants and the dihedral angles, which were also consistent with those obtained for (−)-syringaresinol (**6.129**; Ar = 4-hydroxy-3,5-dimethoxyphenyl) by X-ray analysis (Bryan and Fallon, 1976). However, oxy-substitution of aliphatic hydrogens leads to torsional changes and it is evident from ^{13}C shift differences (p. 244) that α- and β-linked aryl substituents exert different torsional effects on the ring system (pp. 215–16). Further, the rings are flexible and either or both oxygen atoms may invert and adopt an *exo*- or *endo*-conformation. Caution is therefore needed in assigning configuration and the use of NOE spectra is again recommended in this context; an example is provided by the work of Stevens and Whiting (1986) on lactones of the styraxin type (**2.289**).

Chemical correlation of furofurans and furans

The value of this approach is highlighted by the difficulties encountered when using only spectral methods.

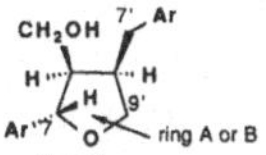

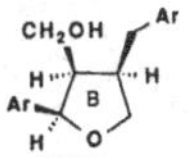

It should be recalled that dihydrofurofurans have recently been shown (MacRae and Towers, 1985) to occur naturally, although they have long been known as the products of hydrogenation or sodium/ammonia reduction (Klemm, 1978). Thus symmetrical furofurans like eudesmin (**6.129**; Ar = 3,4-dimethoxyphenyl) yield a single cleavage product (dihydro-eudesmin, **6.135**), whereas both **6.135** and **6.136** are possible products of the reduction of epieudesmin (**6.130**); the isolation of two distinct dihydro-compounds proves the precursor to be an unsymmetrical epi-isomer. The shielding effects in these molecules and other criteria have been discussed (pp. 206–7) and additional comments on the relation between structure and chemical shifts in these alcohols have been made (Banerji *et al.*, 1984; MacRae and Towers, 1985; Shen and Theander, 1985). When the methylol group is *cis*- to the aryl group at C7-, the methylene peak is at higher field ($\delta = 3.22$) relative to that in a *trans*-isomer, where the signal is partly coincident with the C9′-H multiplet in the region of $\delta = 3.6$–4.0. In **6.135**, C7-H lies at $\delta = 4.80$ but it is deshielded to $\delta = 5.20$ in the *cis*-isomer sylvone (7′-carbonyl in **6.136**).

An extension of the chemical method to the assignment of structures to the gmelinols (Birch *et al.*, 1967) provides a good example of the technique. It will be recalled (p. 183–4) that the structure of neogmelinol was first thought to include a C7-C7′ link but that mass spectral evidence was not in accord with this conclusion.

On reduction gmelinol itself (**6.137**; Scheme 6.17) gives two different compounds, dihydrogmelinol I (**6.138**) and dihydrogmelinol II (**6.139**). This is consistent with the structure (**6.137**), where a 7′-aryl group is retained as a *cis*-substituent in **6.138** and a 7-aryl group as a *trans*-substituent in **6.139**. Isogmelinol (**6.140**) gave **6.139** as a common reduction product. Periodate fission showed that this was a *vic*-diol and hence that both gmelinol and isogmelinol had an equatorial aryl group *cis*- to the bridgehead hydroxyl group; hence isogmelinol must differ in the B ring

Scheme 6.17 Chemical correlations of the gmelinols

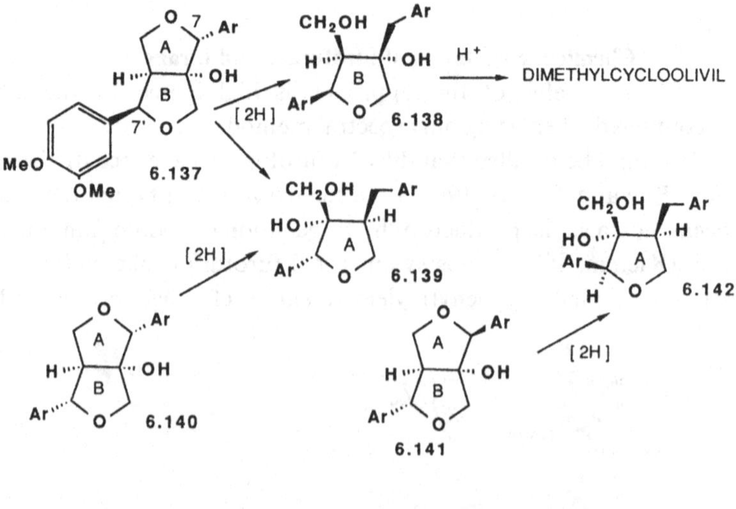

6.137 —[2H]→ 6.138 —H^+→ DIMETHYLCYCLOOLIVIL

6.137 —[2H]→ 6.139

6.140 —[2H]→ 6.139

6.141 —[2H]→ 6.142

and have two equatorial substituents (**6.140**). Reduction of **6.140** could in theory produce dimethyl olivil (**6.122**) although it was not isolable here.

Neogmelinol was now identified as **6.141** because the PMR spectrum of its dihydroproduct was consistent only with that of a tetrahydrofuran (**6.142**), in which the A ring survived and where the aryl substituent was *trans*- to the bridgehead —OH group. The acid-catalysed cyclisation of **6.138** yielded cyclo-olivil dimethyl ether (cf. **2.318** and **6.120**) and so provides a link to the tetralin class.

A comparison (Figure 6.1, p. 210) of the PMR spectra of pinoresinol (**6.129**) and isogmelinol (**6.140**) points to diequatorial (α-) aryl substitution in both. The latter has a more complex spectrum but only exhibits notable chemical shift differences in the 9α-H adjacent to the tertiary hydroxyl group, now located at $\delta = 4.08$; whereas in pinoresinol the equivalent signal falls at $\delta = 4.27$. It is significant that although 7-H has simplified to a singlet its chemical shift is still close to the doublet arising from 7′-H; therefore neither of these protons can be shielded by an axial (β)-aryl group. The assignments are confirmed by taking the highest field 8,8′-signal as a marker in the 2D-COSY spectrum of pinoresinol (Figure 6.2), when one sees an off-diagonal response from the higher field 9α-H pair ($\delta = 4.27$) and also that the 9β-H signals are close in their shift ($\delta = 3.9$) to the ether peaks.

In both gmelinol and neogmelinol all seven aliphatic protons are clear of the ether signals (Figure 6.1). Gmelinol (7′-β-aryl) has one shielded 9′-H ($\delta = 3.32$) and 7-H ($\delta = 4.60$) is at high field relative to 7′-H ($\delta = 5.23$). In accord with the assigned structures (Scheme 6.17) the shift of 7-H ($\delta = 4.74$) relative to 7′-H ($\delta = 4.49$) is reversed in neogmelinol and one of the 9-H pair is shielded ($\delta = 3.52$). Again the 2D—COSY spectrum of neogmelinol allows a firm assignment of chemical shifts based on off–diagonal relationships with the highest field 8′-H peak.

The coincidence of the 8′-H signals ($\delta = 3.15$) in gmelinol and isogmelinol runs contrary to the above analysis, because one expects to find evidence of the shielding of this proton in isogmelinol by the *cis*-aryl group. However, NOE experiments (Ayres and Haycock, 1988) established an interaction between 8′-H and 7′-H in gmelinol which was absent in iso- and neogmelinol. 8′-H, 9′α-H interactions were found for all three isomers, but consistent with the assigned structures the 7β-H, 9β-H response was shown only by gmelinol and isogmelinol. NOE assignments for hydroxypinoresinol (**2.255**) and hydroxymedioresinol (**2.251**) have been published (Abe and Yamauchi, 1988).

The apparent absence of a shielding effect by the C7′-aryl group in isogmelinol as compared to gmelinol must arise from counterbalancing torsional effects of this bulky substituent in the two models. A detailed analysis of this effect in tetrahydrofuran lignans has been made by Biftu *et al.* (1986) and an X-ray study (Ghisalberti *et al.*, 1987) provides evidence of this in furofurans. Their CMR spectra were also seen above to typify torsional changes (p. 213).

The sequence of acid-catalysed structural changes can be explained in that, although neogmelinol is the thermodynamically stable end product, initial isomerisation at C-7 is disfavoured in gmelinol when C-7′ is axially substituted. Once change has occurred here to produce the C-7′ substitution of isogmelinol then the isomersation at C-7 may proceed to yield neogmelinol. The apparent anomaly of a stable axial group at C-7 in neogmelinol must be due to the interaction of the equatorial with the *cis*-hydroxyl group at C-8 in isogmelinol.

The CMR shifts of the C1 and C1′-ring atoms of gmelinol are 127.8 and 133.1 ppm and are typical of one axial and one equatorial substituent. This is also true for neogmelinol with shifts of 129.6 and 132.9 ppm but for isogmelinol the relation does not hold since the observed shifts of 128.1 and 130.7 ppm are akin to those of axial substituents. Since isogmelinol (**6.140**) is a strained compound and gives neogmelinol (**6.141**) in acid, it is probable that torsional strain is transmitted to the C1-position of the adjacent ring.

It is worthy of note that a trihydroxyfuran (**6.144**) related to massoniresinol (**6.125**) is produced on reduction of kigeliol (**6.143**; Inouye, 1981), and also that the acyclic triol (**6.146**) was obtained from the furofuran (**6.145**) in a one-pot reaction (Tsukamoto *et al.*, 1984a,b).

Aryltetrahydronaphthalenes

The configuration and conformation of the reduced rings of this group govern their physiological activity and hence are of the first importance. The relationship between coupling constants and geometry has been discussed (Ayres, 1978); difficulties may arise for the following reasons:

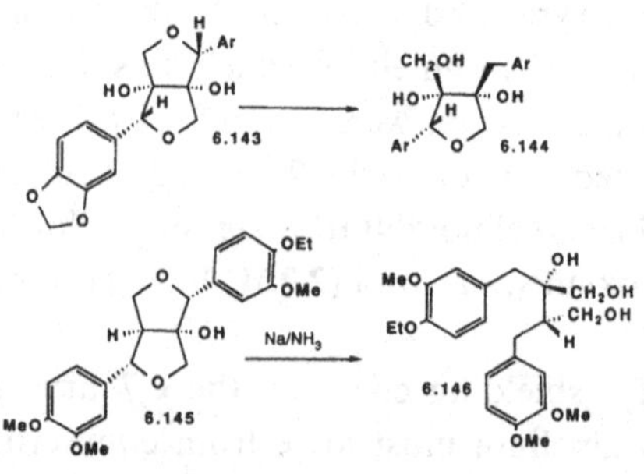

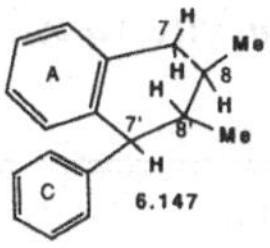

(a) In many instances only the doublet arising from the C-7′ proton has been resolved and the Karplus relation applied.
(b) Two quasi-chair conformers are possible and depending on the aromatic substitution pattern they may be no more favoured than one of two possible boat forms.
(c) An unfavourable feature of the chair form (**6.147**) is the interaction between the C7′-aryl substituent (C) and the fused aromatic ring (A).
(d) In this cyclohexene model an axially placed group experiences only one destabilising 1,3-interaction.

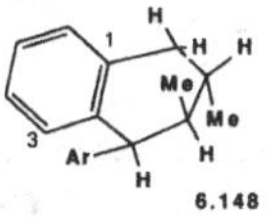

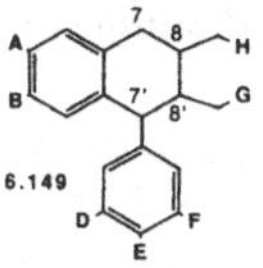

The data (Table 6.16, **6.149**, p. 218) shows that *trans, trans*-substitution results in a substantial value for $J_{7,8}$ (9–10 Hz) with one constant $J_{7,8}$ of like magnitude and the other small. In *cis, trans*-substituted compounds the record is less clear, although all authors report $J_{7',8'}$ as small (4.5–5 Hz).

The chemical shifts of the C-Me groups are a guide to conformation. The all *trans*-model is almost certainly a chair form (**6.147**) with all substituents equatorial. The group remote from ring C is consistently reported to lie at lower field ($\delta = 1.10$) than C8′-Me ($\delta = 0.9$) which is shielded by the pendent ring (C) in keeping with conformation **6.147**.

In the *cis, trans*-model the record is confused. A detailed study at 500 MHz by Nemethy *et al.* (1986) confirmed J values given by earlier workers (Murphy *et al.*, 1975; Snider and Jackson, 1983) and assigned equal shifts to C8-Me and C8′-Me. This is inconsistent with a single conformer (e.g. **6.147**) and averaging must occur, most probably about a form such as **6.148** in which all C-H dihedral angles are small (cf. Table 6.16). The last result reported by Takeya *et al.* (1983) is not in keeping since the methyl groups have quite different chemical shifts and the value for $J_{7,8}$ is comparatively high. This compound was reported to have isomerised to the less stable *cis*-isomer during treatment with palladium/charcoal and it seems probable that this interpretation is wrong.

The effect of the ring C–C3-interaction is clearly seen on examination of the spectrum of lyoniresinol (**6.150**; Kato, 1966) which is substituted at this position. This has a *trans, trans*-configuration yet $J_{7',8'}$ is only 5.5 Hz and it is evident that a significant contribution is made by the conformer (**6.151**)

Table 6.16. *Chemical shifts and coupling constants in aryltetrahydronaphthalene lignans*

	Shifts		Coupling			
6.149 subst.	9-Me	9′-Me	$J_{7',8'}$	$J_{7,8}$	$J_{7,8}$	Config.
None			4.5	9.2	4.5	*cis, trans*
None			4.5	9.8	5.0	*cis, trans*
None			5.8	6.9	5.0	*trans, cis*
B=E=OH	1.05	0.85	9.0	—	—	*trans, trans*
A, D, F=OMe B, E, H=OAc, G=Ac.xylosyl			9.0			*trans, trans*
G, H=OH	—	—	10.0 5.0			*trans, trans* *cis, trans?*
A, B=O—CH_2—O; E=OH	0.99	0.78	10	11	4*	*trans, trans*
A, B, D, E=OMe	0.9	0.9	5.5	5.4	7.0**	*cis, trans*
A, B=O—CH_2—O, D, E=OMe	0.9	0.9	5.5	5.3	7.3	*cis, trans*
A, B, D, E=OMe	1.1	0.9	10	10	4.2	*trans, trans*
A, B=O—CH_2—O D, E = OMe	1.08	0.9	9.8	10	4.1	*trans, trans*
A, B, D, E=OMe	1.15	0.74	7.4	—	—	*cis, trans*

* $J_{8,8'}$ = 10 Hz; ** $J_{8,8'}$ = 3 Hz. Only substituted positions A–H are listed.
References: (a) Murphy, 1975; (b) Vieira, 1983; (c) Vecchietti, 1979; (d) Snider, 1983; (e) Loriot, 1983; (f) Nemethy, 1986; (g) Takeya, 1983.

in which the dihedral angle C7′-H, C8′-H is small. An aromatic proton at C3- has no comparable effect but lies at high field owing to shielding by ring C. The C3-OMe signal in lyoniresinol is shielded only to the extent that the equatorial ring C conformer (**6.147**) participates; a related structure is that of nirtetralin (**6.152**). When ring C is forced away from the plane of ring A C7′-H comes within the deshielding zone and its chemical shift may fall as low as $\delta = 4.55$ compared to a typical value of $\delta = 4.10$ (see Ward *et al.*, 1979).

The structure originally proposed (Anjaneyulu *et al.*, 1973) for phyltetralin (**6.153**; X = H, Y = OMe) included a C3-OMe group, although none of the four —OMe resonances was at higher field than $\delta = 3.82$ and one aromatic proton was at high field ($\delta = 6.23$); these observations are only compatible with ring C being free to adopt an equatorial conformation. This can be correct only if the conformation **6.151** predominates. It was later shown (Stevenson and Williams, 1977) that the *trans*-lactone conidendrin (**6.154**) gave the enantiomer of phyltetralin by reduction and etherification and that therefore its structure is **6.153** (X = OMe, Y = H).

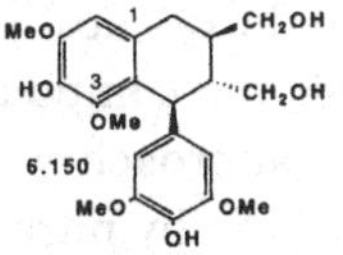

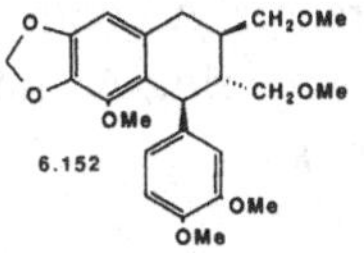

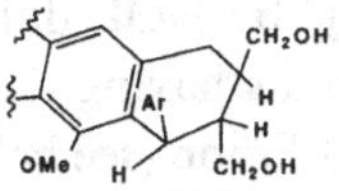

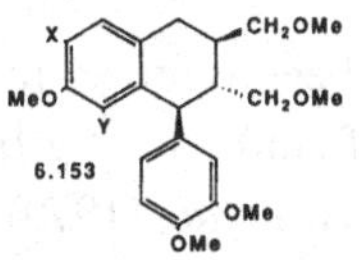

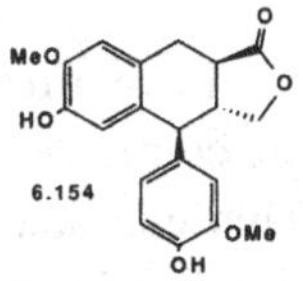

(i) *Orientation of aromatic substituents*

Some relevant material is given in an earlier section (pp. 195–6) and the following is intended to broaden the discussion but with special reference to the tetrahydronaphthalene lignans.

With a few exceptions *ortho*-oxysubstituents are found in both aromatic rings and no naturally occurring compound is known in which either ring is completely substituted. If two Ar-H bonds remain in ring A, they will be *ortho*- or *para*-coupled (J = *ca.* 9 Hz or <1 Hz); a proton at C3- or C6- may well be distinguished by neighbouring group effects (p. 227).

In ring C the most common pattern is one of 3′, 4′-substitution. At 250 or 500 MHz the *cis*- (8.2 Hz) and *meta*- (2.2 Hz) coupling constants can be defined (Nemethy *et al.*, 1986). Should a substituent be present, or be inserted at C-2′, then the favoured conformation is one with this group outside the shielding zone of ring A and the highly shielded C6′-H within it. This is evident in derivatives of podophyllotoxin (Ayres and Lim, 1972b), where 3′,4′,5′-trisubstitution is characterised by a singlet (2H) due to the equivalent pair of residual protons at $\delta = 6.29$ whilst in the 2′-chloroderivative the remaining singlet (1H) is shifted upfield to $\delta = 5.88$.

A methylenedioxy group forms an A,B pair (**2.365**: Hokanson, 1979). A pronounced shielding effect by the C ring on this group is apparent when its shift of $\delta = 5.57$ in otobain (**2.304**; Wallis *et al.*, 1963) is compared with that in attenuol (**2.290**; Joshi *et al.*, 1978; $\delta = 5.81$).

A difficulty arises when different dioxy-substituents are similarly placed on the aromatic rings, for example the compounds represented in **6.155** (X = H) and **6.156** (X = H). One approach would be to identify unresolved protons coupled to position (X) by nitration or bromination, and then to test for adjacent methoxyl by NOE differences. A comparison of CMR shifts with known compounds having the same ring A and ring C orientation could also provide a solution (see below).

(ii) *Use of CMR spectra*

A number of compilations have already been referred to, but of special relevance to tetralins is the comparison of shifts arising from the patterns (A–D; Scheme **6.18**). This enabled Ward *et al.* (1979) to distinguish nirtetralin (**6.152**) as type (A) and hypophyllanthin (**6.157**) as type (B).

Note here that the quaternary carbon shifts are similar only in (D), which otherwise resembles (C) in that both have a high-field PMR MeO- peak and two MeO- groups which show a NOE with a common proton. The C3-H peak makes (B) unique and here the CMR quaternary shifts are farthest apart.

In disubstituted benzenes the CMR shifts due to individual groups can be evaluated, but this is not possible for more highly substituted compounds because *ortho*-steric interactions perturb the electronic effects (Craik, 1983).

Typical CMR shifts (Fonseca *et al.*, 1978) are shown in Scheme 6.19 for the aryltetrahydronaphthalene skeleton (R = H or Me). There will be

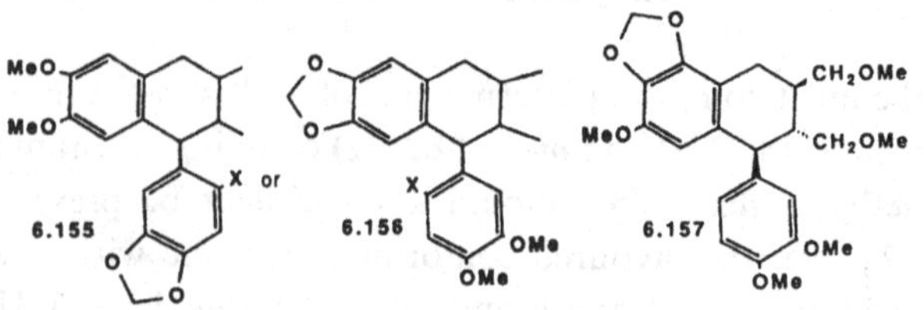

MeO MeO X or 6.155 O O

O O X 6.156 OMe OMe

O O MeO CH_2OMe CH_2OMe 6.157 OMe OMe

Scheme 6.18 Ring A CMR shifts in trioxy - tetralins (WARD, 1979)

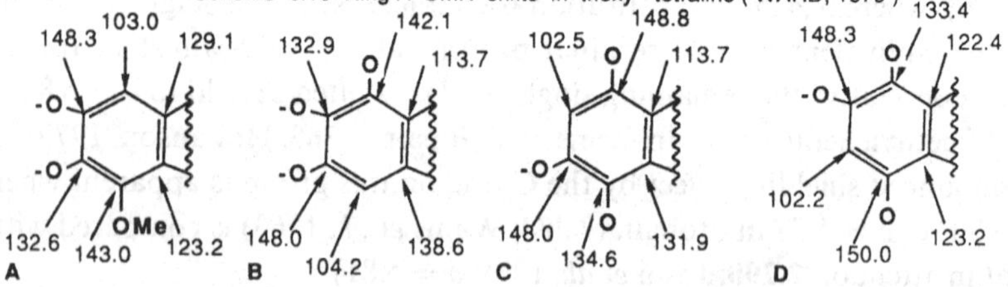

A: 148.3, 103.0, 129.1, -O, -O, OMe, 132.6, 143.0, 123.2

B: 132.9, 142.1, O, 113.7, -O, -O, 148.0, 104.2, 138.6

C: 102.5, 148.8, O, 113.7, -O, O, 148.0, 134.6, 131.9

D: 148.3, 133.4, O, 122.4, -O, 102.2, O, 150.0, 123.2

Scheme 6.19 Chemical shifts of some aryltetrahydronaphthalenes

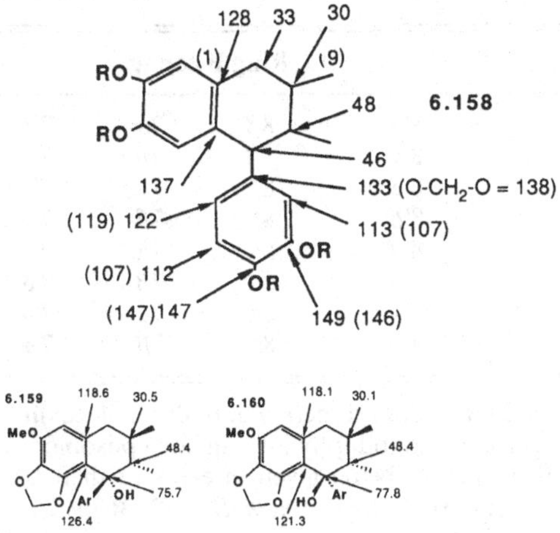

variations for methylenedioxy ethers and these values are given in parentheses (Martinez *et al.*, 1985, and Scheme 6.18). Free phenolic —OH groups may be detected by the downfield shifts which occur in solution at pH 12.5 (Lapper *et al.*, 1975) and in *O*-acetyl derivatives. Off-resonance decoupling has been used to observe the multiplicity of these peaks and hence to confirm the assignments.

Oxy-substitution at aliphatic carbon atoms causes large downfield shifts (see p. 200) and typical values can be obtained from papers cited above and also by Vecchietti *et al.* (1979) and Garzon *et al.* (1987). There are no significant variations in chemical shift in *cis*- or *trans*-isomers of **6.157**. However, it is important to record that in the isomeric hydroxymethoxyotobains (**6.159, 6.160**), interaction between congested substituents led to a sensible change in the quaternary C-2 resonance and that this is relevant to assignments of the kind illustrated in Scheme 6.18.

Details of the CMR spectra of a group of arylnaphthalene lignans have been published (Ardullaev *et al.*, 1987).

(iii) *Aryltetrahydronaphthalene lactones*

Those which are *trans*-linked have rigid structures which cannot undergo conformational inversion (cf. pp. 216–17) and therefore coupling constants relate to configuration. In lignans of the podophyllotoxin subgroup (**6.161**; Table 6.17) this rigidity of *trans*-lactones ensures that the dihedral angle between the two aromatic rings is retained *in vivo*. This is essential for the interaction with protein and the inhibition of tubulin synthesis which is the cause of its activity in cell division (pp. 91–2). In the

Table 6.17. *Podophyllotoxin and its diastereoisomers*

Compound		Ring B config.		
1 Podophyllotoxin	8′α	8β	7α	7′α
2 Epipodophyllotoxin	8′α	8β	7β	7′α
3 Picropodophyllin	8′β	8β	7α	7′α
4 Epipicropodophyllin	8′β	8β	7β	7′α
5 Isopodophyllotoxin	8′β	8α	7α	7′α
6 Epiisopodophyllotoxin	8′β	8α	7β	7′α
7 Isopicropodophyllin	8′α	8α	7α	7′α
8 Epiisopicropodophyllin	8′α	8α	7β	7′α

The Table describes the eight diastereoisomers of L-podophyllotoxin (7R, 7′R, 8R, 8′R) but the trivial names for epiisopodophyllotoxin and epiisopicrodophyllin are inconsistent owing to these names being given in error to their C7-epimers Gensler, W.J. and Johnson, F. (1963), *J. Amer. Chem. Soc.* **85**, 3670–3.

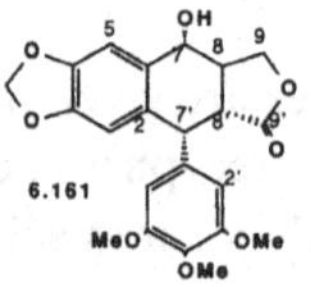

anti-cancer drugs Etoposide and Teniposide (p. 117) this geometrical requirement persists, although their mode of action differs and requires interaction with topoisomerase enzymes (Chen, G.L., 1984). Once the strain in these molecules is relieved through the action of base (p. 363) the fully flexible inactive *cis*-lactones are obtained. It might be thought that their flexibility would permit them to adopt the critical geometry but evidently the energy to activate this change is unavailable *in vivo*.

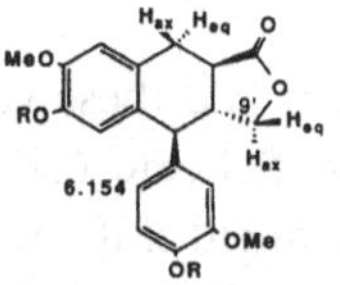

The podophyllotoxins and retrolactones like α-conidendrin (**6.154**) and the butyrolactones provide reference points for the allocation of structure including absolute configuration. One example (Stevenson and Williams, 1977) involving α-conidendrin (**6.154**) and its *cis*-analogue β-conidendrin has been mentioned (p. 219). Another illustration is provided by Kato (1963, 1966) who showed that (+)-lyoniresinol (**2.329**) could be oxidised by chromium trioxide/pyridine to correlate with (−)-conidendrin (**6.154**) which was identified by PMR and other spectra. It was also shown that (+)-lyoniresinol could be regenerated by the hydride reduction of (−)-conidendrin. The possible variations (cf. Hartwell and Schrecker, 1952) in the ring B substitution pattern are listed in Table 6.17.

Closely related to this group are the α- and β-peltatins (**2.354, 2.355**) and the 7-ketone podophyllotoxone (**2.364**).

FIGURE 6.3. 400 MHz ($CDCl_3$) spectrum of podophyllotoxin with expanded detail of aliphatic protons (6.3A, 6.3B overleaf).

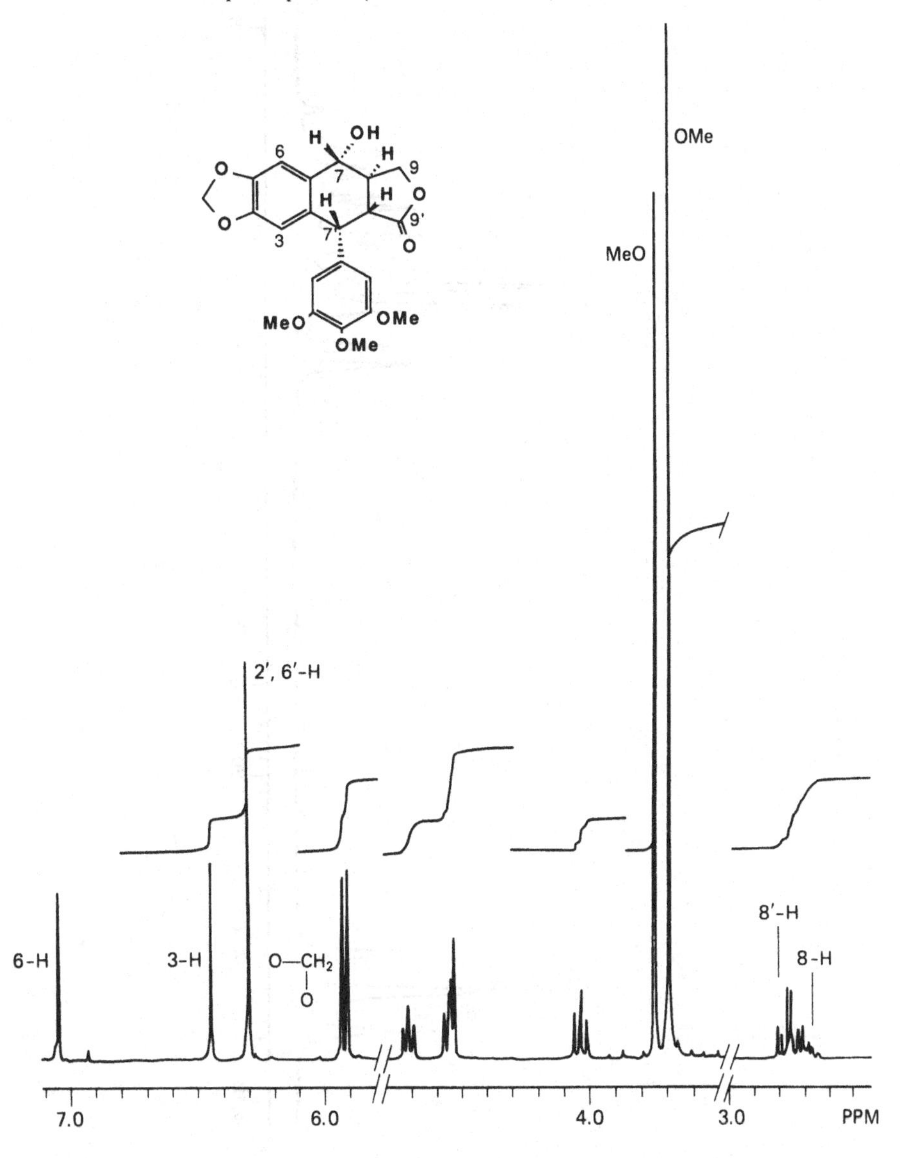

On comparing chemical shifts (Table 6.18 and Figures 6.3, 6.4) C8′-H and C8-H are very close in both the 9-methylene and 9′-methylene lactones and second-order character is found here. There are however several features which distinguish between the two lactone types. In podophyllotoxin the aromatic singlet (2′,6′-H) is at lower field and C7′-H is markedly

Fig. 6.3A. Details of signals for lactone β-H, C_8-H and C'_8-H.

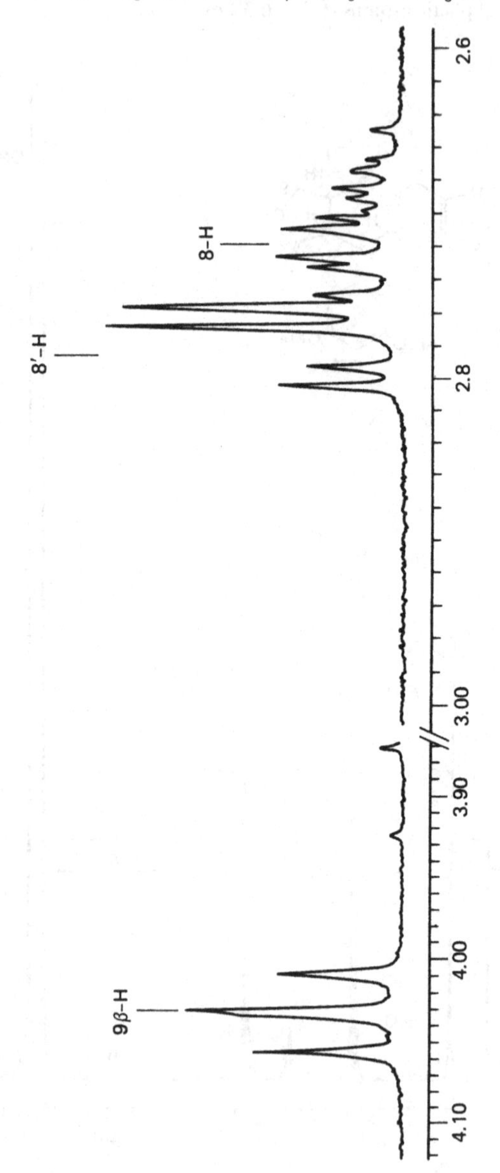

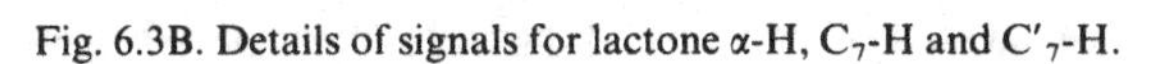

Fig. 6.3B. Details of signals for lactone α-H, C_7-H and C'_7-H.

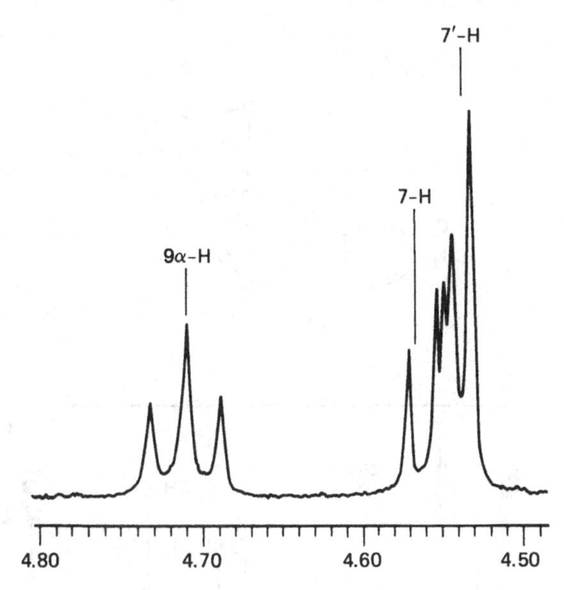

Table 6.18. *PMR spectra of tetralin lactones* (δ, $CDCl_3$)

Podophyllotoxin (360 MHz; Brewer, 1979)	Position	α-Conidendrin, di-Bz (400 MHz; Dahl, 1986)
4.60 4.09	lactone	4.17(eq) 3.97(ax)
2.83	8′	2.48
4.59	7′	3.80
2.77	8	2.55
4.77	7	2.95(ax) 3.18(eq)
7.11	6	6.68
6.51	3	6.28
6.37	2′ 6′	6.42 6.57

FIGURE 6.4. 2D-COSY spectrum of podophyllotoxin-β-D-thioglucoside at 400 MHz.

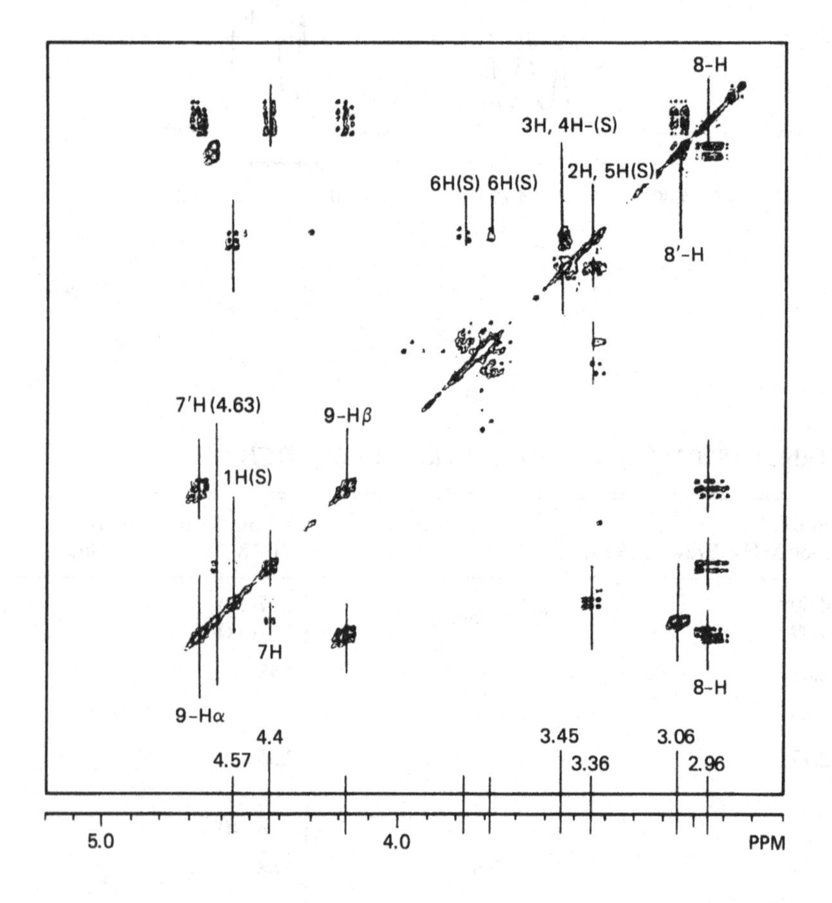

Table 6.19. *Coupling constants (Hz) of lactones.*

Position	*trans*-8,8′		*cis*-8,8′	
	6.161	**6.154**	**6.154**(8α)	**6.161**(8′-β)*
8,8′	14–15	13.5	9	9.0
7′,8′	3.5	10	9.8	6.75
7,8	8.0	—	—	9

* for D_6-DMSO

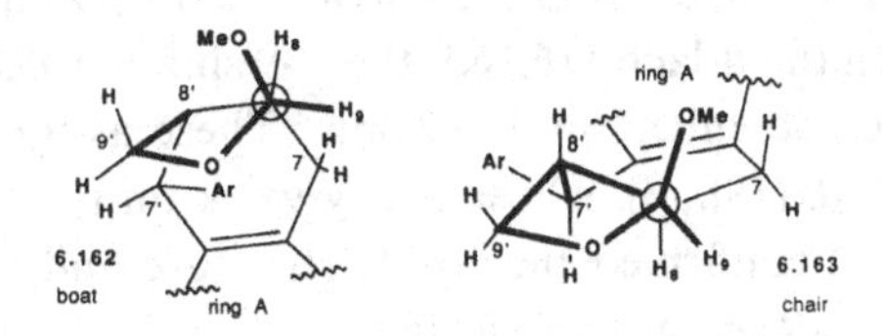

deshielded by the carbonyl group. The equatorial proton at C7 lies within the deshielding zone of the 9-carbonyl group, but an exact comparison is not possible since only a partial assignment has been made (Ayres *et al.*, 1972a) for the analogous retrolactone desoxypodophyllotoxin. Another point of interest is that in α-conidendrin (**6.154**) the aryl substituent at C7′- is equatorial, because C3-H lies at high field ($\delta = 6.28$) weakly coupled ($J = 0.75$ Hz) to C7′-H, whereas in podophyllotoxin (**6.161**) this resonance is at $\delta = 6.46$ consistent only with an axial aryl group at this position.

The C6-H resonance is sensitive to structural variation at C-7. Thus it is shifted upfield by *ca.* 20 ppm in epipodophyllotoxin and a similar distinction may be made (Kuhn and Wartburg, 1969) between podophyllotoxin-β-D-glucoside ($\delta = 7.38$, DMSO) and the epiglucoside ($\delta = 7.07$, DMSO). The crowded spectra of the glucosides required 2D experiments for their resolution and an annotated diagram is shown in Figure 6.4.

The *trans*-8,8′-configuration is reliably characterised by J values in the range of 13.5–15 Hz in these lactones (**6.161, 6.154**, Table 6.19). The two *cis*-8,8′-lactones exhibit typical couplings which are smaller than those found for the *trans*-isomers.

The magnitude of J7′8′ (9.8 Hz) and the chemical shift of C3-H ($\delta =$ 6.24) point to a boat conformation in β-conidendrin. It is also likely that the naturally occurring lactol of β-conidendrin includes a major contribution from a boat form (**6.162**) since two reports (Cambie, 1979; Dahl, 1986) assign a zero value to $J_{8,9}$ in this substance, consistent with a dihedral angle close to 90°C for C8-H, C9-H.

The use of the Karplus relation led to error in the assignment of configuration to epimeric lactols. Thus Dahl suggested that the original assignment (Cambie *et al.*, 1979) of a C-9β-configuration to the analogous lactol from α-conidendrin was incorrect. The confusion arose because of very similar values for the coupling constants: $J_{8,9} = 4.5$ Hz for a β-lactol and *ca.* 6 Hz for the α-epimer. Since the dihedral angles are very different it is clear that the coupling is affected by the electronegative oxysubstituents on C-9 (p. 211). This feature is also evident in cubebin (**2.021**) where the β-lactol has a value for $J_{8,9}$ of 4.6 Hz, whilst in the α-epimer this coupling constant is only 1.8 Hz (Wei-Ming, 1987). A reliable assignment of configuration to these lactols can be made from the CMR spectra (Dahl *et al.*, 1986; Cambie *et al.*, 1985). In the β-lactol (**6.163**) the C-9 shift is 105.0 ppm and in the α-epimer this lies downfield at 110.2 ppm. These assignments were confirmed by Fang *et al.* (1985b), whose X-ray work on the related compound tsugacetal (**2.311**) confirmed the whole structure and further showed that the higher field C-9 signal arose from the β-lactol.

When reliable comparison spectra are available conformational analysis is possible using 1-H coupling constants. From an examination of models, and assignment of $J_{7,8} = 9.5$ Hz (eq, eq) in a boat form and of 1.5 Hz (ax, ax) in a chair form, estimates of the two conformers can be made (Ayres *et al.*, 1972a) for the hydrated form of picropodophyllin. In DMSO this compound has $J_{7',8'} = 6.75$ Hz which represents a weighted mean of the boat form (65%) and the axial form (35%).

(iv) *Absolute configuration*

In their classic paper Schrecker and Hartwell (1953) interrelated the structures of the whole subgroup (Table 6.17) by synthesis of all four diastereoisomers with the absolute configuration shown in Scheme 6.20. Deoxypodophyllotoxin (**6.164**) was obtained by hydrogenolysis of the 7-chloroderivative and readily isomerised to deoxypicropodophyllin (**6.165**). α- and β-apopicropodophyllin (**6.166, 6.167**) were obtained by dehydration of picropodophyllin and have been characterised by PMR (Ayres, 1969). Hydrogenation of α-apopicropodophyllin yielded desoxypicropodophyllin (**6.165**) and also isodeoxypodophyllotoxin (**6.168**), whilst a fourth deoxylactone was obtained by *cis*-addition of hydrogen to β-apopicropodophyllin (**6.167**). It was then argued that, since **6.165** and **6.168** had the same configuration at C7′-C8′ and yet differed in themselves and also from **6.169**, that they must commonly be *trans*-7′,8′. Deoxypodophyllotoxin (**6.164**) must differ from the apocompounds and have a common *cis*-7′, 8′-configuration with isodeoxypicropodophyllin (**6.169**), their difference arising at the C8, C8′-lactone fusion.

Scheme 6.20 Configuration of aryltetrahydronaphthalenes

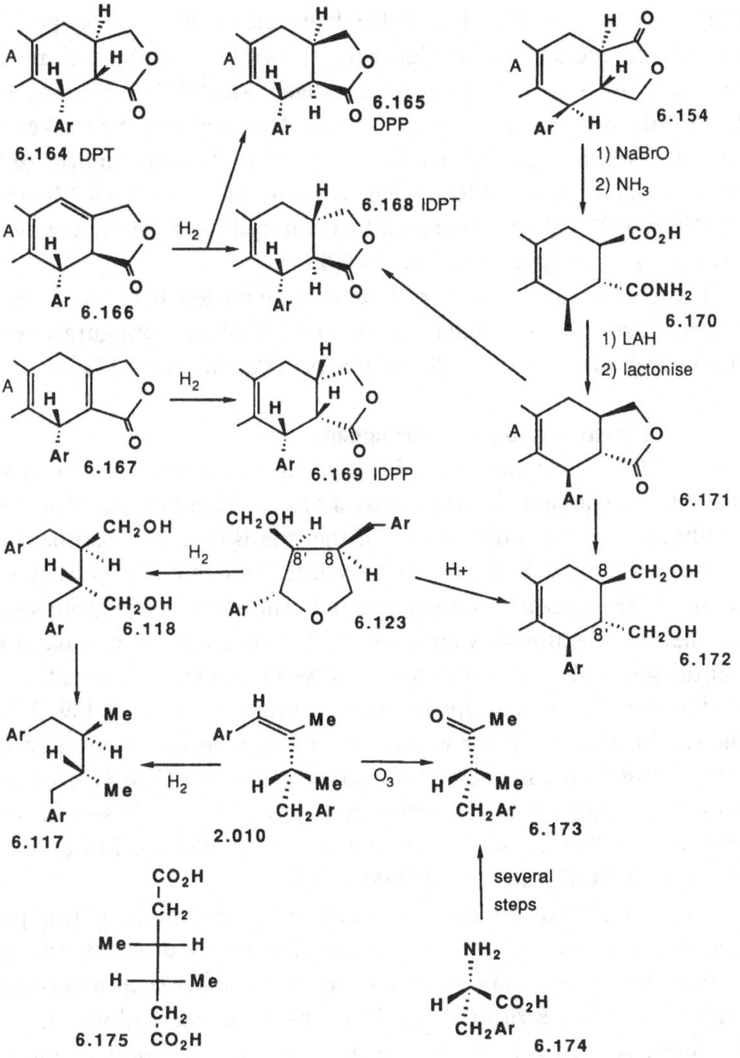

Further evidence was obtained (Cisney *et al.*, 1954) by the conversion of (−)-dimethyl-α-conidendrin (as **6.154**) via the acid amide (**6.170**) into (−)-retrodendrin (**6.171**), which had the *ent*-configuration to that of iso-deoxypodophyllotoxin (**6.168**).

Reduction of (−)-retrodendrin (**6.171**) gave (+)-cyclolariciresinol (**6.172**) and provided a link with other groups of lignans, because the latter is also the product of acid-catalysed cyclisation of (+)-lariciresinol (**6.123**) and retains configuration at C8, C8′. By its reduction (+)-lariciresinol affords (−)-secoisolariciresinol (**6.118**) whose correlation with (−)-dihydroguaiaretic acid (**6.117**) has been mentioned above (pp. 205–6). (−)-Guaiaretic acid (**2.010**) has been shown to have the L-configuration by the synthesis of its cleavage product, 3-methylbutan-2-one (**6.173**), from L-dihydroxyphenylalanine (**6.174**; Hartwell and Schrecker, 1958). Evidence of the absolute configuration of (−)-α-conidendrin (**6.154**) is also available (Kato, 1966) from its relationship with (+)-lyoniresinol (**2.329**, p. 222), which was shown by ozonolysis to have the configuration of (−)-(3*R*, 4*R*)-dimethyladipic acid (**6.175**).

The chemically based correlations summarised here have been confirmed through determination of the absolute configuration of 2′-bromopodophyllotoxin by X-ray analysis (Petcher *et al.*, 1973).

Dibenzocyclo-octadiene lignans

The first compound of this group was isolated by Kochetkov *et al.* (1961) and assigned the structure **6.176** in which the optical activity was attributed to a *trans*-disposition of the ring B methyl substituents. Later workers (Chen, Y-Y. *et al.*, 1976) pointed out that this activity was the result of the asymmetry or atropoisomerism of the biaryl groups and that the methyl substituents were *cis*-related. This geometry is usual throughout the series of compounds unless an 8α-OH group is inserted, when the C9,9′-methyl groups adopt the *trans*-configuration (cf. **2.429**, **2.430**). In the earlier work an error was made in assigning the orientation of aromatic substituents in the related lignan wuweizisu C (**6.177**) through an invalid comparison of low-intensity peaks in the mass spectra of these compounds. The structure was corrected to **6.178** by Liu *et al.* (1978) through an NOE study (pp. 234–5).

The inclusion of the biaryl group within a cyclo-octane ring prevents free rotation and the system becomes inherently dissymmetric. In this respect the lignans may be compared to alkaloids such as neodihydrothebaine (**6.179**; Ferris *et al.*, 1971) in which the biaryl adopts the *S*-configuration. A secure basis for the absolute configuration of the lignans comes from an X-ray study (Ikeya *et al.*, 1979a) of the dibromoderivative

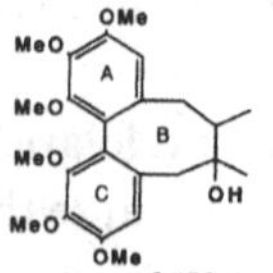

6.176 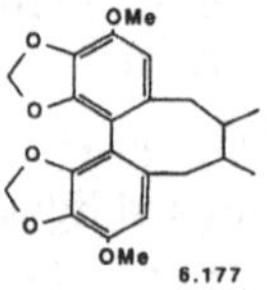 6.177 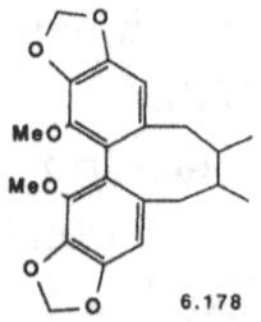 6.178

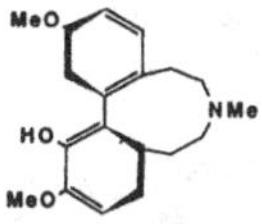

6.179

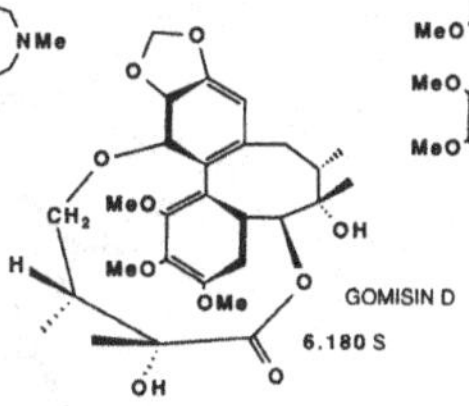

GOMISIN D

6.180 S

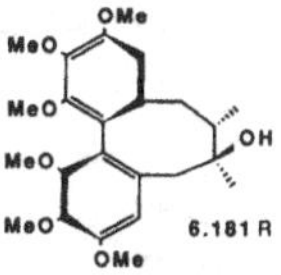

6.181 R

of gomisin D (**6.180**) in which the biaryl is also in the *S*-configuration. CD comparisons are of value in assigning configuration (Ikeya *et al.*, 1979b) and this is discussed on p. 257, although it may be noted that the configuration of schizandrin (**6.181**) was shown to be *R*- and that both types are well represented in nature.

With the atropoisomerism defined we must also distinguish between two possible conformations of ring B in each. It was shown (Anet and Yavari, 1975) that the twist-boat form (**6.182**) is flexible but the twist-boat-chair (**6.183**) is rigid. The two conformers may best be distinguished by viewing them as drawn, when the dihedral angle between the C2,2′ and C8,8′ bonds is almost ninety degrees in the twist-boat (TB) and only small in the twist-boat-chair (TBC). The transition between these two forms is marked by strong non-bonded interactions between one of the C7-H and C7′-H atoms and the transannular aromatic ring. In the boat form in the absolute configuration shown (**6.182**, *R*) the unfavourable interactions are those between 8*β*-H and Ar′ and 8′*α*-H and Ar. In a lignan one of a pair of *cis*-methyl groups must replace one of these C—H bonds and be placed so as to interact with the *transannular* ring and to be shielded by it. In the rigid TBC form 8-*α* H and 8′-*β* H interact with *adjacent* Ar groups but to a lesser extent and there is also an interaction between 7-*β* H and 8′-*β* H. Here also in a *cis*-model one of the methyl substituents will be shielded, but by the *adjacent* aryl group.

Provided that the skeleton is not additionally substituted in ring B, then the methyl groups are *cis*-related in all lignans so far described and this ring adopts a TBC conformation. As shown (Scheme 6.21; **6.184**) the PMR spectrum will include two clearly separated peaks – one arising from a

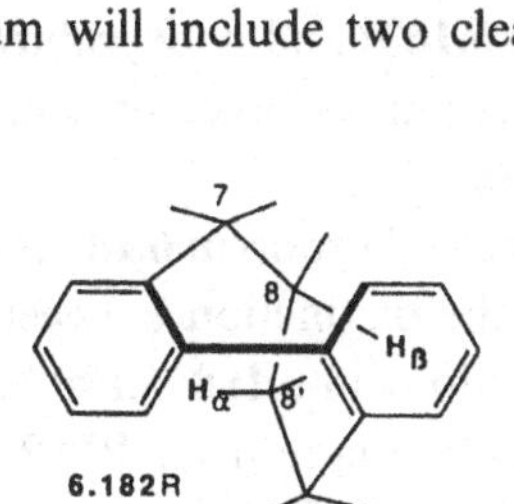

6.182R

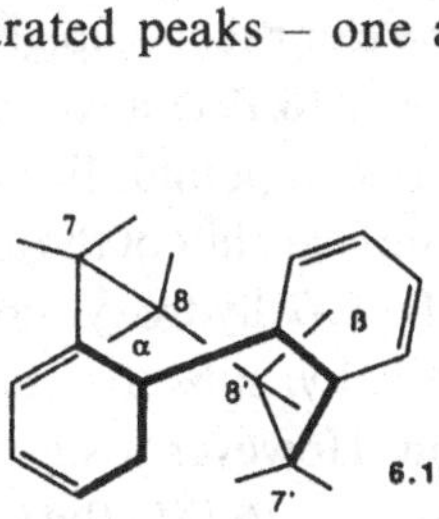

6.183 R

Scheme 6.21 ^{1}H- chemical shifts of schizandrin & deoxyschizandrin (∂)

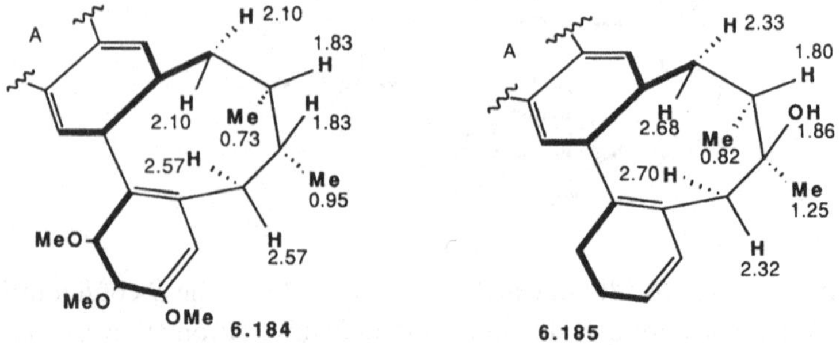

6.184 **6.185**

shielded axial methyl group ($\delta = 0.73$) and the other from an equatorial methyl substituent ($\delta = 0.95$). A synthetic *trans*-C8,C8′-dimethyl analogue is known (Takeya *et al.*, 1985) and because of higher symmetry it has a simpler spectrum in which equivalent methyl groups resonate at $\delta = 0.84$. The same workers also prepared the boat form bearing equatorial *trans*-substituents with common $\delta = 1.01$; these substituents would be eclipsed were it not for the flexing of the boat. Gomisin O (**6.186**) was the first of these lignans to be isolated in the boat form (Ikeya *et al.*, 1979d) and it is of interest to note that on hydrogenolysis it yields the *cis*-TBC form. This is a practical pointer to destabilisation of the boat through the unfavourable non-bonded interaction of an 8_{axial}-H and an aromatic ring (cf. **6.182**).

As seen above (p. 206) the introduction of further substituents lowers the symmetry and facilitates the assignment of peaks. If for example one examines the chemical shifts of ring B protons (Scheme 6.21), then a full assignment of benzyl protons is not possible for deoxyschizandrin (**6.184**) but in schizandrin (**6.185**) these peaks are resolved (Ikeya *et al.*, 1979*b*).

CMR spectra (Ikeya *et al.*, 1980a) are of value in placing ring B substituents and this is shown by the allocation of carbon shifts in a group of alcohols (Scheme 6.22) which, with the exception of gomisin O (**6.186**), all have the TNBC conformation.

It is evident that the shifts of C7′, 8′, and 9′ differ in the boat form (**6.186**) from those in the chair forms epigomisin O (**6.187**) and deoxyschizandrin (**6.184**). Gomisin R (**6.191**) also has the uncommon boat form and is a valuable reference compound (Ikeya *et al.*, 1982b).

The marked upfield shift of the axial C9-methyl group in **6.187** is due to the γ-effect of the C-7 hydroxyl group and the distinction between axial and equatorial methyl substituents is less obvious in **6.188**–**6.190** because of hydroxylation. However assignments can be made from PMR shifts, since in the boat form (**6.186**) the C9- and C9′-methyl groups are equival-

Scheme 6.22 ^{13}C- chemical shifts (∂) in ring B of hydroxycyclooctadienes

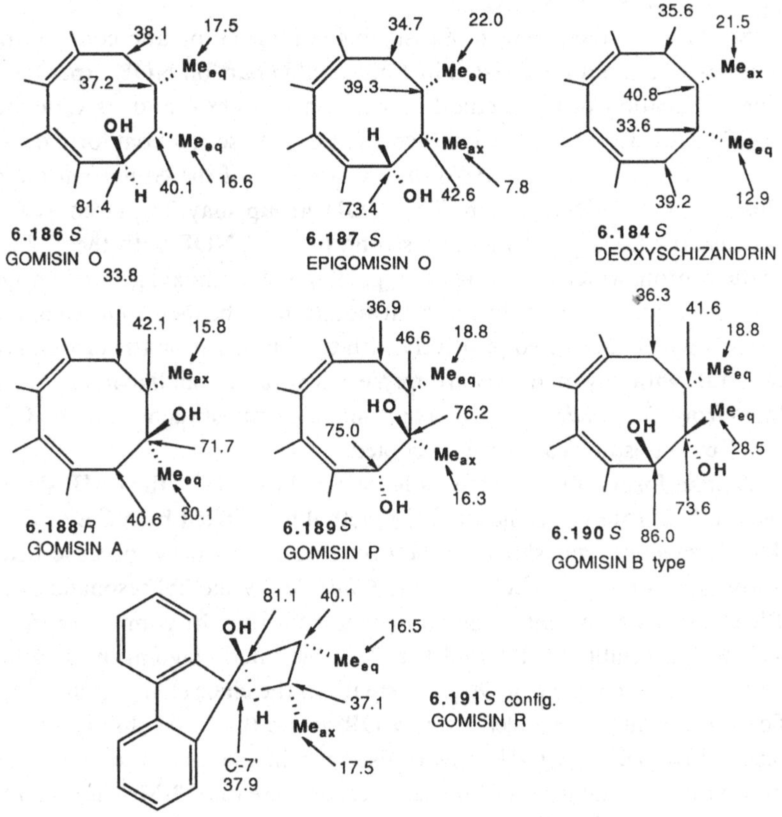

6.186 *S*
GOMISIN O

6.187 *S*
EPIGOMISIN O

6.184 *S*
DEOXYSCHIZANDRIN

6.188 *R*
GOMISIN A

6.189 *S*
GOMISIN P

6.190 *S*
GOMISIN B type

6.191 *S* config.
GOMISIN R

N.B. aromatic substituents omitted

Scheme 6.23 NOE experiments on gomisin O & epigomisin O

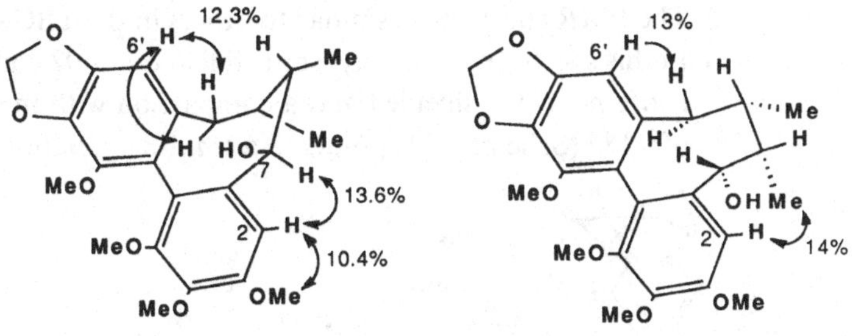

6.186
GOMISIN O (boat)

6.187
ISOGOMISIN O (chair)

ent ($\delta = 0.92$) whereas in the chair (**6.187**) axial ($\delta = 0.70$) and equatorial ($\delta = 1.00$) peaks are resolved.

Solutions to the problem of assigning conformation and configuration in the whole series have rested in considerable part on NOE experiments. The application of the method to gomisin O (**6.186**) and its C7-epimer (**6.187**) provides an illustration (Scheme 6.23). As seen a boat form may be allocated to gomisin O (**6.186**) on the strength of the equivalent methyl shifts in the PMR spectrum. The —OH group may be placed at C-7β because the C-7α proton gives a strong positive NOE with the C-2 aromatic proton, which is itself related (p. 199) to a methoxyl group in ring C. This orientation was confirmed by irradiation of the methylene signal due to C-7′, which interacted positively with C6′-H: this aromatic proton gives no NOE with any methoxyl group. Very fortunately all lignans so far isolated from *Schizandra* sp. bear six aromatic oxysubstituents and the C2-H and C6′-H positions are never occupied.

A chair form is indicated for epigomisin O (**6.187**) by the CMR shifts of the methyl groups and the axial C9-methyl was linked by NOE to C2-H. The 7α-position was shown to be occupied by hydroxyl because irradiation of 7-H had no effect on either C2-H or C9-methyl resonances. The PMR coupling constants are consistent with this. In gomisin O (**6.186**) $J_{7,8}$ is appreciable (8 Hz) whilst in 7α-substituted epigomisin O (**6.187**), with a dihedral angle near 90°, it is small. In the chair form it is also significant that when deshielded by a 7α-OR group the C2-H shift is typically 0.80 δ downfield of C6′-H. This is also seen in the α effect of a hydroxyl group on aliphatic protons for example, compare the C9-Me shifts in **6.185** and **6.184** (Scheme 6.21).

The study of the complex molecule gomisin D (**6.192**, **6.193**; Ikeya *et al.*, 1979a), provides an illustrative review of the techniques which have been mentioned. Here the C7- and C2′-positions are bridged by a 2,4-dihydroxy-2,3-dimethylbutanoic residue (**6.193**) and the criteria for assigning the structure may be summarised:

(a) The PMR spectrum is similar to others in the TBC series.
(b) In this spectrum the methyl shifts fall at $\delta = 1.02$ and $\delta = 1.20$, hence neither is shielded and a comparison with the spectrum of **6.185** (Scheme 6.21) points to a *trans*-8,8′-relationship.

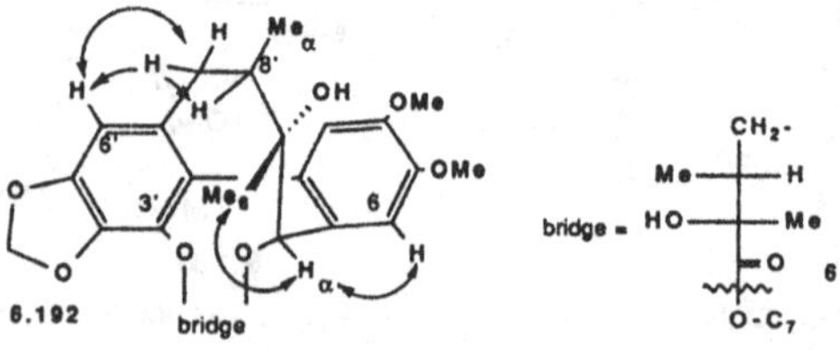

6.192

6.193

(c) After hydrolysis of the esterified C-7 terminal of the bridge with retention of the ether linkage, the revealed hydroxyl group was shown to be vicinal to the tertiary —OH by periodate cleavage.

(d) The methyl group deshielded ($\delta = 1.20$) by adjacent —OH was linked by NOE with C7-Hα, which in turn gave signal enhancement on irradiation of C6-H.

(e) The geometry was confirmed when further NOE experiments showed that C8′-H, coupled to C9′-Me, was close to C6′-H. It was also shown that the C7′-benzyl pair included one proton which was close to C6′-H.

The bridging residue is almost certainly linked biogenetically with angeloyl and tigloyl esters which also occur in these lignans (**2.431**, **2.472**).

(i) *Orientation of aromatic substituents*

There has been previous mention of this problem (p. 219). In order to provide an 'anchor' for NOE experiments it is of first importance to relate ^{1}H-shifts to the two unsubstituted positions.

Early on in these studies (Chen, Y-Y. *et al.*, 1976) oxidative cleavage was shown to yield *meso*-3,4-dimethyladipic acid and trioxyphthallic acids (Liu *et al.*, 1978). For example, schizantherinol A (Scheme 6.24: **6.194**) afforded the keto-alcohol (**6.195**) and 3,4,5-trimethoxyphthallic acid (**6.196**). These diacids are generally available as reference compounds (Beilstein, 1971, 1972) and diphenic acids may also be obtained by cleavage of ring B (Ikeya *et al.*, 1979b).

With one position always free in each ring no problem arises if the substituents are the same since no room remains for permutations. Should

Scheme 6.24 Oxidative degradation of schizantherinol A

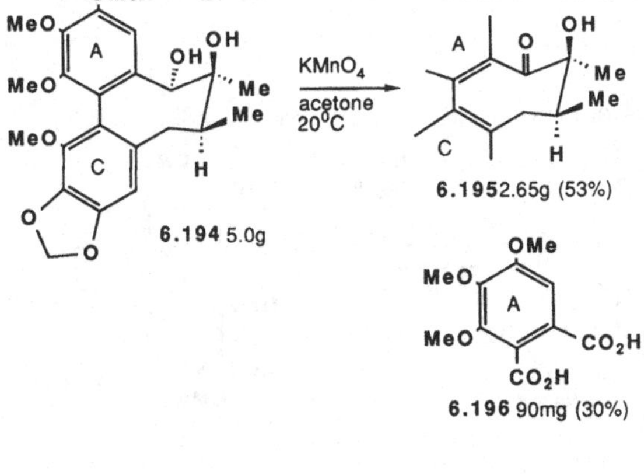

OMe
MeO
MeO
MeO
A
C
OH
OH
Me
Me
H
O
O
6.194 5.0g

$KMnO_4$
acetone
20°C

A
C
O
OH
Me
Me
H
6.195 2.65g (53%)

OMe
MeO
MeO
A
CO_2H
CO_2H
6.196 90mg (30%)

Scheme 6.25 Orientation of substituents in binankadsurin A

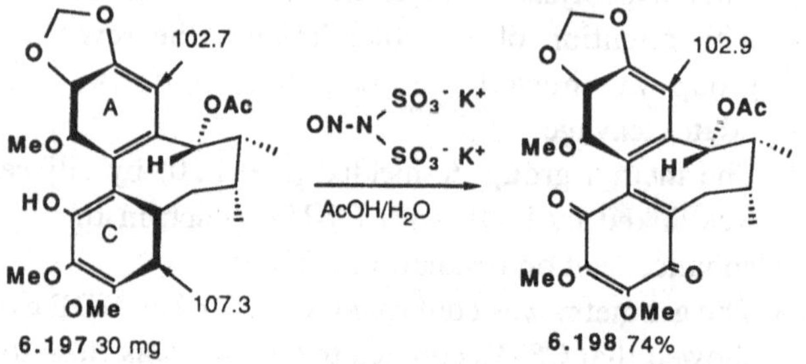

6.197 30 mg → ($ON\text{-}N(SO_3^-K^+)_2$, $AcOH/H_2O$) → **6.198** 74%

one ring include a phenolic —OH group the associated proton may be found by the effect of alkali on its PMR shift or through off-resonance decoupling (p. 201) of the one proton associated with an alkali sensitive ^{13}C-shift. This approach is illustrated by work on binankadsurin A (**6.197**; Ookawa, 1981) which depended on existing data on CMR shifts of aromatic systems in these lignans (p. 233) and the use of Fremy's salt (Moser and Howie, 1968). This reagent will only oxidise phenols to *para*-quinones when this position is unsubstituted (Scheme 6.25). Here the CMR peak at $\delta = 102.7$ was assigned to ring A and that at 107.3 to ring C. It was evident that the latter arose from a C—H position since on oxidation it was replaced by a low-field carbonyl peak. The absorptions due to the aromatic protons were close together at $\delta = 6.41$ and 6.44 and were replaced in the quinone (**6.198**) by a single peak at $\delta = 6.44$.

Another approach to the problem of placing phenolic substituents is by reduction followed by determination of the characteristic -^{1}H coupling constant in the dihydrobenzene obtained. The structure of Gomisin K_3 (**6.199**) was established in this way (Ikeya *et al.*, 1980b; Scheme 6.26). The sodium salt was first formed using sodium hydride and this was then combined with 2,4-dinitrofluorobenzene. The powerful electron-attracting groups in the intermediate render the C—O bond in the original phenol

Scheme 6.26 Reduction of a phenolic -OH group

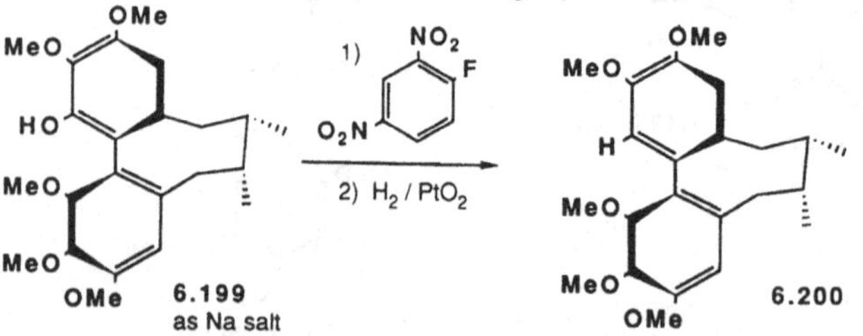

6.199 as Na salt → (1) 2,4-dinitrofluorobenzene; 2) H_2 / PtO_2) → **6.200**

Table 6.20. *Aliphatic -*[1]*H shifts* ($\delta = 250\,MHz$) *of biphenyl cyclo-octadienes* (*Taafrout, 1983a*) *and comparison with acetyl podophyllotoxin* (*Ayres and Lim, 1972*)

	Steganacin	Ac podophyllotoxin	Steganangin
7-H	2.59 3.03	4.43 (C7′)	2.57 Hα 3.07Hβ
8-H	2.51	3.0 (C8′)	2.55
8-H′	2.53	2.7 (C8)	2.52
9-H′	—	4.26 (9-Hα or 4.00 9-Hβ)	4.26 Hα 4.00 Hβ
7-H′	5.82	5.80 (C7)	5.98

liable to cleavage by hydrogenolysis (step 2). This leads to a product (**6.200**, 42% overall yield) in which three *singlet* PMR peaks appear showing that the replaced phenolic —OH group was in a *para*-relationship.

(ii) *The steganacin subgroup*

These compounds differ from the gomisins in that the ring B is rendered inflexible by a *trans*-lactone fusion. The first of these lignans to be isolated (Kupchan *et al.*, 1973) was (−)-steganacin (**6.201**; R = Ac) from *Steganotaenia araliacea* (0.4% yield) together with steganangin (**6.201**; R = angeloyl, 0.1%), steganol (**6.201**; R = —OH, 0.001%) and the 7′-ketone, steganone. In this pioneering work the configuration at C-7′ in steganacin was wrongly assigned and the absolute configuration, based on an X-ray analysis without reference to a heavy atom, was also incorrect. The reader is cautioned that this has led to errors in the literature which even persist after a correction was made in 1980 (Robin *et al.*; Tomioka *et al.*, 1980).

The structure of steganacin was originally defined by a comparison of its spectra with those of podophyllotoxin (**6.161**; Table 6.20). The presence of a five-membered lactone ring was indicated by the IR spectrum ($\nu = 1770\,cm^{-1}$) and a *trans*-configuration by the large coupling between 8-H and 8′-H ($J = 13\,Hz$). Significant differences are the downfield shifts of these two protons (compare Scheme 6.21) and the absence of a diphenylmethane proton signal. These features are consistent with a cyclo-octadiene ring in steganacin in the form of a twist-boat distorted by the nearly coplanar atoms of the lactone ring.

The geometry of ring B in the steganacin group is thus well characterised by the PMR spectra. Epimeric variation at C7′ is shown by twisting of ring B and a small value (1.2 Hz) for $J_{7a,8}$.

Table 6.21. *Coupling constants for araliangine* (**6.202**) (Hz at *250 MHz*)

Paired protons	J value	Paired protons	J value
7α,8	11.0	8′,9′α	7.5
7β,8	7.2	8′,9′β	12.0
7α,7β	15.0	9′α,9′β	9.0
8,8′	13.5	8′,7′α	8.2

Table 6.22. *Aromatic proton shifts in biarylcyclo-octadienes*

	6.201 (7′, C═O)	**6.201** (R = Ac)	**6.201** (R = Angeloyl)	**6.202** (R = Angeloyl)
6H	6.56	6.58	6.45	6.60
6′H	7.45	6.89	6.87	7.1

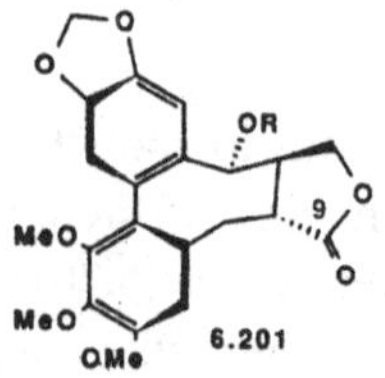

6.201 (refers to Table 6.22)

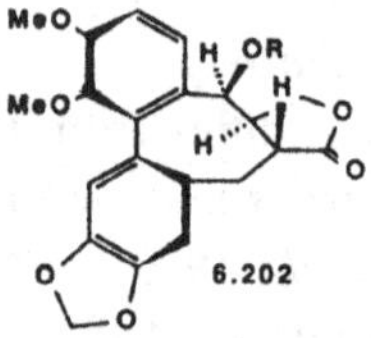

6.202 (refers to Tables 6.21 & 6.22)

In this group NOE has not been used to confirm structures although it has been shown (Taafrout *et al.*, 1983a) that weak coupling (0.5–0.9 Hz) occurs between 7-H, 7′-H in ring B and any adjacent aromatic proton at C6 or C6′. The chemical shift of the C6′-proton responds to substitution at C7′ and the C6-H provides a point of reference (Table 6.22). A weak *para*-coupling can be demonstrated for $J_{3',6'}$. In general, problem solving by NOE would relate the adjacent methoxyl group to 6-H which is always at higher field than 6′-H owing to deshielding of the latter by 7′-oxysubstituents.

With the relative configuration of these compounds established it was pointed out (Zavala *et al.*, 1980) that since steganacin and podophyllotoxin both showed strong anti-tubulin activity they should have the same absolute configuration: podophyllotoxin (pp. 228–30) has the *ent*-form of that then accepted for steganacin. These doubts were resolved in the same year by the stereoselective synthesis of (−)-steganone (Robin *et al.*, 1980) and also by the synthesis of (+)-*ent*-steganacin from L-glutamic acid (Tomioka *et al.*, 1980; see p. 346). As a result of this work the absolute configuration of steganacin was revised to that of **6.201** shown above; this has been further confirmed by X-ray analysis (Houlbert, 1985).

An *S*-configuration in steganacin and steganone was inferred from its laevorotation and comparison of the circular dichroism of the gomisin group (p. 256). It is worth noting that only a twist-boat form in ring B can accomodate a *trans*-lactone fused at C-8′β–C-8α, since in a TBC model these bonds would be *trans*-anti-*trans*. Nevertheless it is known (Kende *et al.*, 1976) that the *R*-isomer (+)-isosteganone (**6.203**) gives the atropoisomer (**6.205**) on heating in acetic acid/acetate (Scheme 6.27). Prolonged treatment is needed to effect this change in isostegane which then affords a mixture of products (see p. 240). This is explained by hydrolysis of the lactone in the presence of a trace of acid when isoteganone with a TBC ring B can flip on heating into a twist-boat form and re-lactonise. The driving force for the more facile keto-isomerisation is thought to be that in the intermediate (**6.204**) the keto- and exocyclic methylene groups come into conjugation in the plane of ring A. In *S*-steganone, which predominates, the keto group retains this near planar relationship but in *R*-isosteganone it is almost orthogonal to ring A.

An important feature of the chemistry of this class is the effect of conformational change in ring B on reactivity. For example, in the *R*-diketone (**6.206**), written in the twist-boat conformation, the repulsion of the axial methyl group by ring A forces the 7-carbonyl group almost into this ring plane but the 7′-carbonyl group lies well out of the plane of ring C.

Scheme 6.27 Isomerisation of R-(+)-isosteganone

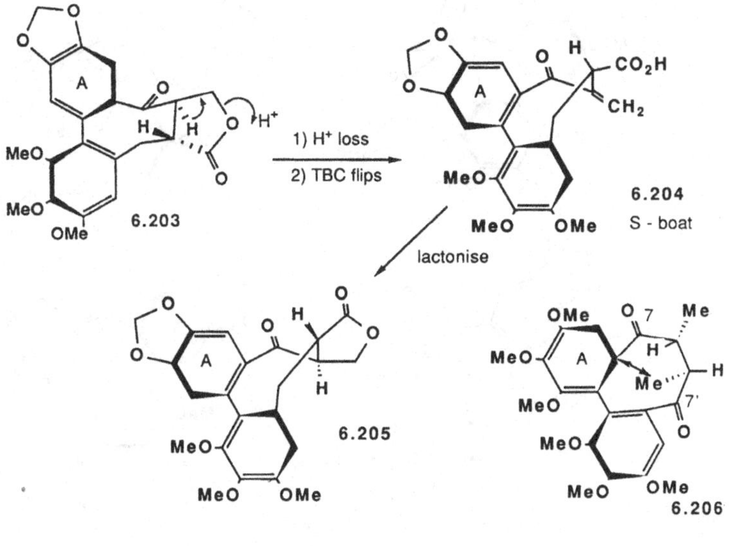

6.203 — 1) H^+ loss, 2) TBC flips → 6.204 S - boat — lactonise → 6.205

6.206

Although neither of these groups will enolise (Ghera and Ben-David, 1978) the near-planar function may be hydrogenated to a secondary alcohol, whilst the out-of-plane group is more electrophilic and is preferentially reduced by hydride donors. It must be pointed out that in this symmetrically substituted example both carbonyl groups are equivalent and out-of-plane in the TBC form. However under these mild conditions the *R*-configuration is unchanged and therefore only one carbonyl can be hydrogenated in the twist-boat conformation.

Of the four possible steganes (Scheme 6.28; Robin *et al.*, 1984) the picro-isomers are characterised by a typical *cis*-coupling with $J_{8,8'} = 8$ Hz (cf. picropodophyllin): in stegane and isostegane this value is 13 Hz. The iso-series may also be distinguished (Taafrout *et al.*, 1983a) by small or zero coupling constants for the proton pairs 7′α-8′ and 7β-8 (cf. Table 6.21) which in this *R*-configuration have a 90° dihedral relation. The CMR shifts (Scheme 6.22) distinguish the (*S*)-series of stegane from the (*R*)-series of isostegane.

As mentioned above the conformations of the stegane group are 'locked' by the *trans*-fused lactone ring, but given sufficient energy input these atropoisomers can be interconverted (Tomioka *et al.*, 1979b). On heating in nitrogen at 195°C isostegane (**6.208**) gave a binary mixture with stegane (**6.207**) in the relative proportions of 67:28 parts. When stegane (**6.207**) was isomerised in this way it gave a similar mixture including iso-

Scheme 6.28 The steganes showing CMR data

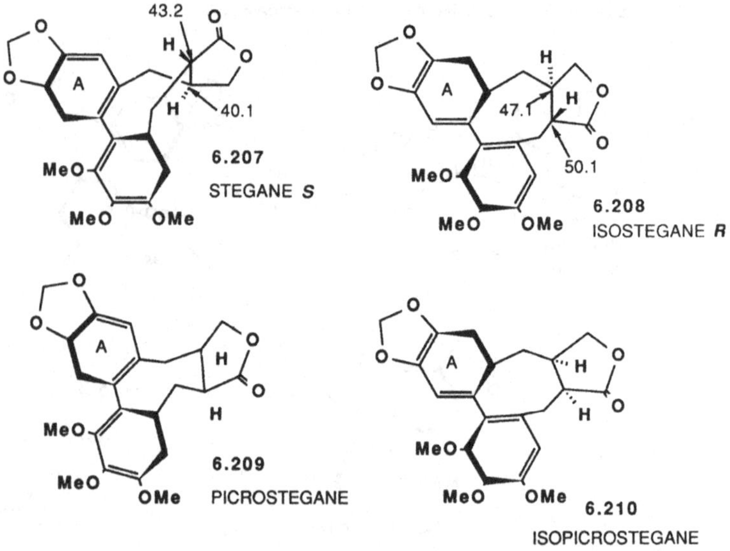

43.2 H O O H 40.1 A O O MeO MeO OMe
6.207 STEGANE *S*

O O A H 47.1 H O O 50.1 MeO MeO OMe
6.208 ISOSTEGANE *R*

O O A H O O H MeO MeO OMe
6.209 PICROSTEGANE

O O A H O H O MeO MeO OMe
6.210 ISOPICROSTEGANE

stegane (60%). When one of the group is heated in acetic acid enolisation of the lactone also occurs and as a result mixtures which include all four isomeric steganes are formed.

Neoisostegane, a naturally occurring pentamethoxy analogue, was recently isolated (Hicks and Sneden, 1983; Taafrout *et al.*, 1983b) and it was correlated with (*R*)-isostegane as shown in Scheme 6.29. The synthetic atropoisomer (**6.212**) had properties including a CMR spectrum which were generally similar to those of neo-isostegane (**6.211**), but analytical TLC showed that they were not identical. The difference was shown to lie in the ring C orientation:

(a) In the PMR spectrum of **6.211** the three methoxyl peaks lay at $\delta = 3.93$, 3.89, and 3.85 and showed no evidence of shielding by ring A. In **6.212** this effect gives rise to one high-field methoxyl peak at $\delta = 3.57$.

(b) 3′-H, 6′-H, and 7′-H had similar shifts in both spectra but in **6.211**, 7β-H had moved downfield to $\delta = 3.63$ ($J_{7\alpha,8} = 0$), which pointed to a methoxy group at C-2.

(c) The CMR shifts at C8 and C8′ were identical in both compounds (**6.211, 6.212**) with those of isostegane (50.1, 47.1) but the C-7 shift upfield in **6.211** to $\delta = 24.3$ (cf. p. 233) provided further evidence of C2-methoxylation.

In concluding this section one draws attention to two recent detailed studies. In the first (Houlbert, 1985) the constants recorded above

Scheme 6.29 Comparison of neoisostegane with isostegane

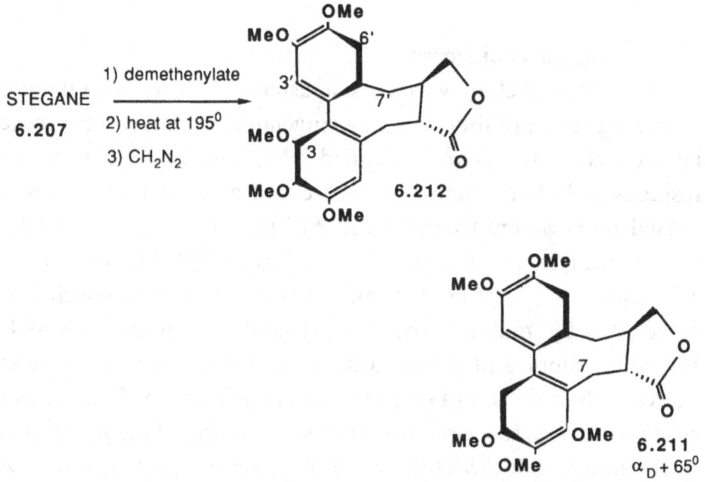

STEGANE 6.207 → 1) demethenylate 2) heat at 195° 3) CH_2N_2 → 6.212

6.211 α_D + 65°

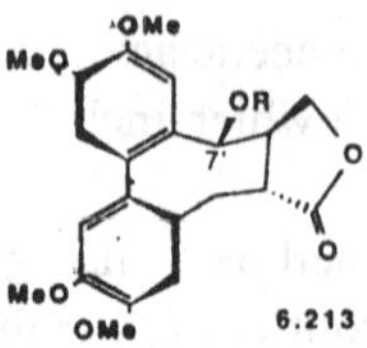

6.213

(Scheme 6.22, Table 6.20) were confirmed by a 2D-COSY spectrum of a steganol glucoside, a physiologically inactive analogue of Etoposide (p. 113). In the second study, Robin *et al.* (1986) records a 2D-correlation of the ^{13}C- and ^{1}H- resonances of steganacin (**6.201**; R = 7′-α-acetyl), steganangine (**6.201**; R = 7′-α-angeloyl), araliangine (**6.202**; 7′-β-angeloyl), and steganolide B (**6.213**; 7′-β-angeloyl). In the 7′-α-oxygenated pair this peak has a CMR shift of 77.2 ppm compared to that in the 7′-β-substituted pair where this peak falls at $\delta = 70.5$ ppm. There is also a characteristic difference of about 5 ppm in the lactone methylene resonance of these 7′-epimeric esters. This work provides further evidence of conformational change in ring B arising from a change in the configuration at C-7′. Thus 7′-α-substituents occupy a quasi-axial position and the dihedral angle between C7-αH and C8-H is close to 90°C ($J_{7\alpha,8} = 1.2$ Hz), but 7′-β substituents interact with ring A (cf. p. 227) so inducing a dihedral angle close to 180° for C7-αH and C8-H with a related increase in the coupling constant ($J_{7\alpha,8} = 10$–11 Hz).

The interesting structural variation shown in **6.211** leads one to consider it likely that other bicyclo-octadiene lactones await discovery, possibly in the numbers uncovered in the *Schizandrin* subgroup by the Japanese workers.

Lignan conjugates

These include various acid esters that have been mentioned in the foregoing text, notably, acetates, angelates, benzoates, caproates, and tiglates which all have clearly defined NMR characteristics; combined sugar residues include D-glucose, D-xylose and disaccharides. They are characterised by cleavage fragments in MS (pp. 193–4) and may be identified after hydrolysis by chromatography (pp. 151, 155). Hydrolysis can be by acid catalysis (pp. 144, 155) or through the use of specific enzymes: α-glycosides by maltase and β-glycosides by emulsin (Marshall, 1974). Phenolic esters and glycosides are more susceptible to acid-catalysed cleavage than those linked to the lignan side-chain (Stanek, 1963) and the emphasis in this section is on the determination of the point of attachment.

In nudiposide (**6.214**) the β-D-xylopyranosyl residue was placed

Scheme 6.30 Nature of the conjugation in nudiposide(6.214) & in an olivil diglucoside

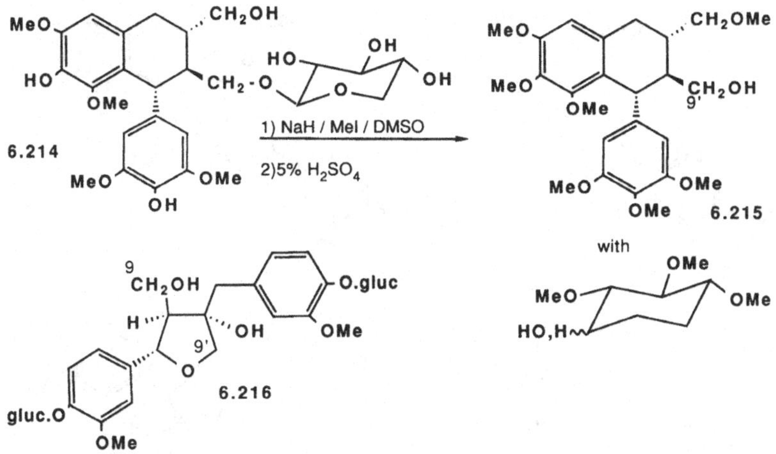

1) NaH / MeI / DMSO

2)5% H_2SO_4

(Ogawa and Ogihara, 1976) by complete methylation using the Hakomori method (Scheme 6.28) followed by hydrolysis to the monohydric alcohol (**6.215**) and the reducing sugar trimethylxylose, which was characterised by GLC (pp. 155–6). In this instance oxidation of the primary alcohol (**6.215**) yielded a known C9′-carboxylic acid, otherwise it is a straightforward exercise to compare the spectra of this product with those of the parent aglycone.

An alternative method of marking hydroxyl groups is by acetylation with acetic anhydride. If the product mixture is allowed to become acidic during working up then any phenolic groups will be isolated unchanged; however in the presence of base (pyridine) phenol acetates will be obtained. Free phenolic —OH groups may be identified before or after hydrolysis by the IR spectra, base-induced shifts in the UV (p. 168), or by the downfield chemical shift in the PMR spectrum (cf. p. 196).

An example is provided by the work of Deyama *et al.* (1985) who showed that in olivil-di-*O*-*β*-D-glucopyranoside (**6.216**) one hydroxyl group was resistant to acetylation, since a nona-acetate with a free aliphatic —OH group was obtained together with a deca-acetate. Evidently no free phenolic —OH group was present in **6.216** and hence the glucosyl residues were not combined in a disaccharide but each was effective as a blocking group. The more resistant —OH group was confirmed as tertiary by its chemical shift in the deca-acetate of $\delta = 1.62$ ($CHCl_3$), downfield of the C9-*O*-acetate which was not resolved from the eight acetyl signals of the glucosyl elements ($\delta = 2.04, 2.08$). Acetates of benzylic secondary alcohols as they occur in podophyllotoxins and steganacin have

Scheme 6.31 ^{13}C- Shifts in glycosides

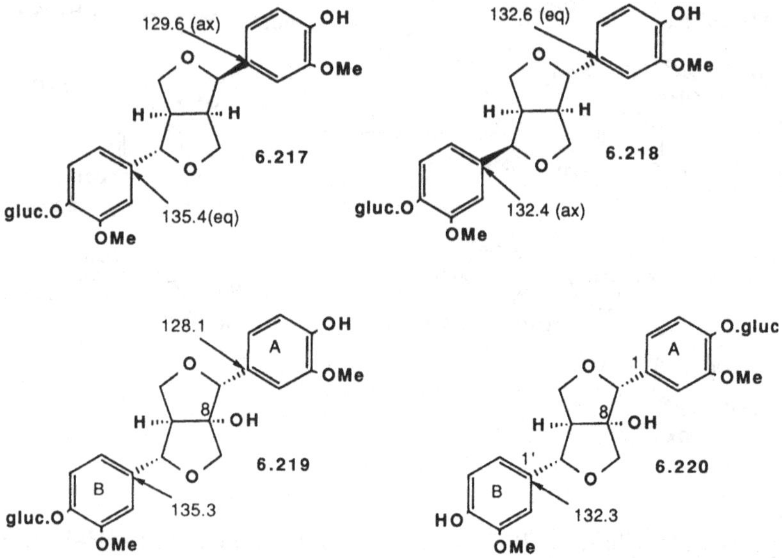

6.217 6.218 6.219 6.220

chemical shifts in the range $\delta = 1.55$–1.92. Typical changes in PMR and CMR shifts which occur on acetylation have been mentioned.

The range of monosaccharides known to occur in natural glycosides includes the hexoses – D-glucose, D- and L-galactose, D-mannose, D-fructose, and L-rhamnose. Conjugated pentoses include D- and L-arabinose, D-xylose, and D-ribose. In addition to the chromatographic techniques used for the identification of these substances there exists an extensive collection of CMR spectral data (Bock, 1983). This can be used for the identification of glycosides in the lignan conjugates before hydrolysis, but these have crowded spectra which may well require 2D-NMR techniques for their resolution (p. 226).

The most difficult problem in the placement is that which arises in asymmetrical glycosides of furofurans (7,9′:7′9-diepoxylignans). As seen above (p. 211) the chemical shift of the aromatic ring carbon at the point of attachment to the furofuran is sensitive to the variation in torsional response and can be used to distinguish between axial and equatorial substitution. A similar approach was made by Chiba *et al.* (1980) in establishing the point of glucosylation of isomeric aglycones in (−)-simplocosin (**6.217**; Scheme 6.31) and in (+)-epipinoresinol glucoside (**6.218**). In (**6.217**) the shift (129.6 ppm) of the reference atom in the axial aryl group falls within the typical range (p. 211), but that of the equatorial C2-substituent (135.4 ppm) is downfield of that typical of a guaiacyl residue. The

Table 6.23. *CMR shifts (ppm) in pinoresinol derivatives*

Compound	C-1	C-1′
(6.219)	128.1	135.3
(6.220)	131.1	132.3
8-Hydroxypinoresinol	128.1	132.4
Pinoresinol	132.4	132.4

attachment of the bulky glucosyl group has caused a further shift in this frequency. The isomeric model **(6.218)** *known to include a monoglucoside* has a different set of shifts – one (132.6 ppm) is typical of an unmodified equatorial substituent, but the other (132.4 ppm) can only be assigned to an axial substituent shifted about 3 ppm downfield by glucosylation.

The placing of glycosides is more complex if the furofuran is further substituted as for example in (+)-8-hydroxypinoresinol (Deyama *et al.*, 1986); this substance was isolated as a monoglucoside **(6.219)** together with the position isomer **(6.220)**. The presence of one tertiary and four other alcoholic groups was shown by acetylation, since the former gave a peak (3H) at $\delta = 1.96$ resolved from that (12H) at $\delta = 2.00$. The presence of one free phenolic —OH group in **6.219** was marked by the phenylacetate peak at $\delta = 2.24$. There is here evidence of shielding of the tertiary acetate by an equatorial ring A and the proximity of these two groups is confirmed by the CMR spectrum. Here the γ-related —OH leads to a downfield shift ($\delta = 3.7$ ppm) in ^{13}C-1 of the acetate, whereas the change in ^{13}C-1′ was minimal ($\delta = 0.9$ ppm).

The placing of the glucosyl residues can be explained by reference to Table 6.23. In **6.220** one whole set of aromatic shifts is identical to those in pinoresinol itself and so characterises the equatorial ring B with a free phenolic —OH group. When the glucoside **(6.220)** was hydrolysed the aglycone retained the equatorial characteristic (132.4 ppm) but the shift of the other reference atom changed (from 131.1 to 128.1 ppm); therefore the glucosyl attachment was to ring A.

Hydrolysis of the glucoside **(6.219)** gave the same aglycone but here the change was in the C1′-shift, which moved by a similar amount from 135.3 to 132.4 ppm; therefore the glucosyl attachment in this isomer was to ring B (Deyama *et al.*, 1986). One recalls that PMR evidence existed for the equatorial position of ring A and one can record that an adjacent C8-OH group shifts the C1-reference 4 ppm downfield but no data on such an axially placed ring are available.

A very significant structural feature is the recently reported isolation

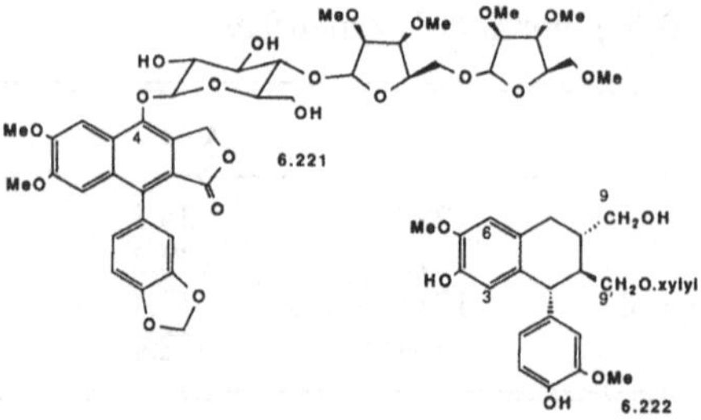

(Abe and Yamauchi, 1986) of the 5′-C-β-D-glucoside of matairesinol (**2.087**) when CMR spectroscopy was used to locate the glucosyl residue.

Higher glycosides

Disaccharide conjugates are rare in lignans but the 4′-*O*-[β-D-xylosyl-(1-6)-β-D-glucosyl] derivative of the furofuran phillygenol (**2.216**) has been described (Zheng and Wang, 1985). The mass spectrum of cleistanthin E, an interesting glycoside of an arylnaphthalene has been discussed (cf. p. 194); it is 4-*O*-[2,3,5-tri-*O*-methyl-D-xylofuranosyl-2,3-di-*O*-methyl-D-xylofuranosyl-(1–4)-D-glucopyranosyl-diphyllin (**6.221**). This substance afforded the monohydric phenol diphyllin on hydrolysis with hot methanolic hydrogen chloride together with β-D-glucose, β-D-2,3-di-*O*-methylxylose and β-D-2,3,5-tri-*O*-methylxylose. The last named was evidently the terminal unit in the trisaccharide since after permethylation it was the only *mono*-hydroxylated fragment. Mild hydrolysis of cleistanthin E with sulphuric acid (0.05N) liberated only 2,3,5-tri-*O*-methylxylose and 2,3-di-*O*-methylxylose; therefore the latter, with two free —OH groups, is centrally placed and D-glucose is directly linked to the lignan.

In a related study Batirov *et al.* (1985) placed rhamnose as the terminal residue in versicoside (**2.278**), 4-[α-L-rhamnopyranosyl-(1-6)-β-D-glucopyranosyl]-epipinoresinol, where the disaccharide was shown by CMR to be linked to the axial aryl group of the lignan.

An example of a conjugated aryltetralin is that of schizandriside (**6.222**; Takani *et al.*, 1979), the 9′-β-D-xyloside of (+)-cyclolariciresinol, which gave D-xylose on treatment with emulsin. Acetylation of schizandriside yielded a hexa-acetate with four aliphatic primary/secondary ^{1}H-acetyl resonances in the range $\delta = 2.04$–2.07 ppm with *two* phenyl acetates ($\delta = 2.21, 2.32$). Hence as the aglycone was identified by its spectra as a *bis*-primary diol, the blocking xylose residue must be located at one of the terminal primary positions. A comparison (Table 6.24) of the CMR shifts for isolariciresinol tetra-acetate with those of the tri-*O*-acetyl

Table 6.24. *CMR shifts of cyclolariciresinol derivatives*

Carbon	Cyclolariciresinol tetraacetate	Schizandriside hexacetate
7′	47.2	46.8
8′	43.5	44.1
8	35.2	34.6
7	33.1	33.1
9′	63 — — — — — — — — —	67.1
9	66	65.9

schizandriside (**6.222**) was made, and since the only divergence in shift is that of C-9′ this must be the point of glycosylation.

Optically active conjugated molecules may sometimes be identified by the difference in the molecular rotation of the conjugated and the parent lignan. This is illustrated for schizandriside (**6.222**) in the following section.

There is no doubt that other glycosides await discovery in plants, for example no glucosides of the dibenzocyclo-octadienes have been reported. The isolation (Tanahashi *et al.*, 1987) of sambacolignoside (**2.274**) where hydroxypinoresinol is combined with a secoiridoid glucoside illustrates the wide range of possibilities. This complex structural assignment was based on 2D COSY spectra.

Optical rotatory dispersion and circular dichroism

It is to be regretted that all too often only the optical rotation at the sodium D line – $[\alpha]_d^t$ has been recorded and this for a family of compounds where absolute configuration is often associated with physiological activity. It is also significant that optical activity may not be detected when measurements are restricted to this one wavelength. An example of this is provided by anhydropodophyllol, a reduction product of podophyllotoxin, which only develops its ORD curve (see below) at shorter wavelength (Ayres and Pauwels, 1962). This feature also emerged very recently (L-Niang *et al.*, 1988) in neokadsuranin (**2.450**), which was also inactive at the D-line but gave a negative Cotton effect in its CD curve typical of its *S*-configuration (cf. Scheme 6.39).

(i) *Optical rotatory dispersion*

The calculation of the molecular rotation, $[\phi]$, at a given wavelength

$$[\phi] = M.a^\circ/lc$$

Scheme 6.32 Optical rotatory dispersion curves of helianthoidin and related lignans.

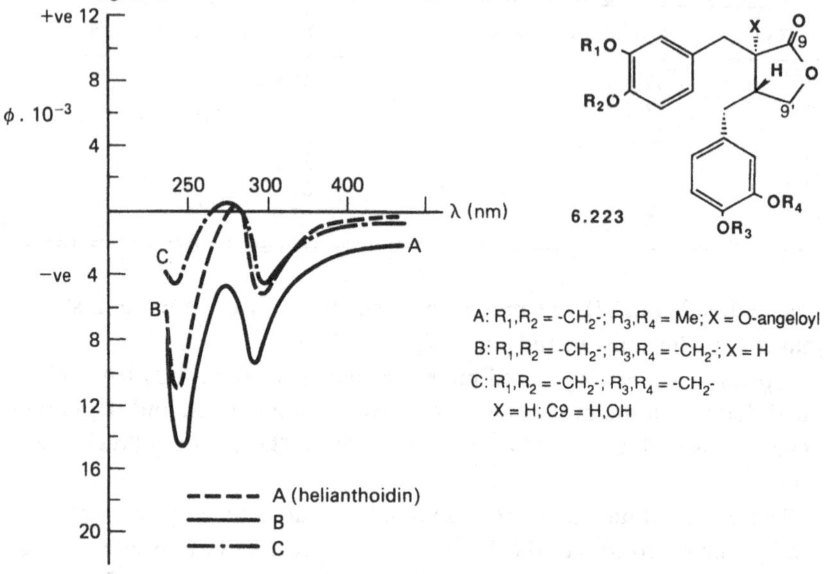

where M = molecular weight; l = path length (mm); c = solution conc. (g/ml); and $a°$ is the rotation at the wavelength, permits significant comparisons between different compounds on a molar basis. Thus the molecular rotation of schizandriside (**6.222**) is $+148 \times 10^2$; the specific rotation of (+)-isolariciresinol (M = 360), $[\alpha]_d$ (acetone) = +69.5, which gives a value for $[\phi]$ of 250×10^2. Subtraction of these two molecular rotations gives a difference of -102×10^2, which is close to that (-108×10^2) calculated for methyl β-D-xyloside.

When the rotation is measured at a wavelength where light absorption occurs, $[\phi]$ diverges from linearity and rises to a maximum in the region of λ_{max} for a given chromophore. This is known as the Cotton effect. It has an amplitude which is the difference between the first and second extremum divided by one hundred, $[\phi]_1 - [\phi]_2 \times 10^{-2}$. Examples of these curves are given above (Scheme 6.32; Burden *et al.*, 1969). They provide evidence for the common absolute stereochemistry of (−)-helianthoidin (curve A), (−)-hinokinin (curve B) and of (−)-cubebin (curve C). The insertion of oxygen in helianthoidin has little effect on the form of the curve, but the preference rules require that it is designated (8*S*, 8′*S*) whilst (−)-hinokinin is designated (8*R*, 8′*R*). More recently (−)-hinokinin has been related (Lopes *et al.*, 1983) with two lactones from *Virola sebifera* which have the same absolute configuration (**6.223**). In hinokinin the first peak is negative ($\phi = -10.9 \times 10^3$, 250 nm), the amplitude is zero at the absorption maximum (282 nm) and the next extremum is positive ($\phi = +400$, 285 nm). ORD measurements are of value in assigning absolute configuration and in that they show an enhancement at a distance from the wavelength maximum. However, the curves can be complicated when, as in lignans, several peaks occur in a narrow wavelength band (p. 167) and for this reason the measurement of circular dichroism is now preferred.

(ii) *Circular dichroism*

An optically active substance absorbs left-hand and right-hand circularly polarised light to different extents and this is the phenomenon of circular dichroism. It is measured as $\Delta\varepsilon$ between ε_1 and ε_2 which are the molar extinction coefficients for LH and RH circularly polarised light. The results are sometimes quoted in terms of the ellipticity, $[\theta] = 3300 \times \varepsilon$. In the region of an absorption band $\Delta\varepsilon$ follows a Gaussian curve with a positive or negative maximum coincident with λ_{max} (see Table 6.25 below), which is dependent on the chirality. The shape of the curve may not coincide with that of the absorption band and may be modified further by the interaction of closely related maxima. With a modern instrument such as the Jouan Dichrograph the strong UV response of lignans allows measurement at concentrations as low as 0.1 mg/ml. A clear account of the basic theory has been given (Bayley, 1980).

(a) *Dibenzylbutyrolactones*

(−)-Hinokinin (**6.223B**) has also been isolated from *Piper cubeba* and six other lignans of this group were obtained later (Badheka *et al.*, 1986). Typical of these was isoyatein (**6.223**; R_1, R_2 = —CH_2—; R_3, R_4 = Me; C5′ = OMe; X = H) which also had the (8*R*, 8*R*′) configuration and which showed negative Cotton effects in the CD curves at 234 and 276 nm.

(−)-Clusin (**6.224A**; X, Y = H, OH) and (−)-cubebinin (**6.224B**; X, Y = H, OH) have also been characterised as having the (8*R*, 8*R*′) configuration with negative CD responses near 244 and 270 nm. The antipodal form (8*S*, 8*S*′) of these compounds is found in (+)-arctigenin (**6.225**; Suzuki *et al.*, 1982) in which the relationship was shown by $[\phi]_{230} = +24180$. The typical form of CD response is shown (Scheme 6.33) below for three dibenzylbutyrolactones (Nishibe *et al.*, 1981) with the antipodal curves of (+)- and (−)-nortrachelogenin.

(b) *Furans and furofurans*

A definitive paper has been published (Hofer *et al.*, 1981; 1988) which describes the CD of sesamin and related lignans with the same abso-

Scheme 6.33 Circular dichroism of (A) dibenylbutyrolactones and (B) the nortrachelogenins.

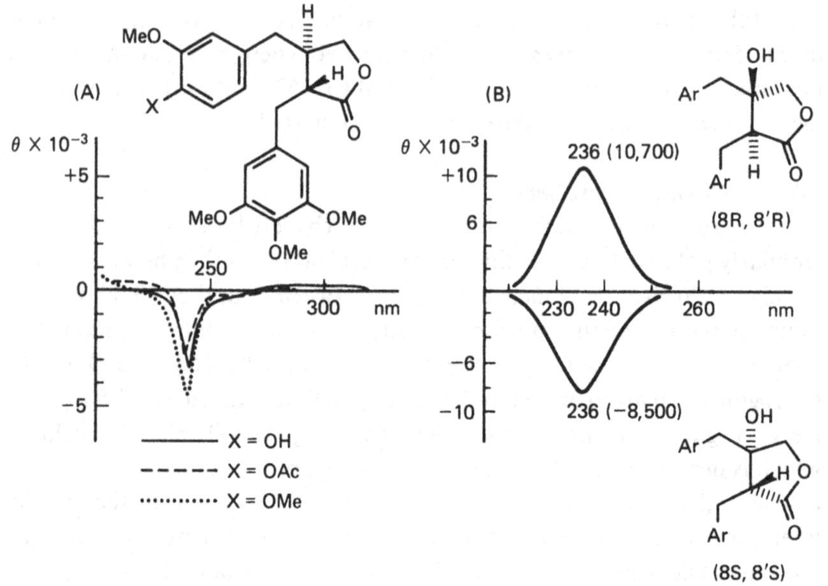

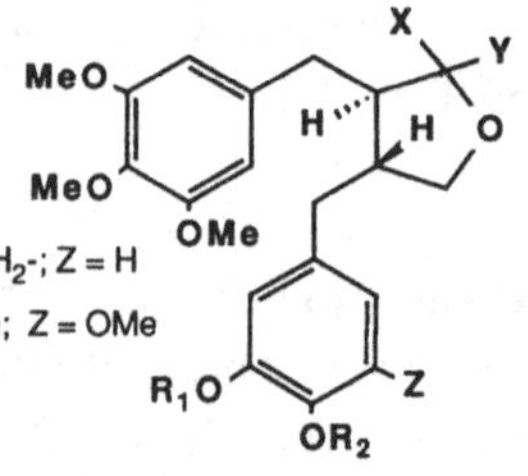

6.224
A: R_1,R_2 = -CH_2-; Z = H
B: R_1,R_2 = Me; Z = OMe

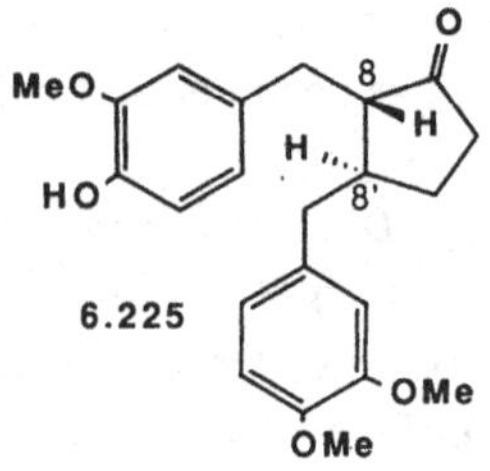

6.225

lute configuration (**6.226–8**) and which shows how the curves are affected by variation in the aromatic substitution pattern. In (+)-sesamin itself with equivalent di-equatorial 3,4-methylenedioxyphenyl groups and in three others with this configuration (Ar = 3,4-dimethoxy, or 3-methoxy-4,5-methylenedioxy, or 3,4,5-trimethoxyphenyl) very similar curves were obtained with the maxima shown in Table 6.25.

In compounds with axial-equatorial substituents (Ar = 3-methoxy-4,5-methylenedioxy- or 3,4,5-trimethoxyphenyl) three well-separated Cotton effects appear. These same maxima are seen for (+)-epieudesmin (**6.226**; Ar = 3,4-dimethoxyphenyl) and also for **6.229** but for **6.230** (p. 252) no reliable assignment can be made as the curve consists of weak negative Cotton effects which result from interaction between the transitions.

In the two diaxial and symmetrically substituted lignans [**6.228**; Ar = 3,4,5-trimethoxyphenyl (A) and 3-methoxy-4,5-methylenedioxyphenyl

Table 6.25. *Circular dichroism in the sesamin subgroup*

Wavelength range (nm)	CD (Δε)
275–280 (^1L_B)	+0.8 to +1.6
230–235 (^1L_A)	+2.2 to +4.8
ca. 205 (^1B_B)	+20.5 to +21.0

Scheme 6.34 Circular dichroism of furofurans.

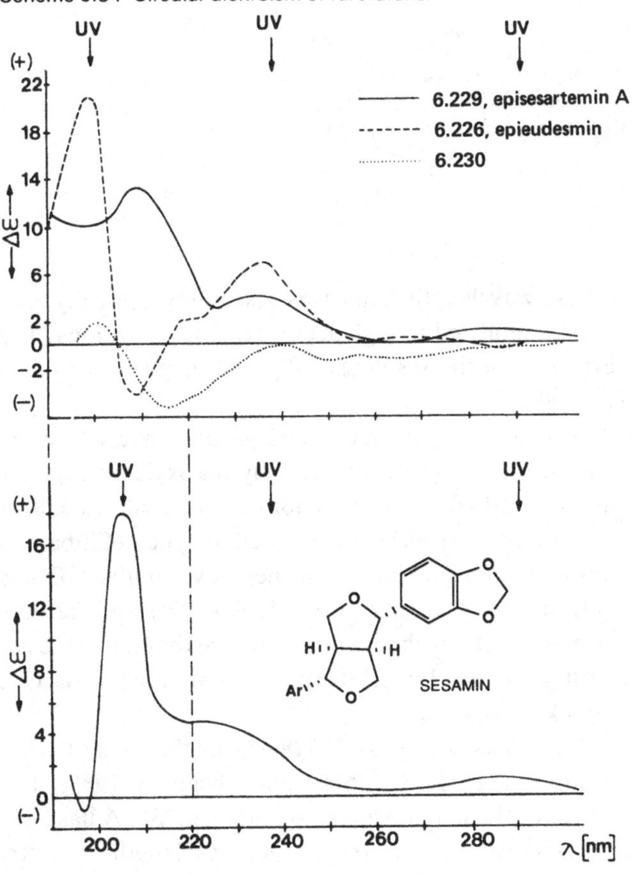

Ar, H, H, O, Ar — **6.226** (eq, eq)

Ar, H, H, O, Ar — **6.227** (ax, eq)

Ar, H, H, O, Ar — **6.228** (ax, ax)

Table 6.26. *CD of symmetrically substituted furofurans*

Wavelength (nm)	Δε (**6.228A**)	Δε (**6.228B**)
280	—	—
240	+22	+14
215	+78	+63
207	0	0
204	−35	−23

6.229

6.230

(B)] the longest wavelength band was essentially optically inactive but strong dipole coupling led to high activity in ^1B_B transitions (Table 6.26). Chemical evidence for the absolute configuration of (+)-sesamin has been presented (p. 000).

Furofurans with free phenolic —OH groups have CD curves which differ significantly from those of the fully alkoxylated type so far discussed. Thus (+)-1-hydroxypinoresinol-β-D-glucoside (**6.220**) and (+)-pinoresinol-β-D-glucoside (**6.129**) have been assigned (Chiba *et al.*, 1979) the same absolute configuration since they have similar CD curves with negative Cotton effects at 225 ($\theta = -5.56 \times 10^3$) and 223 nm ($\theta = -5.31 \times 10^3$) respectively. It should be noted here that the curve is similar in overall intensity to that for **6.230** since the values for Δε (−2.41 and −2.31) are of like magnitude.

The CD of (−)-massoniresinol (**2.128**) shows that it has the same absolute configuration as (−)-olivil (Shen and Theander, 1985). The relative stereochemistry of the magnostellins A and C (**6.231**; A has $Ar_1 = Ar_2 =$ 3,4-dimethoxyphenyl; C has Ar_1 = 3,4-methylenedioxy-, Ar_2 = 3,4-dimethoxyphenyl) was established by Kato *et al.* (1986). Their absolute stereochemistry was then found by the acid-catalysed cyclisation of the hydrogenolysis product (**6.232**) from magnostellin A (see Scheme 6.35). In the aryltetralin product **6.233** the relative configurations were known and the ORD showed that there was an *S*-configuration at the 7′-position because the first Cotton effect (at 284 nm) was negative (Hulbert *et al.*, 1981), hence the tetralin had the (*R*)-configuration at C-8. It therefore fol-

Scheme 6.35 Correlation of the magnostellins with aryltetralins

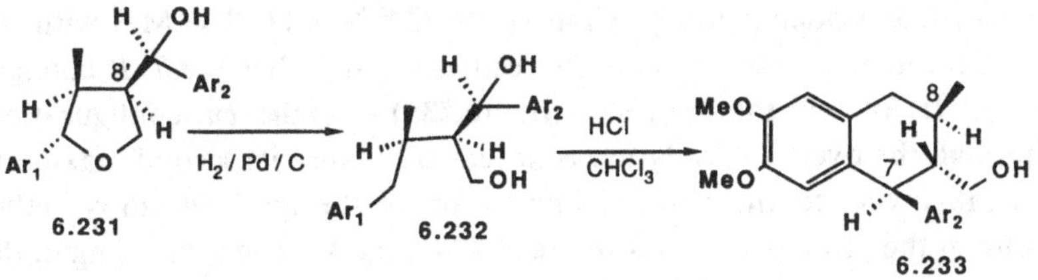

lowed that the magnostellins have the 7*S*, 8*S*, and 8′*R* chiralities. Note here that the cyclisation to **6.233** is not stereoselective since the least-compressed isomer is preferred, but this does not invalidate the comparison which gives the absolute configuration at C-8′ in the tetralin nor does returning this as a point of reference in the furan (**6.231**). It is also pointed out that the selection rules require that C-8 in the tetralin be given the 8*R*-configuration although in absolute terms it is the same as in the precursor (**6.231**).

(c) *Aryltetrahydronaphthalenes*

In making comparisons like the last it must be borne in mind that the dominant chromophores are the aryl groups and that the others are secondary. Lactones may be seen as a weak secondary peak at 223 nm but the CD spectra of glycosides differ only in amplitude from those of the aglycone. It follows from this that confusion may arise as a result of vari-

Scheme 6.36 CD curves of tetrahydronaphthalenes showing the enantiomeric relation and the effect of variations in the aryl substituents.

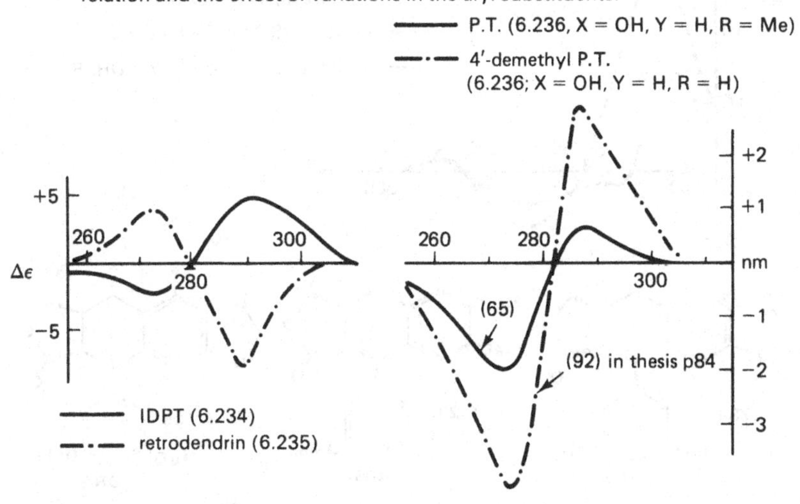

ation of the aryl substituents. This is illustrated in Scheme 6.36 where it is evident that isodeoxypodophyllotoxin (**6.234**; X = H, R = Me) with the first Cotton effect positive and the second negative has the 7′-*R* configuration. It follows that retrodendrin (**6.235**) has the *ent*-configuration, although the overall correlation is not ideal because its second maximum is weak owing to the reduced interaction of the methylenedioxy ether group in the principal chromophore (**6.234**, ring A). The weakening of the second maximum (near 270 nm) is also to be seen in derivatives of plicatic acid (**2.368**).

The matching of curves is also affected when functional group variation occurs in ring C, as between podophyllotoxin (**6.236**; X = OH, Y = H, R = Me) and 4′-demethylpodophyllotoxin (**6.236**; X = OH, R = H); the latter shows a positive first Cotton effect and a negative second one as does podophyllotoxin (Scheme 6.36). However, there is a disparity in the amplitudes, for those of the demethyl compound are much larger (+2.8, 287 nm and −4.2, 274 nm) than those of podophyllotoxin (+0.7, 287 nm and −2.0, 274 nm).

The α- and β-peltatins resemble podophyllotoxin in their physiological activity. The former (**6.236**; Y = OH, X = R = H) has a curve like that of podophyllotoxin itself and the two compounds clearly have the same absolute configuration; in common with all phenols in this subgroup both have a large third maximum in the region of 240 nm. However, when the

Scheme 6.37 CD curves of podophyllotoxin and the peltatins showing the correlation with absolute configuration and of substituent variation.

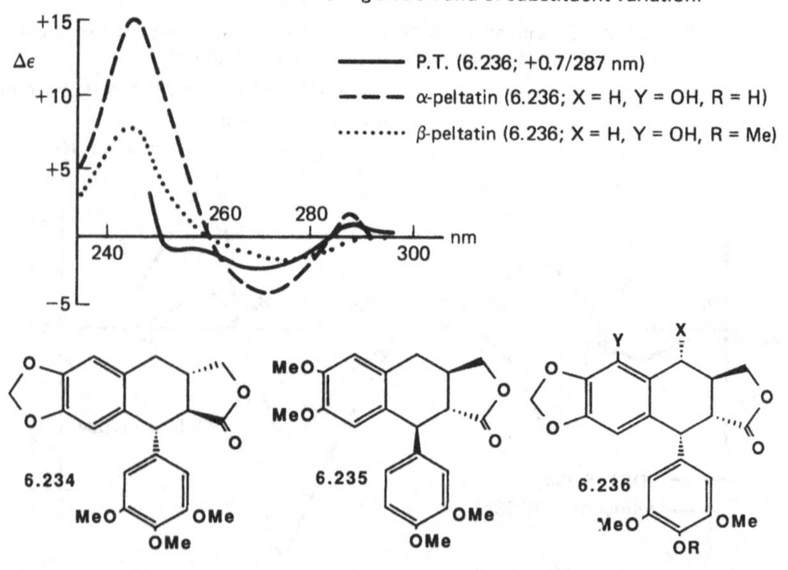

Scheme 6.38 CD curves for picropodophyllin and apocompounds.

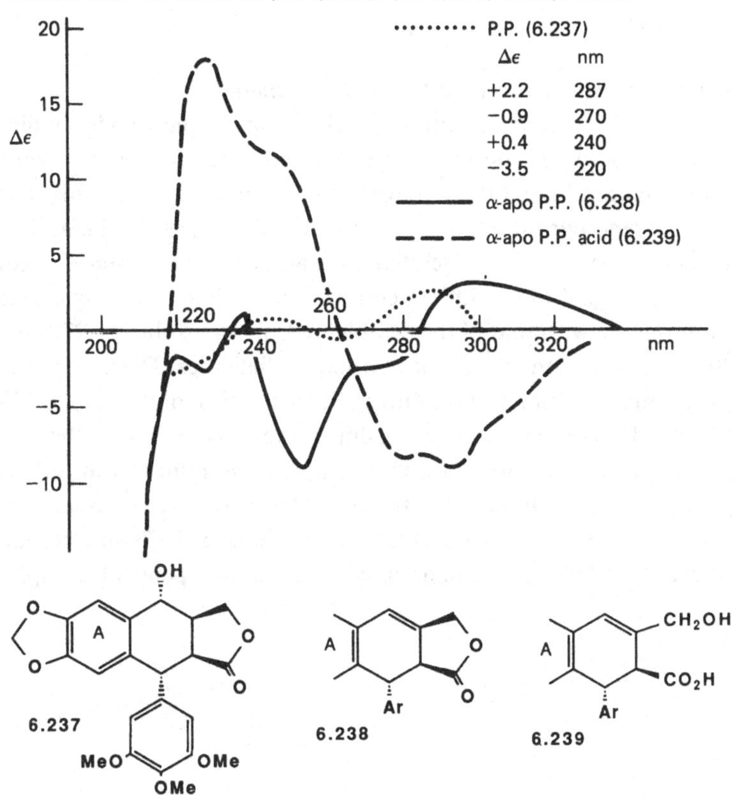

ring C hydroxyl group is blocked in β-peltatin (**6.236**; X = H, Y = OH, R = Me) the sterochemical relationship is less evident as the first maximum is lost (Scheme 6.37).

As has been mentioned (NMR section) conformational effects can be important in this class of lignans and ORD/CD curves are sensitive to this, since the relative positions of the interacting aryl chromophores are affected. An important example is that of picropodophyllin (**6.237**), the physiologically inactive C-8′ epimer of podophyllotoxin, which has frequently been obtained as a base-induced artefact. The active lignan has a rigid structure in which ring C takes up a quasi-axial position, but picropodophyllin is flexible and exists as a conformational equilibrium in which the contribution of the equatorial component is about twice that of the axial. This difference results in a CD curve for picropodophyllin (Scheme 6.38) in which the first maximum is now larger than the second. When the axial component is further reduced by the attachment of bulky groups such as acetyl at the C-7 position then the second maximum may be lost altogether.

(d) *Compounds with inherent dissymmetry*

Elimination of the C7-OH group in picropodophyllin (**6.237**) gives α-apopicropodophyllin (**6.238**) and this molecule resembles the biaryl lignans in that it has a form which is itself chiral (Ayres *et al.*, 1971). The styrene chromophore is dominant; the curves (Scheme 6.38) differ from those of the other cyclolignans and give clear evidence of conformational change. For example, a comparison of that for the rigid α-apopicropodophyllin with that of the flexible hydrolysis product (**6.239**) shows that the styrene chromophores have opposite helicities. There could have been no change in configuration during the formation of the acid (**6.239**) so the observed 'enantiomeric' relationship in the CD curves must be the result of its adopting a conformation with ring C predominantly in the axial position. Another example of a skewed styrene with a right-handed helix is that of sequirin-D (**6.240**; Hatam and Whiting, 1978) where both PMR and CD evidence point to an axially substituted pendent group.

OMe
6.240
MeO

Scheme 6.39 CD curves of atropoisomeric cyclo-octadienes.

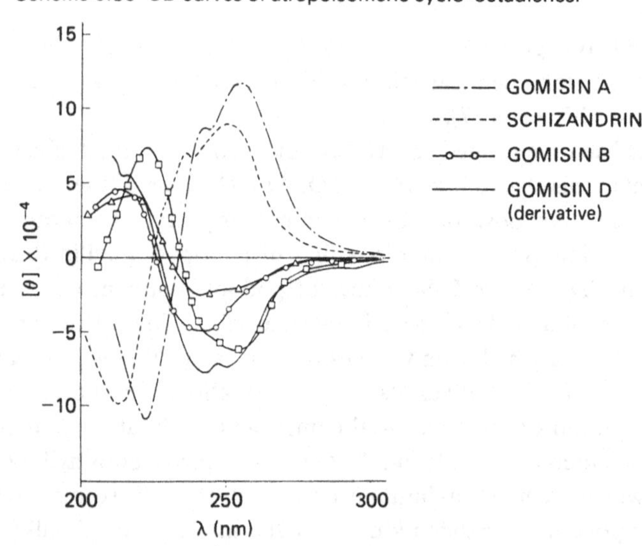

(e) *Dibenzocyclo-octadienes*

The CD curves of these atropoisomers relate to those of α-apopicropodophyllin (**6.238**) and their absolute configuration is securely based on the X-ray analysis of gomisin D (**6.180**; Ikeya *et al.*, 1979a) which has the *S*-biphenyl configuration. X-ray studies of schizandrol B (**2.459**; L-Niang *et al.*, 1985, and Spencer and Flippen-Anderson, 1981) have also been reported. The enantiomeric relationship is clearly shown in the CD curves (Scheme 6.39) where compounds with the *R*-configuration – schizandrin, gomisin A – show positive Cotton effects at 235–255 nm and a negative one near 220 nm. In contrast gomisins B and D with the *S*-configuration give curves which are mirrored by the *X*-axis.

No full CD curve has been published for any of the stegane group, but the data (Raphael, 1982) for (−)-steganone (**6.205**) show that it has the *S*-configuration with major maxima at 244 nm ($\Delta\varepsilon = -41.0$) and 218 nm ($\Delta\varepsilon = +39.6$) matching those shown in Scheme 6.39. A small positive maximum in steganone at 276 nm is attributed to absorption by the keto-group, but typically the curve is dominated by the intense response of the inherently disymmetric biaryl chromophore.

In concluding this section one emphasises the value of CD spectroscopy in relating absolute configurations. The potential of the method is enhanced by the wide range of reference lignans now available. These measurements are of especial value when crystals suitable for X-ray analysis are not available.

References

Abdullaev, N.D., Yagudaev, R., Batirov, E.K. and Malikov, V.M. (1987). CMR spectra of arylnaphthalene lignans. *Chem. Nat. Compd.* **23**, 63–74.

Abe, F. and Yamauchi, T. (1986). Lignans & lignan glucosides from *Trachelospermum asiaticum*. *Chem. Pharm. Bull.* **34**, 4340–5.

Achenbach, H., Waibel, R. and Addae-Mensah, I. (1983). Lignans & other constituents from *Carissa edulis*. *Phytochemistry* **22**, 749–53.

Adler, E. and Hernestam, S. (1955). Estimation of phenolic hydroxyl groups in lignin. Part 1. Periodate oxidation of guaiacol compounds. *Acta Chem. Scand.* **9**, 319–34.

Agrawal, P.K. and Rastogi, R.P. (1982). Two lignans from *Cedrus deodara*. *Phytochemistry* **21**, 1459–61.

Agrawal, P.K. and Thakur, R.S. (1985). CMR spectroscopy of lignan & neolignan derivatives. *Magn. Reson. Chem.* **23**, 389–418.

Aiyar, V.N. and Chang, F.C. (1975). New diastereoisomers of podophyllotoxin: related hydroxyacids. *J. Org. Chem.* **40**, 2384–7.

Andersson, R., Popoff, T. and Theander, O. (1975). A new lignan from Norway spruce. *Acta Chem. Scand.* **29B**, 835–7.

Anet, F.A.L. and Yavari, I. (1975). Conformational properties of *cis,cis*-1,3-cyclooctadiene. *Tetrahedron Letters* **19**, 1567–70.

Anjaneyulu, A.S.R., Rao, K.J., Row, L.R. and Subrahmanyam, O. (1973). Crystalline constituents of *Euphorbiaceae*. Part XII. *Tetrahedron* **29**, 1291–8.

Anjaneyulu, A.S.R. and Murty, V.S. (1981a). Chemical examination of *Guazuma tomentosa* Kunth. *Indian J. Chem.* **20B**, 85–7.

Anjaneyulu, A.S.R., Ang, A.S.P., Ramiah, P.A., Row, L.R. and Venkateswarlu, V. (1981b). New lignans from the heartwood of *Cleistanthus collinus*. *Tetrahedron* **37**, 3641–52.

Ayres, D.C. and Pauwels, P.J.S. (1962). Lignans. Part 2. Reduction of lactones of the podophyllotoxin group with lithium aluminium hydride. *J. Chem. Soc.* 5025–30.

Ayres, D.C. and Mundy, J.W. (1968). The base-catalysed aromatisation of dihydronaphthalenes. *J. Chem. Soc. Chem. Commun.* 1134.

Ayres, D.C. (1969). Lignans & related phenols. Part IX. Characterisation & conformational analysis of apocompounds of the aryltetrahydronaphthalene class. *Canad. J. Chem.* **47**, 2075–80.

Ayres, D.C., Harris, J.A. and Hulbert, P.B. (1971). Lignans & related phenols. Part XI. Circular dichroism of the arylnaphthalene class. *J. Chem. Soc.* (*C*), 1111–14.

Ayres, D.C., Harris, J.A., Jenkins, P.B. and Phillips, L. (1972a). Lignans & related phenols. Part XII. Application of nuclear magnetic double resonance to the aryltetrahydronaphthalene class. *J. Chem. Soc. Perkin I*, 1343–50.

Ayres, D.C. and Lim, C.K. (1972b). Lignans & related phenols. Part XIII. Halogenated derivatives of podophyllotoxin. *J. Chem. Soc. Perkin I*, 1351–5.

Ayres, D.C. and Harris, J.A. (1973). Lignans & related phenols. Part XIV. Selective oxidation of aryltetrahydronaphthalenes. *J. Chem. Soc. Perkin I*, 2059–63.

Ayres, D.C. (1978). In *Chemistry of Lignans*, ed. Rao, C.B.S., Andhra Univ., pp. 123–73.

Ayres, D.C. and Evans, D.M. (1987), unpublished results.

Ayres, D.C. and Haycock, P. (1988), unpublished results.

Bhacca, N.J. and Stevenson, R. (1963). The constitution of otobain. *J. Org. Chem.* **28**, 1638–42.

Badheka, L.P., Pradhu, B.R. and Mulchandani, N.B. (1986). Dibenzylbutyrolactone lignans from *Piper cubeba*. *Phytochemistry* **25**, 487–9.

Banerji, A. and Pal, S. (1982). Constituents of *Piper sylvaticum*. Structure of sylvatesmin. *J. Nat. Prod.* **45**, 672–5.

Banerji, A., Sarkar, M., Ghosal, T. and Pal, S.C. (1984). Sylvone, a new furanoid lignan from *Piper sylvaticum*. *Tetrahedron* **40**, 5047–52.

Barger, G. (1908). The action of thionyl chloride & phosphorus pentachloride on the methylene ethers of catechol derivatives. *J. Chem. Soc.* **93**, 563–73.

Batirov, E.K., Matkarimov, A.D., Malikov, V.M. and Yagudaev, M.R. (1985). Versicoside – a new lignan glycoside from *Haplophyllum versicolor*. *Chem. Nat. Compds.* **21**, 584–7.

Batterbee, J.E., Burden, R.S., Crombie, L. and Whiting, D.A. (1969). Chemistry & synthesis of the lignan (−)-cubebin. *J. Chem. Soc.* (*C*) 2470–7.

Bayley, P. (1980). In *An Introduction to Spectroscopy for Biochemists*, ed. Brown, S.P. Acad. Press: London, pp. 148–234.

Beilstein, F.K. (1971). *Handbook of Organic Chemistry*, **10**, part 3, syst. no. 1182, p. 2555.

Biftu, T., Gamble, N.F., Doebber, T., Hwang, S-B., Shen, T-W., Snyder, J., Springer, J.P. and Stevenson, R. (1986). Conformation and activity of tetrahydrofuran lignans as specific platelet activating factor antagonists. *J. Med. Chem.* **29**, 1917–21.

Birch, A.J., Moore, B., Smith, Mrs E. and Smith, M. (1964). The conversion of gmelinol into neogmelinol. *J. Chem. Soc.* 2709–12.

Birch, A.J., MacDonald, P.L. and Pelter, A. (1967). A revised structure for neogmelinol: determinations of configurations in tetrahydrofuranoid lignans. *J. Chem. Soc.* (*C*) 1968–72.

Blears, J.G. and Haworth, R.D. (1958). The constituents of natural phenolic resins. Part 24. A synthesis of galgravin. *J. Chem. Soc.* 1985–7.

Bock, K. and Pedersen, C. (1978). 13C-NMR spectroscopy of monosaccharides. *Adv. in Carbo. Chem. & Biochem.* **41**, 27–67.

Brieskorn, C.H. and Huber, H. (1976). New lignans from *Aptosimum* species. *Tetrahedron Letters* 2221–4.

Breitmaier, E. and Voelter, W. (1987). Carbon-13 NMR spectroscopy, 3rd edn. VCH Verlag: Weinheim.

Brewer, C.F., Loike, J.D. and Horwitz, S.B. (1979). Conformational analysis of podophyllotoxin & its congeners. Structure–activity relationship in microtubule assembly. *J. Med. Chem.* **22**, 215–21.

Briggs, L.H., Colebrook, L.D., Fales, H.M. and Wildman, W.C. (1957). Infrared absorption spectra of methylenedioxy and aryl ether groups. *Anal. Chem.* **29**, 904–11.

Bryan, R.F. and Fallon, L. (1976). Crystal structure of syringaresinol. *J. Chem. Soc. Perkin II*, 341–5.

Burden, R.S., Crombie, L. and Whiting, D.A. (1969). The extractives of *Heliopsis scabra*: constitution of two new lignans. *J. Chem. Soc.* (*C*) 693–701.

Cambie, R.C., Pang, G.T.H., Parnell, J.C., Rodrigo, R. and Weston, R.J. (1979). Chemistry of the Podocarpacea. Part 54. Lignans from the wood of *Dacrydium intermedium*. *Aust. J. Chem.* **32**, 2741–51.

Cambie, R.C., Clark, G.R., Craw, P.A., Jones, T.C., Rutledge, P.S. and Woodgate, P.D. (1985). Chemistry of the Podocarpacea. Part 69. Further lignans from the wood of *Dacrydium intermedium*. *Aust. J. Chem.* **38**, 1631–45.

Chandel, R.S. and Rastogi, R.P. (1980). Pygeoside a new lignan xyloside from *Pygeum acuminatum*. *Indian J. Chem.* **19B**, 279–82.

Chen, C-L. and Chang, H.M. (1978). Lignans and aporphine alkaloids in bark of *Liriodendron tulipfera*. *Phytochemistry* **17**, 779–82.

Chen, G.L. (1984). Non-intercalative anti-tumour drugs interfere with the breakage–reunion reaction of mammalian DNA topoisomerase II. *J. Biol. Chem.* **259**, 13560–6.

Chen, Y-Y., Shu, Z-B. and Li, L-N. (1976). Studies on *Schizandrae* fruits. Part IV. Isolation & determination of active compounds (in lowering high SGPT levels) of *Schizandra chinensis* Baill. *Sci. Sin.* **19**, 276–90.

Chen, Y-P., Liu, R., Hsu, H-Y., Yamamura, S., Shizuri, Y. and Hirata, Y. (1977). The new schizandrin type lignans, kadsurin & kadsurarin. *Bull. Chem. Soc. Japan* **50**, 1824–6.

Chiba, M., Okabe, K., Hisada, S., Shima, K., Takemoto, T. and Nishibe, S. (1979). Elucidation & structure of a new lignan glucoside from *Olea europa* by CMR spectroscopy. *Chem. Pharm. Bull.* **27**, 2868–73.

Chiba, M., Hisada, S., Nishibe, S. and Thieme, H. (1980). ^{13}C-NMR analysis of symplocosin and L(+)-epipinoresinol glucoside. *Phytochemistry* **19**, 335–6.

Cisney, M.E., Shilling, W., Hearon, W.M. and Goheen, D.W. (1954). Conidendrin. Part II. The stereochemistry and reactions of the lactone ring. *J. Amer. Chem. Soc.* **76**, 5083–7.

Cole, J.R., Bianchi, E. and Trumbull, E.R. (1969). Antitumour agents from *Bursera microphylla*. Part 2. Isolation of a new lignan – burseran. *J. Pharm. Sci.* **58**, 175–6.

Colombo, T., D'Incalci, M., Donelli, M.G., Bartosek, I., Benfenati, E., Farina, P.

and Guatini, A. (1985). Metabolic studies of a podophyllotoxin derivative (VP-16) in the isolated perfused liver. *Xenobiotica* **15**, 343–50.

Colthup, N.B., Daly, L.H. and Wiberley, S.E. (1975). *Introduction to Infrared & Raman Spectroscopy*, 2nd edn. Acad. Press: New York.

Cooley, G., Farrant, R.D., Kirk, D.N., Patel, S., Wynn, S., Buckingham M.J., Hawkes, G.E., Hursthouse, M.B., Galas, A.M.R., Lawson, A.M. and Setchell, K.D.R. (1984). Structural analysis of the urinary lignan enterolactone. A 400 MHz NMR study of the solution state and X-ray study for the solid state. *J. Chem. Soc. Perkin II*, 489–97.

Cooper, R., Gottlieb, H.E., Lavie, D. and Levy, E.C. (1979). Lignans from *Aegilops ovata* L. Synthesis of a 2,4- & 2,6-diarylmonoepoxylignolide. *Tetrahedron* **35**, 861–8.

Corrie, J.E.T., Green, G.H., Ritchie, E. and Taylor, W.C. (1970). The chemical constituents of Australian *Zanthoxylum* species. Part 5. The constituents of *Z. pluviatile* Hartley. *Aust. J. Chem.* **23**, 133–45.

Craik, D.J. (1983). Substituent effects on nuclear shielding. *Ann. Rep. NMR Spec.* **15**, 17–75.

Crossley, N.S. and Djerassi, C. (1962). Naturally occurring oxygen heterocycles. Part XI. Veraguensin. *J. Chem. Soc.* 1459–62.

Deyama, T. (1983). The constituents of *Eucommia ulmoides* Oliv. Part 1. Isolation of (+)-medioresinol-*O*-*β*-D-glucopyranoside. *Chem. Pharm. Bull.* **31**, 2993–7.

Deyama, T., Ikawa, T. and Nishibe, S. (1985). The constituents of *Eucommia ulmoides* Oliv. Part II. Isolation & structures of three new lignan glycosides. *Chem. Pharm. Bull.* **33**, 3651–7.

Deyama, T., Ikawa, T., Kitagawa, S. and Nishibe, S. (1986). The constituents of *Eucommia ulmoides* Oliv. Part 3. Isolation & structure determination of a new lignan glucoside. *Chem. Pharm. Bull.* **34**, 523–7.

Dhal, R., Nabi, Y. and Brown, E. (1986). Lignan studies. Part 7. Total synthesis of (±)-*α*- & *β*-conidendrin & methyl (±)-*α*- & *β*-conidendrals, *Tetrahedron* **42**, 2005–16.

Duffield, A.M. (1967). Mass spectrometric fragmentation of some lignans. *J. Hetero. Chem.* **4**, 16–22.

Duh, C-Y. *et al.* (1986). Plant anti-cancer agents XLII. Cytotoxic constituents of *Wikstroemia elliptica. J. Nat. Prod.* **49**, 706–9.

Emsley, J.W., Feeney, J. and Sutcliffe, L.H. (1965). High resolution NMR spectroscopy, Vol. 1, p. 280. Pergamon: Oxford.

Evans, D.M. (1986). Unpublished observations.

Evcim, U., Gozler, B., Freyer, A.J. and Shamma, M. (1986). Haplomyrtin and (−)-haplomyrfolin: two lignans from *Haplophyllum myrtifolium. Phytochemistry* **25**, 1949–51.

Fang, J-M., Jan, S-T. and Cheng, Y-S. (1985a). (±)-Calocedrin, a lignan dihydroanhydride from *Calocedrus formosana. Phytochemistry* **24**, 1863–4.

Fang, J.M., Wei, K.M., Cheng, Y-S., Cheng, M-C. and Wang, Y. (1985b). (±)-Tsugacetal, a lignan acetal from Taiwan hemlock. *Phytochemistry* **24**, 1363–5.

Ferris, J.,P., Boyce, C.B., Briner, R.C., Weiss, U., Qureshi, I.H. and Sharpless, N.E. (1971). *Lythraceae* alkaloids. Part X. Assignment of absolute stereochemistry on the basis of chiral-optical effects. *J. Amer. Chem. Soc.* **93**, 2964–8.

Fonseca, S.F., Campello, J. de P., Barata, L.E.S. and Ruveda, E.A. (1978). CMR spectral analysis of lignans from *Araucaria angustifolia. Phytochemistry* **17**, 499–502.

Fonseca, S.F., Barata, L.E.S., Ruveda, E.A. and Baker, P.M. (1979). CMR spectral and conformational analysis of naturally occurring tetrahydrofuran lignans. *Canad. J. Chem.* **57**, 441–3.

Garzon, L., Cuca, L.E., Martinez, J.C., Yoshida, M. and Gottlieb, O.R. (1987). Flavonolignoid from the fruit of *Iryanthea laevis*. *Phytochemistry* **26**, 2835–7.

Ghera, E. and Ben-David, Y. (1978). Total synthesis of the lignan (+)-schizandrin. *J. Chem. Soc. Chem. Commun.* 480–1.

Ghisalberti, E.L., Jefferies, P.R., Skelton, B.W. and White, A.H. (1987). The chemistry of *Eremophila* sp. Part 27. A new lignan from *E. dalyana*. *Aust. J. Chem.* **40**, 405–11.

Ghosal, S., Banerjee, S. and Srivastava, R.S. (1979). Simplexolin, a new lignan from *Justicia simplex*. *Phytochemistry* **18**, 503–5.

Gisvold, O. and Thaker, E. (1974). Lignans from *Larrea divaricata*. *J. Pharm. Sci.* **63**, 1905–7.

Harada, N. and Nakanishi, K. (1983). *Circular Dichroic Spectroscopy*. Oxford, pp. 4–9.

Harmatha, J., Budesinsky, M. and Trka, A. (1982). The structure of yatein, determination of the positions and configurations of benzyl groups in lignans of the 2,3-dibenzylbutyrolactone type. *Collect. Czech. Chem. Commun.* **47**, 644–63.

Hartwell, J.L. and Schrecker, A.W. (1952). Components of podophyllin. Part 9. The structure of apopicropodophyllins. *J. Amer. Chem. Soc.* **74**, 5676–83.

Hartwell, J.L. and Schrecker, A.W. (1958). The chemistry of podophyllum. In *Prog. in the Chem. of Org. Nat. Prod.*, ed. L. Zechmeister. Springer: Vienna, **15**, 83–166.

Haslam, E. (1970). The stereochemistry of sesamolin. *J. Chem. Soc.* (*C*) 2332–4.

Hatam, N.A.R. and Whiting, D.A. (1978). The absolute configuration of sequirin D a biogenetically novel norlignan; synthesis of (±)-dimethylsequirin D, *Tetrahedron Lett.* 5145–6.

Hayashi, T. and Thomson, R.H. (1975). New lignans in *Conocarpus erectus*. *Phytochemistry* **14**, 1085–7.

Hicks, R.P. and Sneden, A.T. (1983). Neoisostegane, a new bisbenzocyclooctadiene lignan lactone from *Steganotaenia araliaceae* Hochst., *Tetrahedron Lett.* 2987–90.

Hofer, O. and Scholm, R. (1981). Stereochemistry of tetrahydrofurofuran derivatives – CD & absolute configuration, *Tetrahedron* **37**, 1181–6.

Hofer, O., Wagner, U.G. and Greger, H. (1988). X-ray structural analysis of tetrahydrofuran lignans. *Monatsh.*, **119**, 1143–53.

Hokanson, G.C. (1979). The lignans of *Polygala polygama* (Polygalaceae): deoxypodophyllotoxin and three new lignan lactones. *J. Nat. Prod.* **42**, 378–84.

Holloway, D. and Schienmann, F. (1974). Two lignans from *Litsea grandis* and *L. gracilipes*. *Phytochemistry*, **13**, 1233–6.

Holthuis, J. (1985). *Bioanalysis, electrochemistry and pharmacokinetics of Etoposide, Teniposide and metabolites*, pp. 105–116, Kempenaer.

Houlbert, N., Brown, E., Robin, J.P., Davoust, D., Chiaroni, A., Prange, T. and Riche, C. (1985). Preparation, stereochemical study & evaluation of the anti-leukaemic activity of glucosides of steganol. *J. Nat. Prod.* **48**, 345–6.

Hulbert, P.B., Klyne, W. and Scopes, P.M. (1981). Chiroptical studies Part 100. Lignans, *J. Chem. Res.* (*S*), **27**; *J. Chem. Res.* (*M*), 0401–49.

Hunter, W.H. and Levine, A.A. (1926). The oxidation of the tribromo- & trichloroderivatives of pyrogallol-1,3-dimethyl ether. *J. Amer. Chem. Soc.* **48**, 1608–14.

Iida, T., Noro, Y. and Ito, K. (1983). Magnostellin A & B. Novel lignans from *Magnolia stellata. Phytochemistry* **23**, 211–13.

Ikeya, Y., Taguchi, H., Yosioka, I., Iitaka, Y. and Kobayashi, H. (1979a). Constituents of *Schizandra chinensis* Baill. Part II. The structure of a new lignan, gomisin D. *Chem. Pharm. Bull.* **27**, 1395–401.

Ikeya, Y., Taguchi, H. and Yosioka, I. (1979b). The constituents of *Schizandra chinensis* Baill. Part IV. The cleavage of the methylenedioxy moiety with lead tetra-acetate in benzene, and the structure of angeloyl gomisin Q. *Chem. Pharm. Bull.* **27**, 2536–8.

Ikeya, T., Taguchi, H., Yosioka, I. and Kobayashi, H. (1979c). Constituents of *Schizandra chinensis* Baill. Part I. Isolation & structure determination of five new lignans gomisin A, B, C, F, & G and the absolute configuration of schizandrin, *Chem. Pharm. Bull.* **27**, 1383–94.

Ikeya, Y., Taguchi, H., Yosioka, I. and Kobayashi, H. (1979d). The constituents of *Schizandra chinensis* Baill. Part III. The structure of four new lignans; gomisin H and its derivatives angeloyl, tigloyl and benzoyl-gomisin H, *Chem. Pharm. Bull.* **27**, 1576–82.

Ikeya, Y., Taguchi, H., Yosioka, I. and Kobayashi, H. (1979e). The constituents of *Schizandra chinensis* Baill. Part V. The structure of four new lignans. *Chem. Pharm. Bull.* **27**, 2695–709.

Ikeya, Y., Taguchi, H., Sasaki, H., Nakajima, K. and Yosioka, I. (1980a). The constituents of *Schizandrfa chinensis* Baill. Part VI. CMR spectroscopy of dibenzocyclo-octadiene lignans. *Chem. Pharm. Bull.* **28**, 2414–21.

Ikeya, Y., Taguchi, H. and Yosioka, I. (1980b). The constituents of *Schizandra chinensis* Baill. Part VII. The structure of three new lignans: (−)-gomisin K_1, (+)-gomisin K_2 & K_3. *Chem. Pharm. Bull.* **28**, 2422–7.

Ikeya, Y., Taguchi, H. and Yosioka, I. (1982a). The constituents of *Schizandra chinensis* Baill. Part X. The structures of γ-schizandrin and four new lignans, (−)-gomisin L1 and L2, (±)-gomisin M1 and gomisin M2, *Chem. Pharm. Bull* **30**, 132–9.

Ikeya, Y., Taguchi, H. and Yosioka, I. (1982b). The constituents of *Schizandra chinensis* Baill. Part XII. Isolation and structure of a new lignan, Gomisin R. The absolute structure of Wuweizisu C and isolation of Schisantherin D. *Chem. Pharm. Bull.* **30**, 3207–11.

Inoue, K., Inouye, H. and Chen, C-C. (1981). A naphthoquinone & a lignan from the wood of *Kigelia pinnata. Phytochemistry* **20**, 2271–6.

Jackson, D.E. and Dewick, P.M. (1985). Tumour-inhibitory aryltetralin lignans from *Podophyllum pleianthum. Phytochemistry* **24**, 2407–9.

Jakupovic, J., Schuster, A., Bohlmann, F., King, R.M. and Robinson, H. (1986). New lignans and isobutylamides from *Heliopsis buphthalmoides. Planta. Med.* **52**, 18–20.

Joshi, B.S., Ravindranath, K.R. and Viswanathan, N. (1978). Structure & stereochemistry of attenud, a new lignan from *Knema attenuata* (Wall.) Warb., *Experientia* **34**, 422–3.

Kato, A., Hashimoto, Y. and Kidikuro, M. (1979). (+)-nortrachelogenin, a new physiologically active lignan from *Wikstroemia indica. J. Nat. Prod.* **42**, 159–62.

Kato, M.J., Fo, H.P.F., Yoshida, M. and Goittlieb, O.R. (1986). Neolignans from the fruits of *Virola elongata. Phytochemistry*, **25**, 279–90.

Kato, Y. (1963). Structure of lyoniresinol (dimethoxyisolariciresinol). *Chem. Pharm. Bull.* **11**, 823–7.

Kato, Y. (1966). The absolute configuration of lyoniresinol. *Chem. Pharm. Bull.* **14**, 1438–9.

Kemp, W. (1986). *Nuclear Magnetic Resonance in Chemistry, a Multinuclear Introduction.* Macmillan: Basingstoke.

Kende, A.S., Liebeskind, L.S., Kubiak, C. and Eisenberg, R. (1976). Isosteganacin. *J. Amer. Chem. Soc.* **98**, 6389–91.

Khalid, S.A. and Waterman, P.G. (1981). Alkaloid, lignan & flavanoid constituents of *Haplophyllum tuberculatum* from Sudan. *Planta Med.* **43**, 148–52.

Khan, K.A. and Shoeb, A. (1985). A lignan from *Lonicera hypoleuca*. *Phytochemistry* **24**, 628–30.

Kilidhar, S.B., Parthasarathy, M.R. and Sharma, P. (1982). Prinsepiol, a lignan from stems of *Prinsepia utilis*. *Phytochemistry* **21**, 796–7.

King, F. and Wilson, J.G. (1964). The chemistry of extractives from hardwoods. Part 36. The lignans of *Guaiacum officinale* L. *J. Chem. Soc.* 4011–24.

Kirk, D.N. (1987). Personal communication.

Klemm, L.H. (1978). In *Chemistry of Lignans*, ed. Rao, C.B.S. Andhra Univ., pp. 175–226.

Kochetkov, N.K., Khorlin, A., Chizhov, O.S. and Sheichenko, V.T. (1961). Schizandrin lignan of unusual structure. *Tetrahedron Lett.*, 730–4.

Kohen, F., MacLean, I. and Stevenson, R. (1966). The constitution of otobaphenol, *J. Chem. Soc. (C)* 1775–80.

Kowaleski, V.J. (1969). The INDOR technique in high resolution NMR. *Progr. NMR Spec.* **5**, 1–31.

Kudo, K., Nohara, T., Komori, T., Kawasaki, T. and Schulten, H-R. (1980). Lignan glucosides from the bark of *Ligstrum japonicum*. *Planta. Med.* **40**, 250–61.

Kuhn, M. and von Wartburg, A. (1969). Synthesis of glycosides. Part 2. Glycosides of 4′-demethylpodophyllotoxin. *Helv. Chim. Acta* **52**, 948–55.

Kuo, Y.H., Tsung Lin, Y. and Tang Lin, T. (1985). Tawanin H, a new lignan from the bark of *Tawania cryptomeroides* Hayata. *J. Chin. Chem. Soc.* **47**, 457–60.

Kupchan, S.M., Britton, R.W., Ziegler, M.F., Gilmore, C.J., Restivo, R.J. and Bryan, R.F. (1973). Steganacin & steganangin, novel anti-leukemic lignan lactones from *Steganotaenia araliacea*. *J. Amer. Chem. Soc.* **95**, 1335–6.

Lapper, R.D., Mantsch, H.H. and Smith, I.C.P. (1975). CMR studies of compounds of vitamin B_6 group & related pyridine derivatives. *Canad. J. Chem.* **53**, 2406–12.

Larson, E.R. and Raphael, R.A. (1982). Synthesis of (−)-steganone. *J. Chem. Soc. Perkin Trans I*, 521–5.

Lavie, D., Levy, E.C., Cohen, A., Evenari, M. and Gutterman, Y. (1974). New germination inhibitors from *Aegilops ovata* L. *Nature* **249**, 388.

Le Quesne, P.W., Larrahondo, J.E. and Raffauf, R.E. (1980). Anti-tumour plants Part X. Constituents of *Nectandra rigida*. *J. Nat. Prod.* **43**, 353–9.

Liu, C.S., Fang, S-D., Huang, M-F., Kao, Y-L. and Hsu, J-S. (1978). Studies on the active principles of *Schizandra spenanthera*. *Sci. Sin.* **21**, 483–502.

L-Niang, L., Hung, X. and Rui, T. (1985). Bicyclo-octadiene lignans from roots & stems of *Kadsura coccinea*. *Planta Med.* **51**, 297–300.

L-Niang, L., Xiang, Q., Da-Lun, G. and Man, K. (1988). Neokadsuranin, a tetrahydrodibenzocyclo-octadiene lignan from stems of *Kadsura coccinea*. *Planta Med.* **54**, 45–6.

Lopes, L.M.X., Yoshida, M. and Gottlieb, O.R. (1983). Dibenzylbutyrolactone lignans from *Virola sebifera*. *Phytochemistry* **22**, 1516–18.

Loriot, M., Brown, E. and Robin, J-P. (1983). Total synthesis of (±)-attenuol. *Tetrahedron* **39**, 2795–8.

Ludemann, H.D. and Nimz, H. (1974). ^{13}C-Resonance spectra of lignines. *Macromol. Chem.* **175**, 2393–422.

Mabry, T.J. and Markham, K.R. (1975). Mass spectroscopy of flavonoids. In *The Flavonols*, ed Harborne, J.B., Mabry, T.J. and Mabry, H. Chapman & Hall: London.

McAlpine, J.B., Riggs, N.V. and Gordon, P.G. (1968). Absolute stereochemistry of calopiptin. *Aust. J. Chem.* **21**, 2095–106.

McKechnie, J.S. and Paul, I.C. (1969). Molecular structure of nordihydroguaiaretic acid. *J. Chem. Soc.* (*B*) 699–702.

MacLean, H. and MacDonald, B.F. (1969). Lignans of Western red cedar (*Thuja plicata* D. Don). Part VIII. Plicatinaphthol. *Can. J. Chem.* **47**, 457–60.

MacRae, W.D. and Towers, G.N.N. (1985). Non-alkaloidal constituents of *Virola elongata* bark. *Phytochemistry* **24**, 561–6.

Majumder, P.L., Chatterjee, A. and Sengupta, G.C. (1972). Lignans from *Machilus edulis*. *Phytochemistry* **11**, 811–14.

Marshall, J.J. (1974). Application of enzymic methods to the structural analysis of polysaccharides. *Adv. in Carbohydrate Chem. & Biochem.* **30**, 257–79.

Martinez, J.C., Cuca, L.E., Santana, A.J., Pombo-Villar, M.E. and Golding, B.T. (1985). Neolignans from *Virola elongata*. *Phytochemistry* **24**, 1612–14.

Matlin, S.A., Bittner, M. and Silva, A. M. (1984). Lignan & norditerpene dilactone constituents of *Podocarpus saligna*. *Phytochemistry* **23**, 2867–70.

Mikaya, G.A., Turabelidze, D.G., Kemertelidze, E.P. and Wulfson, N.S. (1981). Kaerophyllin, a new lignan from *Chaerophyllum maculatum*. *Planta Med.* **43**, 378–80.

Milne, G.W.A., Fales., H.M. and Axenrod, T. (1971). Identification of dangerous drugs by isobutane C.I.M.S. *Anal. Chem.* **43**, 1815–20.

Momose, T., Kanai, K.I. and Hayashi, K. (1978). Synthetic studies on lignans & related compounds. Part 8. Synthesis of justicidin B, diphyllin & of tawanin C & E. *Chem. Pharm. Bull.* **26**, 3195–8.

Morris, G.A. (1986). Modern NMR techniques for structure elucidation. *Mag. Res. in Chem.* **24**, 371–403.

Moser, W. and Howie, R.A. (1968). Nitrosodisulphonates. Part 1. Fremy's salt (potassium nitrosodisulphonate). *J. Chem. Soc.* (*A*) 3039–43.

Mujumdar, R.B., Srinavasan, R. and Venkataraman, K. (1972). Taxiresinol, a new lignan in the heartwood of *Taxus baccata*. *Indian J. Chem.* **10**, 677–80.

Murphy, S.T., Ritchie, E. and Taylor, W.C. (1975). Some constituents of *Austrobaileya scandens*: structures of seven new lignans. *Aust. J. Chem.* **28**, 81–90.

Murrell, J.N. (1971). *The Theory of the Electronic Spectra of Organic Molecules*. Chapman & Hall: London, pp. 91–100.

Nemethy, E.K., Lago, R., Hawkins, D. and Calvin, M. (1986). Lignans of *Myristica otoba*. *Phytochemistry* **25**, 959–60.

Nishibe, S., Chiba, M., Sakushima, A., Hisada, S., Yamagouchi, S., Takido, M., Sankawa, U. and Sakakibara, A. (1980). Introduction of an alcoholic hydroxyl group into 2,3-dibenzyl- butyrolactone lignans with oxidising agents and CMR of the oxidation products. *Chem. Pharm. Bull.* **28**, 850–60.

Nishibe, S., Okabe, K. and Hisada, S. (1981). Isolation of phenolic compounds and spectroscopic analysis of a new lignan from *Trachelospermum asiaticum* var. *intermedium*. *Chem. Pharm. Bull.* **29**, 2078–82.

Nishibe, S., Tsukamoto, H. and Hisada, S. (1984). Effects of *O*-methylation and *O*-glycosylation on CMR chemical shifts of matairesinol, (+)-pinoresinol, and epipinoresinol. *Chem. Pharm. Bull.* **32**, 4653–7.

Noggle, J.H. and Schirmer, R.E. (1971). The nuclear Overhauser effect: chemical applications. Acad. Press: New York.

Noguchi, K. and Kawanami, M. (1940). Study of the active components of *Anthriscus sylvestris* Hoffm. *J. Pharm. Soc. Japan* **60**, 629–33.

Ogawa, M. and Ogihara, Y. (1976). Studies on the constituents of *Enkianthus nudipes*. Part V. A new lignan xyloside from the stems. *Chem. Pharm. Bull.* **24**, 2102–5.

Ookawa, N., Ikeya, Y., Taguchi, M. and Yosioka, I. (1981). The constituents of *Kadsura japonica* Dunal. Part I. The structure of 3 new lignans: acetyl, angeloyl & caproyl-binankadsurin A. *Chem. Pharm. Bull.* **29**, 123–7.

Pelter, A. (1967). The mass spectra of oxygen heterocycles. Part IV. The mass spectra of some complex lignans. *J. Chem. Soc.* 1376–80.

Pelter, A. (1968). The mass spectra of some lignans of the 1-phenyl-1,2,3,4-tetrahydronaphthalene class. *J. Chem. Soc.* (*C*), 74–9.

Pelter, A., Ward, R.S., Rao, E.V. and Sastry, K.V. (1976). Revised structure for pluviatilol, methyl pluviatilol and xanthoxylol. *Tetrahedron* **32**, 2783–8.

Pelter, A. and Ward, R.S. (1978). In *Chemistry of Lignans*, ed. Rao, C.B.S. Andhra Univ. pp. 227–75.

Petcher, T.J., Weber, H.P., Kohn, M. and von Wartburg, A. (1973). Crystal structure and absolute configuration of 2′-bromopodophyllotoxin – 0.5 ethyl acetate, *J. Chem. Soc. Perkin Trans. II*, 288–92.

Platt, J.R. (1949). Classification of spectra of catacondensed hydrocarbons. *J. Chem. Phys.* **17**, 484–95.

Powell, R.G. and Plattner, R.D. (1976). Structure of a secoisolariciresinol diester from *Salvia plebeia* seed. *Phytochemistry* **15**, 1963–5.

Prabhu, B.R. and Mulchandani, N.B. (1985). Lignans from *Piper cubeba*. *Phytochemistry* **24**, 329–31.

Rao, M.N. and Lavie, D. (1974). The constituents of *Ecballium elaterium* L. *Tetrahedron* **30**, 3309–13.

Rao, K.V. and Wu, W.N. (1978). Glycosides of *Magnolia*, Part III. Structural elucidation of magnolenin C, *Lloydia* **41**, 56–62.

Ritchie, T.J. (1986). *The Chemistry of Lignans*, Ph.D. Thesis, London.

Robin, J.P., Gringore, O. and Brown, E. (1980). Asymmetric total synthesis of the anti-leukemic lignan precursor (−)-steganone & revision of its absolute configuration. *Tetrahedron Lett.* **21**, 2709–12.

Robin, J.P., Dhal, R. and Brown, E. (1984). Total synthesis of biologically active lignans. Part III. First synthesis of picrosteganes and formal synthesis of (±)-steganacin. *Tetrahedron* **40**, 3509–20.

Robin, J.P., Davost, R. and Taafrout, M. (1986). Steganolides B & C, new lignans analogous to episteganacin, from *Steganotaenia araliacea* Hochst. -2D correlation of ^{13}C–^{1}H in NMR. *Tetrahedron Lett.* 2871–4.

Rose, M.E. and Johnstone, R.A.W. (1982). *Mass Spectroscopy for Chemists and Biochemists*. Camb. Univ. Press: Cambridge.

Sadtler Handbook of Ultraviolet Spectra (*1979*). Sadtler-Heyden: Philadelphia – London.

Sanders, J.K.M. and Hunter, B.K. (1987). *Modern NMR Spectroscopy, a Guide for Chemists*. Oxford Univ. Press: Oxford.

Sarkanen, K.V. and Wallis, A.F.A. (1973a). PMR analysis and conformation of 2,5-bis-(3,4,5-trimethoxyphenyl)-3,4-dimethyl THF isomers. *J. Hetero. Chem.* **10**, 1025–7.

Sarkanen, K.V. and Wallis, A.F.A. (1973b). Oxidative dimerisation of (*E*)- and (*Z*)-isoeugenol and (*E*)- and (*Z*)-2,6-dimethoxy-4-propenylphenol. *J. Chem. Soc. Perkin I* 1869–78.

Sastry, K.V. and Rao, E.V. (1983). Isolation and structure of cleistanthoside A. *Planta Med.* **47**, 227–9.

Satyanarayana, P., Rao, P.K., Ward, R.S. and Pelter, A. (1986). Arborone & 7-oxodihydrogmelinol from the heartwood of *Gmelina arborea. J. Nat. Prod.* **49**, 1061–4.

Schrecker, A.W. and Hartwell, J.L. (1953). Components of podophyllin. Part XII. The configuration of podophyllotoxin. *J. Amer. Chem. Soc.* **75**, 5916–24.

Schrecker, A.W. and Hartwell, J.L. (1955). Application of tosylate reductions and molecular rotations to the stereochemistry of lignans. *J. Amer. Chem. Soc.* **77**, 432–7.

Schrier, E. (1964). The structure of sikkimotoxin. Part 2. Partial synthesis of the 6,7-dimethoxyanalogue of podophyllotoxin, epi-, neo- & deoxypodophyllotoxin, *Helv. Chim. Acta* **47**, 1529–55.

Setchell, K.D.R., Lawson, A.M., Mitchell, F.L., Adlercreutz, H., Kirk, D.N. and Axelson, M. (1980). Lignans in man and animal species. *Nature* **287**, 740–2.

Setchell, K.D.R., Lawson, A.M., Conway, E., Taylor, N.F., Kirk, D.N., Cooley, G., Farrant, R.P., Wynn, S. and Axelson, M. (1981). The definitive identification of the lignans *trans*-2,3-*bis*(3-hydroxybenzyl)-γ-butyrolactone & 2,3-*bis*(3,hydroxybenzyl)-butane-1,4-diol. *Biochem. J.* **197**, 447–000.

Shen, Z. and Theander, O. (1985). (−)-Massoniresinol, a lignan from *Pinus massoniana. Phytochemistry* **24**, 364–7.

Sheriha, J.M. and Amer, K.M.A. (1984). Lignans of *Haplophyllum tuberculatum. Phytochemistry* **24**, 364–7.

Shimomura, H., Sashida, Y. and Oohara, M. (1988). Lignans from *Machilus thunbergii. Phytochemistry* **27**, 634–6.

Smith, M. (1963). The structure of (−)-olivil. *Tetrahedron Lett.* **15**,) 991–2.

Snider, B.B. and Jackson, A.C. (1983). Synthesis of the aryltetralin lignan skeleton via the Prins reaction. *J. Org. Chem.* **48**, 1471–4.

Spencer, G.F. and Flippen-Anderson, J.L. (1981). Isolation & X-ray structure determination of a neolignan from *Clerodendron inerme* seeds. *Phytochemistry* **20**, 2757–9.

Stanek, J. (1963). *The Monosaccharides.* Academic Press; London, New York, p. 272.

Stevens, D.R. and Whiting, D.A. (1986). Synthetic methods for (exo, exo)- and (exo, endo)-2,6-diarylbicyclo-[3:3:0]-octane lignans: synthesis of aptosimon, (±)-styraxin, (±)-asarinin, and pluviatolide. *Tetrahedron Lett.* **27**, 4629–32.

Stevenson, R. and Williams, J.R. (1977). Concerning phyltetralin: synthesis of lignan aryltetralin isomers. *Tetrahedron* **33**, 2913–17.

Suzuki, H., Lee, K-H., Haruna, M., Iida, T, Ito, K. and Huang, H-C. (1982). (+)-Arctigenin, a lignan from *Wikstroemia indica. Phytochemistry* **21**, 1824–5; with details supplied by Professor K-H. Lee.

Taafrout, M., Rouessac, F. and Robin, J-P. (1983a). Arialangin, a new dibenzocyclo-octadiene lignan. *Tetrahedron Lett.* **24**, 197–200.

Taafrout, M., Rouessac, F. and Robin, J-P. (1983b). Neoisostegane, a new *bis*-benzylcyclo-octadienolactonic lignan from *Steganotaenia araliaceae* Hochst. *Tetrahedron Lett.* 2983–6.

Taafrout, M., Rouessac, F. and Robin, J-P. (1983c). Prestegane from *Steganotaenia araliacea* Hochst – a short synthesis of *cis*- & *trans*-(±)-arctigenin. *Tetrahedron Lett.* **24**, 3237–8.

Takani, M., Ohya, K. and Takahashi, K. (1979). Studies on the constituents of medicinal plants. Part XXI. Constituents of *Schizandra nigra* Max. *Chem. Pharm. Bull.* **27**, 1422–5.

Takaoka, D., Watanabe, K. and Hirui, M. (1976). Studies on lignoids in lauraceae. Part II. Studies on lignans in the leaves of *Machilus japonica* Sieb. et Zucc. *Bull. Chem. Soc. Japan* **49**, 3564–6.

Takeya, T., Matsumoto, H., Kotani, E. and Tobinaga, S. (1983). New reagent systems containing chromium trioxide provide precursors for synthesis of neolignans. *Chem. Pharm. Bull.* **31**, 4364–7.

Takeya, T., Okubo, T., Nishida, S. and Tobinaga, S. (1985). A practical synthesis of D,L-dibenzo-cyclooctadiene lignans. D,L-deoxyschizandrin, D,L-Wuweizisu C and their stereoisomers. *Chem. Pharm. Bull.* **33**, 3599–607.

Tanahashi, T., Nagakura, N., Inoue, K., Inoue, H. and Shingu, T. (1987). Sambacolignoside, a new lignan-secoiridoid glucoside from *Jasminium sambac. Chem. Pharm. Bull.* **35**, 5032–5.

Toda, F. *et al.* (1981). *Handbook of CMR Spectra.* Sankyo: Tokyo.

Tomioka, K. and Koga, K. (1979a). Stereoselective total synthesis of optically active *cis*- and *trans*-burseran. *Heterocycles* **12**, 1523–8.

Tomioka, K., Ishiguro, T. and Koga, K. (1979b). Novel isomerisation of dibenzocyclo-octadiene lignan lactones: first synthesis of (±)-stegane. *Tetrahedron Lett.* 1409–10.

Tomioka, K., Ishiguro, T. and Koga, K. (1980). First asymmetric total synthesis of (+)-steganacin – determination of absolute stereochemistry. *Tetrahedron Lett.* **21**, 2973–6.

Tomioka, K., Ishiguro, T. and Koga, K. (1985). Stereoselective reactions. Part X. Total synthesis of the optically pure lignan, burseran. *Chem. Pharm. Bull.* **33**, 4333–7.

Tsukamoto, H., Hisada, S. and Nishibe, S. (1984a). Lignans from the bark of *Olea* plants. Part I. *Chem. Pharm. Bull.* **32**, 2730–5.

Tsukamoto, H., Hisada, S. and Nishibe, S. (1984b). Lignans from the bark of *Fraxinus mandshurica* var. *japonica* & *F. japonica. Chem. Pharm. Bull.* **32**, 4482–9.

Turabelidze, D.G., Mikaya, G.A., Kemertelidze, E.P. and Vu'lfson, N.S. (1982). Lignan lactones from beak-chervil *Anthriscus nemerosa* Bieb. *Soviet J. Bioorg. Chem.* **8**, 374–9.

Ululeben, A., Saiki, Y., Lotter, H., Char, V.M. and Wagner, H. (1978). Chemical components of *Styrax officinalis*. The structure of a new lignan styraxin, *Planta. Med.*, **34**, 403–7.

Urzua, A., Freyer, A.J. and Shamma, M. (1987). 2,5-diaryl-3,4-dimethyltetrahydrofuranoid lignans. *Phytochemistry* **26**, 1509–11.

Vande Velde, V., Lavie, D., Gottlieb, H.E., Perold, G.W. and Scheinmann, F. (1984). Synthesis and PMR spectroscopic analysis of some 3,7-dioxabicyclo-[3,3,0]-octane lignans. *J. Chem. Soc. Perkin Trans. 1*, 1159–63.

Vecchietti, V., Ferrari, G., Orsini, F. and Pelizzoni, F. (1979). Alkaloid and lignan constituents of *Cinnamosmum Madagascariensis. Phytochemistry* **18**, 1847–9.

Vieira, P.C., Gottlieb, O.R. and Gottlieb, H.E. (1983). Tocotrienols from *Iryanthera grandis. Phytochemistry* **22**, 2281–6.

Wallis, R., Porte, A.L. and Hodges, R. (1963). Lignans from *Myristica otoba.* The structure of hydroxyotobain & isootobain. *J. Chem. Soc.*, 1445–9.

Ward, R.S., Satyanarayana, P., Row, L.R. and Rao, B.V.G. (1979). The case for a revised structure for hypophyllanthin – an analysis of the CMR spectra of aryltetralins. *Tetrahedron Lett.* 3043–6.

Ward, R.S., Satyanarayana, P. and Rao, B.V.G. (1981). Reactions of aryltetralin lignans with DDQ – an example of DDQ oxidation of an allylic ether group. *Tetrahedron Lett.* **22**, 3021–4.

Wei-Ming, C., Mayer, R. and Rucker, G. (1987). Configuration of the hydroxy group of (−)-cubebin & its epimer. *Arch. Pharm.* **320**, 374–6.

Weinges, K. and Spanig, R. (1967). In *Oxidative Coupling of Phenols*, ed. Taylor, W.I. and Battersby, A.R. Arnold: London.

Yamaguchi, H. (1970). *Spectral Data of Natural Products*, Vol. 1. Elsevier: Amsterdam.

Yamaguchi, H., Arimoto, M., Tanoguchi, M., Ishida, T. and Inoue, M. (1982). Studies on the constituents of the seeds of *Hernandia ovigera* L. Part III. Structures of two new lignans. *Chem. Pharm. Bull.* **30**, 3212–18.

Zavala, F., Guenard, D., Robin, J-P. and Brown, E. (1980). Structure–anti-tubulin activity relationships in steganacin congeners & analogues. Inhibition of tubulin polymerisation *in vitro* by (±)-isodeoxypodophyllotoxin. *J. Med. Chem.* **23**, 546–9.

Zheng, R. and Wang, M. (1985). Study on the chemical constituents of *Lancan tibetica* Hook. *Zhiwu Xuebao* **27**, 402–6.

Zhuang, L. (1984). CMR spectral analysis of lignans from *Daphne tangutica*. *Chem. Abs.* **100**, 44219.

7

Biosynthesis

The dimeric lignans can be related to lignin, the three-dimensional polymer which is intercalated with cellulose and hemicelluloses so as to rigidify the structure of vascular plants. In the absence of direct evidence of lignan biogenesis it was assumed that dimerisation involved similar precursors and processes to those that had been demonstrated for lignin. Although some direct evidence of the formation of lignans from cinnamyl precursors is now available, much still depends on the lignin analogy. There is also some indication that experiments designed to evaluate pathways in the lignan pool are affected by the co-existence *in vivo* of a pool of intermediates for lignin synthesis and a brief account of its chemistry is therefore given by way of introduction.

The chemistry of lignin

One-third of the world's land surface is covered by forest containing 3×10^5 million cubic metres of timber and some 2.6×10^9 cubic metres are harvested annually. This amounts to a production of 1.3×10^9 metric tons or twice the world production of steel. Cellulose and hemicelluloses form over half of this but lignin is the remaining bulk constituent and is therefore the second largest natural source of organic material: it is restricted to vascular plants (Fengel and Wegener, 1984).

A diagramatic representation (Pettersen, 1984) of a lignin structure is shown (Scheme 7.1). The structure is constructed by the random linking of the three *trans*-(E)-monolignols (**7.1**) through free radicals (**7.2A–C**) generated by peroxidases bound to the cell wall. (Grisebach, 1981). The ratio of these components is dependent on the plant species (Gross, 1979) and on the morphological origin (Whiting and Goring, 1982).

p-Coumaryl units (**7.1**; $R_1 = R_2 = H$) are of minor occurrence, coniferyl alcohol (**7.1**; $R_1 = MeO$, $R_2 = H$) is the principal unit in softwood

Scheme 7.1 A representation of lignin

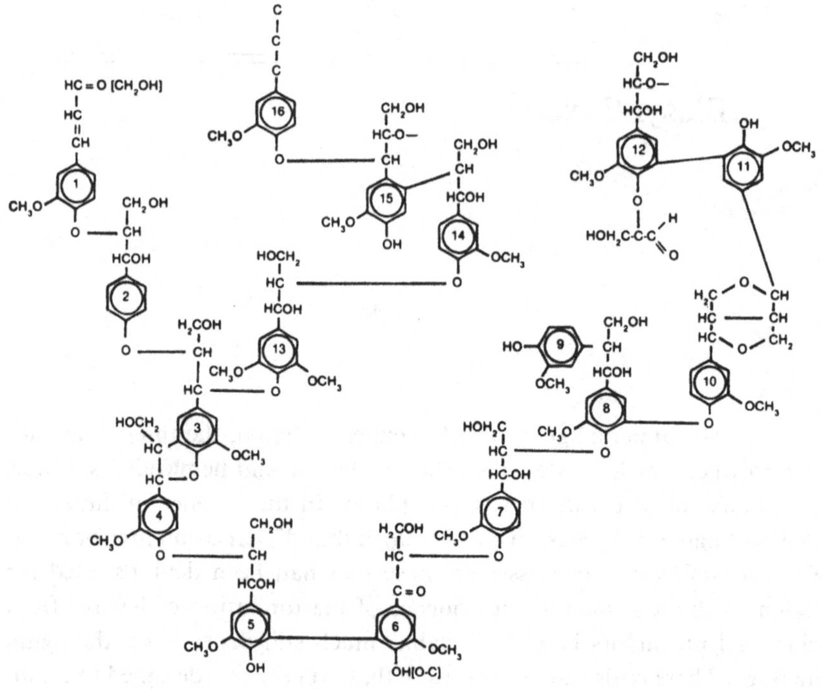

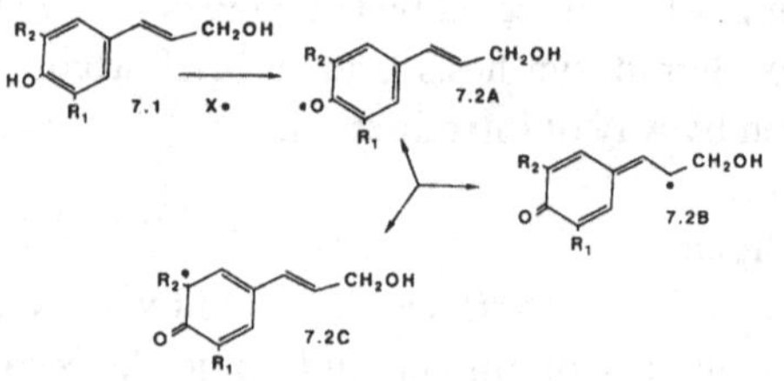

lignin and both coniferyl and sinapyl (**7.1**; $R_1 = R_2 = OMe$) residues contribute to the structure of hardwood lignin. The lignin is deposited principally as thickening proceeds in the secondary cell walls, but it also occurs in cell corners and where cells are joined in the middle lamellae. There are considerable variations in the relative distribution in these parts between different plant species.

The route to lignin depends upon the shikimic acid–cinnamic acid path (Scheme 7.2). The compounds in this pool were identified by the classical work of Davis (1958), who showed that carbohydrates were converted into shikimic acid (**7.3**) with subsequent branching to give either phenylalanine or tyrosine. It is significant to all work on lignan biosynthesis that these two branches are distinct and interchange between their components does not occur. Tyrosine may give rise to *p*-coumaryl alcohol (**7.1**; $R_1 = R_2 = H$) by the action of TAL (Scheme 7.2) but this route to lignin is not general in plants and appears to be limited to grasses. The lignin of grasses also includes *p*-coumaryl units. The principal route to lignin and lignans follows the action of a lyase on phenylalanine to give cinnamic acid; there is a branch point here also and the aminoacid may be taken up for the biosynthesis of alkaloids. The development of cinnamic acid by the action of phenolases (McCalla and Neish, 1959) and of methyl transferase is outlined (Scheme 7.2) but the hydroxylation level at which the pool is linked

Scheme 7.2 The shikimic acid – cinnamic acid path

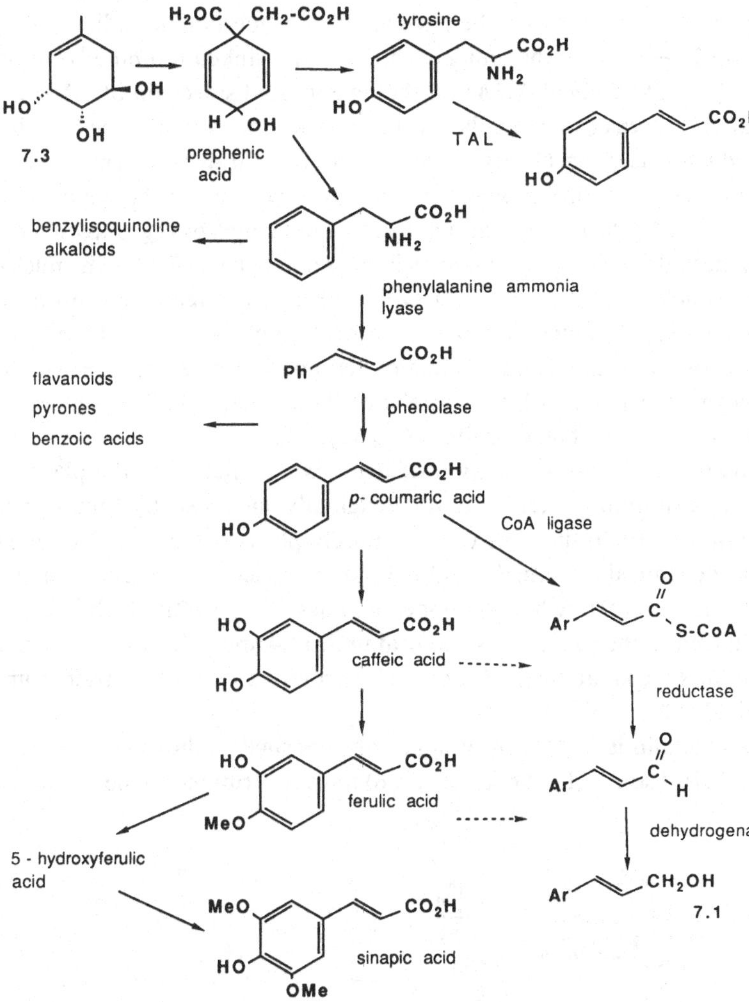

to lignin and lignan synthesis is not known. However, these acids may be stored in the plant as glucosides and converted into lignols (**7.1**) by the action of dehydrogenase (Scheme 7.2). Once they have been transported to the cell wall the blocking glucosyl group is removed and polymerisation is induced by the action of peroxidase. A detailed account of these processes has been given (Higuchi, 1985) and the mechanism for the reduction of the terminal carboxyl group to alcohols and then to alkenes is referred to in an earlier review of lignan biosynthesis (Pelter, 1986). The enzymic interconversion of *E*- and *Z*-monolignols has been discussed in relation to the biosynthesis of coniferyl alcohol in *Fagus grandiflora* (Lewis *et al.*, 1987).

Reference to the Scheme 7.1 shows that the nature of the growth of the polymer is consistent with the participation of mesomeric radicals of the types (**7.2A–C**). Thus the unit 2 (Scheme 7.1) is linked to unit 3 by attack of the phenoxy radical (**7.2A**) on the unsaturated sidechain of **7.1**. Other bonding patterns can be explained in this way for example, the 8,8′-linkage which arises via (**7.2B**) is seen in the furofuran fragment (unit 10; Scheme 7.1). The formation of ether linkages such as that between units 3 and 13 may be the result of attack by a phenolic hydroxyl group on a quinone methide; a similar mechanism is shown in Scheme 7.3 for the nucleophilic attack of the methide (**7.7**). Other schemes for lignin have proposed (Nimz *et al.*, 1984) including the incorporation of tetralins and the lignan conidendrin is a significant product of degradation (p. 274); note also that growth may also occur by the attack of the radical type **7.2C** on an aromatic ring to give a biaryl (units 5–6; Scheme 7.1).

The nature of the association of lignin with cellulose in the plant is a topic of continuing interest. It was originally suggested by Freudenberg and Harkin (1960) that this was not purely physical but included an element of chemical bonding brought about by nucleophilic attack of hydroxyl groups on intermediate quinone methides of type **7.2B**. Such a process would lead to the suggested structural units (**7.4** and **7.5**), since the former would arise from attack at C7 by a glucuronic acid and the latter from a terminal xylose residue.

Some lignin is soluble in organic solvents such as dioxan and acetone and NMR studies (Hawkes *et al.*, 1986) have contributed to the evaluation

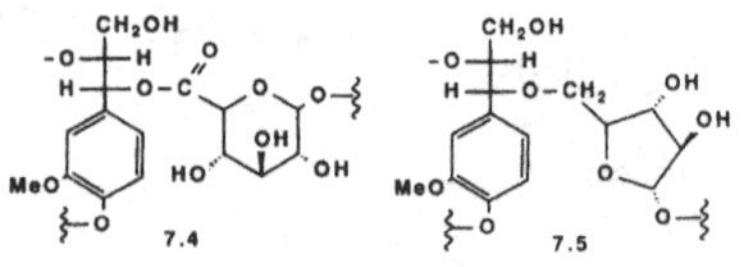

7.4 7.5

of structure. The problem is eased if the carbohydrate components are first removed by biological degradation. The application of CMR is then also considerably enhanced by the availability of higher-field spectrometers – at 25.2 kHz only five or six clear peaks are resolved and off-resonance techniques (p. 201) are ineffective because of multiplet overlap. At 90.7 MHz between twenty and thirty signals are resolved and the weighting of the aromatic substitution patterns can be determined. The incidence of particular linkages such as the Cβ—*O*—C4 (Scheme 7.1 – 1,2; 4,5; 7,8) can also be estimated (Nimz *et al.*, 1984). Structural details have also been obtained recently (Hatcher, 1987) by the application of the dipolar CMR to solid lignins, when the shifts typical of the aromatic rings and of the side chains were distinguished.

Cleavage of lignin

(i) *Acidic cleavage*

There is evidence of the participation of quinone methides (**7.7**) in the acid-induced cleavage of lignin which proceeds in part by the reversal of Freudenberg's proposed mechanism of carbohydrate conjugation (Scheme 7.3). In the preparation of cellulose pulp for paper making it is necessary to remove the lignin in order to achieve a planar form and also to ensure stability to light. This may be done by the sulphite process in which lignin is cleaved (see **7.6**) and solubilised by the addition of sulphite

Scheme 7.3 Acidic cleavage of lignin

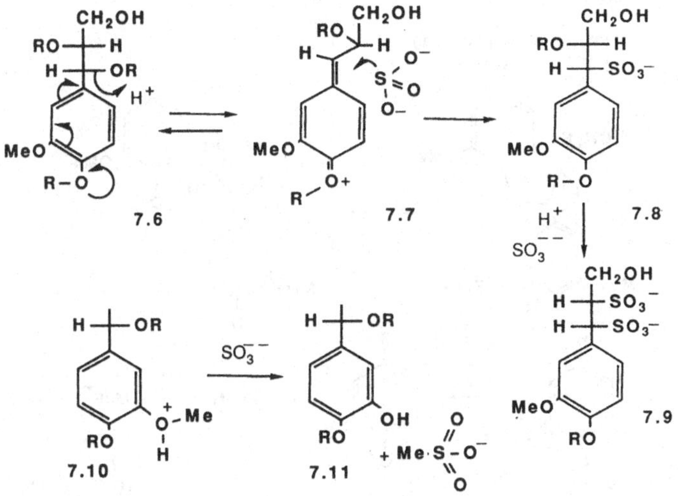

ions to **7.7** giving partially degraded material (**7.8**). The process will continue to give **7.9** and sulphite ion will also remove methyl or cleave other phenyl ether groups by the additional mode of attack (**7.10** to **7.11**).

Conidendrin (**2.348**) has been described as 'sulphite liquors lactone' since it occurs in liquors from cooking many coniferous woods (Hearon and MacGregor, 1955). A commercial process for its separation from western hemlock was devised and with renewed interest in these compounds this could yet provide a useful source of material for pharmaceutical research (cf. Table 5.1).

Fragments may also be cleaved from lignin by treatment with 2% acetic acid, with thioacetic acid or by alcoholysis. These fragments include furofuran lignans of the type shown in Scheme 7.1 (structure 10) and milder treatments of this kind yield oligomers of the type **7.12** and **7.13**.

(ii) *Alkaline cleavage*

Lignin may also be removed from wood by the alkaline pulping process in the presence of the strongly nucleophilic hydrosulphide ion. Scheme 7.4 illustrates the type of attack initiated by a free phenolic hydro-

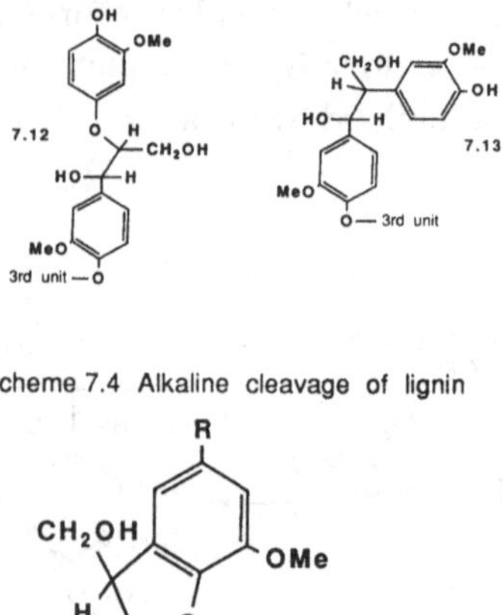

Scheme 7.4 Alkaline cleavage of lignin

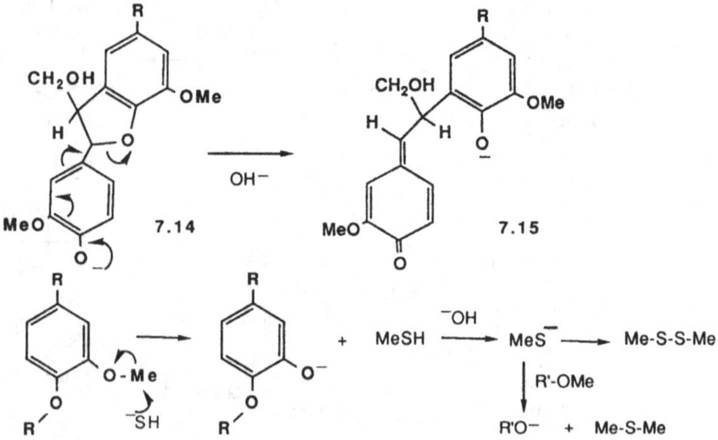

Scheme 7.5 Oxidative cleavage of terminal chains by white rot enzymes (Fengel, 1984)

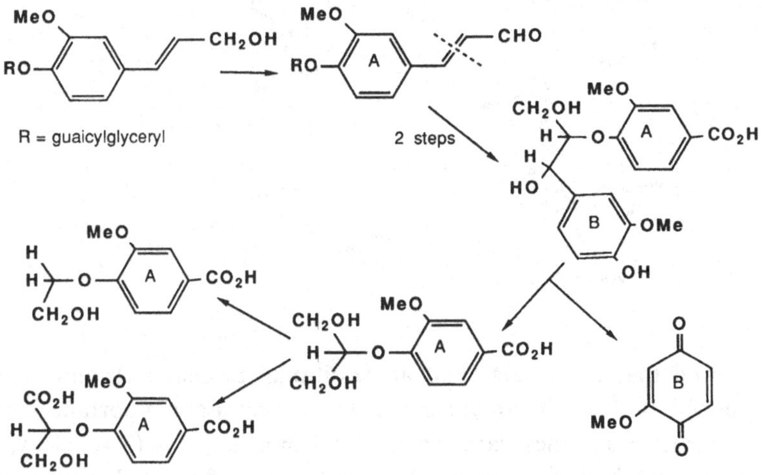

R = guaicylglyceryl

xyl group. The process of solution of the lignin continues through the quinone methide (**7.15**) in a similar manner to that shown above (Scheme 7.3). The hydrosulphide ions are capable of cleaving benzyl ether groups (Scheme 7.1, structure 7.15) so as to reveal other phenoxide centres. They also remove blocking methyl groups so increasing solubility, with the formation of methyl thioether and dimethyldisulphide (Scheme 7.4).

(iii) *Enzymic cleavage*

These processes are related to the formation of phytoalexins from lignans which was mentioned in chapter three (p. 104). The principal organisms are the brown and white rot fungi. The brown rots principally attack the polysaccharide material of wood, whilst the white rots (*Basidiomycetes*) attack both this and lignin; the soft rots which include the *Ascomycetes* and the *fungi imperfecti* are also effective (Scheme 7.5) (Fengel and Wegener, 1984).

Oligomers of cinnamic acid

The sesquilignans lappaol A (**7.16**) and lappaol B (**7.17**), with three cinnamyl residues, were isolated (Ichihara *et al.*, 1976) from *Arctium lappa* L. together with matairesinol (**7.18**) and arctigenin (**7.19**). A comparison of the aromatic substituents points to a biogenetic relationship between lappaol A and matairesinol. A similar relationship is likely for lappaol B and arctigenin. The molecular extinction coefficients at 283 nm in the UV spectra of lappaol A ($\varepsilon = 7800$) and arctigenin ($\varepsilon = 5200$)

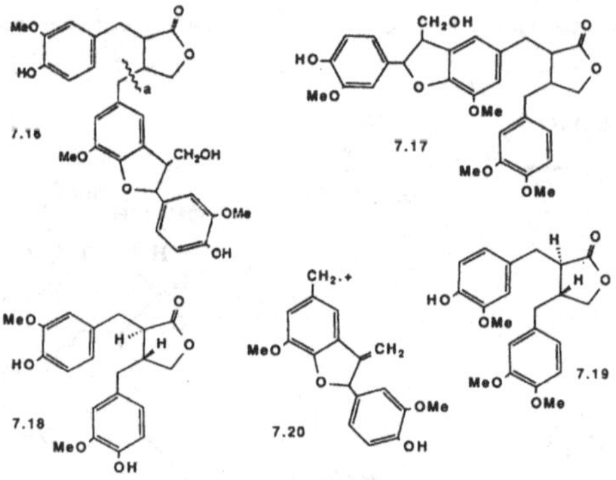

showed that the former had an additional aromatic chromophore included in a third cinnamyl residue. Mass spectra are important in assigning structures to these large molecules. For example in (**7.16**) cleavage (a) leads to the loss of a fragment (m/z = 137) including the lactone ring and the formation of the ion (**7.20**) characteristic of two further aromatic groups.

The β-syringylglyceryl ether of syringaresinol (**7.21**), a sesquilignan with the β—*O*—C4 linkage, was isolated from the wood of *Fraxinus mandshurica* (Omori and Sakakibara, 1979). Four other sesquilignans were obtained (Kikuchi *et al.*, 1985) as their acetyl derivatives. Typical of these compounds is hedyotol A (**7.22**; R = H). It is significant that the triacetate was optically active with [α_D] +29°C and that it also co-occurs with (+)-pinoresinol (**2.217**). The biogenetic link is further implied by the relationship between hedyotol B (**7.22**; R = OMe) and medioresinol (**2.214**) which are both found in this same source plant. A review of sesqui- and di-lignans is given in the Kikuchi *et al.* (1985) paper.

Extraction of *Saururus cernuus* yielded (Rao and Alvarez, 1983) the sesquilignan saucerneol (**7.23**; R = H) together with two dilignans, manussantin A (**7.23**; R = A) and B (R = B). Both of these compounds were optically active with $[\alpha]_D$ −100° and −99° respectively. It is possible that

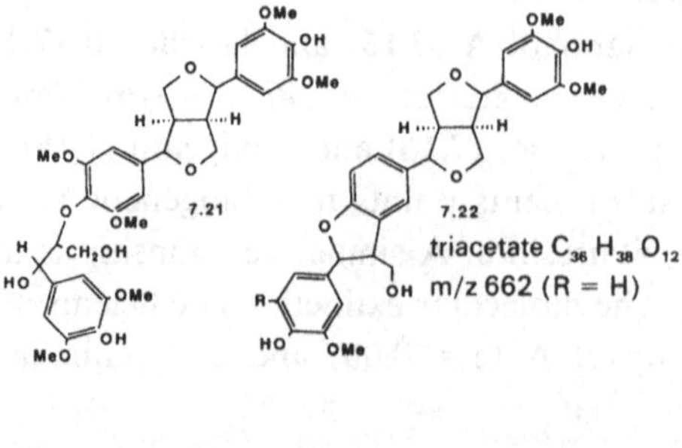

7.22 triacetate $C_{36}H_{38}O_{12}$ m/z 662 (R = H)

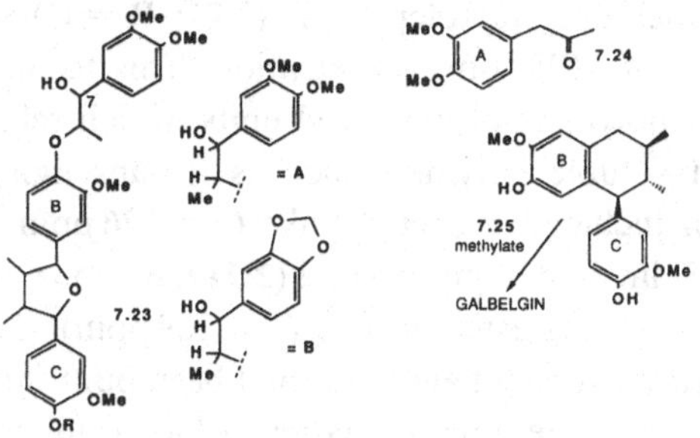

the further extension of these structures is blocked by the alkylation of the C4—OH group. They decomposed in the mass spectrometer but structural indications follow their treatment with acid. Thus cleavage of ring A in manassantin A gives dimethoxybenzyl methyl ketone (**7.24**), presumably via the enol ether obtained on elimination of the C7—OH group, and rearrangement of the furanoid lignan fragment affords the tetralin (**7.25**), which on further methylation gives galbelgin (**2.154**). Manassantin A is a neuroleptic agent.

The dilignans lappaol F and H have been isolated (Ichihara *et al.*, 1978) from the seeds of *Arctium lappa*. The isolation of di-, tri-, and tetralignans from *Larix leptolepis* (Sakakibara *et al.*, 1987) lends further support to the hypothesis of their formation by the sequential addition of individual cinnamyl units. In all, 42 compounds were obtained from this source including ten flavanoids. The lignans were found in the wood and bark but there was evidence of biogenetic differences, for example sesquilignans such as leptolepisol A (**7.26**) and leptolepisol B (**7.27**; R = H) were located in the inner bark rather than the wood. In contrast, neolignans (e.g. **7.28**) with side chains dehydrogenated by the action of TAL (Scheme 7.2) occurred in heartwood or sapwood but not in the bark.

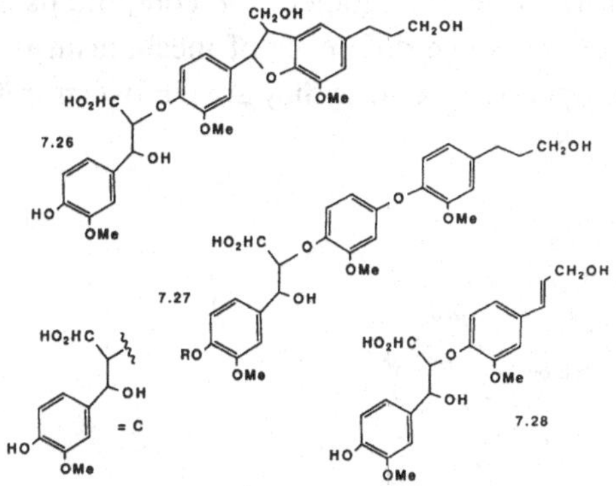

Tetrameric lignans including leptolepisol E (**7.27**; R = C) were also obtained (Sakakibara *et al.*, 1987) by gel filtration. This technique also gave evidence of higher polymers of coumaryl units, which related these oligomers with solvent–soluble polymers such as Brauns lignin. CMR showed that this lignin included biphenyl links ($\delta = 126$ ppm, guaiacyl groups), and the C—8,8′ linkage of pinoresinol (**2.217**, $\delta = 53$ ppm), there was also evidence of coumaryl ester groups ($\delta = 169$ ppm). The CMR spectra indicated that co-occurring flavonoids had been built into the lignin structure; this is also consistent with a pattern which leads to the biosynthesis of flavonolignans. Silybin (**7.29**), the first compound of this class, was described in 1968 (Pelter and Hansel) and a group of coumarinolignans including cleomiscusin A (**7.30**) has also been identified (Lin and Cordell, 1984). Foo and Karchesy (1989) have isolated pseudotuganol from Douglas fir where pinoresinol is linked through a biphenyl to dihydroquercetin.

The development of flavanones by the combination of cinnamate with three molecules of malonyl coenzyme A was established by Grisebach (1957, see Neish, 1964) and the relationship of this pool to that linked to neolignans has been described (Gottlieb and Kubitzi, 1981). The ability of plants of the *Lauraceae* to synthesise flavones in this way was attributed to enzymes that have *evolved* with development of individual species. In contrast the cinnamate path to neolignans was attributed to an *inherited* or primitive characteristic (cf. Estabrook, 1978). This view is consistent with the limited range of individual chemicals produced by primitive plants such as *Podophyllum* sp. where the dominant path is one of lignan synthesis.

Lignan biosynthesis

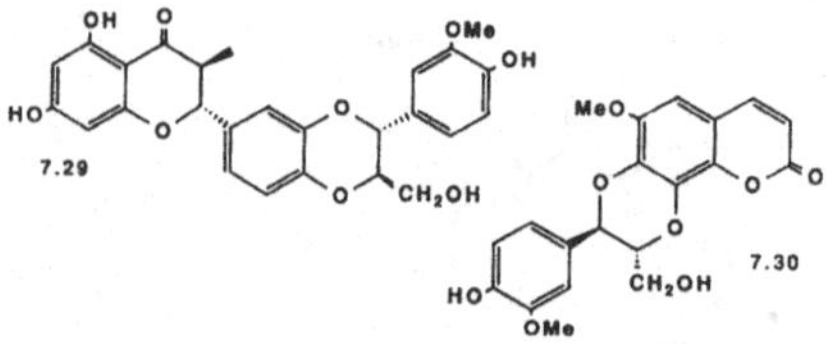

The evidence briefly reviewed above for the participation of lignols such as (**7.1**) in lignin and flavone biosynthesis is compelling. They must be involved in the formation of oligomers and compounds of mixed origin such as flavonolignans. Some difference of mechanism for lignan synthesis is believed to be operative because they are almost entirely optically active compounds whereas no optical activity has ever been found in a lignin, which is therefore regarded as the product of random radical reactions. Lignan biosynthesis is thought to take place under enzymic control. No evidence to the contrary is available concerning this implied difference but one should not dismiss the possibility that lignin does have asymmetric centres which because of their number and diversity are internally compensated. One can make a number of relevant observations:

(a) The overall similarity in conditions for limited growth and also for polymerisation makes it unlikely that a fundamental difference in mechanism should exist between them.
(b) The very intractability of lignin makes it very difficult to retain configuration during its degradation.
(c) Evidence is coming to light (Sakakibara *et al.*, 1987) of a growth continuum linking lignans with lignin.
(d) The optically active cyclolignan todolactol B has been obtained by the mild hydrolysis of Brauns lignin (Ozawa and Sasaya, 1987).
(e) Sesqui- and dilignans are themselves optically active: a striking example is the isolation from *Cerbera* sp. of three such dimers of (−)-olivil (Abe *et al.*, 1988).
(f) The magnitude of the optical rotation in oligomers of *Larix leptolepis* decreases to very low values of $[\alpha_D] = -3.0$ in the larger molecules.

(i) *The quinone methide mechanism*

This was originally formulated by Erdtmann (1933) and has been discussed in detail (Birch and Liepa, 1978); an example of its application was given in the Introduction (Scheme 1.2) which showed a possible mechanism for hydroxylation. Its operation may also be seen in the reversal of the alkaline cleavage (Scheme 7.4) and in the acceptance of nucleophiles as in the attack of sulphite on structure **7.7** (Scheme 7.3). Cyclisation to tetralins is illustrated by the conversion of **7.46–7.47** and in the formation of **7.52** (Scheme 7.8). Quinone methides may also participate in the biosynthesis of sequirin D (Scheme 7.14) and modification of intermediates such as **7.84** is a likely route to furofurans. The point has been made (Gottlieb, 1978) that this mechanism could account for the formation of 116 out of the 120 neolignans then known. It can also account for all the 8,8′-linked lignan dimers, besides its probable involvement in degradation of lignin (Scheme 7.3) and as an explanation of the formation of lignol oligomers such as **7.21** and **7.22**.

(ii) *Biomimetic synthesis*

These studies give general support to the current view of synthesis *in vivo* and may as a result lead to efficient connections – a case in point is the direct approach to carpanone (Scheme 7.11) which is discussed later in this section. However these results can be misleading as to detail since in accord with good practice *in vitro* the individual cinnamate precursors have their principal functions in place, whereas *in vivo* processes may well assemble these parts in a stepwise manner.

One of the key lignols (**7.1**; $R_1 = R_2 =$ OMe) has been obtained (Zanarotti, 1982) by oxidation of **7.31** (Scheme 7.6) with silver oxide to give the vinyl methide (**7.32**) followed by 1,8-addition of acetate and cleavage of the product (**7.33**) by reduction. This work was developed (Zanarotti, 1983) so as to mimic the formation of the C*β*—*O*—4 linkage found in neolignans and which occurs widely in lignin. Thus (*E*)-2,6-dimethoxy-4-(prop-1-enyl)phenol (**7.34**) dimerises through nucleophilic attack by the phenolic hydroxyl group with re-oxidation to the methide (**7.35**, quantitative yield). Subsequently further nucleophilic addition is possible – water

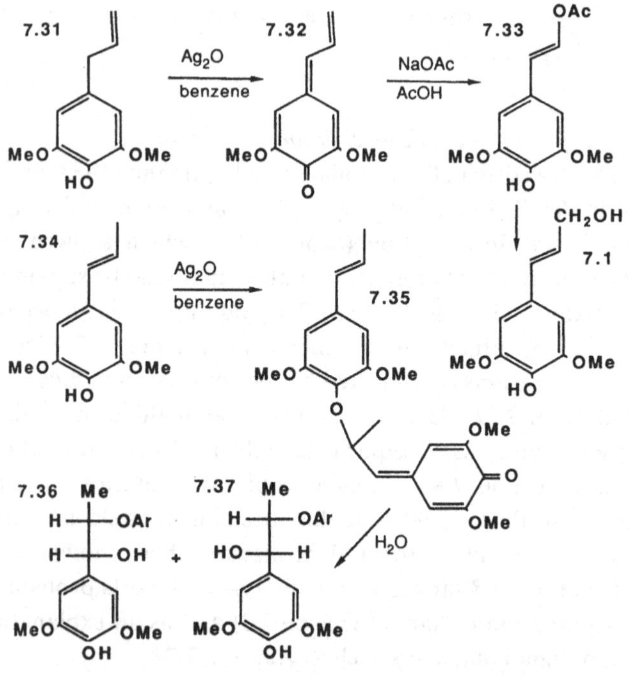

Scheme 7.6 The role of quinone methides in the synthesis & reactions of sinapyl derivatives

for example adds in the 1,6-manner to give a mixture of erythro-(**7.36**) and threo- (**7.37**, Scheme 7.6) products.

Similar results were obtained (Nishiyama, 1983a, b, c) by the electrochemical oxidation of a mixture of (*E*) and (*Z*)-isoeugenols (Scheme 7.7, structure **7.38**). A two-stage process has been written for this involving the reaction of the carbocation (**7.40**) with *either* one molecule of the precursor phenol and one of solvent to give **7.41** *or* with two molecules of solvent

Scheme 7.7 Electrochemistry of isoeugenol

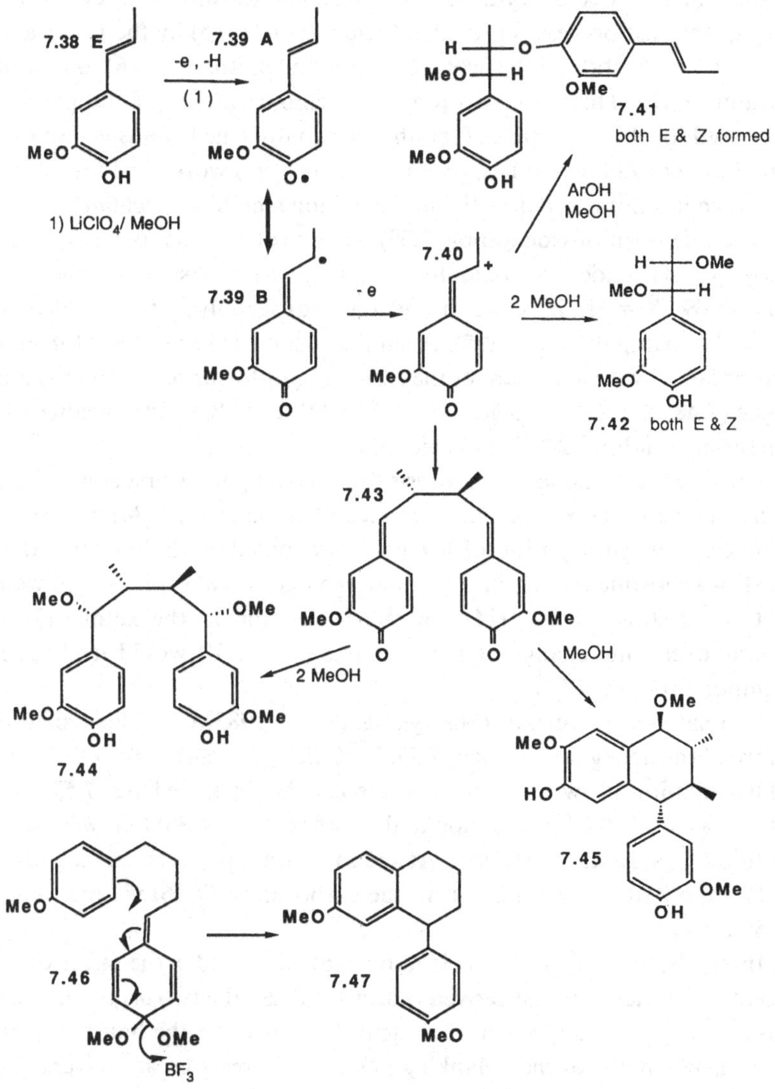

to give **7.42**. It is important to note that in the scheme discussed nucleophilic attack by a molecule of precursor upon a cation (**7.40**) is an alternative to the assumption that radical coupling occurs. This proviso extends to schemes for the dimerisation of cinnamyl precursors for lignans; it follows that the further product of this electrosynthesis (**7.43**) could be formed either by the coupling of the radical (**7.39B**) or by the attack of the cation (**7.40**) on isoeugenol (**7.38**). Two further products (**7.44**, **7.45**) were also isolated from this reaction, one (**7.44**) by the reductive 1,6-addition of solvent to the *bis*-quinone methide (**7.43**) and the other by a similar reduction of one ring followed by intramolecular aromatic substitution. A close analogy to this last process is provided (Pelter *et al.*, 1985) by the Lewis acid-catalysed cyclisation of the acetal of a quinone methide (**7.46**) to give the tetralin (**7.47**). These are two practical demonstrations of the pathway proposed for the formation of tetrahydronaphthalene lignans *in vivo* from lignol dimers and here it is appropriate to mention work which is related, although it is not based directly on the quinone methide mechanism.

The (*E*)-form of isoeugenol (**7.38**) when protected as its methyl ether (**7.48**; X = H) underwent oxidative coupling with cyclisation to the tetralone (**7.49**; X = H, 16% yield) and on hydrogenolysis this yielded the natural lignan galbulin (**2.295**). A similar result (Takeya *et al.*, 1983a, b) was obtained with a mixture of the (*E*)- and (*Z*)-forms of the sinapyl analogue (**7.48**; X = OMe) to give (**7.49**; X = OMe, 14% yield) together with the furan grandisin (**2.157**; 17% yield).

An alternative mode of biogenesis in which a lignol is first combined as a cinnamyl ester (**7.50**) was demonstrated (Takeya *et al.*, 1984) in the same manner. The participation of a β-carbon coupled methide intermediate (**7.51**) was postulated, which could undergo nucleophilic addition of water at C-7′ as shown with oxidation and cyclisation to the ketone (**7.52**). Alternatively interchange of these centres (7 and 7′) would lead to the common product (**7.53**).

In related experiments (Nishiyama *et al.*, 1983b) anodic oxidation afforded the neolignan asatone (**7.58**; 36% yield, Scheme 7.9) by the Diels-Alder addition of two molecules of the isolable intermediate (**7.57**). The latter was produced by oxidation to the carbocation (**7.56**) followed by the addition of solvent (methanol). At the same time products analogous to (**7.1**) are obtained by oxidation to the carbocation (**7.56**) via the radical (**7.55**).

In syntheses of this kind it is important to consider the effect of adjacent substituents on the derived radicals. In **7.55** the two *ortho*-methoxyl groups hinder the approach of reagents and increase the half-life of the radical. When one of these flanking positions is free a greater diversity of

Scheme 7.8 Oxidative cyclisation of isoeugenols

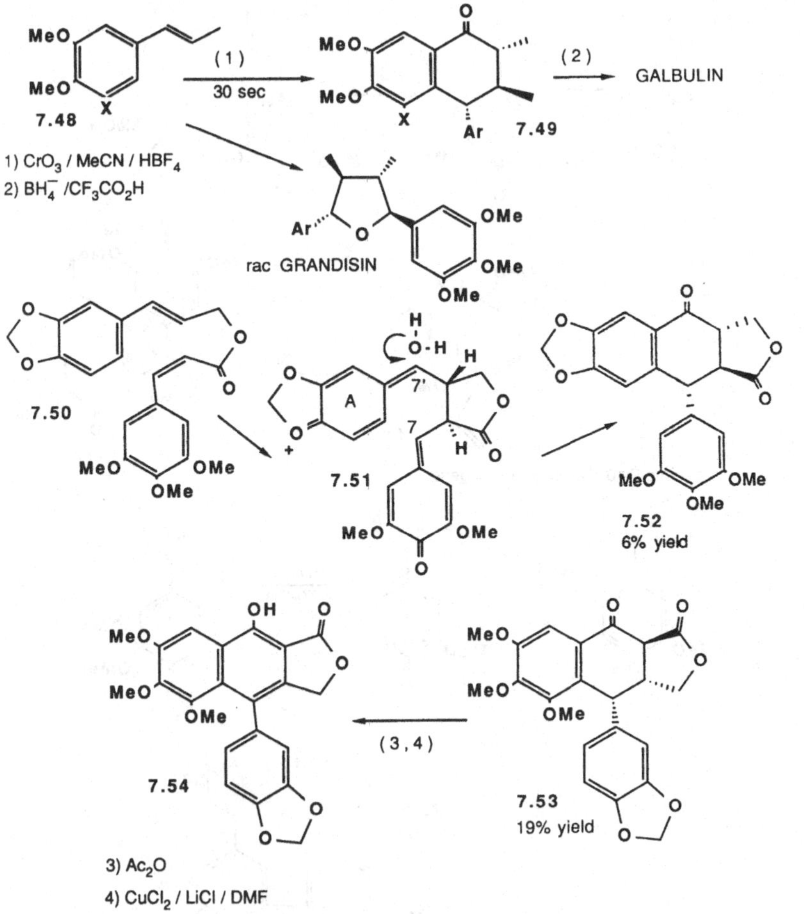

1) CrO_3 / MeCN / HBF_4
2) BH_4^- / CF_3CO_2H

3) Ac_2O
4) $CuCl_2$ / LiCl / DMF

products results. This can be illustrated by the reaction of isoeugenol (**7.59**) where coupling occurs to give the biphenyl (**7.62**). This may have occurred by attack of the radical (**7.60**) on a molecule of the starting phenol (**7.59**) followed by electrochemical modification of one ring. An alternative view is that shown (Scheme 7.10) where the electrochemical change occurs first to give **7.61** which then undergoes radical substitution.

When ferulic acid (**7.63**; X = H) was used (Nishiyama *et al.*, 1983c) as a precursor a compound (**7.64**) of analogous structure to asatone (**7.58**) was obtained (39% yield), but a significant minor product (8% yield) was the bicyclic lactone (**7.65**; X = H) which is related to the furofuran lignans. In closely related work (Ahmed *et al.*, 1973) oxidation with ferric chloride gave the dilactone (**7.65**; X = H, 65% yield). An interesting variation of

Scheme 7.9 Electrochemical synthesis of asatone (7.58)

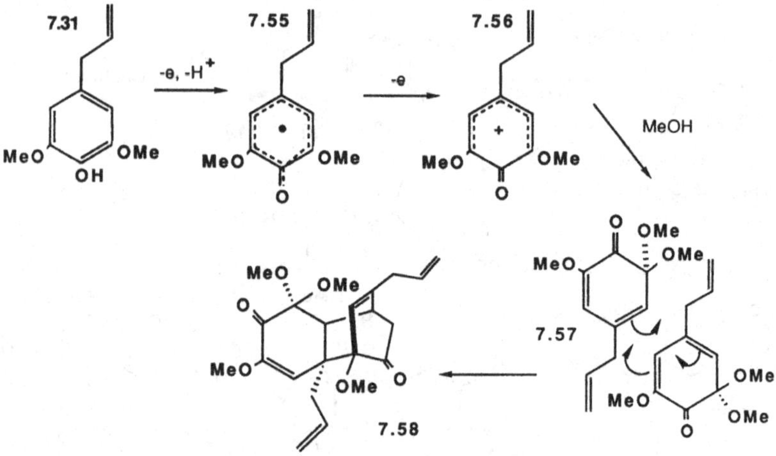

Scheme 7.10 Coupling of eugenol

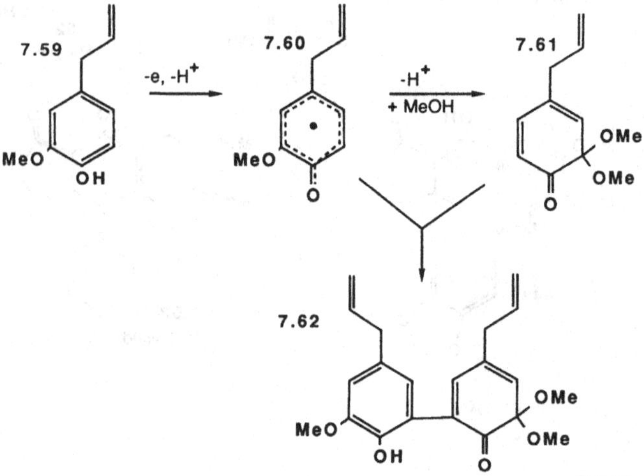

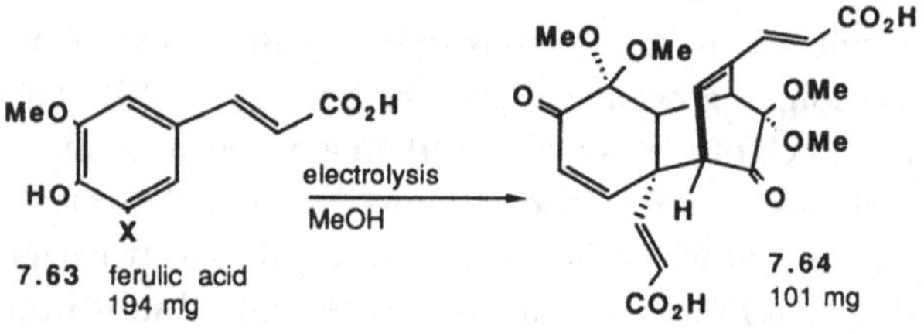

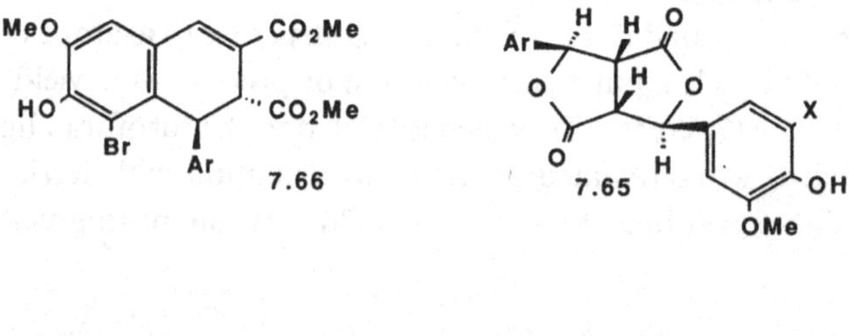

this method (Ahmed *et al.*, 1976) of possible biogenetic significance is that the acid (**7.63**; X = Br) yielded the analogous dilactone (**7.65**; X = Br), but blocking the carboxyl groups as the methyl ester led directly to synthesis of the apo-diester (**7.66**). The cyclisation product is also obtained by rearrangement of dilactones (**7.65**) in acid.

A particularly impressive biogenetic-type synthesis is that of carpanone (Scheme 7.11) by the oxidative dimerisation of the phenol (**7.67**; R = H) with the hydroxyl group *ortho*- to the side chain. This reaction was catalysed by Co(II) and Fe(II) salicylic complexes, when the dimer (**7.68**) gave carpanone in 90% yield by $4\pi + 2\pi$ cyclo-addition (Matsumoto and Kuroda, 1981). A link to possible biosynthesis is the co-occurrence of the *ortho*-methoxyl derivative (**7.67**; R = Me) in the source plant, the carpano tree.

Another phenol with an unusual orientation is prestegane B (**2.076**, Taafrout *et al.*, 1984) with two *meta*-hydroxyl groups. This compound has a common source, *Steganotaenia araliacaea*, with lignans of the dibenzocyclo-octadiene group and this could be of biogenetic significance since their biphenyl structure can be formed by C-6, C-6′ coupling at the two *para*- positions of (**2.076**). A further indicator of a possible route to biphenyls is the eupodienone (**7.69**) isolated (Bowden *et al.*, 1980) from the bark of *Eupomatia laurina*. If this molecule undergoes the dienone–phenol rearrangement then a structure (**7.70**) which is typical of these lignans results. A related norlignan, athrotaxin (**7.71**), has also been described (Beracierta and Whiting, 1978).

Another pointer to modes of biosynthesis is given by the isolation from

Scheme 7.11 Biogenetic - type synthesis of carpanone

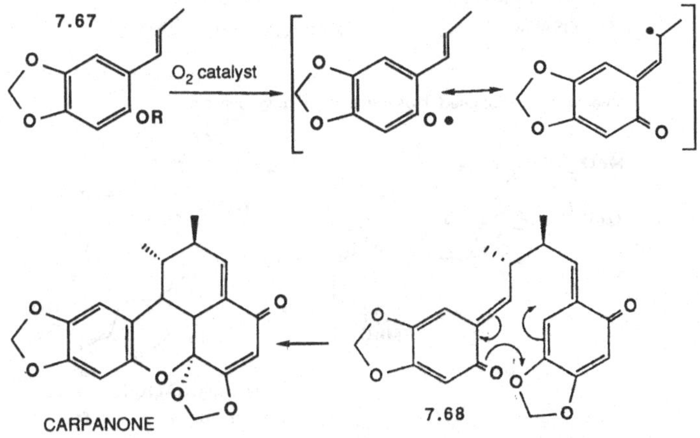

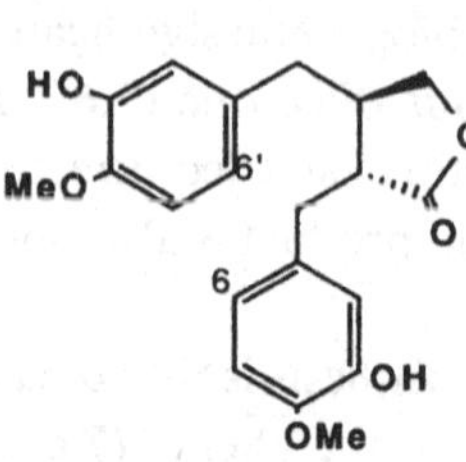

PRESTEGANE B

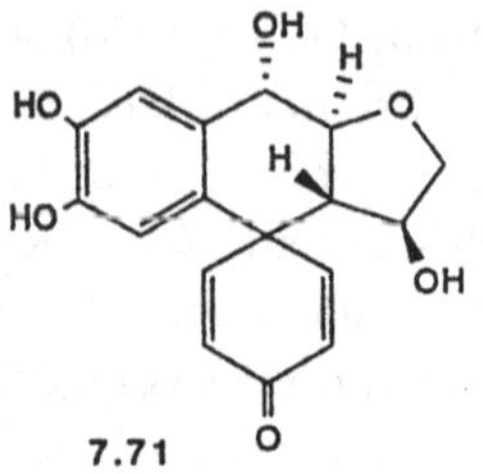

7.71

Scheme 7.12 A dienone - phenol rearrangement

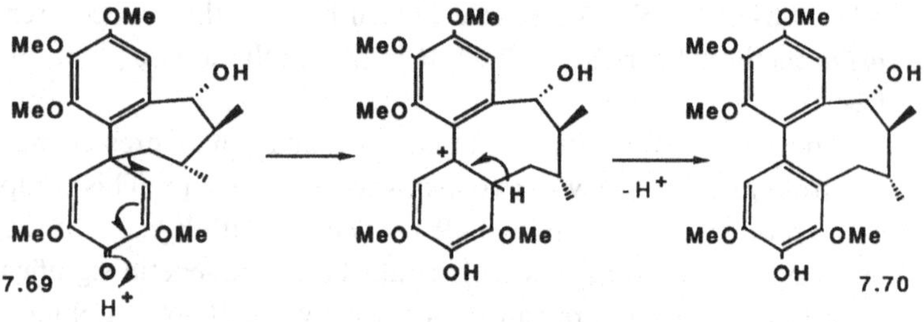

7.69

7.70

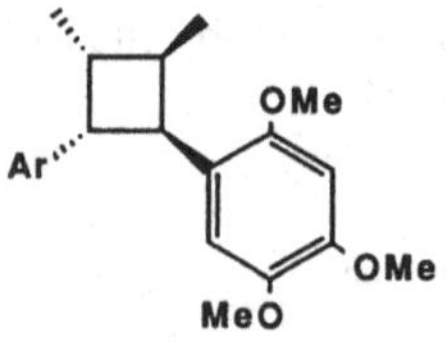

MAGNOSALIN

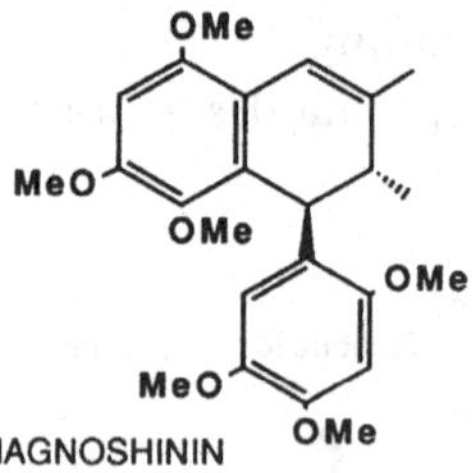

MAGNOSHININ

Scheme 7.13 Proposed biosynthesis of magnosalicin

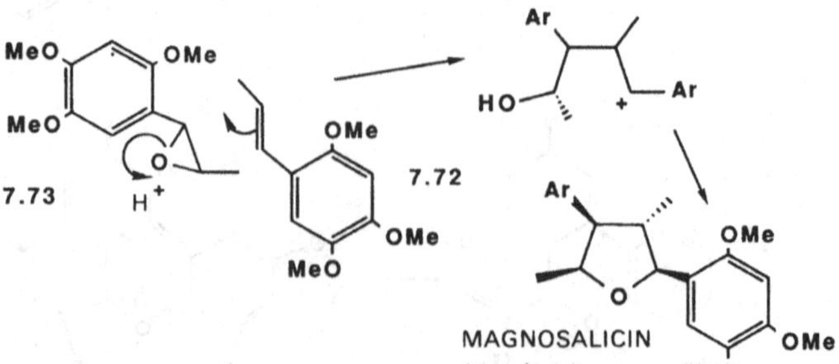

7.73

7.72

MAGNOSALICIN
(*ent*- form)

Magnolia salicifolia of asarene (**7.72**), which is to be compared with carpacin (**7.67**; R = Me), and also with magnosalin (p. 286) and magnoshinin (**2.377**; Kikuchi *et al.*, 1983) which share the same unusual aromatic substitution pattern. Another co-occurrence is that of asarene (**7.72**) with the furanoid lignan magnosalicin (**2.135**). The biogenetic route suggested (Tsuruga *et al.*, 1984; Scheme 7.13) was the condensation of asarene with a molecule of its epoxide (**7.73**).

meta-Hydroxycinnamic acid is believed to be the precursor for the norlignan sequirin D (**7.78**) whose structure has been established by X-ray analysis (Begley *et al.*, 1978) and whose biosynthesis is considered to follow the course shown in Scheme 7.14. Radical coupling of the mesomeric methides (**7.74, 7.75**) could provide an intermediate (**7.76**; R = H), whose pyrophosphate (**7.76**; R = pyrophosphate) is expected to undergo acid-catalysed cyclisation to the transitory carbocation (**7.77**) which decarboxylates to give the norlignan (**7.78**).

The isolation for the first time of *meta*-hydroxylated lignans from mammals in the form of enterolactone (**7.79**; X = H) and the corresponding diol was originally thought to be the result of a distinct mode of biosynthesis. However it has now been established (Adlercreutz, 1984; Bannwart *et al.*, 1984) that these lignans are formed by the action of gut flora on matairesinol (**7.79**; X = OMe) in grain.

(iii) *Experiments in vivo*

(a) *Preliminary studies and structural correlations*

The first confirmation of the link between lignans and compounds of the phenylalanine–cinnamate pool was the demonstration

Scheme 7.14 Biosynthesis of sequirin D

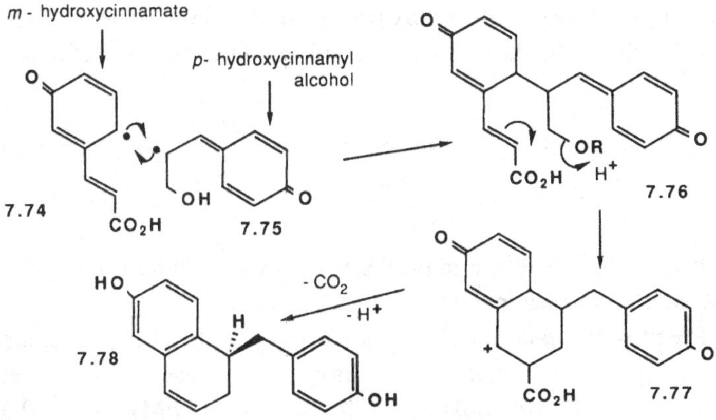

m - hydroxycinnamate

p- hydroxycinnamyl alcohol

7.74 7.75 7.76 7.77 7.78

CO_2H OH OR H^+ HO $-CO_2$ $-H^+$

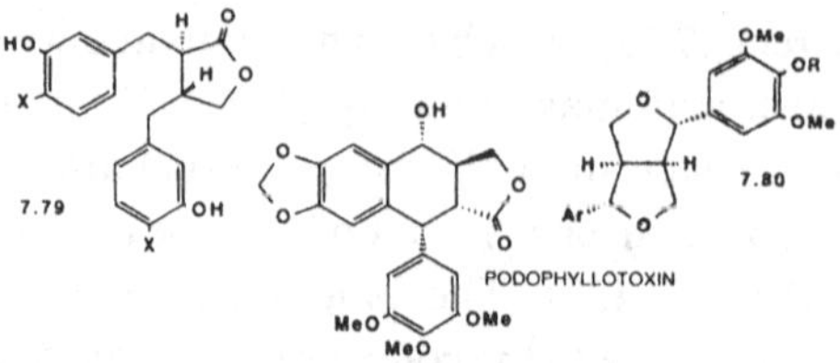

(Ayres, 1969) that [U-^{14}C]phenylalanine was taken up by *Podophyllum hexandrum* Royle (equiv. to *P. emodi* Wall.) plants with an incorporation of 1.4% in podophyllotoxin (**2.357**).

In subsequent work Fujimoto and Higuchi (1977) showed that both U-^{14}C- and 2-^{14}C-sinapyl alcohol were incorporated into syringaresinol (**7.80**; R = H) and into liriodendrin (**7.80**; R = β-D-glucosyl) in the yellow poplar, *Liriodendron tulipfera.* A significant level of incorporation was also found for [U-^{14}C]phenylalanine (dilution value = specific activity of precursor/specific activity of product = 38), but sinapic acid (**7.81**; X = CO_2H) and coniferyl alcohol (**7.82**) were poorly incorporated having dilution values of 334 and 4540 respectively. These low levels were probably the result of their preferential utilisation for the production of tannins and/or lignin. The most efficient precursor for syringaresinol and liriodendrin was 2-^{14}C-sinapyl alcohol (**7.81**; X = CH_2OH); it was concluded that this lignol was the immediate precursor of the lignans and hence that oxygenation was complete before coupling occurred. This conclusion may well be correct but one should draw attention to factors which affect the relative levels of incorporation in plants.

In the first place, variation in the relative rates of metabolism of the interrelated compounds in the pathway leads to their having different concentrations in the pool. The best chance of obtaining a good level of incorporation comes from feeding the active precursor which is least diluted by unlabelled material already present in the plant. Scheme 7.15 (McCalla and Neish, 1959) shows the variation in concentration between the esters

Scheme 7.15. Variation in pool size on sequential hydroxylation of cinnamic acids in *Salvia* sp.

Precursor	*p*-coumaric ester	caffeic ester	ferulic ester	sinapic acid ester
	(0.6 μM)	(20 μM)	(5 μM)	(0.3 μM)

of cinnamic acids in the pool of *Salvia* sp. from which lignin precursors were drawn.

Evidently the limited reservoir of sinapic acid in this plant favours its incorporation and this could have been a factor in the experiments with *Liriodendron*.

A second factor which must be borne in mind is that one is dealing with a dynamic equilibrium and that precursors for lignans are linked competitively with growth of lignin. The conditions prevailing at the time of an experiment may divert a substance into lignin synthesis and so lower the apparent efficiency of its incorporation into lignans. Linked with this uncertainty of metabolism is the possibility that a labelled precursor may not be translocated efficiently under experimental conditions and hence its role in the pathway may not be clear. For success, account must be taken of a number of variables:

(a) Translocation may not always be satisfactory; thus for *Podophyllum hexandrum* feeding is best undertaken through May into later Summer.
(b) Contact times need to be determined and for *P. hexandrum* seven to eight days were required.
(c) In view of the seasonably variable yields of many included lignans it is necessary to ensure that the end product is being *synthesised* and not *degraded* at the time of feeding.
(d) Labelled material may be metabolised with subsequent relocation of the label – an example of this is the demethenylation of methylenedioxycinnamic acid during experiments with *P. hexandrum* (Jackson and Dewick, 1984a).

One unusual feature of the work (Fujimoto and Higuchi, 1977) with the yellow poplar was the synthesis under the experimental conditions of (−)-syringaresinol, although the enantiomer is the normal product. There was some confusion here as the furofuran structures are wrongly represented in the original paper as their (+)-forms. It was suggested that the use of a large excess of the sinapyl alcohol precursor led to its dimerisation by available peroxidase to racemic material, as in lignin biosynthesis, followed by selective glucosidation of the (+)-form under enzymic control to leave a lignan residue with a small negative rotation, $[\alpha]_D$ ca. −4.0°C. The use of circular dichroism in difficult assignments of this kind is much to be desired.

In another study (Stockigt and Klischies, 1977) it was shown that ferulic acid derivatives including coniferin (**7.83**, Scheme 7.16) doubly labelled as shown, or else with $^{14}C-$ at the terminal carbon-9, were incorporated intact into arctiin and phillyrin of *Forsythia suspensa* var. *fortunei.* The precursor was administered in aqueous solution to cut shoots over 20 h. In these experiments glucoferulic acid was incorporated more efficiently than coniferin but again there was evidence of linking to lignin synthesis: of the activity administered 53% was lost in this way. It is also known that *Forsythia* sp. include enzymes capable of reducing ferulic acid to coniferyl alcohol and this is considered to be the oxidation level at which dimerisation takes place to give seco-isolariciresinol (Scheme 7.16). The failure to metabolise 3,4-dimethoxycinnamic acid was further evidence of the nature of this coupling.

The development of lignans of different classes from such an initial dimer is indicated by their common co-occurrence in many evolved plants. In *Forsythia* sp. (Nikaido *et al.*, 1981) one can interrelate arctiin (cf. **2.045**) and matairesinol (**2.062**) with those shown below (Scheme 7.16). In *Olea europa* other furofurans are associated with the furan olivil (**2.129**) and the tetralin cyclo-olivil (**2.318**). The action of dehydrogenase and lactonisation can account for the butyrolactones and a mechanism based on quinone methides is probable for furans (see pp. 279, 282) and similarly for furofurans: since a precursor such as **7.84** may be derived from secoisolariciresinol. An interesting example of diverse co-occurring lignans is found in *Cleistanthus patulus* (Sastry *et al.*, 1987) where tawanin C (**7.85**; R, R = $-CH_2$-, X = H) has been isolated in association with its oxygenated diphyllin analogue (**7.85**; R = Me, X = 3-*O*-methylxylosyl). In this plant

Scheme 7.16 Incorporation of ferulic derivatives into *Forsythia*

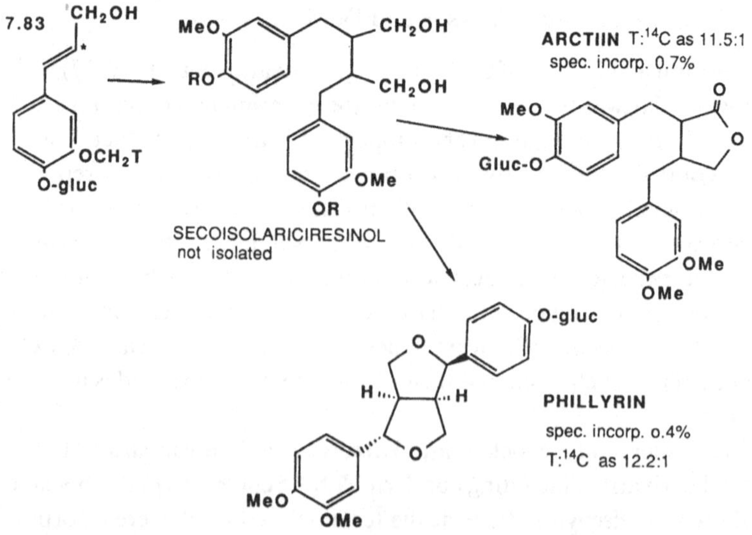

7.83 CH_2OH OCH_2T O-gluc

MeO RO CH_2OH CH_2OH OMe OR

SECOISOLARICIRESINOL
not isolated

ARCTIIN T:^{14}C as 11.5:1
spec. incorp. 0.7%

MeO Gluc-O O O OMe OMe

O-gluc O H H O MeO OMe

PHILLYRIN
spec. incorp. 0.4%
T:^{14}C as 12.2:1

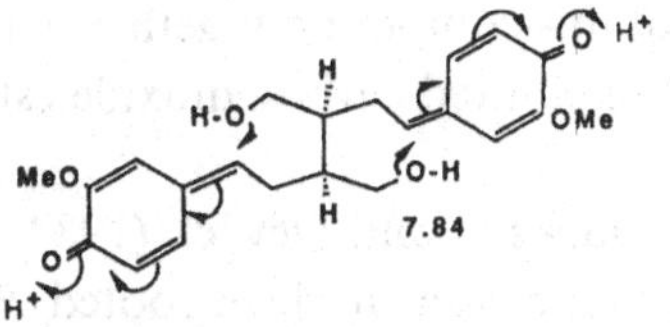

(+)-sesamin (**2.223**) was also found together with its hydroxylation product paulownin (**2.261**).

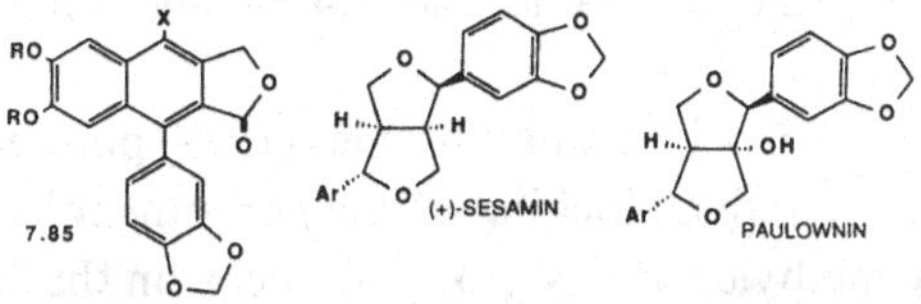

It has been reported (Rahman *et al.*, 1986) that in callus and suspension cultures of *Forsythia* sp. the normal distribution pattern, which includes dibenzylbutyrolactones and furofurans, is not followed since only lignans of the latter class were found. *O*-methylation may also be inhibited in specific instances.

(b) *The tetrahydronaphthalene group*

These compounds are the only lignans to have been studied in detail. It was shown (Ayres *et al.*, 1981) that efficient incorporation was possible by wick feeding of precursors in duplicate to mature *Podophyllum emodi* var. *hexandrum* plants. Initially it was found that acetate (0.042%) and tyrosine (0.023%) were not significantly incorporated but that *p*-coumaric acid (0.57%) and phenylalanine (1.17%) were. The formation of podophyllotoxin from two molecules of the latter was shown by oxidative degradation (Scheme 7.17). The yield of the oxidation product (**7.86**) was

Scheme 7.17 Location of label from DL-[β-^{14}C]-phenylalanine feeding

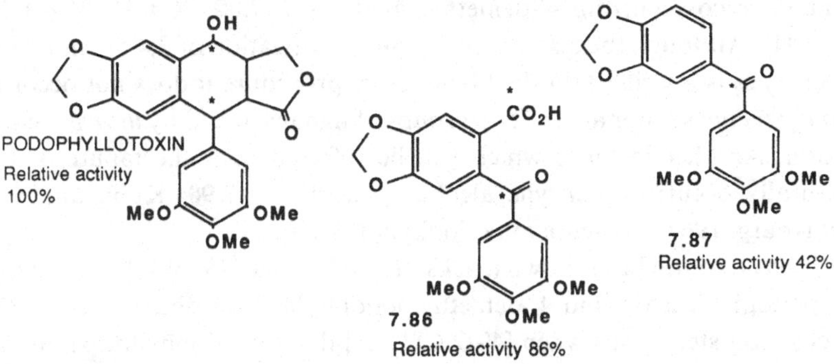

PODOPHYLLOTOXIN
Relative activity 100%

7.86
Relative activity 86%

7.87
Relative activity 42%

low (23%) but the relative activity of this compound together with that of the derived ketone (**7.87**, 73%) and of emitted carbon dioxide established the labelling pattern.

The early study was extended by Jackson and Dewick (1984a, *et seq.*) who fed precursors in solution by immersion of clean-rooted *P. hexandrum* plants, followed by a 7-day growing on period. They showed that cinnamic, ferulic and 3,4-methylenedioxy cinnamic acids were satisfactorily incorporated at levels of 0.17, 0.05 and 1.34% respectively, but that the more highly oxygenated sinapic and 3,4,5-trimethoxycinnamic acids were not.

These results like those of Stockigt and Klischies (1977) pointed to late-stage oxygenation and to effective blocking of the *para*-phenolic position by methyl but not by the methylenedioxy group. A check on the fate of the methylenedioxy ether group was made (Jackson and Dewick, 1984a) using [*O*-$^{14}CH_2$-*O*]-3,4-methylenedioxycinnamic acid. As a result the label was shown to be equally distributed which excludes cleavage of the cyclic ether and incorporation *solely* via labelled ferulic acid. In addition 70% of the label was found in methoxyl methyl groups consistent with incorporation through a demethylenylated precursor, caffeic acid, with ether labelling via *S*-adenosylmethionine. It was further shown that feeding of [3-*O*-$^{14}CH_3$]ferulic acid gave podophyllotoxin in which the extent of labelling in the methylenedioxy ether was almost the same as that in the methyl groups. It is therefore highly probable that the skeleton derives from coupling of two identical ferulyl units and almost certainly excludes a mechanism where an unsymmetrical lactone skeleton (e.g. **7.88**) is first formed.

Many good and philosophically satisfying analogies have been shown to exist between mechanisms established *in vitro* and those paths followed through plant enzymes. However, despite the value of this hypothetical approach reservations must of course be made. An excellent example of this is provided by the podophyllotoxins (**7.89**; X = H, Y = OH, R = Me) and their co-occurring 4′-demethyl analogues (**7.89**; X = H, Y = OH, R = H). Although the acid-catalysed 4-demethylation of 3,4,5-trimethoxybenzenes is a well established laboratory procedure it does not occur in *Podophyllum* sp. nor are the two groups of lignans linked by *in vivo* methylation. Another instance which will be referred to is the failure of the naturally occurring benzylic alcohol podorhizol (**7.90**; Kuhn and von Wartburg, 1967) to cyclise to podophyllotoxin *in vivo*.

It has recently been shown (Jackson and Dewick, 1984b) that the paths to podophyllotoxin and 4′-demethylpodophyllotoxin diverge before the cyclisation step. Thus when [4′-*O*-$CH_2{}^3H$]-desoxypodophyllotoxin is fed to *P. hexandrum* plants it undergoes benzylic hydroxylation at the 7-position to podophyllotoxin, which is converted reversibly into podophyllotoxone but no *in vivo* synthesis of 4′-demethyl compounds takes place. In a series of complementary experiments it was also shown that feeding of [8-^{3}H]-4′-demethyldeoxypodophyllotoxin led to synthesis *in vivo* of labelled 4′-demethylpodophyllotoxin, but with no incorporation into co-occurring podophyllotoxin. The relationships between these lignans are summarised in Scheme 7.18. This pathway is in keeping with the relative natural occurrence of these compounds, because the first in the series is widely occurring (Cole and Weidhopf, 1978); podophyllotoxin is more restricted and podophyllotoxone is found only in *Podophyllum* sp. It also follows that the prior placing of deoxypodophyllotoxin excludes the early insertion of the C7-OH group by a Michael-type hydration of an intermediate quinone methide (cf. **7.84**).

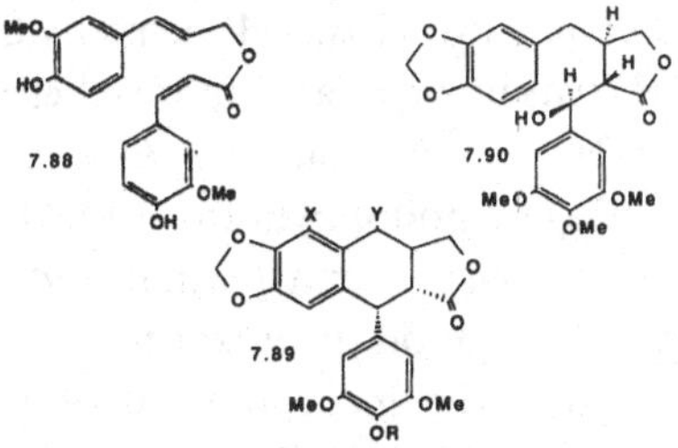

The proposed mechanism for *in vivo* hydroxylation was also supported by studies (Kamil and Dewick, 1986a) on the peltatins (**7.89**; X = OH, Y = H, R = H or Me), which are found only in limited amounts in *P. hexandrum* but which are major metabolites of *P. peltatum* (Jackson and Dewick, 1984c). The annual turnover of the peltatins in the latter plant is subject to seasonal changes (Kamil, 1986a); other examples of this yield variation are the dibenzylbutyrolactones and arylnaphthalenes of *Polygala chinensis* (Ghosal and Banerjee, 1979) and in dibenzocyclo-octadienes of *Schizandra chinensis* (Nakajima *et al.*, 1983). These variations were also referred to with regard to the isolation of lignans (p. 144) and in the present context they show that full analysis is necessary before a feeding programme is undertaken.

Scheme 7.18. Relationships between lignans of *P. hexandrum*

4′-DEMETHYLDEOXYPODOPHYLLOTOXIN → 4′-DEMETHYLPODOPHYLLOTOXIN

DEOXYPODOPHYLLOTOXIN → PODOPHYLLOTOXIN - - → PODOPHYLLOTOXONE

Late-season feeding of *P. peltatum* showed that 4′-demethyldeoxypodophyllotoxin (**7.89**; X, Y, R = H) was incorporated more efficiently into α-peltatin (**7.89**; X = OH, Y, R = H; 0.035%) than into β-peltatin (7.89; X = OH, Y = H, R = Me; 0.0029%) or podophyllotoxin (**7.89**; X = H, Y = OH, R = Me; 0.0076%). It was also found (Kamil and Dewick, 1986a) that deoxypodophyllotoxin was a good precursor for podophyllotoxin (0.33% incorporation) and that the incorporation of β-peltatin (0.023%) exceeded that of α-peltatin (0.0059%). In the low-yielding *P. hexandrum*, β-peltatin (4 mg) had twice the level of specific activity found in podophyllotoxin (62 mg) when the plant was fed with [4′-*O*-methyl-^{3}H]-deoxypodophyllotoxin. These results clearly point to a late-stage route to peltatins from 7-deoxy-precursors.

The widespread co-occurrence of aryltetrahydronaphthalene lignans with the dibenzylbutyrolactones (Cole and Weidhopf, 1978) indicates that they are related biogenetically. Screening of *P. hexandrum* and *P. peltatum* (Kamil and Dewick, 1986b) confirmed reports of podorhizol (**7.90**) and its glucoside in these plants and also showed that anhydropodorhizol (**7.91**) was present (0.0004%). Despite their structural relationship neither compound (**7.90, 7.91**) gives rise to podophyllotoxin by a Friedel–Craft cyclisation *in vivo* because on feeding them to *P. hexandrum* no label was taken up. However podorhizol and anhydropodorhizol are linked structurally to yatein (**7.92**) and although this butyrolactone is *not detectable in the plant* it was incorporated into podophyllotoxin at levels (0.11 and 0.19%) comparable to those found (0.53 and 0.32%) when [4′-*O*-methyl-^{3}H]-deoxypodophyllotoxin was used as a control. This makes it probable that yatein is the immediate precursor of deoxypodophyllotoxin and that it is cyclised via a quinone methide (Scheme 7.19) by a mechanism similar to that

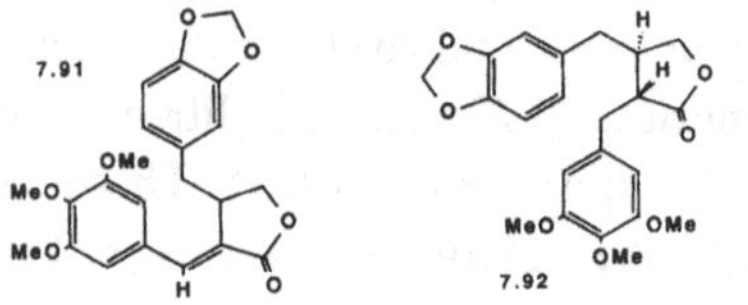

7.91 7.92

Scheme 7.19 Mechanism for the formation of podophyllotoxins from yatein

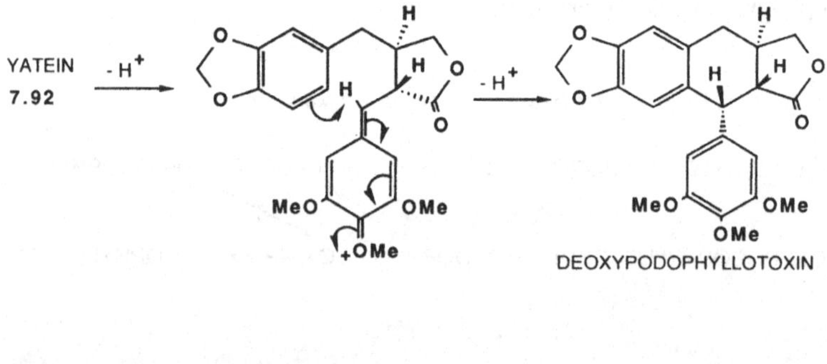

YATEIN 7.92 → DEOXYPODOPHYLLOTOXIN

Scheme 7.20 Biosynthetic pathway to podophyllum lignans

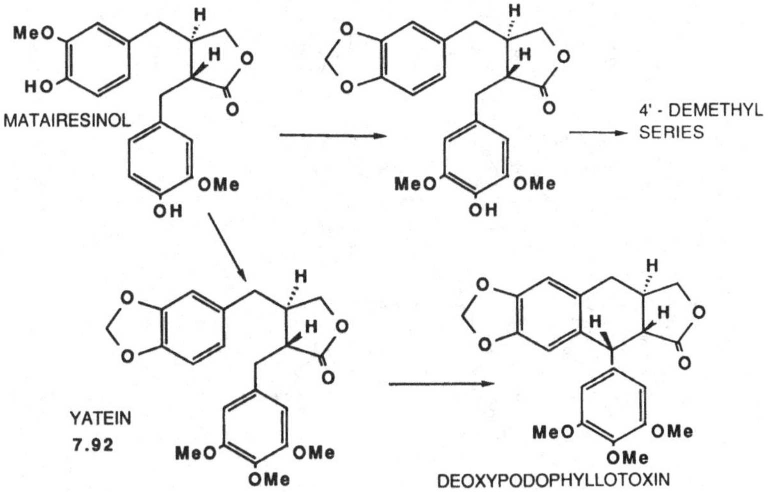

demonstrated in an earlier biomimetic synthesis (Pelter *et al.*, 1985). It is salutary to note that yatein, a critical component of the path, had a pool concentration too small to detect; its precursor remains to be identified but the widely distributed butyrolactone matairesinol is a possibility (Scheme 7.20).

(c) *Apolignans and arylnaphthalenes*

Apolignans have long been known as the products of acid-catalysed elimination of the C7-OH group in picropodophyllins (Hartwell and Schrecker, 1958). The first isolation of β-apoplicatitoxin (**7.93**) from a natural source, *Thuja plicata* Donn., strengthens their postulated link between the arylnaphthalene lignans and the tetrahydronaphthalene group. Thus plicatin (**7.94**; X = H) and plicatinaphthol (**7.94**; X = OH) are also found (McDonald and Barton, 1973) in the heartwood of this cedar tree.

The base-catalysed *in vitro* aromatisation of α- and β-apo-picropodophyllin has been demonstrated (Ayres and Mundy, 1968) and it was suggested (Swan *et al.*, 1969) that this conversion takes place *in vivo* through the action of oxidase enzymes. Further support for this view was provided by the subsequent synthesis (Momose and Kanai, 1978) of the naphthalene (**7.96**; X = H) and the naphthol (**7.96**; X = OH) from the model apolactone (**7.95**) by base-catalysed oxygenation: lignan naphthols and naphthalenes of this type (**7.96**) commonly co-occur naturally.

It is also of note that butyrolactone lignans were found (MacLean and Murakami, 1966; MacLean and MacDonald, 1967) in *Thuja plicata.* The close structural ties between these thujaplicatins (**7.97**) and yatein (**7.92**) reinforces the view that a common butyrolactone cyclisation mechanism operates in all naphthalenic lignans. A possible route to the dihydroxy compound (**7.98**) would be through the epoxidation of a dehydrothujaplicatin, whilst a similar process if followed on a cyclised β-apo-precursor would give rise to plicatic acid (**7.99**). This acid was the subject of an early detailed X-ray analysis (Gardner *et al.*, 1966).

Another highly hydroxylated tetralin lignan is (+)-africanal (**7.100**) which was isolated (Viviers *et al.*, 1979) together with (−)-olivil (**2.129**) and (+)-cycloolivil (**2.318**) from *Olea africana.* A possible pathway to this ketal is the oxidation of the primary alcohol group in cycloolivil to the aldehyde.

(d) *Biogenesis of ether groups*

This work (Kamil and Dewick, 1986b) on the source and sequence of alkoxylation in podophyllotoxin and 4′-demethylpodophyllotoxin provides experimental backing for the role of matairesinol and yatein in a branching pathway to these tetralins.

The origin of the alkyl ether groups in lignans must depend on the methionine path but confirmation was nevertheless obtained by root-feeding [*S*-methyl-^{14}C]-methionine. The incorporation of label (Table 7.1) clearly proves that biosynthesis of phenolic methyl/methylene ethers in lignans is included in this general path. Moreover by further analysis of

Table 7.1. *Labelling of podophyllotoxin (PT) derivatives obtained by feeding [S-methyl-^{14}C]-methionine to* P. hexandrum *(Specific activities dpm/mM × 10^{-5})*

	From PT	From 4′-Demethyl-PT
Podophyllotoxin (PT)	5.15 (Entry 1)	—
4′-Demethyl-(PT)	—	2.15 (Entry 2)
3′,4′-Didemethyl-PT	2.32 (Entry 3A)	1.32 (Entry 3B)
4,5-Demethylene-PT	3.78 (Entry 4)	

these results Kamil and Dewick were able to show that the mechanism of alkoxylation in podophyllotoxin differed from that in 4′-demethylpodophyllotoxin. This followed from the reasonable assumption that coupling of two ferulic residues took place initially and also that no label was included in skeletal carbon atoms.

From the figures given in Table 7.1 (Table 3 of Kamil and Dewick, 1986b) one may calculate the specific activity and relative specific activity of the various ether groups and these are shown in Table 7.2. It is important to note that, although 3′- and 5′-methyl groups are *chemically* equivalent and equally liable to cleavage in the formation of 3′,4′-didemethylpodophyllotoxin, they are not *biochemically* equivalent. In evaluating relative specific activity it is therefore assumed that one of the methyl ether groups has the same origin *in vivo* as the methylenedioxy ether group; hence the total activity of the 3′- and 5′-groups (0.36, column **A**) is allocated (column **B**) so as to equate one of them (3′ – say) with the methylenedioxy activity (0.27).

Examination of the levels of the relative specific activities shows that the patterns in the pendent rings of podophyllotoxin and 4′-demethylpodophyllotoxin are different; for example, one methyl group (C-5′) in the latter compound carries a greater proportion than any other. It is therefore unlikey that the two lignans are linked through 4′-methylation and hence the compound at the branch point will either be 4′-hydroxy-3′-methoxy- or 3′,4′,5′-trimethoxy-substituted in the potential pendent ring.

The assumption (Kamil and Dewick, 1986a, b) that no activity was located on the skeleton is supported by previous work (Jackson and Dewick, 1984a) which showed that loss of label from the methylenedioxy carbon led to incorporation into oxymethyl groups but not into the skeleton. It is therefore possible to elaborate the branching pathway with confidence, as in Scheme 7.20, which is based on matairesinol and was proposed by Kamil and Dewick (1986b).

Table 7.2. *Relative specific activities of alkyl groups*

	Podophyllotoxin		4′-DME-podophyllotoxin	
Group	A	B	A	B
OCH_2O	0.27	0.27	0.22	0.22
3′-OMe	0.18	0.27	0.39	0.22
4′-OMe	0.37	0.37	—	—
5′-OMe	0.18	0.09	0.39	0.56

Despite the demonstrated integrity of the methionine feeding to *P. hexandrum* (Kamil and Dewick, 1986b) it has been claimed (Canonica *et al.*, 1971) that the carboxyl carbon (C-1) of phenylalanine exchanges with methionine during the synthesis of eugenol by *Ocimum basilicum* L. This conclusion was based on a marked rise in the $^3H:^{14}C$ ratio in the eugenol produced when generally tritiated phenylalanine was specifically 1-^{14}C-labelled, but not when this label was at C-2 or C-3. In contrast it was shown (Klischies *et al.*, 1975) that glucoferulic acid and glucoconiferyl alcohol were metabolised by *O. basilicum* to give cinnamyl derivatives with the side chain *intact*. Later work (Canonica *et al.*, 1979) suggests that the apparent contradiction arose because *p*-coumarate and caffeate undergo the exchange with *O. basilicum* but that ferulic derivatives do not.

The oxidation level at which coupling occurs remains uncertain and difficulties are likely to be found when this is further investigated. Thus it has been observed that when a [1-^3H]-cinnamyl alcohol was fed to tissue cultures of *P. hexandrum* it was enriched in tritium. This surprising result may have come about through an enzymic hydride transfer at the aldehyde level and it is probable that difficulties of this kind result from interaction and exchange between the lignan pool and the lignin pool.

References

Abe, F., Yamauchi, T. and Wan, A.S.C. (1988). Lignans related to olivil from genus Cerbera. *Chem. Pharm. Bull.* **36**, 795–9.

Adlercreutz, H. (1984). Does fibre-rich food containing animal lignan precursors protect against both colon & breast cancer? An extension of the fibre hypothesis. *Gastroenterology* **86**, 761–2.

Ahmed, R., Lehrer, M. and Stevenson, R. (1973). Synthesis of thomasic acid. *Tetrahedron* **29**, 3753–9.

Ahmed, R., Schreiber, F.G., Stevenson, R., Williams, J.R. and Yeo, H.M. (1976). Oxidative coupling of bromo- & iodoferulic acid derivatives: synthesis of (±)-veraguensin. *Tetrahedron* **32**, 1339–44.

Ayres, D.C. and Mundy, J.W. (1968). The base-catalysed aromatisation of dihydronaphthalenes. *J. Chem. Soc. Chem. Commun.* 1134.

Ayres D.C. (1969). Incorporation of L-[U-^{14}C]-β-phenylalanine into the lignan podophyllotoxin. *Tetrahedron Lett.* 883–6.

Ayres, D.C., Farrow A. and Carpenter, B.G. (1981). Lignans & related phenols. Part 16. The biogenesis of podophyllotoxin. *J. Chem. Soc. Perkin Trans.* **1** 2134–6.

Bannwart, C., Adlercreutz, H., Fotsis, T., Wahala, K., Hase, T. and Brunow, G. (1984). Identification of *O*-desmethylangolensin, a metabolite of daidzein & of matairesinol one likely plant precursor of the animal lignan enterolactone in human urine. *Finn. Chem. Lett.* 120–4.

Begley, M.J., Davies, R.V., Henley-Smith, P. and Whiting, D.A. (1978). The constitution of (1*R*)-sequirin D from (*Sequoia sempervirens*), a biogenically novel norlignan, by direct X-ray analysis. *J. Chem. Soc. Perkin Trans.* **1**, 750–4.

Beracierta, A.P. and Whiting, D.A. (1978). Stereoselective total synthesis of norlignans of *Coniferae*. *J. Chem. Soc. Perkin Trans.* **1**, 1257–63.

Birch, A.J. and Liepa, A. (1978). Biosynthesis. In *Chemistry of Lignans*, ed. Rao, C.B.S. Andhra Univ., Waltair.

Bowden, B.F. *et al.* (1980). Constituents of *Eupomatia* species. Part 6. The structure of eupodienone-1,2, & 3 *Aust. J. Chem.* **33**, 1823–31.

Canonica, L., Manito, P., Monti, D. and Sanchez, A. (1971). Biosynthesis of allylphenols in *Ocymum basilicum* L. *J. Chem. Soc. Chem. Commun.* 1108–9.

Canonica, L., Gramatica, P., Manito, P. and Monti, D. (1979). Biosynthesis of caffeic acid in *Ocimum basilicum* L. *J. Chem. Soc. Chem. Commun.* 1073–5.

Cole, J.R. and Weidhopf, R.M. (1978). Distribution. In *Chemistry of Lignans*, ed. Rao, C.B.S. Andhra Univ., Waltair, pp. 39–64.

Davis, B.D. (1958). On the importance of being ionised. *Arch. Biochem. Biophys.* **78**, 497–509.

Erdtman, H. (1933). Dehydrogenation in the Coniferyl series. Part 2. Dehydro-diisoeugenol. *Annalen* **503**, 283–94.

Estabrook, G.F. (1978). Some concepts for the estimation of evolutionary relationships in systematic botany. *Syst. Botany* **3**, 146–58.

Fengel, D. and Wegener, G. (1984). *Wood: Chemistry, Ultrastructure, Reactions*, de Gyten, Berlin, p. 1.

Foo, L.Y. and Karchesy, J. (1989). Pseudotsuganol, a [5,5′]-biphenyl linked pinoresinol – dihydroquercetin from Douglas fir bark: isolation of the first true flavonolignan. *J. Chem. Soc. Chem. Commun.* 217–19.

Freudenberg, K. and Harkin, J.M. (1960). Model for the binding of lignin & carbohydrates, *Chem. Ber.* **93**, 2814–19.

Fujimoto, H. and Higuchi, T. (1977). Biosynthesis of liriodendrin by *Liriodendron tulipfera*. *Wood Research* **62**, 1–10.

Gardner, J.A.F., Swan, E.P., Sutherland, S.A. and MacLean, H. (1966). Polyoxyphenols of western red cedar (*Thuja plicata* Donn.). Part 3. Structure of plicatic acid. *Canad, J. Chem.* **44**, 51–8.

Ghosal, S. and Banerjee, S. (1979). Synthesis of retrochinensin; a new naturally occurring 4-aryl-2,3-naphthalide lignan. *J. Chem. Soc. Chem. Commun.* 165–6.

Gottlieb, O.R. (1978). Neolignans. In *Progr. in Chem. of Org. Nat. Prod.*, ed. Herz W. *et al.*, **35**, 1–72.

Gottlieb, O.R. and Kubitzi, K. (1981). Plant chemosystematics and phylogeny. Part 12. Chemosystematics of Aniba. *Biochem. Syst. Ecol.* **9**, 5–12.

Grisebach, H. (1981). In *Biochemistry of plants*', ed. Conn, E.E. Academic Press, New York, 7, p. 457.

Gross, G.G. (1979). Recent advances in the chemistry & biochemistry of lignin. *Rec. Adv. Phytochemistry* **12**, 177–220.

Gross, G.G. (1980). The biochemistry of lignification *Adv. Bot. Res.* **8**, 25–63.

Hartwell, J.L. and Schrecker, A.W. (1958). The chemistry of podophyllum. In *Progr. in the Chem. of Org. Nat. Prod.*, ed. Zechmeister, L. Springer, Vienna, **15**, 83–166.

Hatcher, P.G. (1987), Chemical structure studies of natural lignin by dipolar dephasing solid-state ^{13}C-NMR. *Org.Geochem.* **11**, 312–19.

Hawkes, G.E., Smith, C.Z., Utley, J.H.P. and Chum, H.L. (1986). Key structural features of acetone-soluble phenol-pulping lignins by ^{1}H & ^{13}C NMR spectroscopy. *Holzforschung* **40**, 115–23.

Hearon, W.M. and MacGregor, W.S. (1955). The naturally occurring lignans. *Chem. Rev.* **55**, 1031–2.

Higuchi, T. (1985). Biosynthesis & biodegradation of wood components, ed. Higuchi, T. Academic Press, Orlando, pp. 141–61; see also Gross, G.G., *ibid.* pp. 229–72.

Ichihara, A., Oda, K., Numata, Y. and Sakamura, S. (1976). Lappaol A & B, novel lignans from *Arctium lappa* L. *Tetrahedron Lett.* 3961–4.

Ichihara, A., Kanai, S., Nakamura, Y. and Sakamura, S. (1978). Structures of lappaol F & H, dilignans from *Arctium lappa* L. *Tetrahedron Lett.* 3035–8.

Jackson, D.E. and Dewick, P.M. (1984a), Biosynthesis of *Podophyllum* lignans. Part 1. Cinnamic acid precursors of podophyllotoxin in *Podophyllum hexandrum. Phytochemistry* **23**, 1029–35.

Jackson, D.E. and Dewick, P.M. (1984b). Biosynthesis of *Podophyllum* lignans. Part 2. Interconversions of aryltetralin lignans in *Podophyllum hexandrum. Phytochemistry* **23**, 1037–42.

Jackson, D.E. and Dewick, P.M. (1984c). Aryltetralin lignans from *Podophyllum hexandrum* & *Podophyllum peltatum. Phytochemistry* **23**, 1147–52.

Kamil, W.M. and Dewick, P.M. (1986a). Biosynthesis of the lignans α- & β-peltatin. *Phytochemistry* **25**, 2089–92.

Kamil, W.M. and Dewick, P.M. (1986b). Biosynthetic relationship of aryltetralin lactone lignans to dibenzylbutyrolactone lignans. *Phytochemistry* **25**, 2093–102.

Kikuchi, T., Kadota, S., Yanada, K., Tanaka, K., Watanabe, K., Yoshizaki, M., Yoko, T. and Shingu, T. (1983). Isolation & structure of magnosalin & magnoshinin, new neolignans from *Magnolia salicifolia* Maxim. *Chem. Pharm. Bull.* **31**, 1112–14.

Kikuchi, T., Kadota, S. and Tai, T. (1985). Studies on the constituents of medicinal plants in Sri Lanka. Part 3. Novel sesquilignans from *Hedyotis lawsonia. Chem. Pharm. Bull.* **33**, 1444–51.

Klischies, M., Stockigt, J. and Zenk, M.H. (1975). Biosynthesis of the allylphenols eugenol and methyleugenol in *Ocimum basilicum* L. *J. Chem. Soc. Chem. Commun.* 879–80.

Kuhn, M. and von Wartburg, A. (1967). *Podophyllum* lignans. Structure & absolute configuration of podorhizol β-D-glucoside. *Helv. Chim. Acta* **50**, 1546–65.

Lewis, N.G., Dubelsten, P., Eberhardt, T.L., Yamamoto, E. and Towers, G.H.N (1987). The *E*/*Z* isomerisation step in the biosynthesis of *Z*-coniferyl alcohol in *Fagus grandiflora. Phytochemistry* **26**, 2729–34.

Lin, L-J. and Cordell, G.A. (1984). Synthesis of coumarinolignans through chemical & enzymic oxidation. *J. Chem. Soc. Chem. Commun.* 160–1.

McCalla, D.R. and Neish, A.C. (1959). Metabolism of phenylpropanoid compounds in *Salvia*. Part 2. Biosynthesis of phenolic cinnamic acids. *Can. J. Biochem. Physiol.* **37**, 537–47.

MacLean, H. and Murakami, K. (1966). Lignans of western red cedar. Part 4. Thujaplicatin & thujaplicatin methyl ether. *Canad. J. Chem.* **44**, 1541–5.

MacLean, H. and MacDonald, B.F. (1967). Lignans of western red cedar (*Thuja plicata* Donn.). Part 7. Dihydroxythujaplicatin, *Canad. J. Chem.* **45**, 739–40.

MacDonald, B.F. and Barton, G.M. (1973). Lignans of western red cedar. Part 11. β-apoplicatitoxin. *Canad. J. Chem.* **51**, 482–5.

Matsumoto, M. and Kuroda, K. (1981). Transition metals. Part 2. Schiffs base complex catalysed oxidation of *trans*-2-(1-propenyl)-4,5-methylenedioxy-phenol to carpanone by molecular oxygen. *Tetrahedron Lett.* **22**, 4437–40.

Momose, T. and Kanai, K-I. (1978). A base-catalysed oxygenation of the β-apo-lignans: a biogenetic model. *Heterocycles* **9**, 207–10.

Nakajima, K., Taguchi, T., Ikeya, Y., Endo, T. and Yosioka, I. (1983). The constituents of *Schizandra chinensis* Baill. Part 13. Quantitative analysis of lignans in the fruits of *S. chinensis* by HPLC. *Yakugaku Zasshi* **103**, 743–9; *Chem. Abstr.*, 1983, **99**, 155220.

Neish, A.C. (1964). Major pathways of biosynthesis of phenols. In *Biochemistry of phenolic compounds*, ed. Harborne, J.B., pp. 340–4. Academic Press, London.

Nikaido, T., Ohmoto, T., Kinoshita, T., Sankawa, U., Nishibe, S. and Hisada, S. (1981). Inhibition of cAMP phosphodiesterase by lignans. *Chem Pharm. Bull.* **29**, 3586–92.

Nimz, H., Tshirner, U., Stahle, M., Lehmann, R. and Schlosser, N. (1984). Carbon-13 NMR spectra of lignins. Part 10. Comparison of structural units in spruce & beech lignin. *J. Wood Chem. Technol.* **4**, 265–84.

Nishiyama, A., Eto, H., Terada, Y., Iguchi, M. and Yamamura, S. (1983a). Anodic oxidation of some propenylphenols. Synthesis of physiologically active neolignans. *Chem. Pharm. Bull.* **31**, 2834–44.

Nishiyama, A., Eto, H., Terada, Y., Iguchi, M. and Yamamura, S. (1983b). Anodic oxidation of 4-allyl-2,6-dimethoxyphenol & related compounds. Synthesis of asatone & related neolignans. *Chem. Pharm. Bull.* **31**, 2820–3.

Nishiyama, A., Eto, H., Terada, Y., Iguchi, M. and Yamamura, S. (1983c). Anodic oxidation of 4-hydroxycinnamic acids & related phenols. *Chem. Pharm. Bull.* **31**, 2845–52.

Omori, S. and Sakakibara, A. (1979). *Mokuzai Gakkaishi* **25**, 145–8.

Ozawa, S. and Sasaya, T. (1987). A new cyclolignan containing a lactol ring from *Abies sachaliensis* Masters. *Mokuzai Gakkaishi* **33**, 747–8.

Pelter, A. and Hansel, R. (1968). The structure of silybin (*Silybum* substance E6) the first flavonolignan. *Tetrahedron Lett.* 3961–4.

Pelter, A., Ward, R. S. and Rao, R. R. (1985). An approach to the biomimetic synthesis of aryltetralin lignans. *Tetrahedron* **41**, 2933–8.

Pelter, A. (1986). Lignans: some properties & synthesis. *Recent Adv. Phytochemistry* **20**, 201–41.

Pettersen, R. C. (1984). In *The chemistry of solid wood*, ed. Rowell, R.M. *Amer Chem. Soc. Adv. in Chem.* **207**, 57–126.

Rahman, M., Dewick, P.M., Jackson, D.E. and Lucas, J.A. (1986). Lignans in *Forsythia* leaves and cell cultures. *J. Pharm. & Pharmacol.* **38**, 15P.

Rao, K.V. and Alvarez, F.M. (1983). Manassantins A/B and saucerneol: novel biologically active lignoids from *Saururus cernuus*. *Tetrahedron Lett.* **24**, 4947–50.

Sakakibara, A., Sasaya, T., Miki, K. and Takahashi, H. (1987). Lignans & Brauns' lignins from softwoods. *Holzforschung* **41**, 1–11.
Sastry, K.V., Rao, E.V., Buchanan, J.G. and Sturgeon, R.J. (1987). Cleistanthoside B, a diphyllin glycoside from *Cleistanthus patulus* heartwood. *Phytochemistry* **26**, 1153–4.
Stockigt, J. and Klischies, M. (1977). Biosynthesis of lignans. Part 1. Biosynthesis of arctiin and phillyrin. *Holzforschung* **31**, 41–4.
Swan, E.P., Jiang, K.S. and Gardner, J.A.F. (1969). The lignans of *Thuja plicata* and the sapwood–heartwood transformation. *Phytochemistry* **8**, 345–51.
Taafrout, M., Rouessac, F. and Robin, J-P. (1984). Isolation of prestegane B from *Steganotaenia araliaceae* Hochst. *Tetrahedron Lett.* **25**, 4127–8.
Takeya, T., Kotani, E. and Tobinaga, S. (1983a), New reagent systems containing chromium trioxide and synthesis of lignans. *J. Chem. Soc. Chem. Commun.* 98–9.
Takeya, T., Matsumoto, H., Kotani, E. and Tobinaga, S. (1983b). New reagent systems containing chromium trioxide provide precursors for synthesis of neolignans. *Chem. Pharm. Bull.* **31**, 4364–7.
Takeya, T., Akanabe, Y., Kotani, E. and Tobinaga, S. (1984). Biomimetic synthesis of podophyllum lignans. *Chem. Pharm. Bull.* **32**, 31–7.
Tsuruga, T. Ebizuka, Y., Nakajima, J., Chun, Y-T., Noguchi, H., Iltaka, Y. and Sankawa, U. (1984). Isolation of a new lignan, magnosalicin, from *Magnolia salicifolia. Tetrahedron Lett.* **25**, 4129–32.
Viviers, P.M., Ferreira, D. and Roux, D.G. (1979). (+)-Africanal, a new lignan of the aryltetrahydronaphthalene class. *Tetrahedron Lett.* (39), 3773–6.
Whiting, P. and Goring, D.A.I. (1982). Chemical characterisation of tissue fractions from the middle lamella & secondary wall of black spruce tracheids. *Wood Sci. Technol.* **16**, 261–7.
Zanarotti, A. (1982). Preparation & reactivity of a vinyl quinone methide. A novel synthesis of sinapyl alcohol. *Tetrahedron Lett.* **23**, 3815–18.
Zanarotti, A. (1983), Synthesis & reactivity of lignin model quinone methides. Biomimetic synthesis of 8,*O*,4′-neolignans. *J. Chem. Res. Synop.* 306–7.

8

Synthesis

Useful information is contained in a collection of general articles (Rao *et al.*, 1978) and the synthesis of lignans and neolignans has been reviewed in detail (Ward, 1982). In this review some attention has been paid to old-established methods but the writer has concentrated on developments post 1980.

Owing to the existence of large groups of reference compounds and to the power of modern spectroscopic methods, synthetic techniques are no longer essential to the proof of structure. However, the synthetic approach is vital for the preparation of lignans with useful physiological properties whenever the yields from natural sources are inadequate.

Oxidative coupling

Numerous examples of biomimetic synthesis exist which depend upon this procedure. These were illustrated in the previous chapter (cf. Schemes 7.7, 7.8, 7.9, 7.11) and others which relate to the various classes of lignans are given here.

This method is of interest because of the analogy with biosynthesis. However, as we have seen in oxidative lignin synthesis (p. 272), a number of pathways are usually open and low yields of specific lignans are commonly obtained. An early review (Harkin, 1967) discusses the simulated synthesis of lignin by the oxidative coupling of *p*-hydroxycinnamyl alcohols. Another contribution (Weinges and Spanig, 1967) in the same source describes the limited examples then available of this route to lignans.

In fundamental studies (Freudenberg, 1959, 1965) of lignin synthesis the free radical (**8.1**; X = H) derived from eugenol and the coupled quinone methide (**8.2**; X = H) were identified as intermediates but their half-lives are limited, being only 45 seconds and 1 hour respectively in 1:1 dioxan:water at 20°C. The corresponding radical (**8.1**: X = OMe)

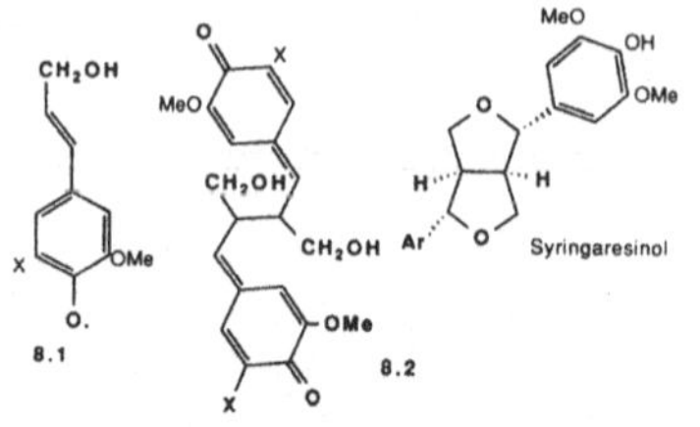

Scheme 8.1 Synthesis of thomasidioic acid

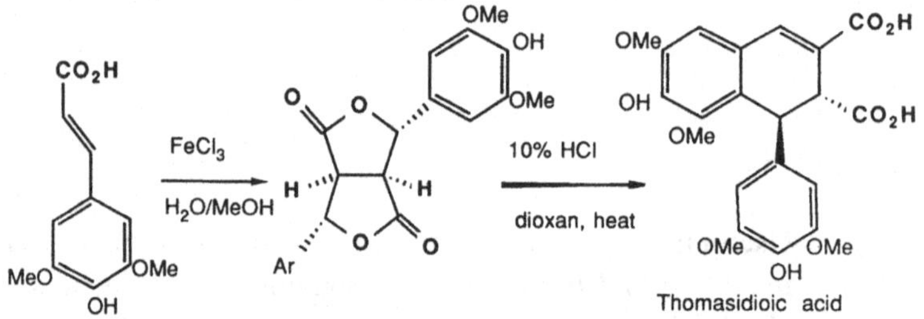

obtained (Freudenberg *et al.*, 1958) by the action of laccase or of copper(II) sulphate/oxygen on sinapyl alcohol is more hindered and therefore less reactive. It dimerises to give an isolable quinone methide (**8.2**; X = OMe, m.p. 255°C) which cyclises to yield syringaresinol (**2.230**; 80%, cf. **7.84**). The same approach was employed (Ahmed *et al.*, 1973a) in a two-step synthesis (Scheme 8.1) of thomasidioic acid (**2.379**) from sinapic acid (**8.3**). No yield was quoted for this first step, but in a parallel study (Wallis, 1973) of the coupling of the isomers of a propenylphenol the (*E*)-form (**8.5**, Scheme 8.2) afforded a 7:3 mixture (56% yield) of the *erythro*- (**8.6**) and *threo*- (**8.7**) ethers. Two minor products were the tetralin (**8.8**; 270 mg, 9%) and its α-apoderivative (**8.9**; 70 mg, 2%), which like the ethers are typical of the participation of an intermediate quinone methide (cf. Scheme 7.6).

It is also of interest that methods for *in vitro* coupling exist where a free phenolic hydroxyl group is not required. Examples of this are provided by the synthesis (Biftu *et al.*, 1978a, Scheme 8.3) of (±)-deoxyschizandrin (**8.11**; R = Me) from the *meso*-diarylbutane (**8.10**, R = Me) and also that of wuweizisu C (**8.11**) from (**8.10**, R, R = —CH_2—) by Schneiders (1980). Vanadium oxyfluoride has also been used (Landais and Robin, 1986) for a synthesis of neoisostegane (**2.466**) from the butyrolactone (**8.12**).

The coupling of phenolic ethers is also initiated (Taafrout *et al.*, 1986) by thallium *tris*(trifluoroacetate), itself obtained by solution of the oxide in trifluoroacetic acid (McKillop *et al.*, 1977). The mechanism of this reac-

Scheme 8.2 Coupling of propenyl phenols

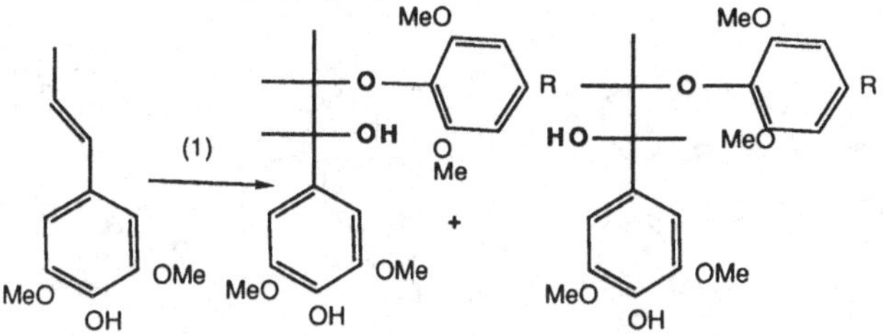

MeO MeO

R R

(1) OH HO MeO

O

Me

+

MeO OMe MeO OMe MeO OMe

OH OH OH

8.5 3.0g 8.6 805mg 8.7 420mg

(1) $FeCl_3$

Acetone/ water

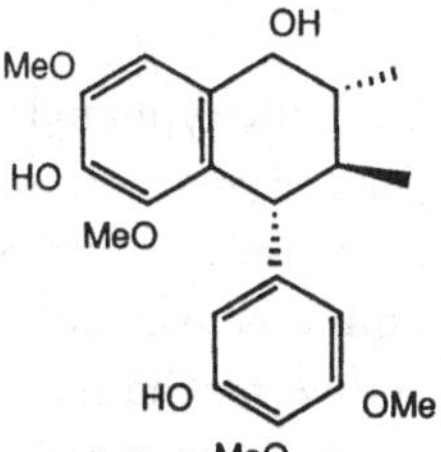

OH

MeO MeO

HO HO

MeO MeO

Ar

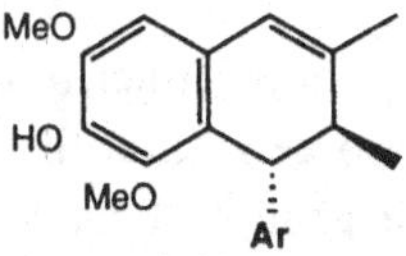

8.9

HO OMe

8.8 MeO

Scheme 8.3 Synthesis of dibenzocyclooctadiene lignans

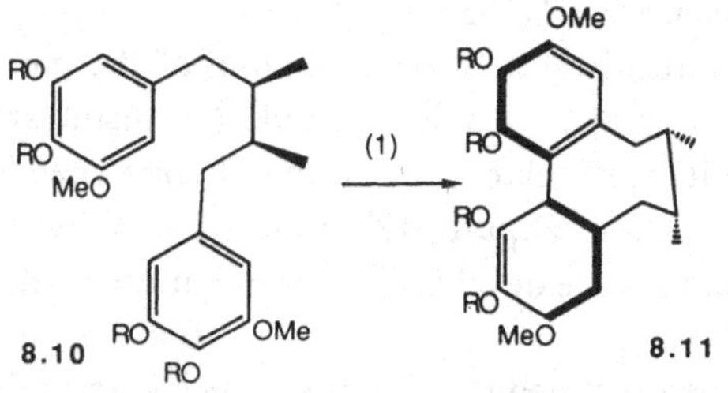

OMe

RO RO

RO (1) RO

MeO

RO

RO

8.10 RO OMe MeO 8.11

RO

(1) VOF_3 R = Me, deoxyschizandrin

CF_3CO_2H R = $-CH_2-$, wuweizisu C

MeO

MeO

O

MeO O

OMe VOF_3 O

O

MeO OMe MeO OMe

OMe MeO

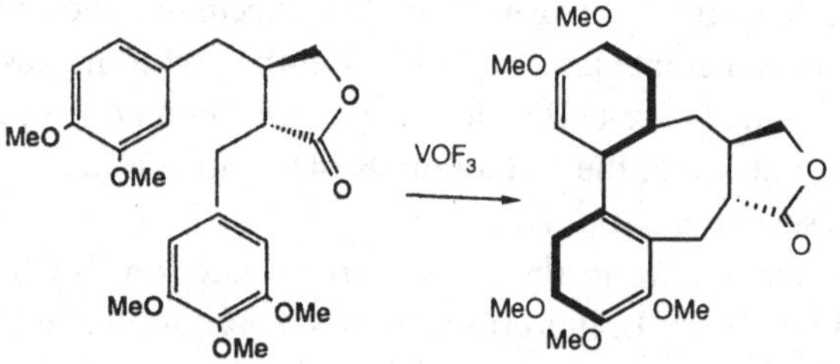

8.12 neoisostegane

Scheme 8.4 Mechanism of radical - cation induced cyclisation of a biaryl

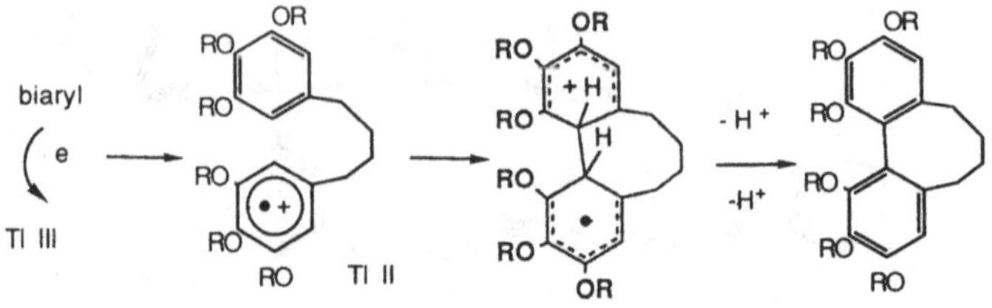

tion is believed (Elson and Kochi, 1973) to involve a single electron transfer to the metal cation followed by electrophilic substitution (Scheme 8.4).

In this context the strongly nucleophilic trimethoxyphenyl group is favoured and reaction proceeds smoothly despite the presence in (**8.12**) of other oxygenated donor groups. This lactone is closely related to yatein (**2.088**, p. 294) which is believed to cyclise via a quinone methide to deoxypodophyllotoxin (Scheme 7.19), but the alternative mechanism (Scheme 8.4) for dibenzocyclo-octadiene lignans has not been tested in plants.

Dibenzylbutanes and dibenzylbutyrolactones

(i) *The Stobbe condensation* (*1893*)

This reaction is ideally suited to assembling of the basic carbon skeleton of lignans. A general review is available (Johnson and Daub, 1951) and summaries of its particular application to lignan chemistry have been made (Bagavant and Ganeshpure, 1978) including those where the simpler open-chain structures are developed to obtain more complex compounds.

In many earlier papers it is stated that mixtures of (*E*)- and (*Z*)-isomers are obtained from the first stage of the Stobbe condensation. It has however been shown (Hart and Heller, 1972) that the preference is for formation of the (*E*)-configuration in both the one- and two-stage reactions. Thus in the condensation between dimethyl succinate and piperonal (Scheme 8.5) the intermediate lactone (**8.13**) is in the less compressed configuration which favours the cyclisation step. This geometry would also result from reprotonation of the carbanion (**8.14A**) which is in equilibrium with **8.14B** as the second stage develops.

The first stage terminates in the stable carboxylate ion (**8.15**) formed from **8.14A** by elimination. In this critical step the interaction between the methoxycarbonyl and aryl groups is minimised by their adopting a *trans-*

Scheme 8.5 The progress of the Stobbe condensation

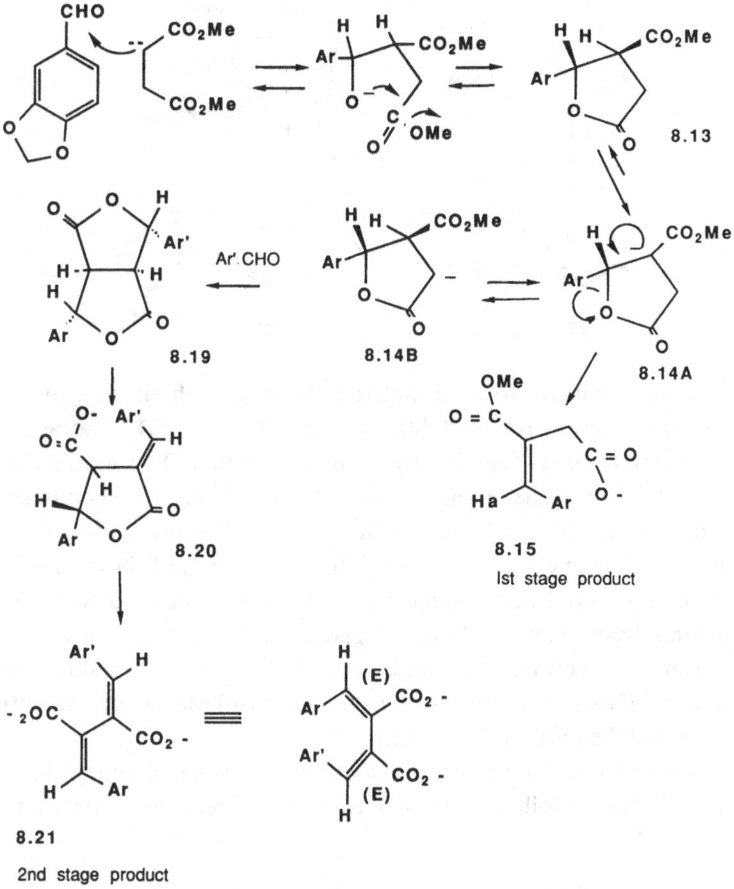

relationship (Hart and Heller, 1972; Anjaneyulu *et al.*, 1984). The (*E*)-configuration (**8.15**) is therefore formed as is shown by the NMR spectrum (p. 31) where the olefinic proton (H_a) is deshielded by the ester carbonyl group, whereas in the (*Z*)-alkene this proton lies at higher field (Banerji *et al.*, 1984).

With the configuration of the product (**8.15**) established, a valuable procedure for stereoselective hydrogenation was developed (Achiwa, 1979). This depends on direction of hydrogen to the least hindered side when the alkene interacts with a chiral pyrrolidinephosphine-rhodium chelating agent (**8.16**, Scheme 8.6). The ester function in the chiral product (**8.17**) is reduced with lithium aluminium hydride to give the lactone (**8.18**). A typical second-stage condensation will now afford optically active butyrolactones.

Scheme 8.6 Enantioselective hydrogenation of itaconates

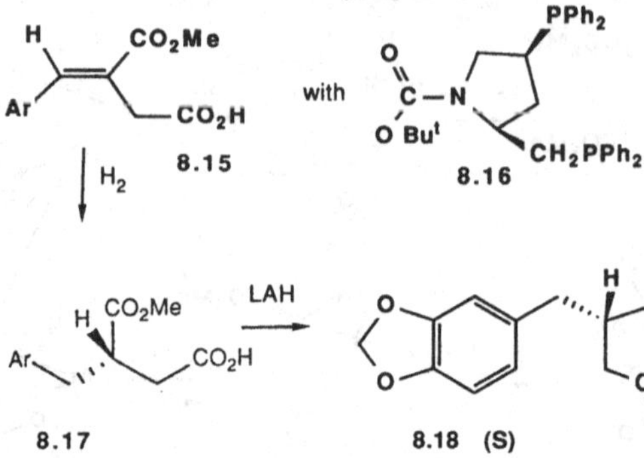

It is important to note that the absolute configuration of the lactone (**8.18**) will not be retained if the carboxylate ion of **8.17** is attacked in preference to the ester function by the hydride donor. This would give the *ent*-form of the lactone (**8.18**) and although this is an unlikely outcome a controlled reduction procedure exists and it is illustrated for a racemic analogue in Scheme 8.7. In this procedure protection of the carboxyl group is ensured by prior and complete conversion into the carboxylate ion with a strong base. The synthesis of gadain (**2.098**), a conjugated lignan of accepted structure, is possible only if the reduction follows the course shown. Other instances of the directed synthesis of unsymmetrical lactones will be referred to (p. 312).

For practical purposes, even for symmetrical dibenzyl lignans, it is usually best to follow a two-step procedure after fully esterifying the inter-

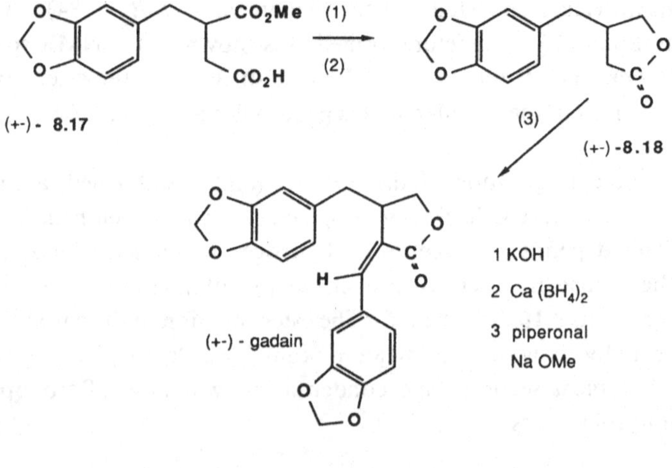

Scheme 8.7 Directed reduction of half-esters from the Stobbe condensation

mediate half-ester (**8.15**). This is because the carbanion (**8.14A**) will be stopped as the precipitated carboxylate salt (of **8.15**) and this will compete with and limit the second-stage reaction of **8.14B** with a second equivalent of aldehyde. The latter probably proceeds through a bicyclic lactone (**8.19**) which will be subject to similar steric constraints as the furofuran lignans (pp. 215–16). Elimination now leads to lactonic acids (**8.20**) and bis(arylmethylene) succinic acids (**8.21**) with the *E*-configuration about both double bonds (Hart and Heller, 1972). This geometry may not be retained in reactions of highly hindered aromatic aldehydes.

It is of interest that the end product of the Stobbe reaction with 3,4,5-trimethoxybenzaldehyde (two mol.) on treatment with methanolic hydrogen chloride gave the dimethyl ester of thomasidioic acid (**2.379**; Bagavant and Ganeshpure, 1978), the same product as that obtained (Ahmed *et al.*, 1973a) from the oxidative coupling of sinapic acid (p. 304).

Precursors for lignan synthesis were obtained (Ayres *et al.*, 1965a) by the two-stage process where piperonal and diethyl succinate afforded the half ester (**8.22**, 76% yield; Scheme 8.8), which was then fully esterified (85%) and combined with 3,4,5-trimethoxybenzaldehyde to give the diarylidene half ester (**8.23**, 81%).

Full experimental details are given in another early paper (Batterbee *et al.*, 1969) which describes the high yield hydrogenation of alkenoic acids of type **8.22** and a procedure for their resolution as quinine salts. These procedures can be followed for the synthesis of lignans with free phenolic hydroxyl groups provided that they are protected by benzylation. This avoids difficulties arising from the precipitation and low reactivity of the enolate salts and the technique is illustrated (Matsuura and Iunama, 1985) by the synthesis of the furan (**2.112**) from *Diospyros kaki* calyces (Scheme 8.9).

Scheme 8.8 Typical two - stage condensation

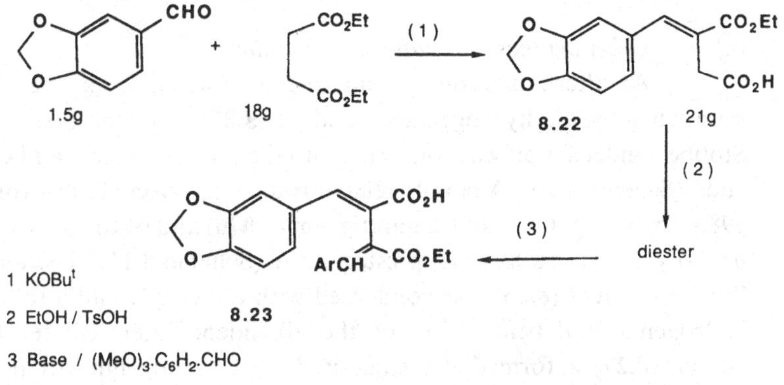

CHO + CO_2Et / CO_2Et —(1)→ **8.22** (CO_2Et, CO_2H)

1.5g 18g 21g

(2) diester —(3)→ **8.23** (CO_2H, ArCH, CO_2Et)

1 $KOBu^t$

2 EtOH / TsOH

3 Base / $(MeO)_3.C_6H_2.CHO$

Scheme 8.9 The use of protecting groups in the Stobbe reaction

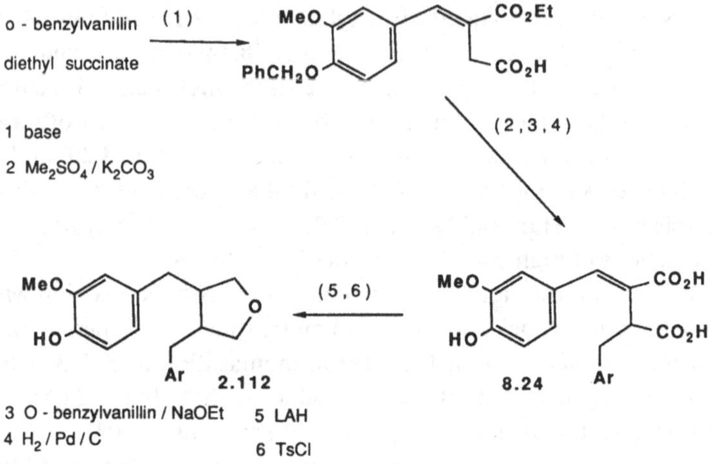

o - benzylvanillin (1)
diethyl succinate

MeO CO_2Et
$PhCH_2O$ CO_2H

1 base
2 Me_2SO_4 / K_2CO_3

(2,3,4)

MeO CO_2H
HO CO_2H
8.24 Ar

(5,6)

MeO
HO
O
Ar 2.112

3 O - benzylvanillin / NaOEt
4 H_2 / Pd / C
5 LAH
6 TsCl

Difficulties often arise in these syntheses owing to the imperfect crystallisation of intermediates such as **8.22** and **8.23**. In view of the mechanistic preference for the (*E*)-configuration this must arise from their isomerisation during working up in strongly acidic media. It has been shown (Banerjee *et al.*, 1984; Banerjee and Das, 1985) that this change may even be effected by traces of acid in chloroform used for spectroscopy and therefore the use of buffers and limiting the time and temperature of contact are recommended.

The Stobbe route was also chosen (Cooley *et al.*, 1981) for the synthesis of enterolactone (**2.049**) and enterodiol (**2.028**), the lignans detected in mammals. Here 3-benzyloxybenzaldehyde (2 mol.) was used in a two-step procedure followed by saturation and debenzylation with hydrogen/palladised charcoal. Another related synthesis was reported by Ganeshpure and Stevenson (1981a).

(ii) *Other syntheses initiated by carbanions*

An alternative route to enterolactone was also given by Ganeshpure whereby the hydrogenated product (**8.25**) from the first stage of a Stobbe condensation was converted into the lithium enolate and C-benzylated (Scheme 8.10). A closely related route was selected (Taafrout *et al.*, 1983) for the synthesis of (±)-arctigenin (**2.046**) and of the unsymmetrical *meta*-hydroxylated lactone, prestegane A (Scheme 8.11). The analogous Stobbe product (**8.26**) was condensed with *O*-benzylvanillin followed by hydrogenation/debenzylation of the alkylidene intermediate. The *cis*-lactone (**8.27**) so formed was isomerised to (±)-arctigenin with potassium

Scheme 8.10 Benzylation of a lithiolactone

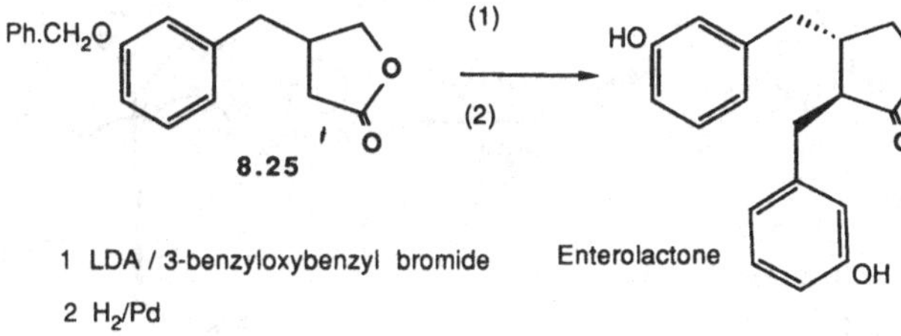

Ph.CH_2O 8.25 (1) (2) HO Enterolactone OH

1 LDA / 3-benzyloxybenzyl bromide

2 H_2/Pd

hydroxide in methanol (43% yield overall). In a detailed study (Tayyeb Hussain *et al.*, 1975) of the base-induced equilibration of 2,3-disubstituted-γ-butyrolactones, the *trans*-isomer was shown to predominate. The same end result was achieved (Taafrout *et al.*, 1983; 68% yield) by the reaction of the enolate (of **8.26**) with *O*-benzylvanillyl bromide. In another synthesis of this type (Ghosal and Banerjee, 1978) the lactone (**8.26**) was condensed with piperonal and after acidification a mixture of isomers was obtained in which the natural (*Z*)-suchilactone (**2.105**) predominated. (*E*)-isosuchilactone (**2.106**) is less polar and was separated from the mixture by preparative layer chromatography. A flexible procedure for the synthesis of symmetrical or unsymmetrical dibenzyl-butyrolactones has been described (Mahalanabis *et al.*, 1982*a*). Here a dimetallated succinamide (**8.28**) may be dibenzylated by the direct addition of one halide (2 equiv.) or by the sequential addition of two different halides (1 equiv. of each). With the first benzyl group (**8.29**. Ar_1) in place, the second entering benzyl group (Ar_2) is directed to the other face and hence the major product has

Scheme 8.11 Synthesis of *rac*-arctigenin

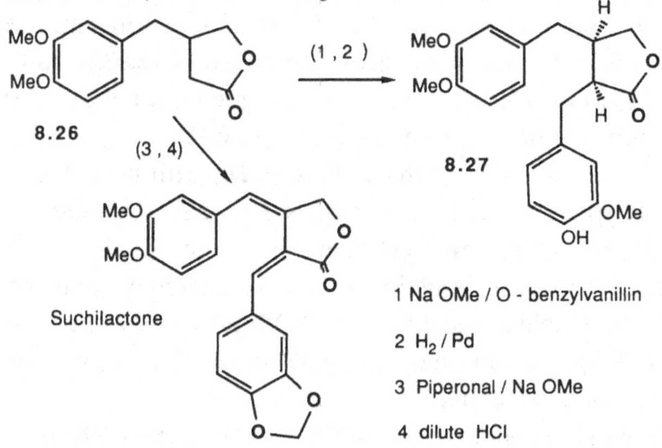

MeO MeO 8.26 (1 , 2) MeO MeO H H 8.27 OMe OH (3 , 4) MeO MeO Suchilactone

1 Na OMe / O - benzylvanillin

2 H_2 / Pd

3 Piperonal / Na OMe

4 dilute HCl

Scheme 8.12 Lignan synthesis from dimetallated succinamides

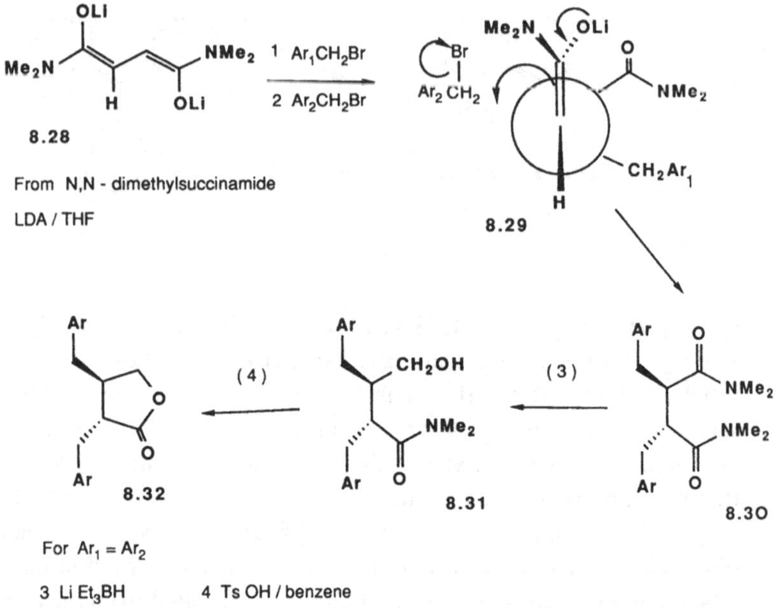

1 Ar_1CH_2Br

2 Ar_2CH_2Br

8.28 8.29 8.30 8.31 8.32

From N,N - dimethylsuccinamide

LDA / THF

For $Ar_1 = Ar_2$

3 Li Et_3BH 4 Ts OH / benzene

the threo-configuration (**8.30**). Partial reduction of symmetrically substituted succinamides (**8.30**; $Ar_1 = Ar_2$) with triethyl borohydride gave amidic alcohols (**8.31**), which formed butyrolactones (**8.32**) on treatment with acid. This sequence led to the synthesis of the symmetrical lignans dimethylmatairesinol (**8.32**; Ar = 3,4-dimethoxyphenyl, 60% overall) and hinokinin (**8.32**; Ar = 3,4-methylenedioxyphenyl, 80% overall). Enterolactone (**2.049**) was also obtained by this route in 32% overall yield from 3-methoxybenzaldehyde with removal of the methyl protecting group by boron tribromide (Mahalanabis *et al.*, 1982b). In the same manner further reduction of **8.32** to the diols secoisolariciresinol (**2.035**) and dihydrocubebin (**2.027**) occurred together with complete reduction of the C9,9′-terminal groups to afford austrobailignan-5 (**2.002**).

The product of reaction of the enolate (**8.28**) with benzaldehyde had a simple NMR spectrum consistent with symmetrical *syn-anti-syn* stereochemistry (**8.33**) and this was confirmed by X-ray analysis. The corresponding product from the reaction with veratraldehyde (Scheme 8.13) on treatment in acid yielded the dilactone (**8.34**) previously obtained by Takaoka *et al.* (1975). The monolactone (**8.35**) in the *Z*-configuration was the major product of this reaction.

β-Coupling of the dianion from 3-(3′,4′-methylenedioxyphenyl)propionic

Scheme 8.13 Reactions of dimetallated succinamides with aldehydes

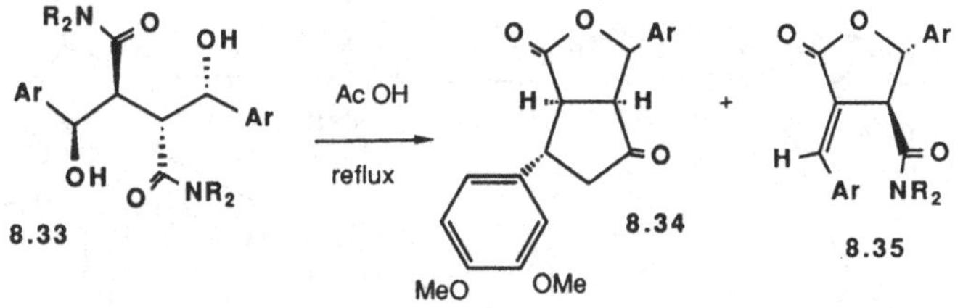

acid with iodine led to a synthesis of hinokinin (**2.057**) in 61% overall yield (Belletire and Fry, 1987).

(iii) *Stereoselective synthesis*

One route to enantiomerically pure material by enantioselective hydrogenation was mentioned above (Scheme 8.6). A clear description of the enantiospecific synthesis of podorhizone (**2.071A**) and deoxypodorhizone (equiv. **2.088**) has been given by Tomioka *et al.* (1982; Scheme 8.14), who employed a route based upon (*S*)-(+)-4-alkoxymethyl-4-butanolides (**8.37**) derived from L-glutamic acid (**8.36**). Other correlations of absolute configuration exist, for this lactone and its enantiomer have also been obtained (Ravid *et al.*, 1978) from L- and D-glutamic acid when used as precursors in pheromone synthesis. L-Glutamic acid has also been linked (Takano *et al.*, 1981a) in this manner to the indole alkaloids.

Condensation of **8.37** (R = benzyl) with piperonal/LDA gave a mixture of isomeric alkenes whose composition varied with the conditions employed to dehydrate the intermediate alcohol. With toluene/*p*-toluenesulphonic acid a yield of 92% was achieved with the *S-trans-* or (*E*)-isomer predominating (*E*:*Z* as 10:1). Asymmetric induction of the hydrogenation of **8.38** (R = benzyl) was inefficient and gave a mixture of isomers in which the *ent-* form of the lactone (**8.39**; R = benzyl) predominated and which on elaboration afforded the *ent-* or (−)-form of the natural (+)-podorhizone (**2.071A**). However a stereoselective route, also based on 1,3-induction, was achieved by alkylation of **8.37** (R = trityl). In this adaption the bulky trityl ether directed the entering methylenedioxybenzyl group to the β-face to give the benzyl lactone (**8.39**) as the major product. Four further steps (Scheme 8.14) were needed to obtain the (+)-secolactone (**8.41**), whence treatment of its conjugate base with the mixed anhydride from ethyl chloroformate and 3,4,5-trimethoxybenzoic acid gave (+)-podorhizone of 57% optical purity. Alternatively treatment with 3,4,5-trimethoxybenzyl bromide yielded (−)-deoxypodorhizone also in the natural configuration.

Scheme 8.14 Enantioselective synthesis of podorhizone and deoxypodorhizone

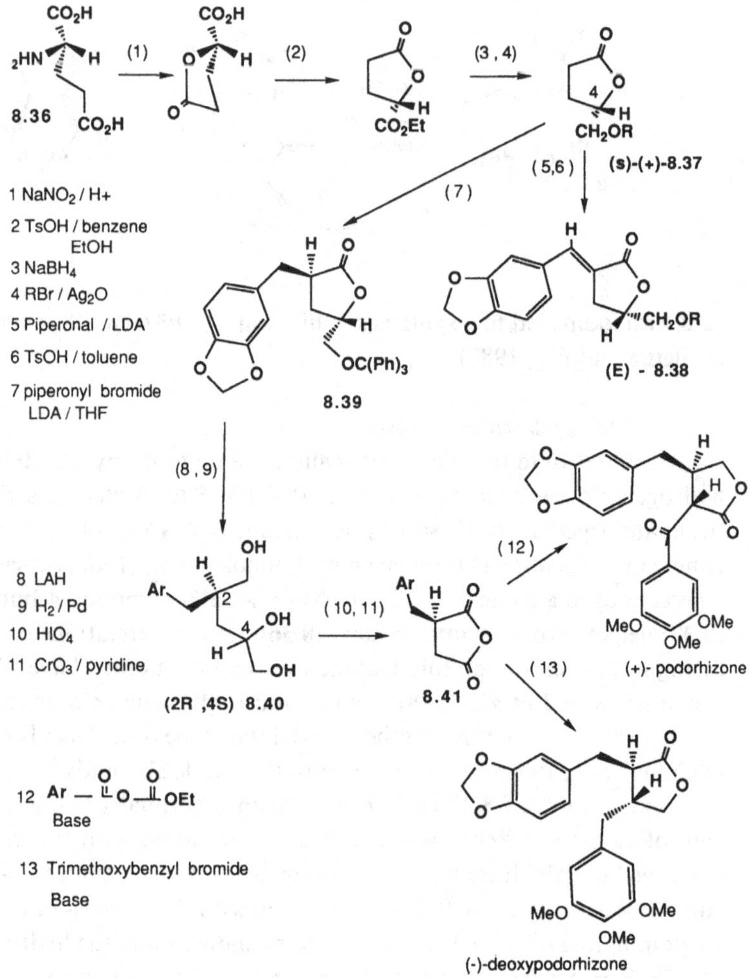
8.36
(1)
(2)
(3, 4)
(S)-(+)-8.37
(5,6)
(7)
1 NaNO2 / H+
2 TsOH / benzene
EtOH
3 NaBH4
4 RBr / Ag2O
5 Piperonal / LDA
6 TsOH / toluene
7 piperonyl bromide
LDA / THF
8.39
(E) - 8.38
(8, 9)
8 LAH
9 H2 / Pd
10 HIO4
11 CrO3 / pyridine
(2R ,4S) 8.40
(10, 11)
8.41
(12)
(13)
(+)- podorhizone
12 Ar — CO—O—CO—OEt
Base
13 Trimethoxybenzyl bromide
Base
(-)-deoxypodorhizone

Some other correlations of absolute configuration arose from the work just described. L-Glutamic acid was converted via the (*S*)-(+)-lactone (**8.37**) into D-ribose and the *ent*-(−)-form was obtained (Takano *et al.*, 1981a) from D-mannitol.

Another procedure of potential value for the enantio-selective synthesis of the lignan skeleton depends upon the modification of D-ribonolactone (Scheme 8.15; Camps *et al.*, 1981; Cardellach *et al.*, 1984) through the orthoformate (**8.42**) to the ene-lactone (**8.43**) and thence to (*S*)-(+)-caprolactones of type **8.44**.

Scheme 8.15 Derivation of (S) - caprolactones from D - ribonolactone

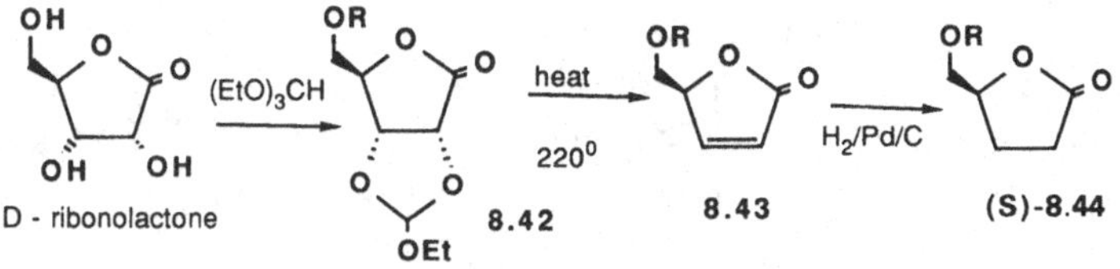

D - ribonolactone; $(EtO)_3CH$; 8.42; heat 220°; 8.43; H_2/Pd/C; (S)-8.44

(iv) *Conjugate additions*

An important study (Tomioka *et al.*, 1980) was based upon the application of this technique to a chiral lactone of type **8.43** and led to the stereoselective synthesis of (+)-steganacin (**8.53**; $[\alpha]_D + 135°C$), which proved to be the *ent*-form of the natural product ($[\alpha]_D - 114°C$) whose absolute configuration was incorrectly assigned (Kupchan *et al.*, 1973) at the time of isolation.

In this later work (Scheme 8.16) the previously described trityl ether (**8.37**) was converted into the selenyl derivative (**8.45**) and thence to the unsaturated chiral lactone (**8.46**). 1,2-Interaction between the bulky trityl ether group and the carbanion from trimethoxybenzaldehyde dithioacetal (**8.47**) now directed conjugate addition to the β-face of (**8.46**). Subsequent desulphurisation of **8.48** and cleavage of the trityl group with hydrochloric acid yielded the hydroxylactone (**8.49**), which was alkylated with piperonyl bromide using two equivalents of lithium diethyl amide to give the homolignan (**8.50**). The skeleton was simplified by the same four steps given in Scheme 8.14 for the conversion of **8.39** into **8.41**, whereby (+)-deoxypodorhizone (**8.51**) was obtained from **8.50**. Non-phenolic coupling (p. 304; Damon *et al.*, 1976) now gave (*S*)-(−)-stegane (**8.52**) which was atropoisomerised at 195°C to give (*R*)-(+)-isostegane followed by acetoxylation to yield (+)-steganacin (**8.53**).

In these selective syntheses it is important to contrast the result of 1,3-induction in the lactone (**8.37**; Scheme 8.14) with 1,2-induction in the lactone (**8.46**). Both depend upon lactones with the same absolute configuration at C-4, but 1,3-induction gives an entering β-benzyl group at C-2. In contrast, 1,2-induction leads directly to β-insertion of trimethoxybenzyl at C-3 of **8.48** and thence to α-benzylation at C-2 of **8.49**. As a consequence of this difference in the approach, one end-product is (−)-deoxypodorhizone and the other is the enantiomer (+)- (**8.51**).

The lignan skeleton can be assembled by addition to butenolides, with concerted trapping of the enolate anion (**8.54**, Scheme 8.17) by a benzyl bromide to give **8.55**. This method was devised (Damon *et al.*, 1976) as a

Scheme 8.16 Enantioselective synthesis of (+)-steganacin

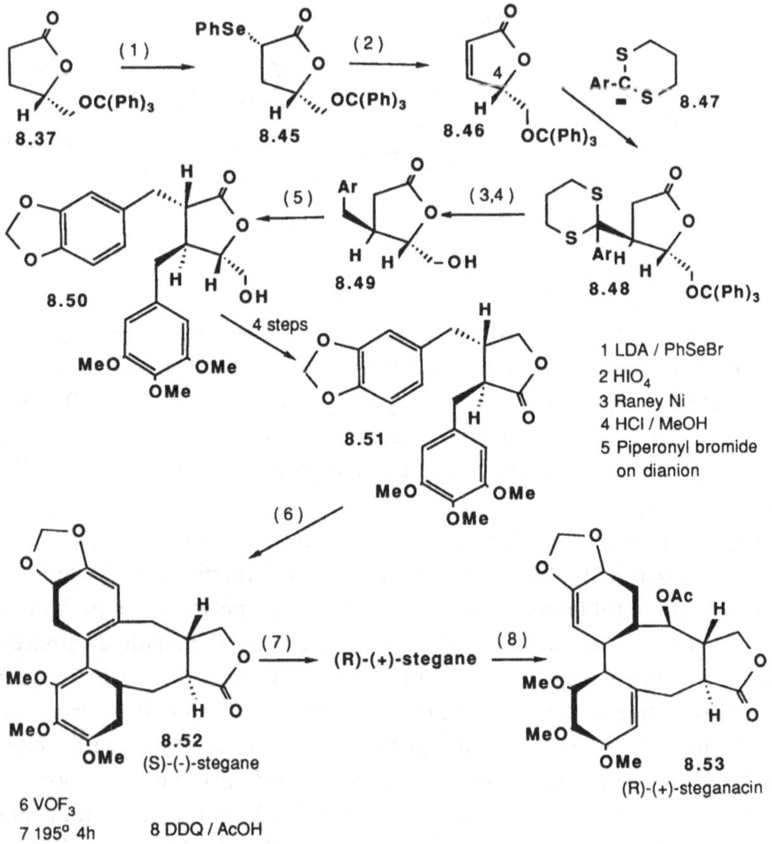

8.37 —(1)→ 8.45 —(2)→ 8.46 → 8.48 (with 8.47) —(3,4)→ 8.49 —(5)→ 8.50 —4 steps→ 8.51 —(6)→ 8.52 (S)-(-)-stegane —(7)→ (R)-(+)-stegane —(8)→ 8.53 (R)-(+)-steganacin

1 LDA / PhSeBr
2 HIO_4
3 Raney Ni
4 HCl / MeOH
5 Piperonyl bromide on dianion
6 VOF_3
7 195° 4h
8 DDQ / AcOH

Note **8.53** has the ent-configuration of natural steganacin (pp. 315, 345–7)

route to precursors of steganes (see above) and aryltetralins. After desulphurisation with Raney nickel the product (**8.55**) may be subjected to non-phenolic coupling or alternatively it may be cleaved with mercuric oxide/boron trifluoride to give the ketone at C-7 (Ziegler *et al.*, 1978a). Oxygen insertion in the side chain may also be affected by trapping the anion with an aromatic aldehyde (Gonzalez *et al.*, 1978) to yield products of type **8.56**. This approach was used (Pelter *et al.*, 1981) in the synthesis of enterolactone (**2.049**) and extended to the synthesis of arylnaphthalenes; it will be discussed further (cf. Schemes 8.57, 8.66) as part of a route to aryltetralins.

Scheme 8.17 Addition of thioacetal carbanions with concerted trapping of the enolate

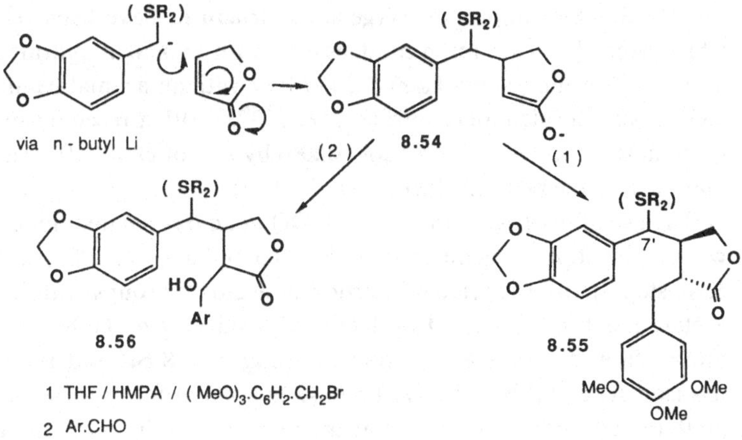

The fundamental lignan skeleton has been assembled (Minato, 1980a) by sequential substitution of thiophene, followed by reductive desulphurisation. This work followed two principal pathways (Scheme 8.18). In (A) 3,4-dibromothiophen (**8.57**) reacted with methyl Grignard and then with bromine to yield 2,5-dibromo-3,4-dimethylthiophen (**8.58**); two equivalent end groups were then inserted by a second Grignard synthesis followed by reductive desulphurisation of this product to give a 4:6 mixture of diastereoisomers of **8.59**. Route (B) led to an unsymmetrical lignan by limiting the displacement by benzylzinc bromides through palladium complexation (Minato, 1980b) to give **8.60** initially, followed by a second insertion catalysed by nickel(II) chloride with a final desulphurisation step to afford **8.62**.

Scheme 8.18 Synthesis of arylbutane lignans from thiophenes

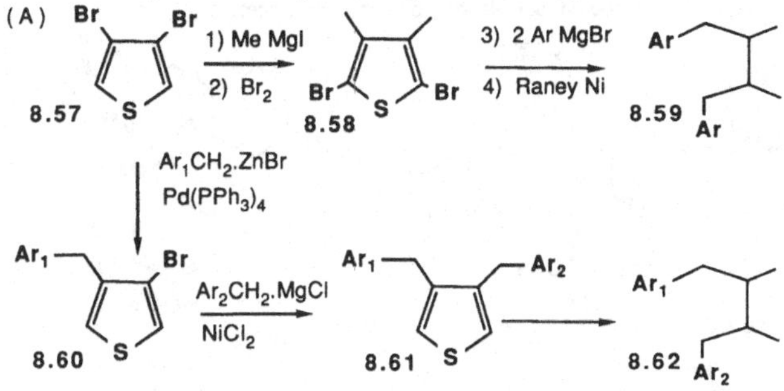

Oxygenation of the side chain

Some examples of oxygenated structures have been referred to which depend on carbanion condensation with an acid anhydride to insert a carbonyl group (Tomioka *et al.*, 1978) or with an aromatic aldehyde to yield a benzylic alcohol (Gonzalez *et al.*, 1978). Other recent examples are given in the synthesis of attenuol (**2.290**) by Loriot *et al.* (1983) and in an approach to the peltatins (Loriot *et al.*, 1984).

The use of hydrogen transfer by DDQ in an oxygenated donor solvent was employed for steganicin synthesis (Tomioka *et al.*, 1980; see Scheme 8.16, step 8) and the related insertion of acetoxyl groups with lead tetra-acetate has been discussed in detail by Nishibe *et al.* (1980). The latter group showed that acetoxylation in arctigenin (**8.64**) and isoarctigenin (*cis*-isomer) could be selective because *O*-acetylation of these precursors inhibited hydrogen abstraction at the related benzylic position. Thus the C-7′ monoacetate (**8.63**) was obtained in good yield from arctigenin. The configuration of the inserted OAc group was established by a comparison of its CMR spectrum with those of the analogous naturally occurring 7′-*allo*-hydroxymatairesinol, 7-hydroxymatairesinol (**2.065**) and parabenz-lactone (**2.068**). The configuration of **8.63** (7′-*S*) was characterised in the derived alcohol by a shift of 73.9 ppm and in the 7′-*R* alcohol parabenz-lactone by one of 75.4 ppm.

Care is needed in applying the lead tetra-acetate reaction to methylene-dioxy lignans, as it has been shown (Ikeya *et al.*, 1979*a*) that on prolonged heating with gomisin A (**8.67**; X = OH, Y = H) in benzene this ether is cleaved. In model compounds such as **8.65** (R = Me) the *bis*-phenol was formed in 40% yield with 13% recovery after seven hours at 60°C. Evidence of hydrogen abstraction with acetyl capture was provided by the isolation of the acetoxymethylene ether (**8.66**) from the reaction with methyl piperonylate. Intermediates of the type **8.66** afford the *bis*-phenols on hydrolysis.

Scheme 8.19 Directed acetylation by lead tetra-acetate

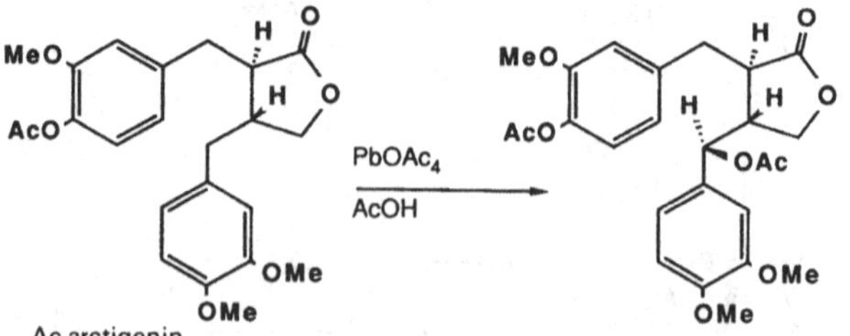

MeO, AcO, H, O, OMe, OMe — $PbOAc_4$ / AcOH → MeO, AcO, H, OAc, OMe, OMe

Ac arctigenin **8.64**

8.63

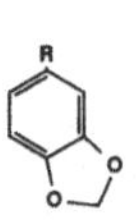

Loss of the methylenedioxy group in this way was also reported (Ikeya, 1978) for gomisin N (**8.67**; X, Y = H). However at a lower temperature (12°C) over 72 hours, acetyl gomisin O (X = H, Y = OAc) was obtained in modest yield from gomisin N without ether cleavage using acetic acid as the donor solvent (Ikeya, 1979b). It was also shown (Mervic *et al.*, 1981) that the highly hindered diol (**8.68**), an intermediate in steganone synthesis, was cleaved to the related diketone (90% yield) by lead tetra-acetate in benzene: pyridine at 25°C over one hour without effect on the ether groups.

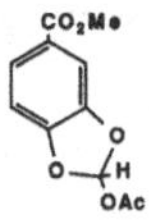

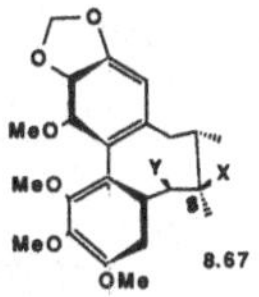

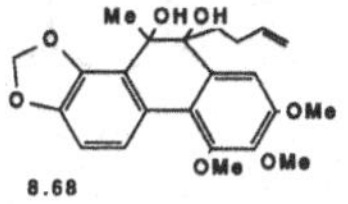

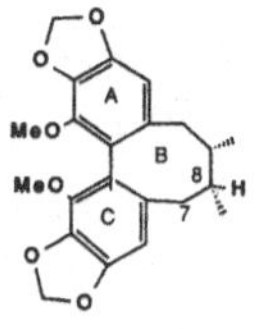

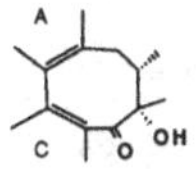

The insertion of oxygen at benzylic positions is difficult to accomplish for the alicyclic rings of bicyclo-octadiene and tetralin lignans, although a high-yielding procedure applicable to the latter has been claimed (Kende *et al.*, 1977).

Traditional reagents such as potassium permanganate have been used in lignan chemistry, principally as a method of allocating the pattern of aromatic alkoxylation through the isolation of piperonylic and alkoxy-benzoic acid fragments (Haworth and Kelley, 1937; Haworth, 1942; Hearon and MacGregor, 1955; Hartwell and Schrecker, 1958; Takaoka *et al.*, 1976). Degradations of this kind are initiated by oxygen insertion in the aliphatic portion of the molecule, but these intermediates are more susceptible to oxidation than the starting materials and hence survive in only low yield. Recent examples of permanganate oxidation are to be found in the extensive work on bicyclo-octadienes by Ikeya and colleagues (1979b, 1982). Thus wuweizisu C on treatment with alkaline permanganate in pyridine at 50°C for 2 hours gave the 7-ketone (6% yield) and the keto-alcohol (**8.69**, 1%) with only 25% recovery. The yields of analogous compounds from gomisin N (**8.67**; X, Y = H) were marginally better and a point of interest was the inversion of configuration which occurred on insertion of —OH at C-8. This implies that reaction occurs through enolisation of a preformed C7-ketone which is attacked from the less-hindered

Scheme 8.20 Oxidation of a primary alcohol by chromium trioxide

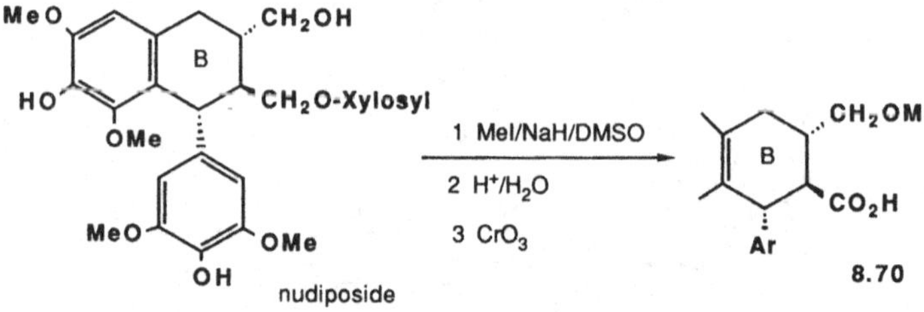

MeO, CH_2OH, B, HO, CH_2O-Xylosyl, OMe, MeO, OMe, OH, nudiposide

1 MeI/NaH/DMSO
2 H^+/H_2O
3 CrO_3

CH_2OMe, B, CO_2H, Ar, 8.70

side. Further oxidation of keto-alcohols of type **8.69** occurs under mild conditions with manganese dioxide/acetone or chromium trioxide/pyridine with cleavage of ring B (Ikeya *et al.*, 1978). Both of these reagents have found additional applications in the selective oxidation of lignan alcohols.

After permethylation of nudiposide (Scheme 8.20 and **2.331**) by Hakomori's method the point of glucosidation was established (Ogawa and Ogihara, 1976) after hydrolysis by oxidation of the revealed primary alcohol to the carboxylic acid (**8.70**). This was identical to the acid derived similarly from the lignan lyoniside, save that CD showed the aglycones to be enantiomers.

The lactol (**8.71**) was converted into the corresponding lactone with Collin's reagent in methylene chloride during Tomioka *et al*'s asymmetric synthesis of podorhizon (1982). This procedure is clearly of value in correlating the unusual lactol lignans with the numerous lactone models. The technique is even applicable (Chakraborty *et al.*, 1979) to sensitive molecules such as the unsaturated lactol podotoxin (**2.104**).

A study of the solvent effect on the oxidation potential of acidic chromium trioxide has been made (Takeya *et al.*, 1983); modest yields of 7-ketones (**8.73**; X = H) are obtained by the *single stage* coupling of 1-arylprop-1-enes (**8.72**) with this reagent. The analogous reaction with the (*E*)- or (*Z*)-3,4,5-trimethoxyphenyl isomers gave the ketone (**8.73**; X = OMe; 14% yield) together with (±)-grandisin (**2.157**, 18%).

Hydroxypicrostegane and its atropoisomer (**8.74**) were both efficiently oxidised (Brown and Robin, 1978) with acidic chromium trioxide to the H-bonded enol (**8.75**), but in pyridine solution ring cleavage of the diol (**8.76**) occurred with formation of the keto-aldehyde (**8.77**; Ikeya *et al.*, 1978). The same reaction took place with manganese dioxide but the *vic*-ditertiary diol from plicatic acid (**2.368**), which lacks a benzylic —OH, was unaffected by the latter reagent.

The 7-position in podophyllotoxin proved extremely difficult to oxidise specifically (Drake and Price, 1951) since chromium trioxide and a range of other oxidising agents caused fragmentation; the required podophyllotoxone was eventually obtained (Gensler and Johnson, 1955) in good yield using manganese dioxide. This reaction is typical of the application of the reagent to the selective oxidation of benzylic and allylic alcohols. Its value in distinguishing between types of primary alcohols is well shown by the reaction (Ahmed *et al.*, 1973b) with the diol (**8.78**) obtained from thomasic acid (**2.378**), which affords the hydroxyaldehyde (**8.79**) and also (Banerji *et al.*, 1984) by its action on the allylic diol from gadain (**2.098**). It was further shown that the cyanohydrin of this aldehyde (**8.79**) was susceptible to manganese dioxide oxidation in methanol by Corey's method to give the 9-carboxylate. This reaction is also applicable to the orientation of lactone groups.

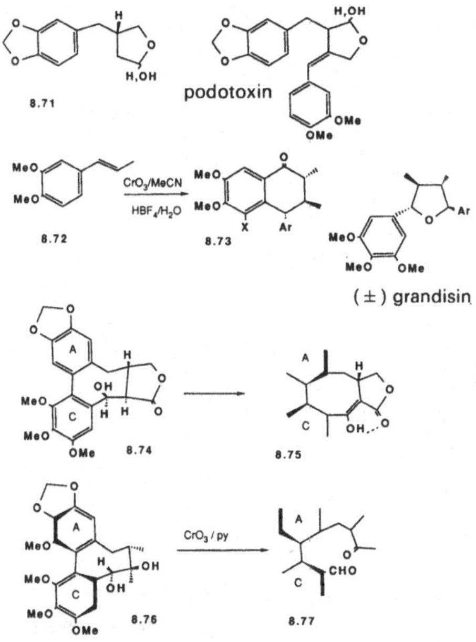

It has been reported (Anjanamurthy and Rai, 1985) that a keto-group was inserted into a β-apopicropodophyllin (**8.80**) on treatment with manganese dioxide in refluxing chloroform for two hours; it is, however, likely that the product was dehydropodophyllotoxin (**2.393**). The same workers also describe the epoxidation of β-apopicropodophyllin with hypochlorite; a reaction of interest in that this change enhances the physiological activity of these compounds. It was not possible to epoxidise the double bond in the schizandrin precursor (**8.81**) with peracid, but this alkene

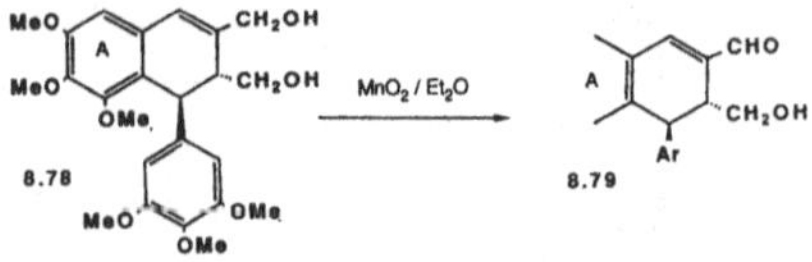

8.78 8.79

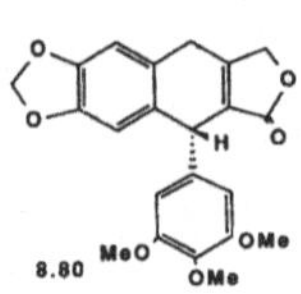

8.80

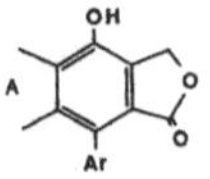

dehydropodophyllotoxin

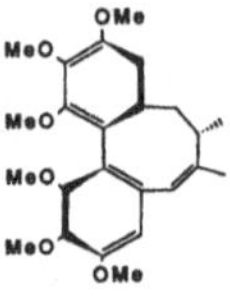

8.81

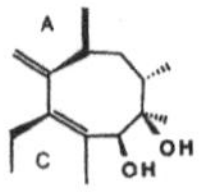

8.82

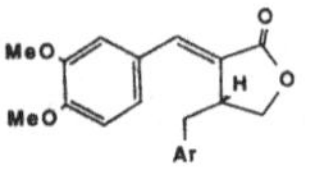

8.83

underwent *cis*-hydroxylation with osmium tetroxide in pyridine to give the diol (**8.82**; 62%) by attack from the less-hindered side. Hydroxylation of open-chain lignans such as jatrophan (**2.100**) and the conjugated lactone (**8.83**) also proceeds in this manner (Nishibe *et al.*, 1980; Banerji *et al.*, 1984).

Reactions with N-bromosuccinimide (NBS) and dichlorodicyano-1,4-benzoquinone (DDQ)

NBS has found only limited application for the benzylic bromination of lignans. The reaction depends (Nonhebel and Walton, 1974) upon hydrogen abstraction by atomic bromine which can also attack the aromatic nuclei and where hydrogen bromide release may also affect the outcome. Steric factors are also of importance (Tedder and Walton, 1980) and this is illustrated by comparing the reactions of the model substance (**8.84**; X = H) with those of conformationally restricted lignans such as dibenzocyclo-octadienes. The monobenzylic lactone (**8.84**) gave the derivative (**8.84**; X = Br) in almost quantitative yield and could also be directly oxygenated using DDQ in acetic acid. However, extension to dibenzocyclo-octadienes was only possible for stegane (**8.52**, $\delta'\alpha$, 8β) in which selective abstraction of an orthogonal hydrogen atom was seen (p. 316) to favour C-7 acetoxylation and where NBS gave a superior yield of the C-7-bromoderivative (Ishiguro *et al.*, 1985).

Although the action of NBS/dioxan has been reported (Kende *et al.*, 1977) to insert a benzyl substituent in high yield on UV irradiation of a flexible precursor (**8.85**) it was disappointing when applied (Yamaguchi *et*

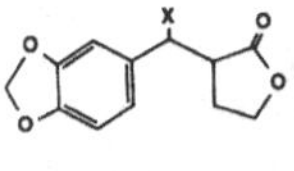

8.84 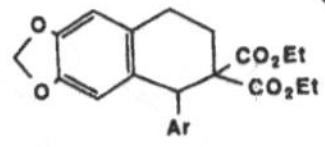 8.85 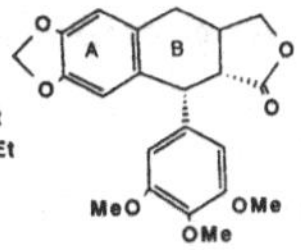 desoxypodophyllotoxin

al., 1984) to deoxypodophyllotoxin (**2.360**), which has a rigid ring B- and no orthogonally directed C7—H bond. In this work as in earlier studies (Kofod and Jorgensen, 1954) bromination occurred at the C2′-position in ring C. On demethenylation of the lignan, aromatic bromination was again the major route only now directed in the free phenol to the C-6 position of ring A; this orientation was established by X-ray analysis. Choice of a polar solvent tends to favour aromatic substitution rather than benzyl-radical reactions, as seen here in Yamaguchi's work where dimethylformamide was the solvent chosen, and in dioxan solution aromatic substitution is also preferred (Rao and Alvarez, 1985).

Benzylic intermediates formed either through hydrogen atom abstraction with NBS or hydride ion abstraction by DDQ may be trapped through interactions of Friedel-Craft type to give arylnaphthalenes. Thus suchilactone (**2.105**, Scheme 8.21) yields justicidin B (**8.86**; X = H) with NBS. The reaction of DDQ gave the diene (**8.87**; Ghosal, 1978) which was cyclised by UV irradiation (pp. 300–1) to give justicidin B together with diphyllin (**8.86**; X = OH). The cyclodehydrogenation of diene anhydrides by DDQ is described by Satyanarayana and Rao (1985).

The point of attack by DDQ-dioxan on isosuchilactone (**2.106**, *E*-form) is marked by oxygen capture to give **8.88** (R = H . DDQ) and an efficient synthesis (85% yield; Ghosal and Banerjee, 1979) of retrochinensin (**2.414**) results from attack in *dry* benzene by the intermediate cation upon ring A. Under these conditions the mode of cyclisation is characteristic of the position of the double bond in the side chain. However this specificity is evidently lacking in the cyclisation of suchilactone by NBS to give justicidin B (**8.86**; X = H) with its pendent methylenedioxyphenyl group and this reaction may well proceed via the diene (**8.87**; pp. 330–1). The synthesis (Banerji, 1984) of justicidin B by the action of NBS on jatrophan (**2.100**), which is *ent*-isosuchilactone, may follow the same route.

The symmetrically etherified (*Z*)-gadain (**2.098**) was dehydrogenated by DDQ and also underwent cyclisation to justicidin E (**2.403**), which may also be obtained (Ganeshpure, 1979) by the action of NBS on apocompounds. The aromatisation of apocompounds by NBS, DDQ and manganese dioxide has been described by Gonzalez *et al.* (1978). Under similar conditions (Ward *et al.*, 1981) tetralin lignans with terminal ether groups

Scheme 8.21 Cyclisation of benzylic intermediates

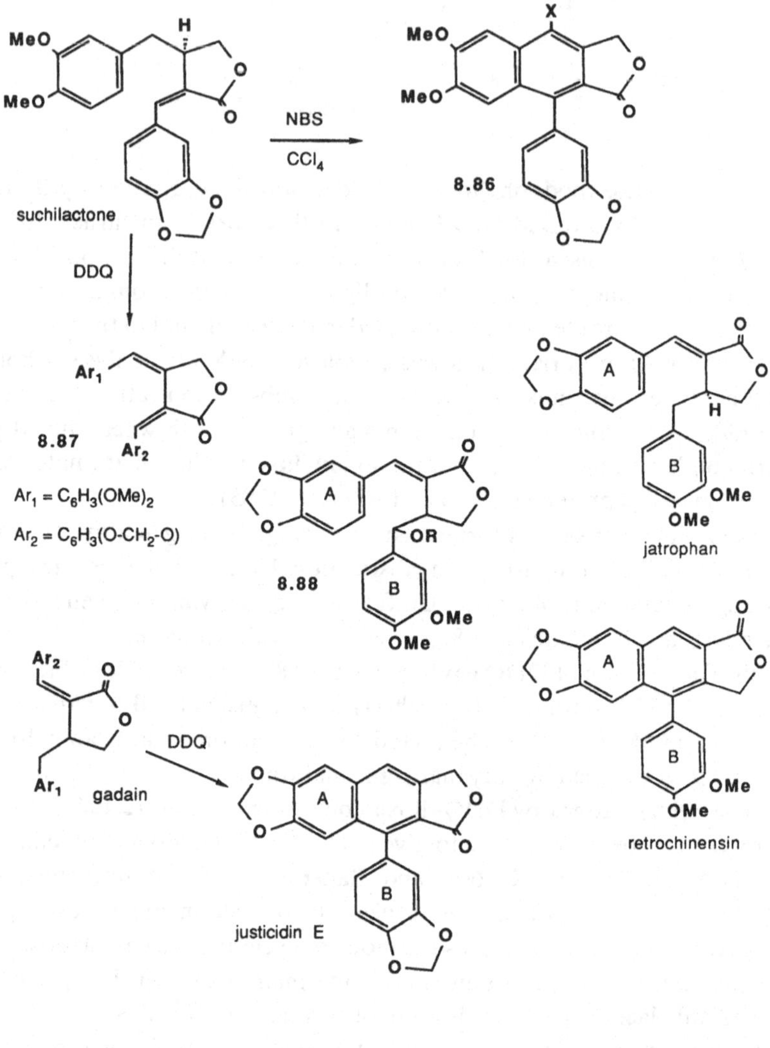

MeO, MeO, H, O, O, suchilactone, NBS, CCl_4, MeO, MeO, X, O, O, 8.86, DDQ, Ar_1, 8.87, Ar_2, O, O

$Ar_1 = C_6H_3(OMe)_2$

$Ar_2 = C_6H_3(O\text{-}CH_2\text{-}O)$

8.88, A, B, OR, OMe, OMe, jatrophan, retrochinensin, Ar_2, Ar_1, gadain, DDQ, justicidin E

(e.g. hypophyllanthin, **2.325**; Scheme 8.22) were dehydrogenated preferentially by DDQ at the less hindered C-7 position to give allylic aldehydes (e.g. **8.89**) with concomittant formation of the naphthalenic aldehydes (e.g. **8.90**). The aromatisation of tetralin lignans by DDQ has also been reported by Yamaguchi *et al.* (1982).

A potentially useful synthesis (Meyers, 1981) of naphthalenic lactones employs NBS to functionalise a methyl group on a model (**8.91**) in which the carboxyl is concealed as an oxazoline, and which yields the lactone **8.92** (Scheme 8.23). The reagent has also been used to convert the lactone

Scheme 8.22 Dehydrogenation of tetralin ethers

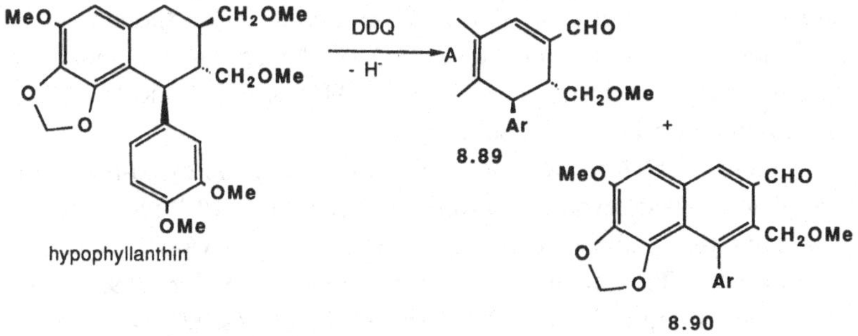

Scheme 8.23 Lactone synthesis and modification by NBS

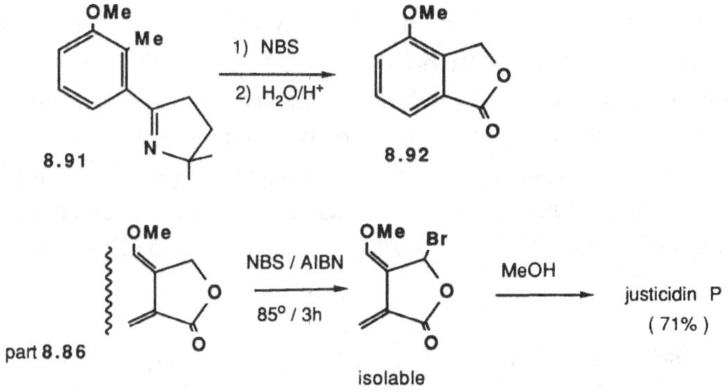

in justicidin A (**8.86**; X = OMe) to the lactol group in justicidin P (**2.405**) through methanolysis of an intermediate bromolactone (Wang, 1983).

Diels–Alder syntheses

This route to the lignan skeleton was pioneered by Klemm *et al.* (1966) and a number of applications have been reviewed (Ward, 1982). The first recorded instance is the dimerisation of phenylpropiolic acid to give 1-phenylnaphthalene-2,3-dicarboxylic acid (Michael and Butcher, 1898) and this was developed by Brown and Stevenson (1965) via cyclisation of the anhydride. Provided a blocking group is inserted (**8.93**; X = Br; Scheme 8.24) the more-hindered ether orientation (**8.94**) can be obtained under mild conditions, otherwise the product is the 4,5-methylenedioxy analogue. Reduction of these anhydrides by, for example, sodium/alcohol proceeds at the C-9 position. The retrolactone may be obtained (Brown and Stevenson, 1965) by non-selective reduction with LAH, which also removes the blocking group, followed by selective oxidation at C-9 with silver carbonate.

C-9′-lactones (e.g., **8.96**) are obtained as major products by the thermal cyclisation of cinnamyl propiolates (Klemm *et al.*, 1976; Gonzalez *et al.*, 1978) when the lowest unoccupied MO of the conjugated 2π-electron system interacts with the highest occupied MO of the 4π-electron system of the cinnamyl portion (**8.95**; Scheme 8.25).

A significant modification (Rodrigo, 1980) was the cyclo-addition of dimethyl acetylenedicarboxylate to the isobenzofuran (**8.97**; Scheme 8.26), followed by base-induced epimerisation at C-8 (of **8.98**) and selective reduction with LTBA at the less-hindered position to give **8.99**. The final hydrogenolysis step yielded methyl epipodophyllate (**8.100**).

The necessary quinone methide intermediates are also generated by extrusion of sulphur dioxide from thiophene-2,2-dioxides at 80–100°C. A route (Charlton and Alauddin, 1986) to cyclolariciresinol (**2.313**) was based on this method where **8.101** (X = OMe, R = Me, Ar = 3,4-dimethoxyphenyl) gave the (*E*,*E*)-methide (**8.102**). In order to maximise the yield (33% overall) stage separations were omitted and this led to an intermediate mixture of diastereoisomers, where NMR pointed to **8.104** as the principal component. The methide was also obtained (Das *et al.*, 1983) by heating the lactone (**8.105**) to 300°C when it was trapped by the addition of

Scheme 8.24 Diels - Alder cyclisation of propiolic anhydrides

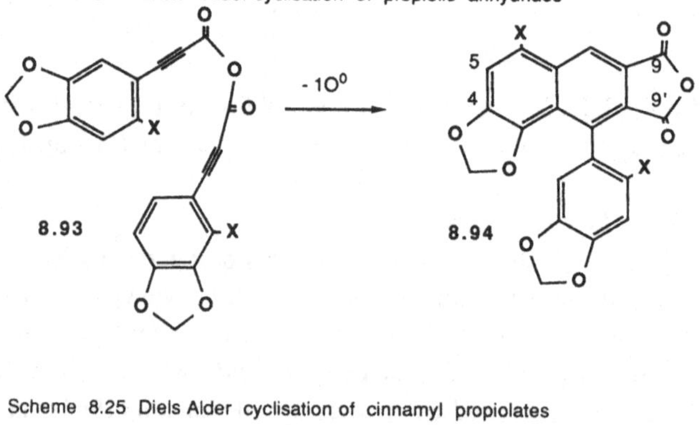

Scheme 8.25 Diels Alder cyclisation of cinnamyl propiolates

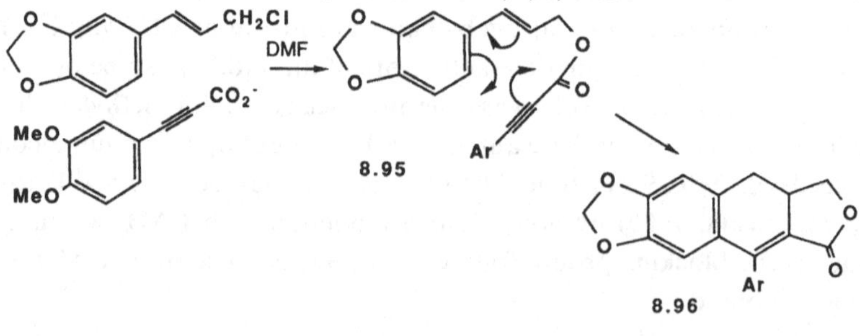

Scheme 8.26 Addition to an isobenzofuran

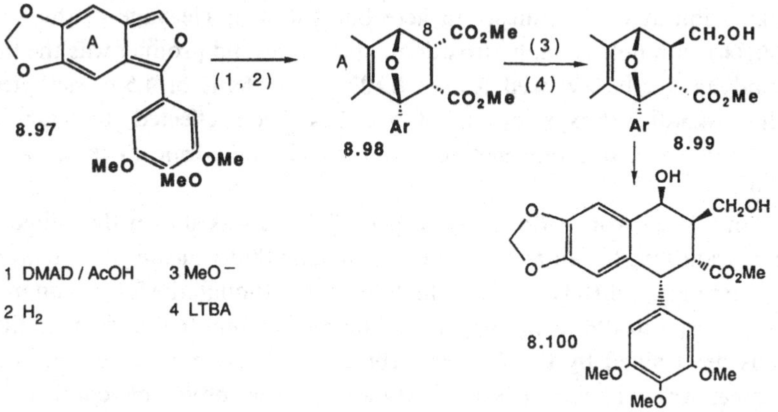

8.97 → (1, 2) → 8.98 → (3) (4) → 8.99 → 8.100

1 DMAD / AcOH 3 MeO^-
2 H_2 4 LTBA

Scheme 8.27 Addition to quinone methides

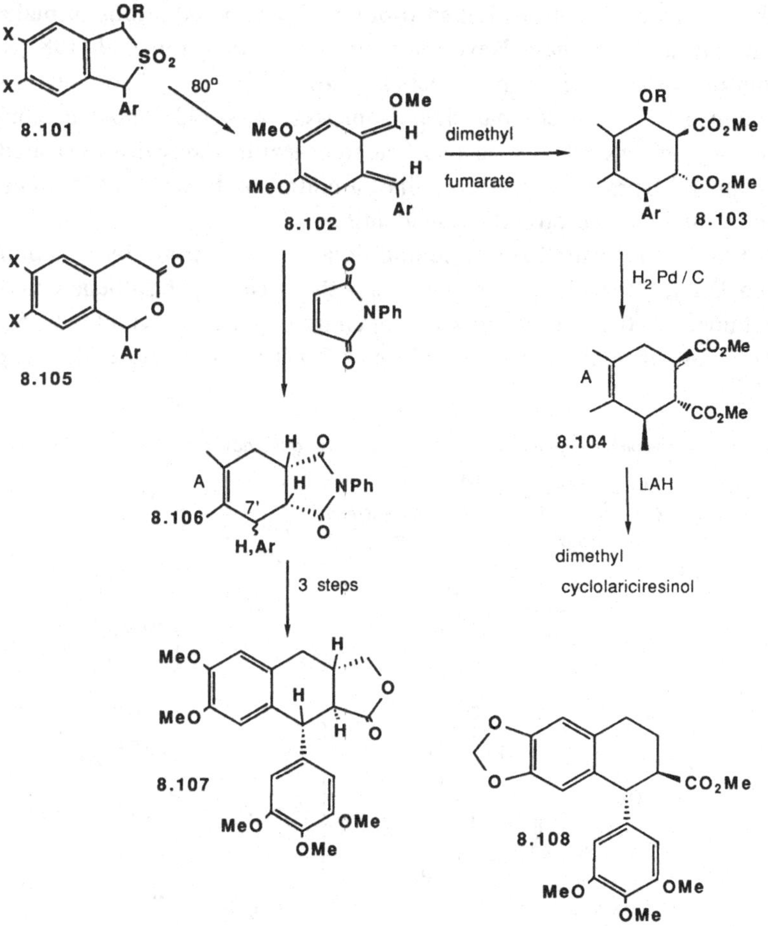

8.101 → 80° → 8.102 → dimethyl fumarate → 8.103 → H_2 Pd / C → 8.104 → LAH → dimethyl cyclolariciresinol

8.105

8.102 → NPh → 8.106 → 3 steps → 8.107

8.108

N-phenylmaleimide. In the absence of a restraining oxygen bridge the configuration at C-7′ is uncertain here but following cleavage of the imide (**8.106**) with potassium hydroxide/methanol the end product was the thermodynamically favoured isomer **8.107**. A synthesis of 4,5-demethylenedioxy-4,5-dimethoxypodophyllotoxin has been effected through the trapping of a quinone methide with maleic anhydride (Takano *et al.*, 1987).

In related work (Mann and Piper, 1982), it was shown that dimethyl fumarate gave the all-*trans*-diester (**8.104**) and that maleate gave an all-*cis*-diester of type **8.104** in good yield. However, although the formation of the mono-ester (**8.108**) with methyl acrylate showed that the mode of addition was determined by the frontier orbital, the predominance of the *trans*-isomer was evidence of reversibility and of thermodynamic control. This conclusion was questioned by Charlton and Durst (1984) who were only able to isolate 7′,8′-*trans*-linked products from the additions of maleate, fumarate and crotonate. Nevertheless in later work Mann *et al.* (1985) substantiated their finding that kinetically controlled *cis*-7′,8′-adducts can be obtained and showed that their isomerisation could follow prolonged heating and/or contact with acid. The same features were demonstrated by Khan and Durst (1987), where minor amounts of the retro-addition products from methyl crotonate were identified.

Equilibration in these cyclo-additions was overcome by Macdonald and Durst (1986a,b) who showed that the necessary *E,E*-diene could be obtained by thermal conrotatory opening of a *trans*-arylbutenol (**8.109**; Ar = 3,4,5-trimethoxyphenyl; Scheme 8.28). This labile molecule was pre-

Scheme 8.28 Stereocontrol by intramolecular cycloaddition

pared from a benzyne intermediate (Jung *et al.*, 1985) and was then linked as its urethane (**8.111**) by reaction with the crotyl isocyanate (**8.110**) under neutral conditions using an acetyltin catalyst. Thermolysis of the free acid then afforded a precursor (**8.113**) for podophyllotoxin (**2.357**) by intramolecular cyclo-addition of the intermediate *E,E*-methide (**8.112**).

In a related approach (Takano *et al.*, 1985a) the skeleton was constructed by protection of an *o*-bromoaldehyde (**8.114**; Scheme 8.29), lithiation and condensation with an aromatic aldehyde to afford the secondary alcohol (**8.115**) which was converted into its maleic half-ester under acidic conditions. This initiated a sequence of changes – the protecting group was cleaved to give a sulphonium intermediate (**8.116**) which generated a methide (**8.117**) by elimination of a proton. Although not suited to intramolecular addition this did give an intermolecular adduct (**8.118**) with excess maleic anhydride; this product gave a naphthalenic anhydride (**8.119**) by thermal elimination of 1,3-propanedithiol and maleic acid. By

Scheme 8.29 Arylnaphthalenes by cycloaddition

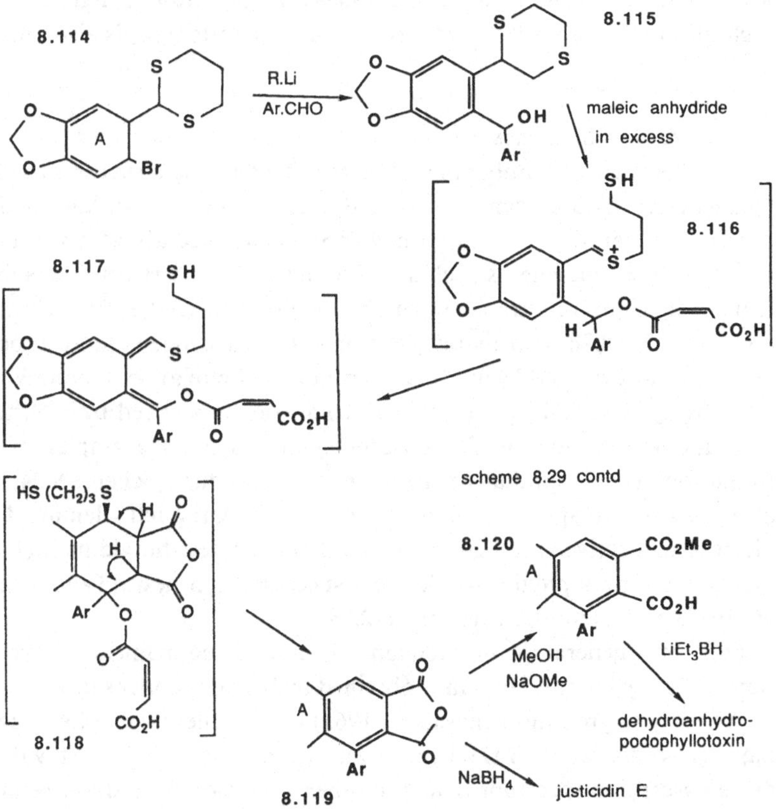

means of selective reduction the anhydrides of this group were converted into six members of the class of arylnaphthalene lactones. This is illustrated for **8.119** (Ar = 3,4,5-trimethoxyphenyl) when methoxide cleavage proceeds at the less-hindered position to give **8.120** in which the ester function is reduced by triethylborohydride to give dehydroanhydropodophyllotoxin (**2.393**; 7-deoxy). Typically where for **8.119** (Ar = methylenedioxyphenyl) direct reduction with borohydride itself led to a mixture of lactones from which justicidin E (**2.403**) was obtained.

Routes to specific classes of lignans

These approaches depend to a considerable extent on the general methods for assembling the skeleton reviewed so far and the following selections from the literature include only brief reference to them, while concentrating on methods for modifying the detail in the chosen examples.

Important variations are methods for conversion between classes for example, the hydrogenolytic sequence linking furofurans – furans – hydroxybutanes. The acid-catalysed cleavage of furanoid lignans, with cyclisation to tetralins by reactions of Friedel–Crafts type, is also important.

Arylnaphthalenes

Owing to a number of early errors in assigning structures to these lignans (Ayres, 1978) synthesis played a significant part in later assignments. The majority of methods have been mentioned above when it was seen that oxidising agents such as NBS and DDQ insert benzylic substituents which enable tetralins to be obtained by Friedel–Crafts cyclisation (Scheme 8.21; **8.88**) and that cyclisation with dehydrogenation can occur directly (Haworth and Kelly, 1936). Aromatisation of apocompounds and tetralins by NBS, DDQ and manganese dioxide is referred to in Scheme 8.22 and the section on Diels–Alder syntheses gives examples of the formation of arylnaphthalenes in a one-step procedure (Scheme 8.24) and also of a route to apocompounds (Scheme 8.25). Although selenium dioxide was ineffective, the action of DDQ on the tert-butyldimethyl-silyl derivative of podophyllotoxin led to a successful synthesis (Tanoguchi *et al.*, 1987) of the tetradehydrolignan (**2.393**).

One other general method remains to be discussed, namely the cyclisation of *bis*-arylidene compounds obtained by Stobbe condensation.

An early (Ayres and Mhasalkar, 1965) and frequently employed technique is irradiation with UV light of wavelength common in part with the diene absorption spectrum. This and other activation procedures result in isomerisation of the dienes although they will initially be in the (*E,E*)-configuration (p. 306). DDQ has been used to initiate this reaction (Satyanarayana and Rao, 1985) which also follows thermal excitation (Anjaneyulyu, 1979b) on heating in a high boiling solvent such as diphenyl ether (b.p. 259°C). In some instances cyclisation occurs merely on prolongation of treatment of the Stobbe product in acetic anhydride. Thus reflux of the *symmetrical* diene (**8.121**; Scheme 8.30) afforded the arylnaphthalene (**8.122**; R_1, R_2 = Me, R_3 = H) in 80% yield as the sole product.

Cyclisation of unsymmetrical dienes is inefficient and as many as four products are possible depending on the aromatic substitution pattern. The precursor (**8.121**; R_1, R_2 = —CH_2—; R_3 = OMe) on UV irradiation gave (Ayres *et al.*, 1965) a mixture of the three possible arylnaphthalenes (**8.122**; and products of 2,7′; 2′,7-cyclisation. By careful exclusion of oxygen and activation with 1,4-diazabicyclo-[2,2,2]-octane, Momose *et al.* (1978) synthesised *β*-apolignans by UV irradiation of the diene lactone (**8.125**; Scheme 8.31). This reaction is therefore identified as of the Diels–Alder type (p. 325). Treatment of these *β*-apolignans with oxygen/base leads to their aromatisation by hydride abstraction – **8.126** gave justicidin B (**8.128**; X = H, 28% yield) with diphyllin (**8.128**; X = OH, 27%).

Other unsymmetrical anhydrides cyclised with oxidation in hot acetic anhydride to give mixtures of arylnaphthalenes (Rao *et al.*, 1978). For example **8.121** (R_1, R_2 = —CH_2—; R_3 = H) gave a mixture from which

Scheme 8.30 Cyclisation of bis - arylidene dienes

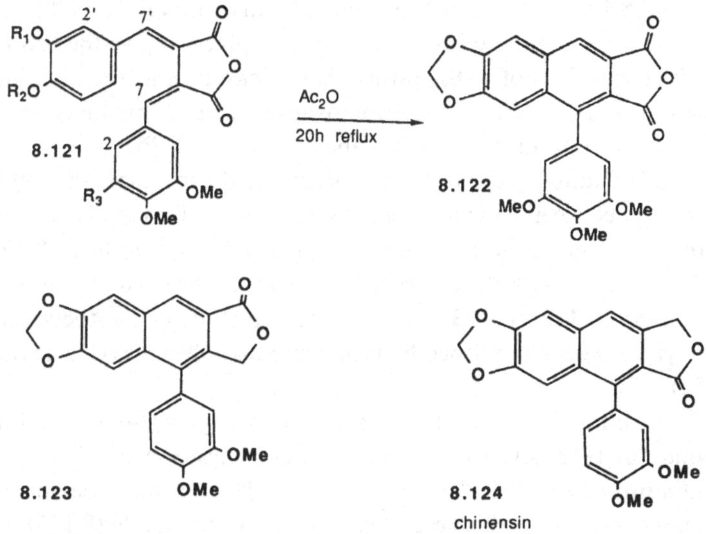

8.121 —(Ac_2O, 20h reflux)→ 8.122

8.123

8.124 chinensin

Scheme 8.31 UV initiated cycloaddition of a diene

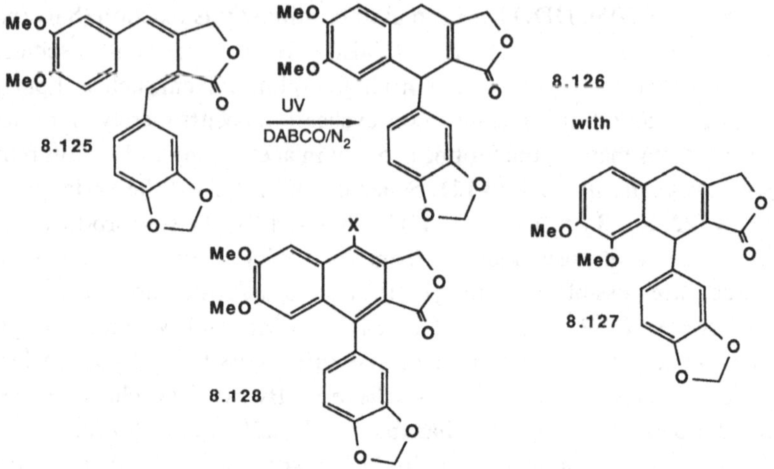

was separated the corresponding anhydride (**8.122**) which was reduced to the diol with LAH and selectively oxidised with silver carbonate to a mixture of lactones; chromatography on silica led to the separation of the major component which was identified as chinensin (**8.124**). It is of interest that the seemingly more polar retrolactone (**8.123**) was the first to be eluted as the minor component.

Some control of thermally induced cyclisations is possible through the introduction of blocking bromosubstituents before Stobbe condensation. Thus the dibromoanhydride (**8.129**; Scheme 8.32) afforded the arylnaphthalene (**8.130**, 80%) as sole product (Anjaneyulu *et al.*, 1978). Although this synthesis is more efficient, attack took place to give the less-hindered product with loss of —Br, rather than cyclisation at the alternative C-6 position so as to retain both bromosubstituents. A methoxyl group may also be displaced in this way (Anjaneyulu *et al.*, 1979b).

The UV-induced cyclisation of brominated precursors (**8.129**) has also been studied (Anjaneyulu *et al.*, 1979a, 1980). This leads to cyclisation with retention of both bromosubstituents in the adduct (**8.131**, 10% yield), together with the corresponding naphthalene which was isolated as the diester (**8.132**; X = H, 20%). A detailed study of photocyclisations of this type has been published by Heller and his colleagues (Hart and Heller, 1972).

An example of direct selective synthesis of arylnaphthalene model compounds of type **8.135** by Friedel–Crafts cyclisation has been described (Agnihotri *et al.*, 1982; Scheme 8.33). The skeleton was assembled by Perkin condensation with a weak base to give a butenolide (**8.133**), followed

Scheme 8.32 Cyclisation of brominated precursors

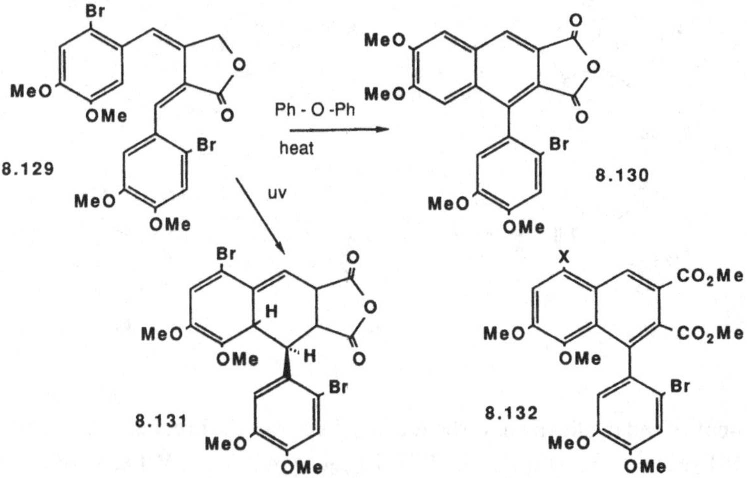

Scheme 8.33 Synthesis of 9-carbonyl lactones

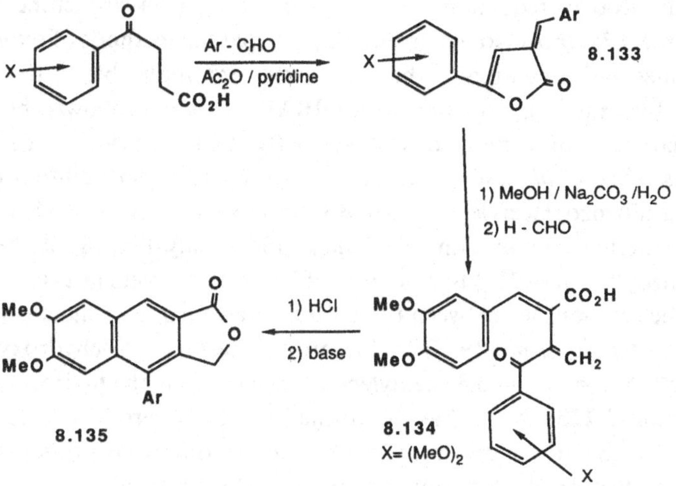

by hydrolysis and reaction of the ketoacid (not isolated) with formaldehyde to yield a methylene derivative (**8.134**) which cyclised in acid and so led to the required lactone (**8.135**) via the chloromethyl precursor.

Furans

(i) *9,9′-Epoxylignans*

This minor group is closely related to the butyrolactones and may be obtained from them by reduction to the *bis*-primary diols followed by

Scheme 8.34 Synthesis of burserans

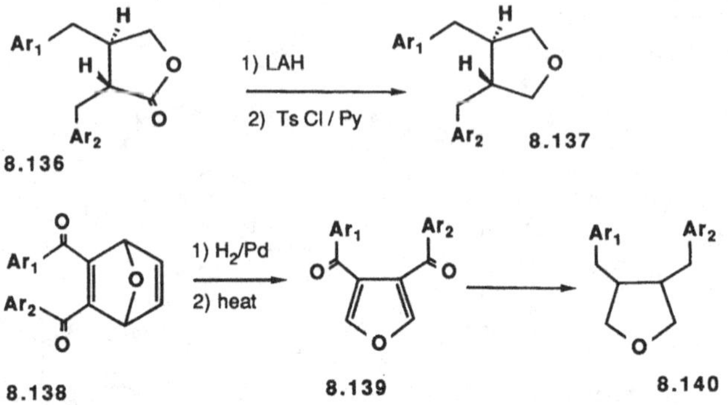

acid-catalysed cyclisation to the tetrahydrofuran (Schrecker and Hartwell, 1955; Freudenberg and Knof, 1957; Briggs *et al.*, 1959; Pelter *et al.*, 1982). A more satisfactory route (Tomioka *et al.*, 1985) is illustrated by the stereoselective reduction of (+)-*trans*- and (−)-*cis*-lactones (**8.136**, Scheme 8.34, *trans*-isomer given). In principal the methods of reduction of lactones with retention of the ring system are applicable to this conversion; for example, reduction with DIBAL to a ketal followed by hydride displacement of its mesylate or tosylate (see Scheme 8.44). An earlier synthesis (Cole *et al.*, 1969) was based upon Diels–Alder addition of furan and a dibenzoylacetylene to give **8.138** followed (Scheme 8.34) by reduction, thermal elimination of ethylene and finally hydrogenation of the dibenzoylfuran (**8.139**) to a mixture of burseran isomers (**8.140**).

The relationship between these furans and furofuran lignans is shown (De Carvalho *et al.*, 1987) by the production of (−)-dehydroxycubebin (**8.137**; $Ar_1 = Ar_2$ = 3,4-methylenedioxyphenyl) by the hydrogenolysis of sesamin (**2.223**). Note that the formation of this product (**8.137**) is the result of the etherification of the initially produced *bis*-primary diol (cf. **8.158A**; Scheme 8.39) during chromatography on silica.

(ii) *7,7′-Epoxylignans*

This subgroup has been identified (Rao and Alvarez, 1983) as a structural element in some dilignans. This was established for isolates from *Saururus cernuus* when hydrogenolysis of manassantin A yielded 3,4-dimethoxyphenylpropane and the monolignan (**8.141**).

An early approach by Blears and Haworth (1958) was the hydrogenation of 2,5-diaryl-3,4-dimethylfurans. These compounds may readily be synthesised via diaryoylbutenes or by the dimerisation of β-ketoesters or α-bromoethyl aryl ketones (Klemm, 1978). They were themselves first

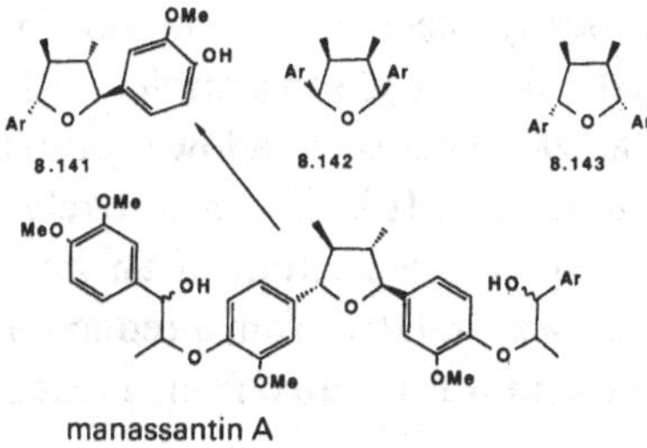

8.141 8.142 8.143

manassantin A

identified as natural products by King and Wilson (1964). The hydrogenation products have the all *cis*-configuration (**8.142**) and it was shown (Blears, King) that they were isomerised to the other less-compressed meso-form, galgravin (**8.143**; Ar = 3,4-dimethoxyphenyl), with acid under mild conditions.

The diketone route was adopted by Biftu *et al.* (1978b, Scheme 8.35), when the sodium enolate of propioveratrone (**8.144**) was alkylated with its α-bromoanalogue (**8.145**) to give largely the racemic product (**8.146**). The isomeric *meso*-compound (**8.147**) may be distinguished from **8.146** by differences in their IR spectra. In the NMR spectra the methyl protons lie at higher field in the *meso*-form due to shielding by the aroyl groups in the favoured conformer (Perry *et al.*, 1972). Unsymmetrical diketones (Ar different) may be obtained in good yield by this route provided that the bromoketone is slowly added to the enolate.

Scheme 8.35 The diketone route to tetrahydrofurans

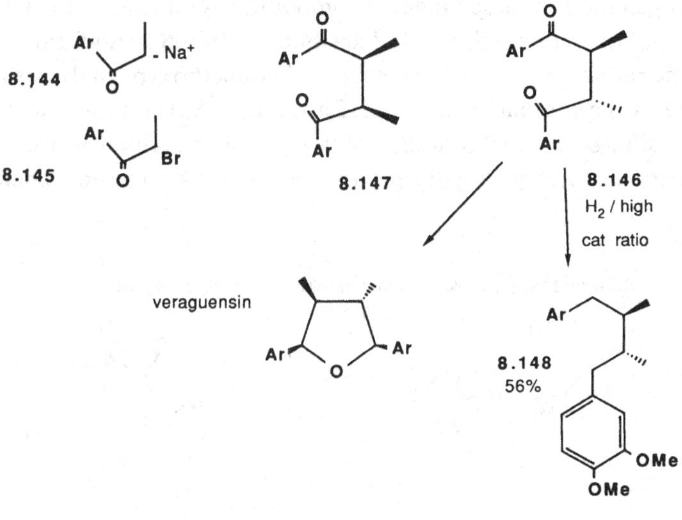

8.144 8.145 8.147 8.146

H_2 / high cat ratio

veraguensin

8.148 56%

Hydrogenation of **8.146** was critically affected by variations in the conditions since a high ratio of catalyst:diketone led to complete reduction to dimethyl dihydroguaiaretic acid (**8.148**), whereas at a lower ratio the principal product was (±)-veraguensin (**2.168**). It is, however, likely that this difference in behaviour is attributable to cyclisation of an intermediate benzyl alcohol, because veraguensin was isolated from a reduction carried out in acetic acid and the guaiaretic acid was isolated from a reaction in an aprotic solvent.

The route shown in Scheme 8.35 is unsatisfactory for unsymmetrical lignans with different aromatic nuclei because in place of veraguensin two isomers will be formed which are inseparable by chromatography (Biftu *et al.*, 1978a,b).

The stereochemical preference for the racemic diketone (**8.146**) was studied by Perry *et al.* (1972) and was explained by them in terms of a transition in which the bromide (**8.145**, Scheme 8.36) approaches orthogonally to the enolate (**8.144**). The aroyl groups are placed furthest apart and the α-H atom is adjacent to the bulky enolate aryl; displacement of bromine with Walden inversion then gives the racemic form of the diketone (**8.146**). The proportion of the *meso*-diketone (**8.147**) in the product mixture rises to 40% when two molecules of the enolate anion (**8.144**) are oxidatively coupled using cupric ion (Biftu *et al.*, 1979). It may also be obtained through the base-induced equilibration of the two enolates in methanol/ether since the *meso*-diketone is the least-soluble component and is precipitated. Reduction of **8.147** (Ar = 3,4,5-trimethoxyphenyl) with LAH gave the racemic diol, as shown by its NMR spectra, which was cyclised via its methanesulphonate to the *meso-trans*-tetrahydrofuran (**8.143**). Catalytic hydrogenation of these hindered compounds is difficult but reaction with sodium/liquid ammonia yields diarylbutanes (**8.148**, *meso*-form).

The racemic diketone (**8.146**; Ar = 3,4-dimethoxyphenyl) was also prepared by the enol/halide reaction (Rao *et al.*, 1985) and was reduced to the diol (Biftu *et al.*, 1978b; Rao, 1985) which was cyclised in various ways (Scheme 8.37) to give galbelgin (**8.149**; Ar = 3,4-dimethoxyphenyl) as a

Scheme 8.36 Sterically preferred transition for enolate alkylation

8.144

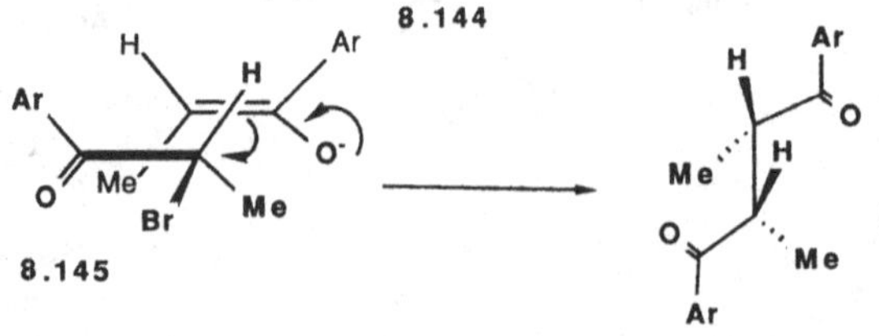

8.145

8.146

Scheme 8.37 Cyclisation of 1,4 - diarylbutane - 1,4 - diols

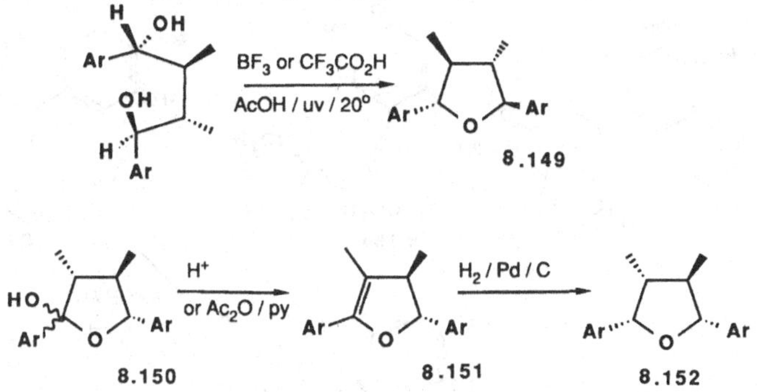

BF_3 or CF_3CO_2H
AcOH / uv / 20°
8.149

H^+
or Ac_2O / py
H_2 / Pd / C
8.150 8.151 8.152

common product. With borohydride an interesting variation occurred (Rao, 1985) leading to the cyclic ketal (**8.150**) of the partially reduced diketone. Acid converted this ketal into the dihydrofuran (**8.151**) which was hydrogenated with *cis*-addition to give (±)-veraguensin (**8.152**; Ar = 3,4-dimethoxyphenyl).

(a) *The bicyclic lactone route*

These compounds (e.g. **8.153**, Scheme 8.38) have been mentioned as deriving from the oxidative coupling of cinnamates (pp. 303–4) and it will be seen that they are important precursors of furofurans. They also provide an entry into the chemistry of tetralins which is discussed in a following section (p. 354), but here one notes that the typical product (Ahmed, 1973a; Takei, 1973) of their acid-catalysed rearrangement is the apotetralin (**8.155**). This arises from the cleavage of one lactone ring to give the carbocation (**8.154A**) via an unsaturated monolactone. Written in the form **8.154B** (X = H) this is seen to be liable to cyclisation by a reaction of Friedel–Crafts type. However the outcome of this reaction may be changed by the insertion of a bulky bromo- or iodo-substituent (Ahmed *et al.*, 1973b, 1976; Stevenson and Williams, 1977) when carbocyclisation is prevented by steric hindrance. For example, in **8.154B** (X = Br), approach is directly obstructed by bromine or in the rotamer by an *ortho*-methoxyl group. In the latter situation there is additional destabilisation by interaction between a 2′-bromosubstituent on a potential pendent ring C (Ayres and Lim, 1972) and as a result the preferred fate of the cation is capture of a water molecule to give **8.156** followed by its cyclisation to give the all *trans*-furan (**8.157**). Typical conditions for the cyclisation to give tetrahydrofurans include treatment of the halo-*bis*-lactone with hydrochloric acid in aqueous dioxan or in methanol. In the latter system no

Scheme 8.38 Acid - catalysed rearrangement of dilactones

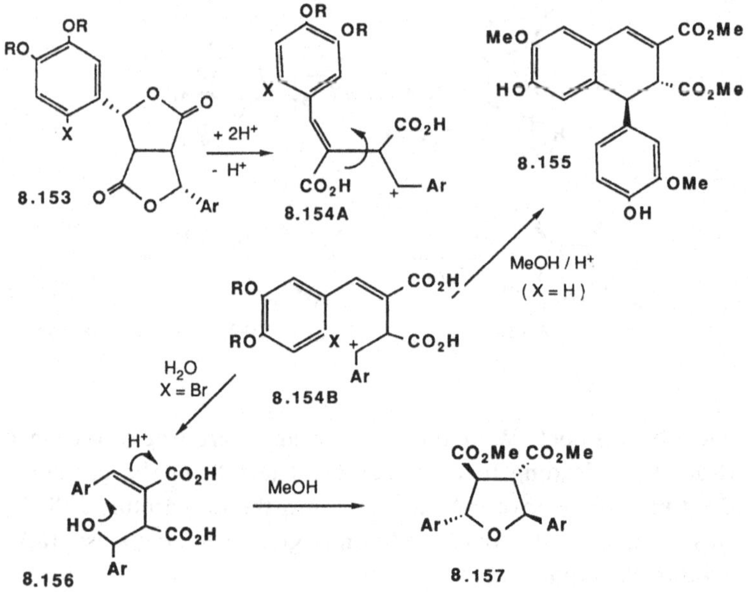

water was deliberately added although the stoichiometry demands it and it must have been introduced by the streaming in of HCl gas and also by its reaction with the solvent alcohol (Ayres and Mundy, 1969):

$$\mathrm{MeOH + HCl \rightarrow MeCl + H_2O}$$

Under these acidic conditions equilibration of the product to this least-congested form (**8.157**) is to be expected. When the alternative method of direct synthesis of a tetrahydrofuran by oxidative coupling of a halocinnamate is adopted this equilibration is not complete. This is shown (Ahmed *et al.*, 1976) by the synthesis of (±)-veraguensin (**8.152**) with the *cis-trans-trans*-configuration. When halosubstituents are used as directing groups they may be removed either by hydrogenolysis (Ahmed *et al.*, 1976) or with LAH (Stevenson and Williams, 1977).

(iii) *7,9′-Epoxylignans*

This is a small subgroup of lower symmetry which has attracted little purely synthetic interest. Their place as possible biosynthetic intermediates between the simple dibenzylbutanes and the bicyclic furofurans is indicated by the co-occurrence of *O*-methyl-lariciresinol (**2.123**) with pinoresinol (**2.217**) and *O*-methyl-secoisolariciresinol (**2.035**). They have

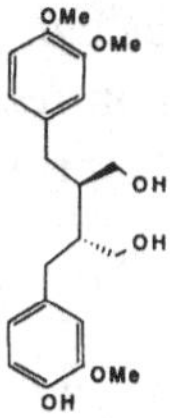

O-methyl-secoisolariciresinol 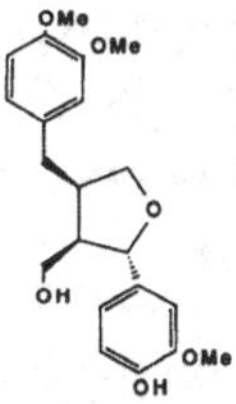　O-methyl-lariciresinol 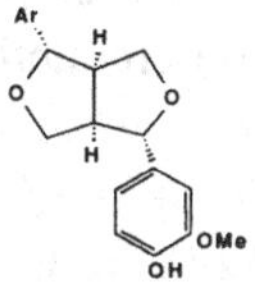　pinoresinol

frequently been obtained *in vitro* by the hemi-hydrogenolysis of members of the larger furofuran group of which pinoresinol is typical (pp. 213–14).

Furofurans (7,9′:7′9-diepoxylignans)

Various biomimetic experiments have been mentioned (p. 283) where cinnamyl alcohols are oxidatively coupled by the action of reagents such as iron(III) chloride and ferricyanide. In principle the intramolecular addition of terminal methylol groups to intermediate quinone methides can give rise to furofurans (cf. **7.84**), but the variety of pathways including oligomerisation through both O- and C- alkylation imposes a limit on the application of these methods. Enzymic coupling of coniferyl alcohol and sinapyl alcohol gives good yields of pinoresinol (**2.217**) and syringaresinol (**2.229**) respectively (Freudenberg and Dietrich, 1953).

It is known that hydroxycinnamic acids undergo oxidative coupling to give *bis*-lactones (**8.4**, Scheme 8.1) in a reaction first investigated by Erdtman (1935) and later by Cartwright and Haworth (1944). One method of extending this sequence is complete reduction to a tetrol (**8.158A** = **8.158B**; Scheme 8.39), but this approach is ambiguous since etherification may proceed through the *bis*-primary hydroxyls to give (**8.159A**) or

Scheme 8.39 Cyclisation modes of acyclic tetrols

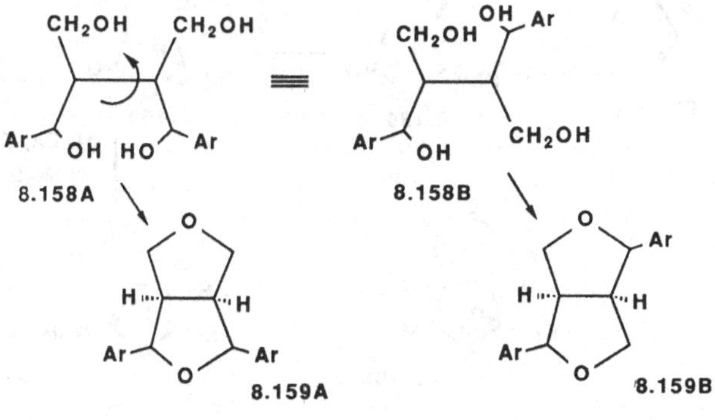

8.158A　8.158B　8.159A　8.159B

through the primary-benzyl hydroxyls to give a true lignan type (**8.159B**). For this reason emphasis will be placed more upon recent work which depends upon reduction of these lactones *without* their rings being opened.

(i) *Synthesis of monolactones*

Work on germination inhibitors (Cooper *et al.*, 1979) illustrates the problems inherent in the synthesis of unsymmetrical coupling products. Here the oxidation of a mixture of ferulic acid (**8.160**; Ar_1 = 4-hydroxy-3-methoxyphenyl; Scheme 8.40) and coniferyl alcohol (**8.161**, Ar_2 = 4-hydroxy-3-methoxyphenyl) gave the required cross-coupled lactone (**8.162**) together with the dilactone (**8.163**) from two molecules of acid and the lignan (**8.164**) from mutual coupling of the alcohol.

A more fruitful route to monolactones of type (**8.162**) has been reported by Till and Whiting (1984), who constructed the lactone ring by the radical addition of acetic acid to *trans*-cinnamyl acetate (Scheme 8.41). A second *different* aryl group was introduced by a base-catalysed reaction between the revealed primary alcohol (from **8.165**) and a 1-ethoxy-1-arylchloromethane; this gave the mixed acetal (**8.166**) which was con-

Scheme 8.40 Mixed oxidative coupling

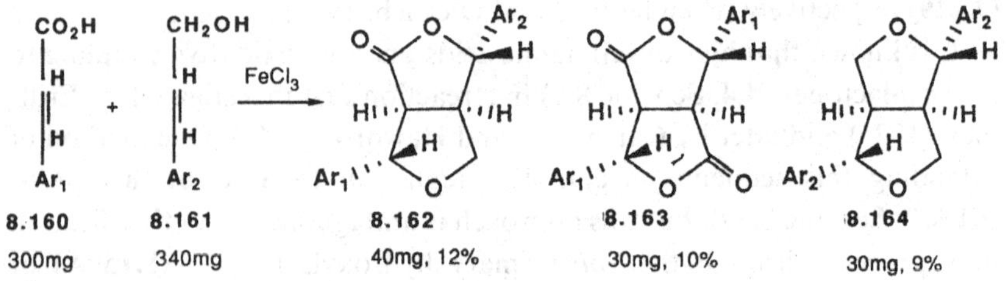

CO_2H / Ar_1 **8.160** 300mg + CH_2OH / Ar_2 **8.161** 340mg —$FeCl_3$→ **8.162** 40mg, 12%; **8.163** 30mg, 10%; **8.164** 30mg, 9%

Scheme 8.41 Intramolecular aldolisation of a monolactone

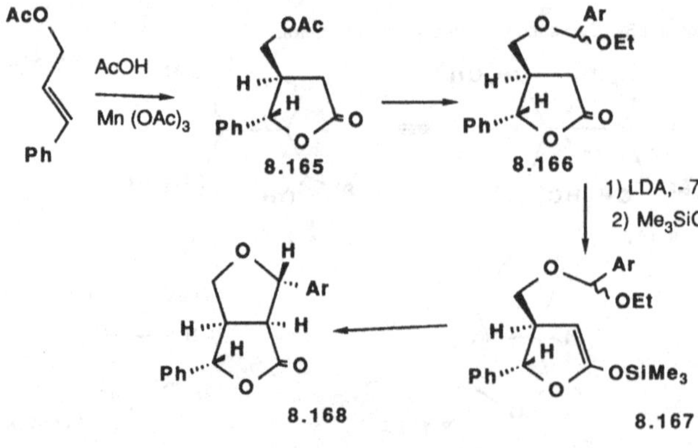

AcO … Ph —AcOH, Mn $(OAc)_3$→ **8.165** → **8.166** —1) LDA, - 70° 2) Me_3SiCl→ **8.167** → **8.168**

verted into its trimethylsilyl enol ether (**8.167**). The final step was an intramolecular aldolisation which gave a single stereoisomer (**8.168**, 40% yield) with both aryl groups in the less-compressed *trans*-configuration. When the cyclisation was induced with trimethylsilyl trifluoromethanesulphonate this product was mixed with the C7-epimer formed under kinetic control.

(ii) *Synthesis of dilactones*

In one route (Brownbridge and Chang, 1980a,b) the structural unit is assembled with ionic intermediates as distinct from the free radical additions previously discussed. It depends upon the conversion of succinic anhydride into the silyloxyfuran (**8.169**, Scheme 8.42), which condenses with two molecules of an aromatic aldehyde so as to eliminate both silyloxy groups and form the dilactone (**8.171**). The mechanism of this process has not been defined but it may depend upon the fact that, under the conditions employed, the silyloxyfuran is equivalent to a *bis*-ketene (**8.170**). The subsequent Lewis acid-catalysed addition of aromatic aldehydes to give intermediate β-lactones is well documented (Hasek, 1964). In this synthesis the ratio of isomers varies with the quantity of titanium tetrachloride used: a deficit (less than 2 equiv.) favours the *cis*:*trans*-compound

Scheme 8.42 Aldehyde condensation with dioxysilylfuran.

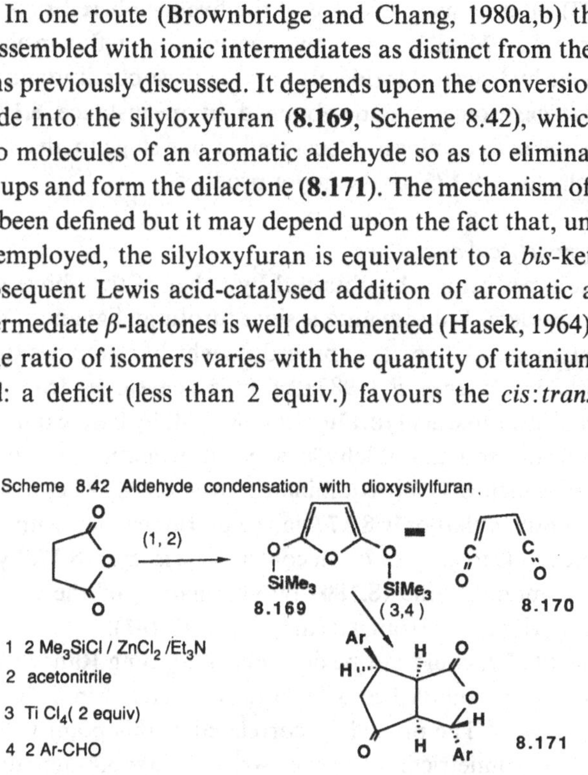

Scheme 8.43 Aldehyde condensation with succindiamides

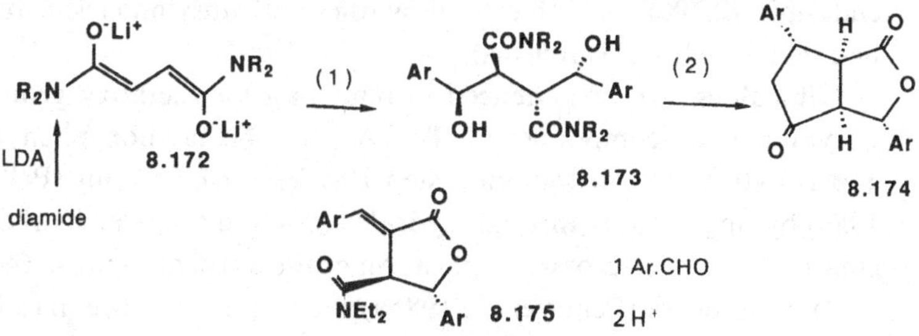

(**8.171**) rather than the fully equilibrated all *trans*-isomer. The former was obtained in 37% yield (when Ar = 3,4,5-trimethoxyphenyl) and the latter in 57% yield (Ar = 3,4-dimethoxyphenyl).

Another approach (Mahalanabis *et al.*, 1982a) depends upon the base-catalysed condensation of aromatic aldehydes with succinamides and this has been mentioned (pp. 311–12) as a route to dibenzylbutyrolactones. The dimetalated species (**8.172**, Scheme 8.43) reacts with two molecules of an aldehyde to give a mixture of isomers in which the *syn-anti-syn* compounds (**8.173**) predominate. Closure of the five-membered rings to give mixed isomers (of **8.174**, all *trans*- shown) occurred on reflux in acetic acid or in methanolic hydrogen chloride. By varying these conditions a proportion of the *cis-trans*-isomer may be obtained. In one instance (**8.173**; Ar = 3,4-dimethoxyphenyl) elimination occurred at one stage to give the unsaturated monolactone (**8.175**) as the major product.

(a) *Unsymmetrical dilactones*

These types cannot be obtained by either of the above methods and this is a considerable disadvantage since furofuran lignans of this kind are commonly found. A two-stage procedure which overcomes this problem was devised by Pelter *et al.* (1983a,b). It was developed from the half-ester of mercaptosuccinic acid (**8.176**, Scheme 8.44) by base-catalysed condensation with an aromatic aldehyde to yield a mixture of isomeric 4-aryllactones in which **8.177** predominated over **8.178**. These two types behaved differently in that only **8.177** gave a useful reaction with a second molecule of aldehyde to give **8.179**. In contrast the isomer (**8.178**) yielded a mixture of the monolactone (**8.180**) by elimination of the methylthio-group, together with the rearranged carboxylate (**8.181**).

The lactone (**8.179**) could not be desulphurised using Raney nickel but this reduction was accomplished after prior conversion into the diacetal (**8.182**, Scheme 8.44). The procedure correlated at this point with earlier work whereby a symmetrical dilactone (Ar_1 = Ar_2 = 3,4-dimethoxyphenyl) was obtained from oxidative coupling and reduced to a lactol of type **8.183** (R = H), whence ditosylation and hydride reduction afforded eudesmin (**2.200**). A viable total synthesis of unsymmetrical furofurans had therefore been established.

Difficulties were experienced in removing the methoxy groups in the unsymmetrical compound (**8.183**; Ar_1 = 3,4-dimethoxyphenyl, Ar_2 = piperonyl), although this final step had been carried out (Pelter *et al.*, 1982) by ring opening through hydride reduction followed by the less-specific acid-catalysed cross-etherification of the tetrol of type (**8.158**, Scheme 8.39). Later work (Pelter *et al.*, 1985a) showed that the acetal (**8.183**; R =

Scheme 8.44 Synthesis of unsymmetrically substituted precursors

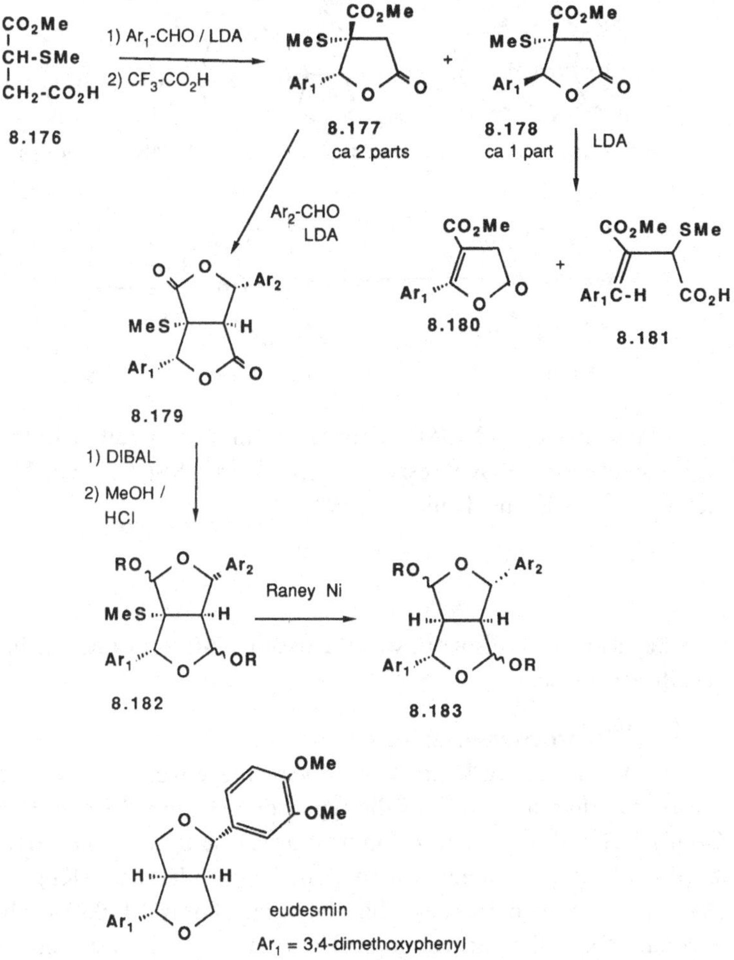

8.176 — 1) Ar_1-CHO / LDA; 2) CF_3-CO_2H → 8.177 (ca 2 parts) + 8.178 (ca 1 part)

8.177 — Ar_2-CHO, LDA → 8.179

8.178 — LDA → 8.180 + 8.181

8.179 — 1) DIBAL; 2) MeOH / HCl → 8.182

8.182 — Raney Ni → 8.183

eudesmin

Ar_1 = 3,4-dimethoxyphenyl

Me, Scheme 8.45) could be reduced *without ring opening* by the use of triethylsilane/boron trifluoride, although these acidic conditions led to a mixture of three diastereoisomers (**8.184–8.186**). This disadvantage was avoided by carrying out the acetal reduction with the bulky methylthio group in place on the precursor (**8.182**), which ensured a *trans*- or *endo*-configuration in the adjacent aryl substituent. Desulphurisation of the principal product (**8.187**) then yielded pluviatilol (**8.188**; Ar_1 = 3,4-dimethoxyphenyl, Ar_2 = piperonyl). This method of controlling stereochemistry is probably applicable more widely within the acid-labile furofuran group.

Scheme 8.45 Reduction with ring - retention and control of stereochemistry

8.183 8.184 35% 8.185 18% 8.186 17%

8.182 8.187 89% 8.188 75%

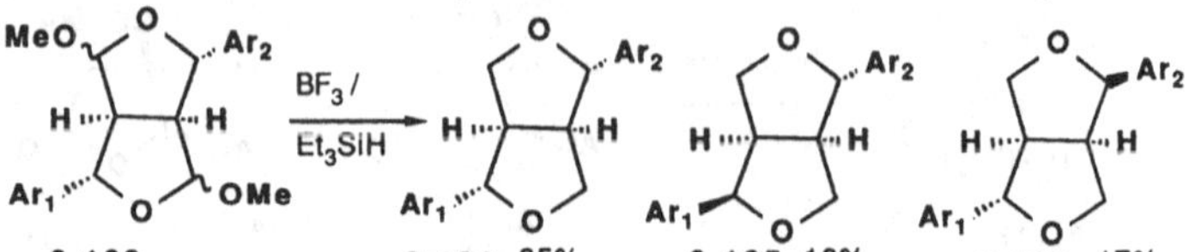

(+)-Phrymarolin (**2.286**) was obtained in 3% overall yield from a 15-step stereoselective synthesis starting from (*S*)-(+)-β-vinyl-γ-butyrolactone (Ishibashi and Taniguchi, 1986).

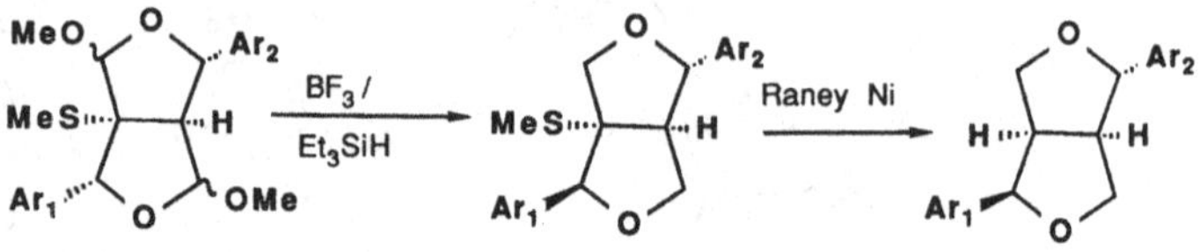

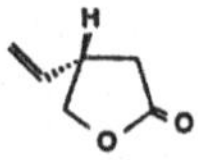

A key feature of this work was the use of a 7-fluoro- intermediate in the acetalisation step.

Dibenzocyclo-octadienes

From the synthetic point of view more work has been related to steganones than to lignans of the type isolated from *Schizandra* spp. (e.g. Gomisin H). This division of interest arises because of the early demonstration of the antileukaemic activity of steganin lactones (Kupchan *et al.*, 1973) in natural occurrence within the range of only 0.1–0.4%, which is inadequate for full pharmacological evaluation. Another factor was the success of Ikeya, Taguchi and colleagues in assigning structures to the *Schizandra* lignans by NOE difference spectroscopy (pp. 233–5) without recourse to total synthesis.

There are four fundamental approaches to these biaryls:

(a) Oxidative coupling of phenols (p. 303).

(S) - steganone

(R) - gomisin H

8.189

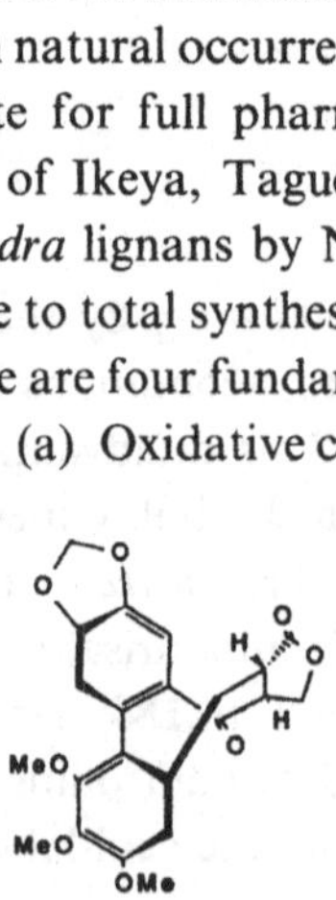

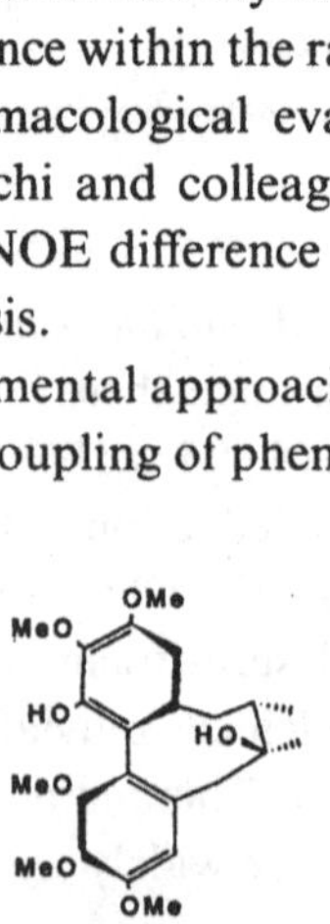

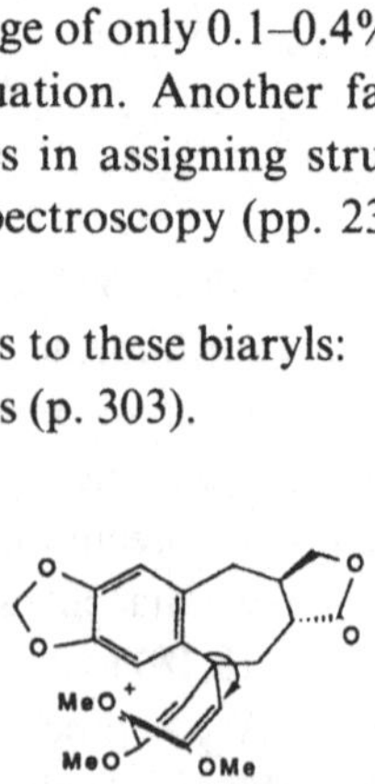

(b) Non-oxidative coupling of their ethers (p. 304).
(c) Classical and modified Ullmann syntheses.
(d) Development of phenanthrenes.

(i) *Oxidative coupling*

Techniques for the coupling of phenols have been mentioned (p. 303) as have the mechanism and methods for non-phenolic coupling (pp. 304–6). The latter procedure is now of first importance for the synthesis of this class. In the work of Biftu *et al.* (1978a, p. 304) deoxyschizandrin was synthesised from a diarylbutane of type **8.10** (Scheme 8.3) where the complete side chain was obtained by reduction of a tetrahydrofuran. A similar approach (Schneiders and Stevenson, 1980) led to wuweizisu C. A key paper is that of Tomioka *et al.* (1980, Scheme 8.16) whose asymmetric total synthesis of (*R*)-(+)-steganacin included the assembly of the complete side chain and established an oxidative link with butyrolactone lignans, besides correcting the wrongly assigned (Kupchan *et al.*, 1973) configuration of the stegane subgroup. The preferred formation of (*R*)-atropoisomers by non-oxidative coupling has been explained in terms of a spirodiene intermediate (**8.189** above; Damon *et al.*, 1976) in which the aryl migration indicated is energetically favoured.

Synthetic details for the unsymmetrical neoisostegane (**8.193**) are shown in Scheme 8.46 (Landais and Robin, 1986). The lactone precursor (**8.190**) was obtained by the Stobbe route and its anion was used to alkylate the benzyl bromide (**8.191**) and the product (**8.192**) cyclised using a ruthenium derivative to induce non-phenolic coupling. This procedure gave a superior yield to those obtained with vanadium oxychloride or vanadium oxyfluoride.

Scheme 8.46 Synthesis of neoisostegane

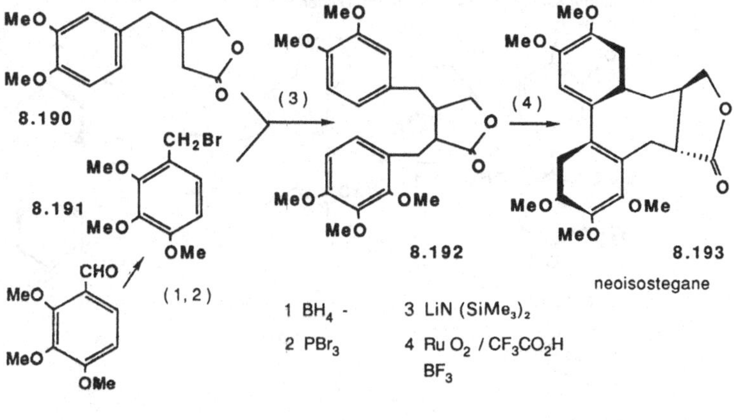

1 BH_4^-
2 PBr_3
3 $LiN(SiMe_3)_2$
4 RuO_2 / CF_3CO_2H BF_3

(ii) *Ullmann synthesis*

Methods for the preparation of relevant aryl iodides are given in papers by Brown and co-workers (1978, 1982) whereby a precursor lactone (**8.190**, but O—CH_2—O ether) was first iodinated and then coupled with 2-bromo-3,4,5-trimethoxybenzaldehyde using copper powder to give (**8.194**; 48% yield, Scheme 8.47). The eight-membered ring was then closed by a Claisen-type procedure (cf. Landais, Scheme 8.46) in almost quantitative yield to an equi-mixture of atropoisomeric alcohols (**8.195**). They were hydrogenolysed with difficulty, when the lactone adopted the *cis*-configuration, leading to syntheses of picrostegane (**8.196**) and isopicrostegane (**8.196**, *R*-form; structure as corrected by Robin *et al.*, 1984).

It was subsequently pointed out (Robin *et al.*, 1980) that there was an anomaly in that the structure originally assigned to (−)-steganone (*R*-form, 8α-H) differed from podophyllotoxin (8β′-H), yet both compounds bind competitively with tubulin protein. Consequently an asymmetric synthesis was carried out starting from the 3-(*R*)-lactone (**8.197**, Scheme 8.48) which was iodinated and treated by a similar procedure, to yield the aldehyde (**8.198**). The mixture of alcohols obtained from its cyclisation by base was oxidised to the enol lactone (**8.199**). This was in its turn decarboxylated, re-oxidised and esterified to yield the ketoester (**8.200**). Condensation with formaldehyde then gave a retrolactone which afforded (*R*)-(+)-isosteganone (**8.201**) on retrieval of the C7′-keto group. The action of heat on this product gave natural (*S*)-(−)-steganone, a result which complimented the synthesis of the *ent*-compound by Tomioka *et al.* (1980) and also the first synthesis of (−)-steganone by Larson and Raphael (1982). Later work by Robin and colleagues (1984), based upon the Claisen-type reaction (Scheme 8.47), led to the epimeric picrosteganols (**8.195**) whence hydrogenolysis yielded (*S*)-picrostegane (**8.196**). The latter was isomerised

Scheme 8.47 Ring closure of an Ullmann product

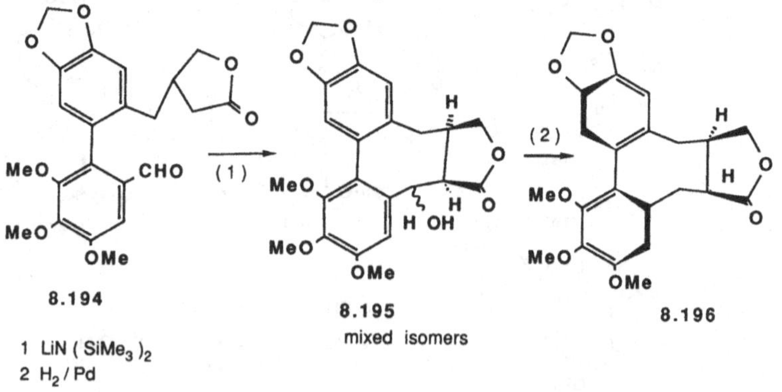

8.194 (1) 8.195 mixed isomers (2) 8.196

1 LiN ($SiMe_3$)$_2$
2 H_2 / Pd

Scheme 8.48 Asymmetric synthesis of (-) - steganone

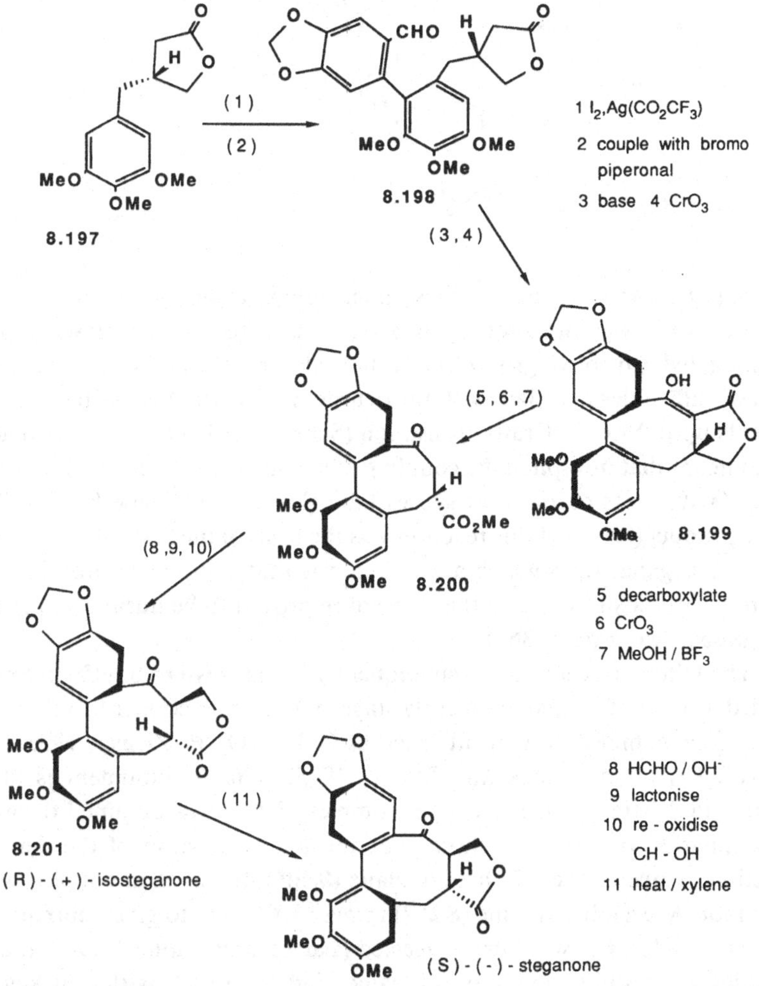

8.197 → (1) (2) → 8.198 → (3,4) → 8.199 → (5,6,7) → 8.200 → (8,9,10) → 8.201 (R) - (+) - isosteganone → (11) → (S) - (-) - steganone

1 I_2,$Ag(CO_2CF_3)$
2 couple with bromo piperonal
3 base 4 CrO_3

5 decarboxylate
6 CrO_3
7 MeOH / BF_3

8 HCHO / OH^-
9 lactonise
10 re - oxidise CH - OH
11 heat / xylene

on heating to (*R*)-isopicrostegane, which was further epimerised by base to give the *trans*-lactone, (*R*)-isostegane.

In contrast to the early-stage coupling (Scheme 8.48, **8.197–8.198**) difficulties were experienced (Ziegler and Schwartz, 1978a; Semmelhack and Ryono, 1975) with the classical Ullmann reaction when late-stage coupling was attempted. A potential precursor (**8.202**; X = I or Br) for steganone gave the monohalogenated iodide as the major product of reduction when treated with copper bronze in dimethylformamide at 160°C. At a lower temperature a symmetrical dimer was obtained by intermolecular coupling at point X. It was concluded that reduction was preferred to intramolecular coupling for these nucleophilic aromatic systems, where one position was hindered by two *ortho*-substituents; the presence of a conjugated carbonyl group also impeded the intramolecular reaction. These difficulties correlate with the determining steric effect of halosubstituents upon Friedel–Crafts cyclisation (Scheme 8.38 and p. 337); here it is significant that non-phenolic coupling of a comparable halogen free structure (**8.203**) was carried out successfully (Kende and Liebeskind, 1976). The product (**8.204**) of this reaction was the basis of the first total synthesis of (±)-steganacin, in which a key step was that of benzylic bromination with N-bromosuccinimide: this procedure proved to be unsatisfactory for large-scale work (cf. p. 365).

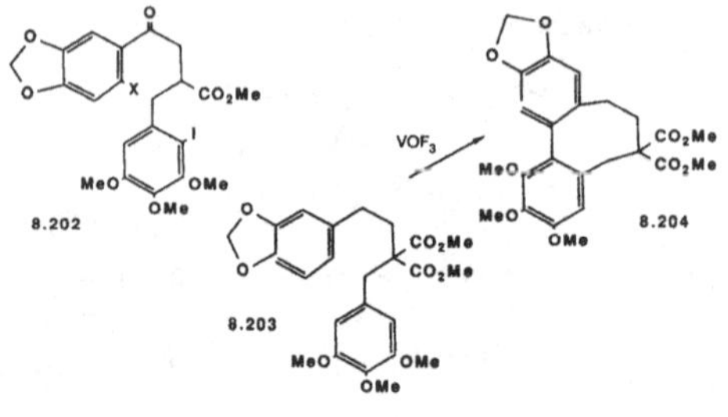

The Ullmann reaction has subsequently been shown to be effective provided: (1) that it is used at an early stage; and (2) that copper bronze is replaced by complexing reagents such as $CuI.(EtO)_3P$ (Ziegler, 1978b) or $Ni(Ph_3P)_4$ (Semmelhack and Ryono, 1975). This development is illustrated in Scheme 8.49. The copper complex (**8.206**) was obtained through bromination of piperonal, homologation and protection of the derived methyl ketone as the 1,3-thiadioxolane (**8.205**). It was then coupled with the iodo-*N*-cyclohexylimine (**8.207**) in over 80% yield to give a mixture of isomers of **8.208**. It was later suggested (Larson and Raphael, 1982) that a similar successful coupling of this same hindered iodide with a nickel(O) catalyst was assisted by chelation of the imino-nitrogen to the arylnickel intermediate.

The Ziegler synthesis (Scheme 8.49) was continued by hydrolysis of the imine and a Knoevenagel condensation of the revealed aldehyde to afford the diester (**8.209**, 87%). Subsequent removal of the thiadioxolane group using Fetizon and Jurion's method (1972), hydrogenation and bromination gave the bromoketodiester (**8.210**). The crucial intramolecular alkylation was then effected in *t*-butanol, without addition of base, to yield the cyclic diester (**8.211**). The diethyl analogue of this compound had been obtained previously by Kende (1976) from the oxidatively coupled precursor (**8.204**). Hydrolysis and decarboxylation of (**8.211**) followed by treat-

Scheme 8.49 Synthesis of (+-) - steganacin (Ziegler, 1978)

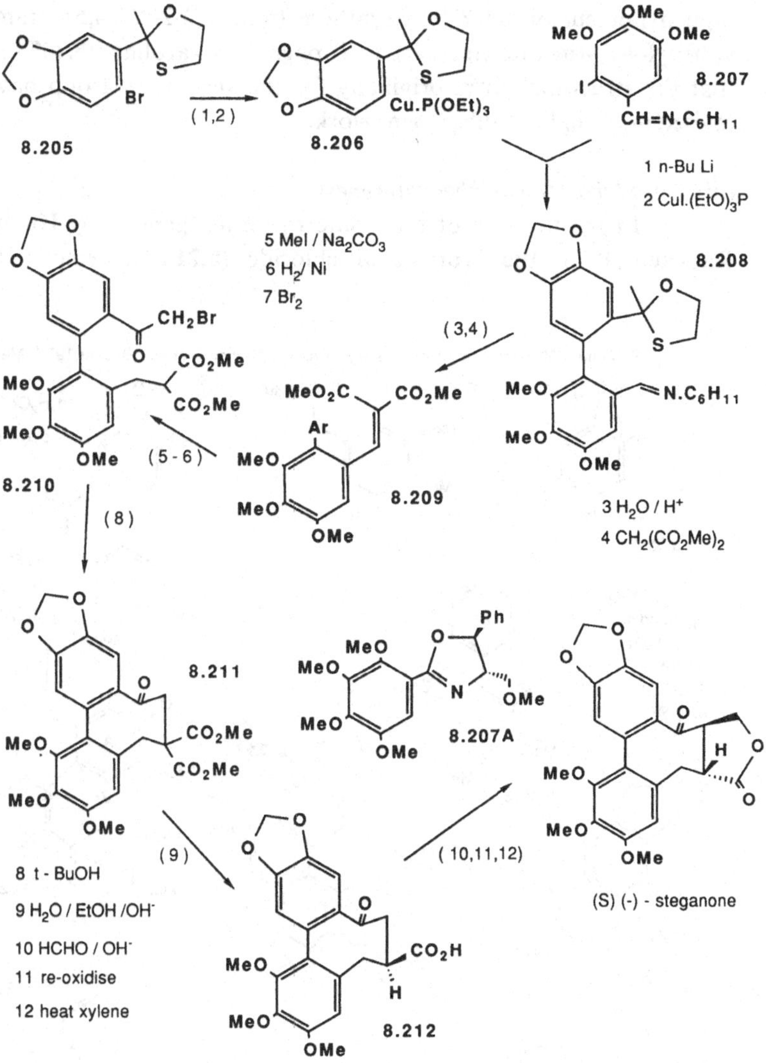

ment with diazomethane led to a mixture of atropoisomeric acids (**8.212**), which were separable as their methyl esters. Reaction of the mixed acids with formaldehyde/base (cf. Kende) and re-oxidation at C-7′ afforded steganone with isosteganone. The more stable steganone was obtained by heating this mixture in xylene, when reduction with sodium borohydride yielded a separable mixture of episteganol with steganol and the latter was acetylated to give (±)-steganacin (**2.465**). Meyers *et al.* (1987) have devised an asymmetric synthesis of (−)-steganone by resolving a chiral biphenyl precursor of **8.210** by synthesis from (+)-2-(2,3,4,5-tetramethoxyphenyl-4(*S*)-(methoxymethyl)-5-(*S*)-phenyl-2-oxazoline (**8.207A**). Note that the structures given originally for the steganone group have been revised in the light of subsequent work.

(iii) *Elaboration of phenanthrenes*

In the first use of this route to (±)-steganone by Hughes and Raphael (1976) the bromo-acid chloride (**8.213**, Scheme 8.50) was

Scheme 8.50 Synthesis of (+-) - steganone via phenanthrenes (Becker,1977; Mervic,1981)

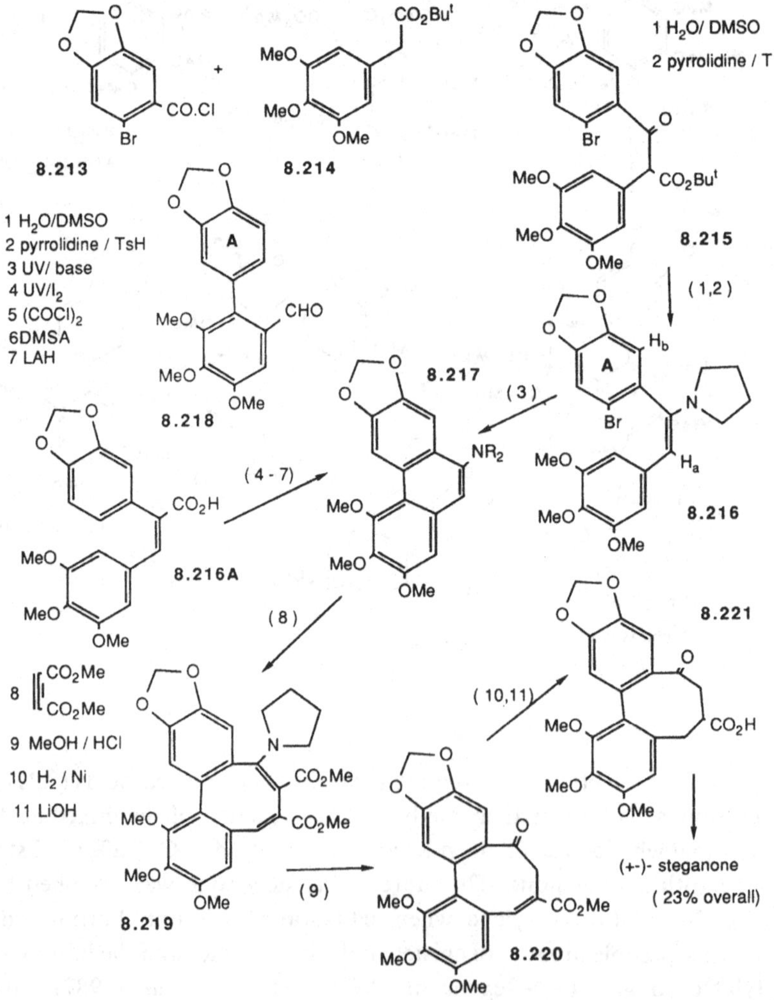
CO.Cl
Br
8.213
MeO
MeO
OMe
CO_2Bu^t
8.214
1 H_2O/ DMSO
2 pyrrolidine / T
8.215
1 H_2O/DMSO
2 pyrrolidine / TsH
3 UV/ base
4 UV/I_2
5 $(COCl)_2$
6DMSA
7 LAH
A
CHO
8.218
8.217
NR_2
(3)
(1,2)
H_b
H_a
8.216
CO_2H
(4 - 7)
8.216A
(8)
8.221
8 CO_2Me CO_2Me
9 MeOH / HCl
10 H_2 / Ni
11 LiOH
CO_2Me
8.219
(9)
(10,11)
8.220
(+-)- steganone
(23% overall)

obtained from piperonal and condensed with the lithio-derivative of the ester (**8.214**), which was best made by a Mannich reaction of 2,6-dimethoxyphenol. The product (**8.215**) was hydrolysed with decarboxylation to afford a ketone which was converted into the enamine (**8.216**): this was a key nucleophilic function which trapped the keto-group and also assisted in the subsequent double ring expansion of the 9-pyrrolidinophenanthrene (**8.217**; R, R = pyrrolidinyl). The latter was probably formed by free radical coupling followed by base-catalysed elimination of hydrogen bromide.

In later large-scale work (Larsen and Raphael, 1982) this biaryl synthesis was supplanted by a modified Ullmann procedure, whereby methylenedioxyphenylzinc chloride coupled with the iodocyclohexylimine (**8.207**, Scheme 8.49) to give the aldehyde (**8.218**). This was then cyclised via its *N*-phenacylpyrrolidine homologue to the enamine (**8.217**).

The eight-membered ring compound (**8.219**, 91% yield) was obtained on double expansion of the donor ring in **8.217** (R, R = $(CH_2)_4$) by heating with dimethyl acetylenedicarboxylate. An analogous precursor (**8.217**; R = Me) was obtained by Krow *et al.* (1978) in a parallel approach shown in Scheme 8.50A. Here the 2,3-diarylacrylic acid (**8.216A**) was coupled by UV irradiation to afford the phenanthrene carboxylic acid (**8.216B**, 81% yield). The derived acid chloride was converted into the isocyanate (**8.216C**) and thence by reduction and *N*-methylation into the *N*-dimethyl analogue of **8.217**.

Acidic hydrolysis of the amino-octatetraene (**8.219**) with partial decarboxylation of the intermediate ketodiester gave the unsaturated ketoester (**8.220**), which was hydrogenated to the saturated ketoester and then

Scheme 8.50A Phenanthrene synthesis

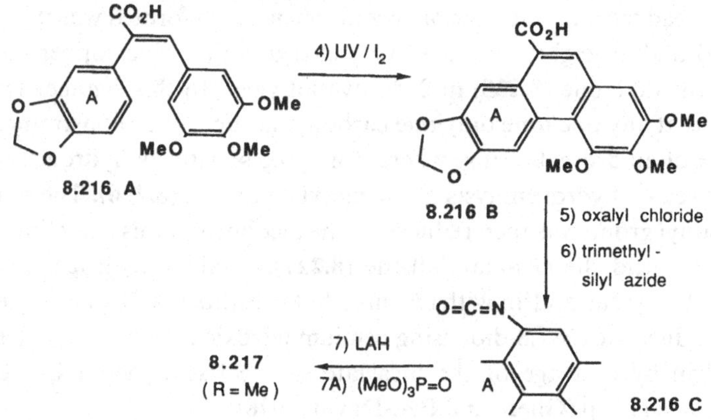

8.216 A → 4) UV / I_2 → 8.216 B

8.216 B → 5) oxalyl chloride; 6) trimethyl-silyl azide → 8.216 C

8.216 C → 7) LAH; 7A) $(MeO)_3P{=}O$ → 8.217 (R = Me)

hydrolysed to the ketoacid (**8.221**). Both epimers of this compound and also their methyl esters were characterised and had properties in common with those obtained by Ziegler (e.g. **8.212**, Scheme 8.49). Raphael found that the acid (**8.221**) of m.p. 146°C yielded (±)-isosteganone after formylation (Scheme 8.49), but that this isomerised to (±)-steganone during measurement of melting point (234°C) or in refluxing xylene.

A point of significance for the characterisation of these ketones is that the carbonyl group in steganone is more fully conjugated with the adjacent aromatic ring than is this group in isostegane. This is shown by its IR absorption at 1665 cm^{-1}, which in isosteganone occurs at the higher frequency of 1710 cm^{-1}. The essentially complete conversion of isosteganone into steganone on heating may be the result of the trend to conjugation of the carbonyl group. In isostegane this interaction is excluded and this compound becomes a major component of the thermally induced equilibrium with stegane (Tomioka *et al.*, 1979).

An intellectually satisfying conclusion to the work was the late-stage resolution of the acid (**8.221**) via its amide with (−)-(*S*)-2-amino-3-phenylpropan-1-ol (Seki *et al.*, 1965) and a synthesis of (−)-(*S*)-steganone, which complemented the approach (Robin *et al.*, 1980) from an asymmetric starting material (**8.197**, Scheme 8.48).

A valuable alternative synthesis of stilbene precursors employs the Wittig reaction (Ghera *et al.*, 1977, Scheme 8.51). Thus combination of trimethoxybenzaldehyde with the phosphonium salt of trimethoxybenzyl bromide gave a mixture of *cis*- and *trans*-styrenes (**8.222**), which cyclised on oxidative UV irradiation (Wood and Mallory, 1964) to the phenanthrene (**8.223**). This was converted into the phenanthraquinone (**8.224**) by the action of osmium tetroxide and oxidation of this product diol with pyridine/sulphur trioxide/DMSO. In four further steps the quinone ring was opened through reaction with ethylmagnesium bromide and cleavage using lead tetra-acetate; the *bis*-propiophenone so formed was then brominated and an eight-membered ring closed with a zinc/copper couple to give the diketone (**8.225**) in 28% overall yield. In this symmetrical compound at any one time only one carbonyl group can be constrained in the plane of an aromatic ring, where it may be selectively hydrogenated and removed by hydrogenolysis of the mesylate to give (**8.226**). The remaining carbonyl group was then reduced to the alcohol and this was eliminated in acid to yield the (*R*)-biarylalkene (**8.227**); catalytic hydrogenation then gave deoxyschizandrin with *cis*-dimethyl substituents. It was also possible to produce the *cis*-7,8-diol using osmium tetroxide and to obtain (±)-schizandrin by cleavage of the mesylate of the less-hindered C-7 benzylic hydroxyl group (Ghera and Ben-David, 1978).

Scheme 8.51 (+-)- Schizandrin and (+-)- steganone from stilbene precursors

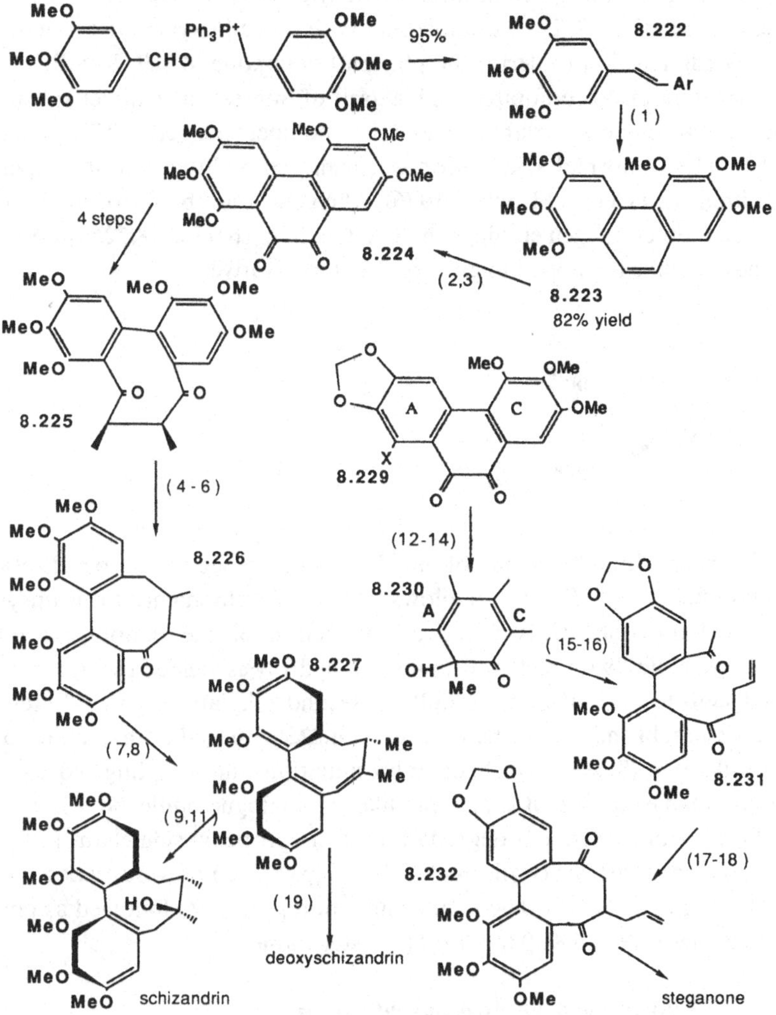

legend for scheme 8.51

1 UV/I_2	6 H_2 / Pd / OH^-	11 $NaBH_4$	16 Pb $(OAc)_4$
2 OsO_4	7 LAH	12 Ketalise	17 Br_2
3 Py/SO_3	8 $KHSO_4$	13 MeLi	18 Zinc
4 H/ Pd / C	9 OsO_4	14 H^+	19 H_2 / Pd
5 $MeSO_2Cl$ / Py	10 $MeSO_2Cl$	15 Grignard	

A variation of the phenanthraquinone synthesis (Mervic *et al.*, 1981) was the combination of 2-bromopiperonal with trimethoxybenzyl bromide in the Wittig reaction to derive the phenanthraquinone (**8.229**, X = Br). The preparation of 2-bromopiperonal is itself of interest since direct substitution by bromine invariably gives 6-bromopiperonal (cf. **8.213**, Scheme 8.50). The 'abnormal' orientation was achieved by lithiation of a piperonal imine (Ziegler and Fowler, 1976), when chelation by nitrogen and the adjacent ether oxygen stabilise the C-2 paired carbanion (**8.228**) and bromine or iodine is trapped to give the 2-haloderivative.

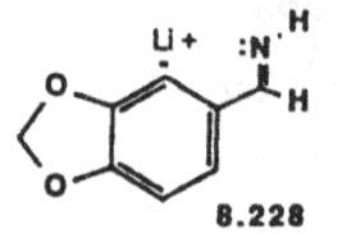

8.228

In this position bromine not only blocks one mode of photocyclisation to give **8.229** (X = Br) it also allows selective development of this unsymmetrical quinone. Thus selective ketalisation of the remote carbonyl group, reaction with methyl lithium and hydrolysis yielded the debrominated keto-tertiary alcohol (**8.230**). A second alkylation with 3-butenylmagnesium bromide and cleavage as for **8.225** with lead tetra-acetate gave the diketone (**8.231**), which on α-bromination and coupling led to the cyclo-octadienedione (**8.232**). This like its analogue could be selectively reduced with subsequent degradation of the propenyl side chain to form the epimer of the carboxylic acid (**8.221**, m.p. 172°C) previously described by Becker *et al.* (1977). The final formylation procedure followed as given in Schemes 8.49 and 8.50 to afford (±)-steganone.

Dihydro- and tetrahydronaphthalenes

In many instances the precursors needed for the synthesis of this important group have been mentioned above since they also feature in routes to lignans of other classes. It is also true that paths to tetralins have whole sequences in common with these as is shown by the following summary of methods from the foregoing text.

(i) *Summary of above procedures relevant to tetralin synthesis*

(a) *Oxidative coupling*

This approach is shown in Scheme 8.1 for the preparation of a dilactone and of its acid-catalysed rearrangement to give thomasidioic acid. Other examples are the route to the α-apodiester (**8.155**, Scheme 8.38) and the direct coupling to the tetralone (**8.73**, p. 321).

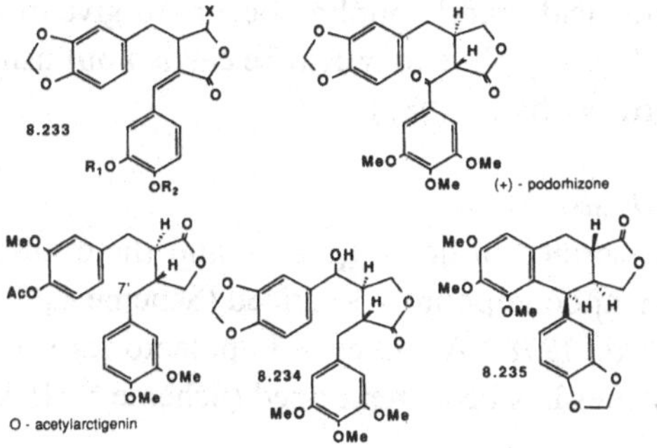

(b) *Development of the Stobbe and carbanion condensation products*

Here unsaturated side chains may be obtained as for gadain (**8.233**; X = H, R_1, R_2 = —CH_2—; Scheme 8.7, p. 308) and this may be followed by acid-catalysed cyclisation of Friedel–Crafts type. Alternatively a scheme typified by the stereoselective synthesis (Tomioka *et al.*, 1982) of (+)-podorhizone via Stobbe condensation and benzoylation of the monolactone (Scheme 8.14) can be employed, which leads to C-7′-alcohols suitable for cyclisation.

This general approach is also illustrated (Scheme 8.19; Nishibe *et al.*, 1980) by the C-7′-acetoxylation of acetylarctigenin, which may also afford a useful concurrent yield of the directly cyclised material (e.g. a β-conidendrin from isoarctigenin). Other examples of this important route via C-7-hydroxyl compounds will be discussed in the sequel.

(c) *Oxidative cyclisation*

A series of intermediates common to the synthesis of both steganes and tetralins has been described (Scheme 8.16; Tomioka *et al.*, 1980). It was shown (Ziegler and Schwartz, 1978a) that biaryl coupling did not occur on treatment with vanadium oxyfluoride or manganese(III)-*tris*-(acetylacetonate) if the C-7 position was oxidisable. In these circumstances tetralins were the preferred products; thus the alcohol (**8.234**) cyclised through attack of the more nucleophilic ring to give (**8.235**). Oxidative cyclisation also gives tetralins when the C-7 position is not functionalised provided that a free phenolic group is present, as in the butyrolactone (**8.27**, Scheme 8.11).

The solvent plays a critical role in these reactions. Thus 1,4-diarylbutanes undergo benzylic acetoxylation on treatment with DDQ in acetic acid, but in trifluoroacetic acid biaryl coupling occurs to give dibenzobicyclo-octadiene lignans. This coupling may also be carried out using manganese(III) acetate (Chattopadhyay, 1987).

(d) *Diels–Alder synthesis*

This has been described in detail (p. 325) and the cyclisation of cinnamyl propiolates to γ-apocompounds described (Scheme 8.25; Klemm *et al.*, 1976; Gonzalez *et al.*, 1978). A route to β-apolactones via the UV initiated cyclisation of dienes has been mentioned (Scheme 8.31; Momose *et al.*, 1977, 1978).

An excellent analysis of the factors which govern the stereochemistry of the hydrogenation products of apocompounds was published by Schrecker and Hartwell (1952, 1958) as part and parcel of their proof of the geometry of podophyllotoxin (**2.357**). They showed that invoking the principle of *cis*- and *syn*-addition of hydrogen (Linstead, 1942) led to a self-consistent structural correlation between the three apocompounds (**8.236**–**8.238**, Scheme 8.52) and their hydrogenation products (**8.239**, **8.240**). Deoxypodophyllotoxin (DPT) and deoxypicropodophyllin (DPP) were available by hydrogenolysis of the parent benzyl alcohols. It was also recorded that:

(a) α-apopicropodophyllin (**8.236**; C-7′, C-8′ then uncertain), with the point of attack twice removed from the pendent ring, was not subject to stereocontrol and both DPP and isodeoxypodophyllotoxin (IDPT) were formed. These two compounds must have retained their common C-7′, C-8′ configuration.

Scheme 8.52 Hydrogenation products of apocompounds

8.236 α-apopicropodophyllin

8.237 β-apopicropodophyllin

8.238 γ-apopicropodophyllin

8.239 isodeoxypodophyllotoxin

8.240 isodeoxypicropodophyllin

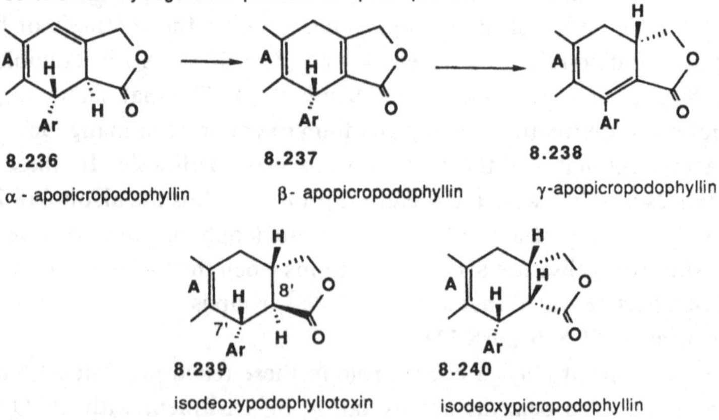

(b) Hydrogenation of both *β*- and *γ*-apopicropodophyllin gave the fourth possible diastereoisomer isodeoxypicropodophyllin (IDPP, **8.240**) as the sole product. This reaction must therefore be stereocontrolled and the common product can only be *cis*-7′,8′; *cis*-8,8′. It follows that DPP and IDPT being derived from α-apoPP have the same configuration at C-7′,8′; this must be *trans*- because only two *cis*-7′,8′-isomers are possible and one has been defined as IDPP (**8.240**).

(c) By a similar argument it follows that DPT cannot be *trans*-7′,8′ and it must therefore be *cis*-7′,8′ yet differ from IDPP (all *cis*-). Hence DPT is *cis*-7′8′; *trans*-8,8′.

(d) As DPP is obtained by base-catalysed epimerisation of DPT at C-8′ its configuration is *trans*-7′,8′; *cis*-8,8′, whence IDPT can only be allocated the all *trans*-configuration. These assignments have all been subsequently confirmed by NMR spectroscopy (pp. 223–8) and the evidence of absolute configuration is discussed in Chapter 6 (pp. 228–30).

The hydrogenation of fully aromatic lignans gives mixtures. It has also been established (Schrecker and Hartwell, 1953) that hydrogen transfer between rings A and B occurs when palladium/charcoal is used as a catalyst.

(ii) *Modification of other lignans by Friedel–Crafts cyclisation*

In addition to the acid-catalysed carbocyclisation of lactones (e.g. Scheme 8.38) the epimerisation of tetrahydrofurans and furofurans via a carbocation has been discussed (p. 158), when it was shown that the trend is towards the least-compressed diastereoisomer. The rate of rearrangement of galgravin (Scheme 8.53) is six times that of galbelgin (Birch *et al.*, 1958), which leads one to the conclusion that ring-opening with strain relief occurs in the transition to give a carbocation of type **8.241**. The furan ring may be re-established through intramolecular attack by the hydroxyl group but the irreversible Friedel–Crafts cyclisation (via **8.241**) will determine the ultimate product. The stereochemistry of these tetralins is of first importance since the method is widely used for their synthesis and physiological activity is bound up with the geometry: witness the structure–activity relation in the podophyllotoxins.

In 7,7′-epoxylignans such as galgravin and galbelgin the revealed benzylic hydroxyl group is labile in the acid conditions and is eliminated, so that the end product is the α-apocompound (**8.242**, Scheme 8.53). The *trans*-7′,8′-relationship was established by hydrogenation of this product

Scheme 8.53 Acid - catalysed rearrangement and cyclisation reactions

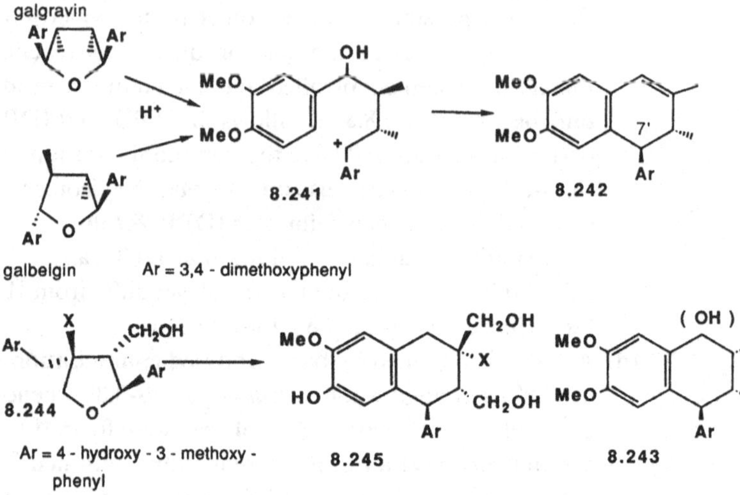

galgravin

galbelgin Ar = 3,4 - dimethoxyphenyl

8.241 8.242

8.244 Ar = 4 - hydroxy - 3 - methoxy - phenyl

8.245 8.243

by the *syn-cis*-path to give galbulin (**2.295**) with the known all *trans*-configuration.

A similar cyclisation was observed (Perry *et al.*, 1972) in the course of the hydrogenation of galgravin in an acidic medium; the product isolated then was isogalbulin (**8.243**) which retained the original *cis*-configuration of the methyl groups. It is evident here that the benzylic hydroxyl group revealed by cyclisation was removed by hydrogenolysis rather than by acid-catalysed elimination.

Another instance of the formation of a tetralin from a tetrahydrofuran is cyclisation of the 7,9-epoxylignan lariciresinol (**8.244**; X = H), where the *bis*-primary diol (**8.245**; X = H) obtained is stable under the conditions. In the congener olivil (**8.244**; X = OH) the reaction takes the same course to give cyclo-olivil (**8.245**; X = OH), but slow elimination of the tertiary hydroxyl group follows (Ayres *et al.*, 1965).

There is a possibility that elimination could occur in the cation (**8.241**) to yield a benzylidene derivative (cf. **8.233**) prior to cyclisation. This would lead to ambiguity in the assignment of geometry at C-8,8′; however the isolation of isogalbulin (**8.243**) with the less stable *cis*-configuration makes this unlikely. It should be noted that in all the above examples the pendent aryl substituent takes up the more stable *trans*-relation. The products of the cyclisation of podorhizol, the 7*S*-alcohol from podorhizone, were analysed in detail by Kuhn and von Wartburg (1967) and although some of the more compressed DPT was formed the major product was the *trans*-isomer IDPT (Scheme 8.52).

A variation is possible by cyclisation of a 7-ketone, such as podorhizone, where as a result of ring closure the keto-oxygen is trapped as a tertiary hydroxyl, which is eliminated to yield a cross-conjugated C-7′,8′ double bond (Munakata *et al.*, 1967). Hydrogenation of this end product will then lead to the all *cis*-compound; a result which should be compared with formation of IDPP (**8.240**, Scheme 8.52).

A convenient route to precursor ketolactones is shown in Scheme 8.54 where an epoxide (**8.246**) undergoes a base-catalysed reaction with a ketoester to give the *secondary* lactone (**8.247**). Protonation of the keto-oxygen now leads to a seven-ring carbocycle. This retains the hydroxyl group, although it is eliminated under more forcing acid conditions with ring contraction and formation of a primary lactone (**8.249**; Ayres and Mundy, 1969).

Recent work based on 7′-hydroxycompounds has shown that yields of cyclisation products of the order of 90–95% can be obtained on treatment with trifluoroacetic acid/methylene chloride. In one synthesis by Mpango and Snieckus (1980) addition of the thioketal anion (**8.250**, Scheme 8.55) to *N*,*N*-dimethylcrotonamide followed by veratraldehyde gave a separable mixture of hydroxyamides. The major *threo*-component (**8.251**) was cyclised and the tetralone (**8.252**) obtained using mercury(II) oxide; the required all *trans*-configuration was then established through enolisation of the *cis*-8,8′-ketone. The sequence of steps required to obtain galcatin from the ketone (**8.252**) are shown in the scheme – note that during reduction of the amide with LTBH some *O*-demethylation occurred. Thioketals also featured in the synthesis of analogues of podophyllotoxone (Gonzalez *et al.*, 1986) and it was shown that with this protecting group in place, the proportion of this *cis*-cyclised product was 40% of the mixture with picropodophyllone.

Scheme 8.54 Synthesis and cyclisation of keto - lactones

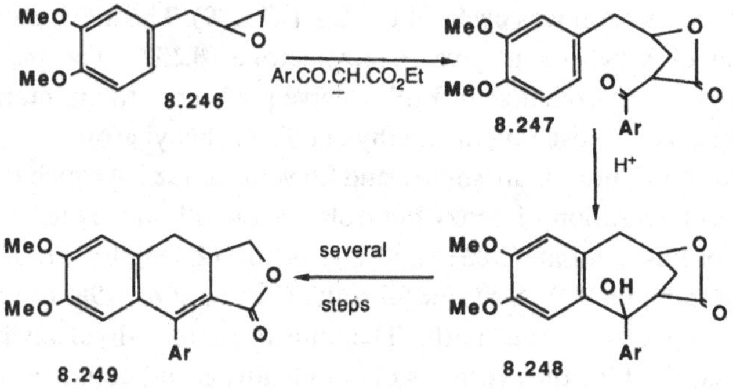

Scheme 8.55 Synthesis of galcatin

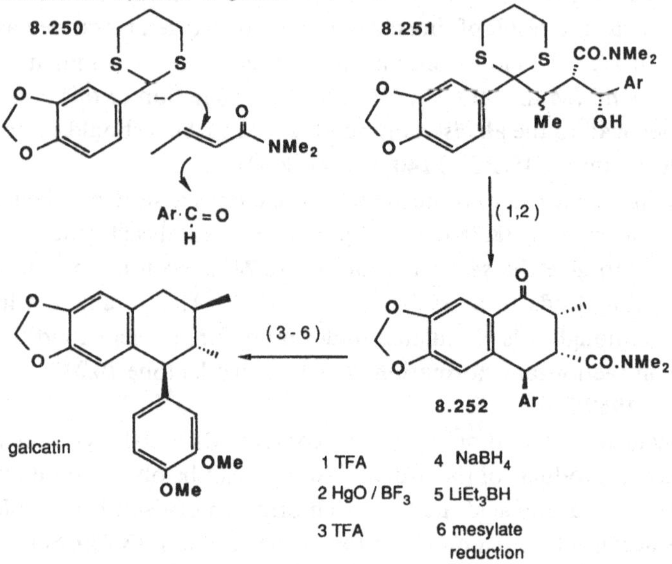

1 TFA
2 HgO / BF_3
3 TFA
4 $NaBH_4$
5 $LiEt_3BH$
6 mesylate reduction

The 9,9′-bismethoxylignan nirtetralin (**2.328A**) was recently synthesised by the cyclisation of a 7-hydroxydibenzylbutane (Satyanarayana *et al.*, 1988). It was noted by Loriot *et al.* (1985) that the more labile protecting *O*-benzyl groups of syringyl residues were liable to be eliminated in trifluoroacetic acid to reveal the phenolic —OH group.

The thioketal route was also used (Pelter *et al.*, 1985b) with capture of the carbanion from a dimethoxyphenyl analogue (of **8.250**) by butenolide/piperonal. Subsequent cyclisation with TFA via the sulphur-stabilised cation yielded retrochinensin (**8.253**), but the alternative cyclisation mode to the retrolactone (**8.254**) was followed if sulphur was first removed by hydrogenolysis. In a similar sequence the monothioanion (**8.255**, Scheme 8.56) gave the hydroxythioether (**8.256**; Ar = 3,4,5-trimethoxyphenyl-) which after reduction was cyclised to IDPT (**8.239**). The alternative direct cyclisation of **8.256** again gave a retrolactone (**8.257**). The mechanism appears to involve exchange of PhS- to give preference to an intermediate cation which is stabilised by the methylenedioxyphenyl group.

Related procedures (Ganeshpure and Stevenson, 1981b) which depended upon the condensation of benzylbutyrolactones with aldehydes led to the synthesis of kusunokinin (**2.061**) and isogalcatin (**2.298**), lintetralin (**2.326**) and phyltetralin (**2.333**). A successful route (Loriot *et al.*, 1983) to attenuol (**2.290**) also followed this path. The intermediate 7′-hydroxydibenzyllactones required for the synthesis of conidendrins and conidendrals were

Scheme 8.56 Alternative cyclisation paths

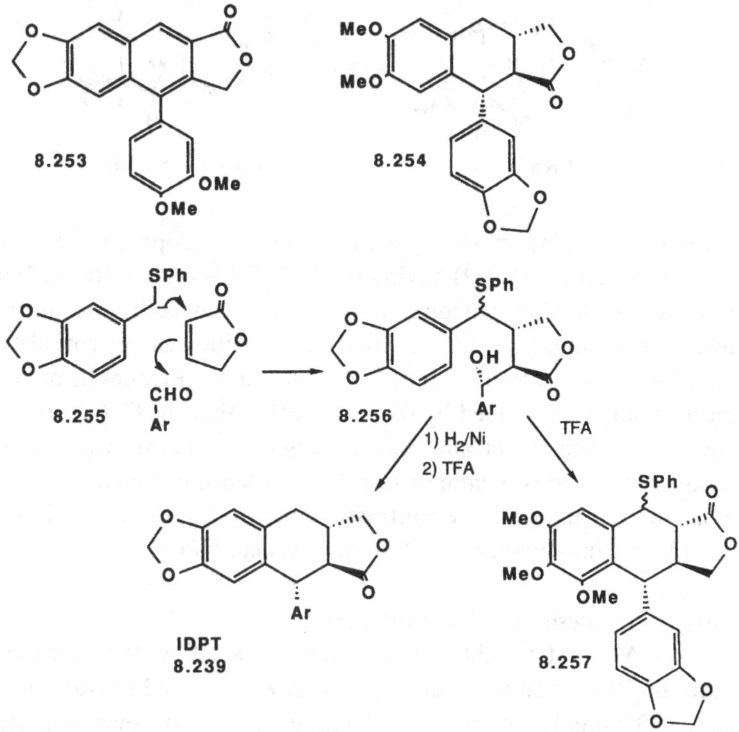

made (Dhal *et al.*, 1986) by Michael addition of thioketal anion to butenolide, followed by alkylation of a benzyl bromide (cf. Scheme 8.17).

In the work which has just been summarised (Ganeshpure and Stevenson, 1981b; Loriot, 1983; Dhal *et al.*, 1986) high yields of C-7′, C-8′-*trans*-products were obtained in the intramolecular alkylation despite the fact that the precursors were mixtures of epimeric 7-hydroxybenzyl-lactones. This argues against any possibility that the reaction is concerted (Scheme 8.57), since the 7-alcohol shown would then give the *cis-trans*-tetralin whilst the all-*trans*-tetralin would be obtained from its 7-epimer. It also follows that this procedure cannot lead to the *cis-trans*-configuration of the physiologically active podophyllotoxins. Confirmation that there is no control over stereochemistry comes from the observation (Robin *et al.*, 1982) that when the separated epimers were subjected to acid-catalysed cyclisation they *both* gave IDPT as the only isolable product.

Steric control over this type of reaction was recently achieved (Van-der-Eycken *et al.*, 1985) following the separation of the partially protected triol (**8.258**; X = OH; Scheme 8.58). This was separated from the epimeric alcohol and converted into the mesylate (**8.258**; X = Me.SO_2—O—) which

Scheme 8.57 Hypothetical concerted reaction at C - 7

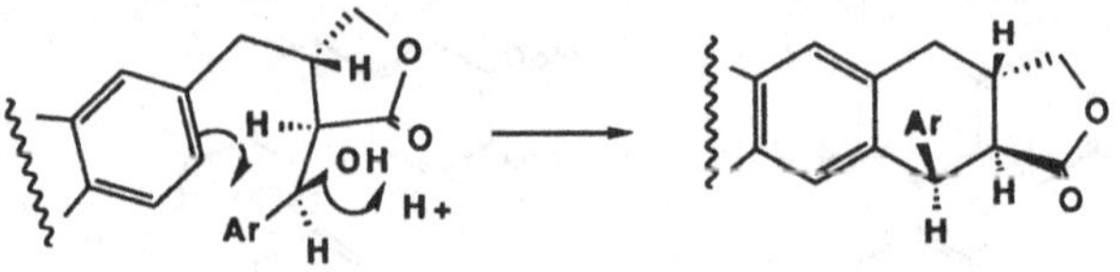

cyclised directly by an SN2-type mechanism, as proposed in Scheme 8.57, to give the tetralin (**8.259**) having the C-7′, C-8′-configuration of epipodophyllotoxin. Such an outcome can only be explained by substitution with retention, although it is unlikely that the transition can be completely concerted as this must lead to considerable strain. Further confirmation of the mechanism was provided by the observation that the C-7-epimer of **8.258** yielded the epimeric tetralin. A surprising feature of this work was that the siladioxan group was labile to both Bronsted and Lewis acids and as a result these conditions by contrast caused cyclisation through attack on Ar_2 (3,4,5-trimethoxyphenyl; Van der Eycken, 1986b).

(iii) *Synthesis of podophyllotoxins*

All eight possible diastereoisomers (**8.260**) of the L-series with the natural (*R*)-configuration at C7′ have been obtained (Aiyar and Chang, 1977). The highly strained *trans*-lactones show anti-cancer activity which is lost on epimerisation at C8′- by base; the equilibrium mixture of podophyllotoxin (**8.260**) and picropodophyllin (**8.261**) includes 97% of the latter.

Gensler *et al.*'s classical synthesis (1960) of picropodophyllin was based on the benzophenone (**8.262**, Scheme 8.59). Owing to the presence of the methylenedioxy ether group, which is readily cleaved by Lewis acids, this ketone is not easily made by the Friedel–Crafts reaction. A good yield was however obtained by the use of stannic chloride to catalyse the trimethoxybenzoylation of methylenedioxybenzene. An alternative route was the action of *N*-(trimethoxybenzoyl)-morpholine on the lithio-derivative of

Scheme 8.58 Cyclisation by an S_n2 type mechanism

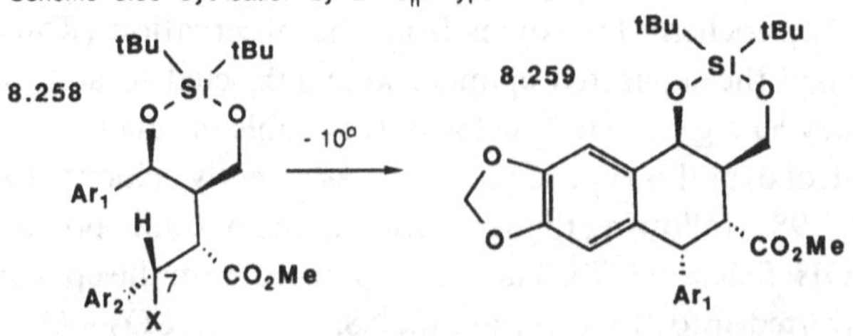

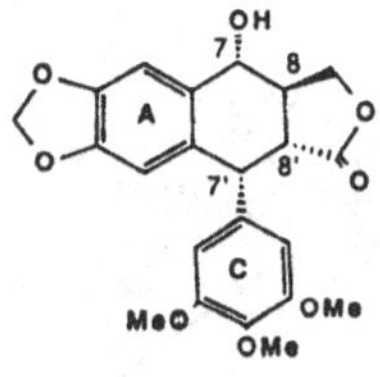

8.260 (7'R, 8'R, 8R, 7R)

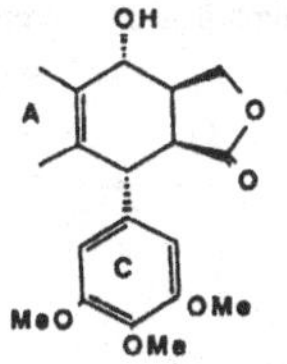

8.261 (7'R, 8'S, 8R, 7R)

this ether. The stability of the methylenedioxy group under acidic conditions is variable, thus although cleaved by boron trichloride it is unaffected by the trifluoride or by zinc chloride (Kuhn and von Wartburg, 1968). The ether groups in podophyllotoxins are stable in the presence of sulphuric and hydrochloric acids (cf. Scheme 8.59), but C4′-demethylation occurs with hydrobromic and nitric acids (Ayres and Lim, 1982). Although polyphosphoric acid provides a superior route to hydroxy and alkoxybenzophenones (Ayres and Denny, 1961) it too cleaves methylenedioxy ethers; fortunately a reliable synthetic method is now available (Bashall and Collins, 1975) which allows re-insertion by the methenylation of catechols.

With a route to the ketone (**8.262**) established Gensler developed his synthesis via a Stobbe condensation followed by hydrogenation to afford the benzylsuccinic acid (**8.263**, Scheme 8.59). This was cyclised in acetyl chloride without loss of the methylene group and the esterified tetralone (**8.264**) was formylated and treated with borohydride to yield the ester of (±)-epiisopodophyllic acid (**8.265**) of uncertain configuration at C8′. This acid reacted with concentrated sulphuric acid to form α-apopicropodophyllin (**8.266**) and this was in turn hydrated by the addition of hydrogen chloride followed by mild alkaline hydrolysis to give picropodophyllin (**8.261**). It was shown that the method was suitable for the preparation of optically active lignans since (±)-apopicropodophyllic acid was resolved as its quinine salt.

The final step in the Gensler synthesis (1962, 1966) was an ingenious but moderately yielding process for the directed protonation of the enolate (**8.267**), produced in an aprotic solvent by the action of a metal alkyl on picropodophyllin protected as its tetrahydropyranyl derivative. Protonation under kinetic control now ensured a significant proportion of podophyllotoxin in the mixture as a result of donation from the less hindered β-face.

An initial study of photoenolisation (Schemes 8.26, 8.27) of *o*-benzylbenzaldehydes led to a synthesis (Arnold *et al.*, 1973, 1974) of arylnaphthalenes by the addition of dienophiles to photoexcited dienols of type **8.268**

Scheme 8.59 Gensler's synthesis of picropodophyllin

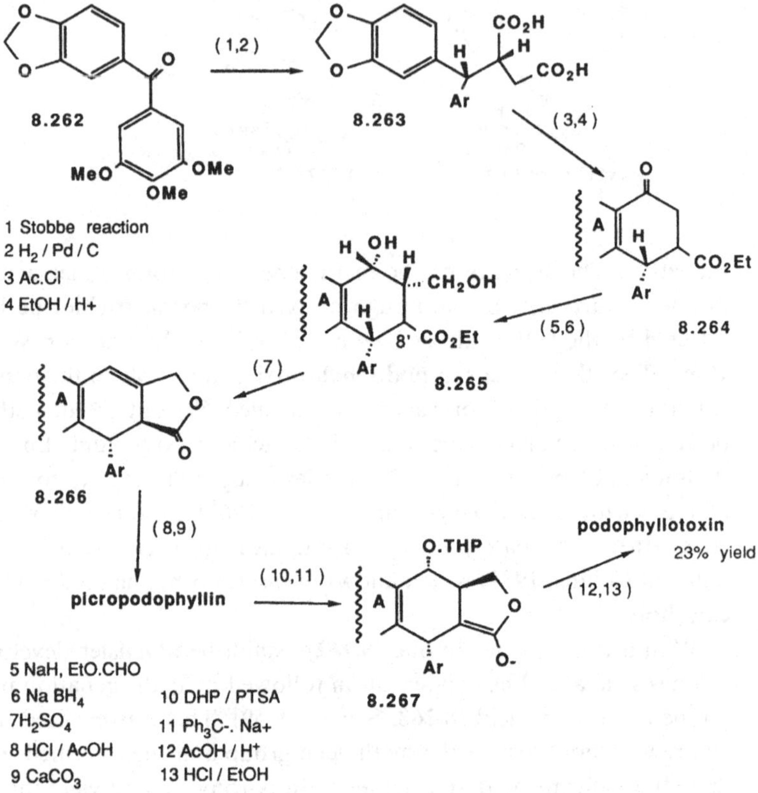

1 Stobbe reaction
2 H_2 / Pd / C
3 Ac.Cl
4 EtOH / H+

5 NaH, EtO.CHO
6 Na BH_4
7 H_2SO_4
8 HCl / AcOH
9 $CaCO_3$

10 DHP / PTSA
11 Ph_3C^-. Na+
12 AcOH / H^+
13 HCl / EtOH

(X = Y = H). Addition of maleic anhydride gave unstable hydroxyanhydrides of type **8.269**. Tetralin lignan models such as **8.270** (R = phenyl) were obtained via **8.269** by endo-addition to the (*E*)-dienol shown (**8.268**).

An improved route to the precursor of **8.268** (X, Y = 0—CH_2—O, R = 3,4,5-trimethoxyphenyl; Glinski and Durst, 1983) was based on coupling between 6-bromopiperonal dimethylacetal and trimethoxybenzyl bromide – this reaction was critically dependent upon catalysis by tri-*n*-butylphosphine copper iodide. Addition of dimethyl fumarate to the photo-enol gave the diester (**8.271**, 47% yield), which was reduced selectively at C-9 and converted into the lactone epiisopodophyllotoxin by a sequence described by Rajapaksa and Rodrigo (1981; Scheme 8.62, p. 367). Another route to (*E*)-dienes was devised by Mann *et al.* (1984) which depends upon the extrusion of sulphur dioxide from sulphoxides; this is illustrated in Scheme 8.27.

Kende and colleagues (1977) showed that the route to steganacin could

Scheme 8.60 Addition to photoenols

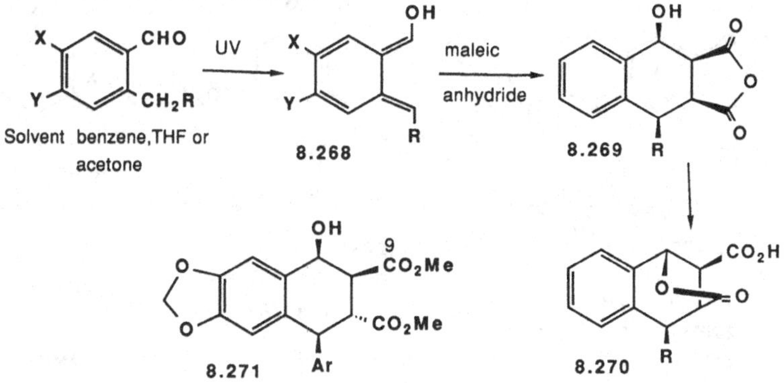

Solvent benzene,THF or acetone

be adapted by synthesis of the 4′-demethyl analogue of **8.203** (p. 348) followed by *oxidative* coupling through an intermediate quinone methide (cf. **7.51**, Scheme 7.8) to the tetralin diester (**8.272**, Scheme 8.61). The corresponding tetralone was then obtained from *O*-methylated material using NBS (cf. p. 322) followed by hydrolysis with decarboxylation to the keto-acid (**8.273**). In view of the failure of the bromination step when attempted on a larger scale an alternative was developed (Kende *et al.*, 1981).

The intermediate lithium salt (**8.274**) was obtained from a dibromo-precursor by the addition of *one equivalent* of methyl lithium, whereupon nucleophilic addition occurred to the diester (**8.275**) with a concerted displacement of the less-reactive residual bromine. The isomeric mixture of tetralol methyl ethers (**8.276**) was then demethylated, the alcohols oxidised and the product ketone hydrolysed to provide a superior route to the ketoacid (**8.273**). Formylation of the acid afforded the *bis*-condensation product (**8.277**) which yielded picropodophyllone on heating through retroaldolisation. This final step overcomes the difficulty of separating a mixture of *mono*- and *bis*-formylation products.

Treatment of picropodophyllone with a hydride donor was followed by an improved (89%) C8′-epimerisation procedure. Here the ketone was protected as its *t*-butyldimethylsilyl ether and the lactone enolate (cf. **8.267**) was protonated using pyridine hydrochloride leading to the synthesis of podophyllotoxin itself.

The eight-stage route to the tetralone (from **8.272**) was shortened by Murphy and Wattanasin (1982) who converted methylenedioxyphenyl-3,4,5-trimethoxycinnamyl ketone, a chalcone, into a mixture of epimeric cyclopropane derivatives (**8.278**, 95% yield) by the addition of ethoxycarbonyldimethylsulphonium methylide ($Me_2S{=}CH.CO_2Et$; Trost and

Scheme 8.61 Tetralone synthesis

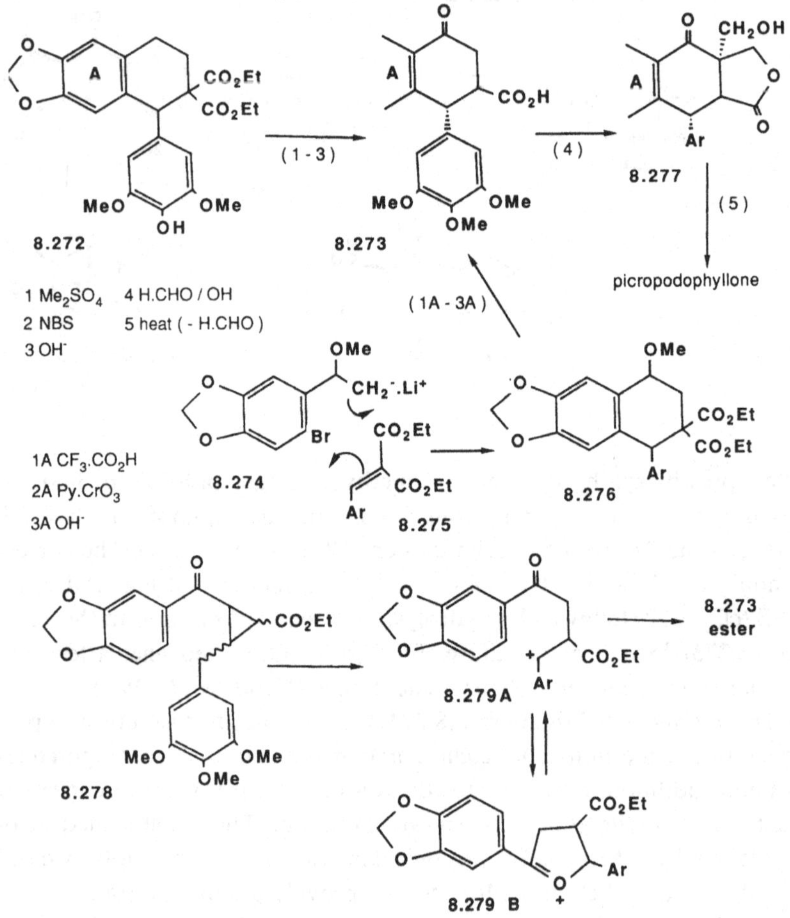

1 Me_2SO_4 4 H.CHO / OH
2 NBS 5 heat (- H.CHO)
3 OH^-

1A $CF_3.CO_2H$
2A $Py.CrO_3$
3A OH^-

Melvin, 1975). The use of nitromethane was essential for success of the cyclisation to the *trans*-tetralone ethyl ester, because it preferentially solvates the benzyl carbocation (**8.279A**) rather than the contributary oxonium ion (**8.279B**). This work also included a valuable assessment of the formylation step.

The use of the Diels–Alder reaction for the synthesis of deoxypodophyllotoxin by Rodrigo (1980) has been mentioned (Scheme 8.26). Later work by Rajapaksa and Rodrigo (1981) led to the epipodophyllic acid ester (**8.280**, Scheme 8.62) through hydrogenolysis of the adduct (**8.99**). An ingenious method of protecting this ester from epimerisation during alkaline hydrolysis was devised whereby it was first converted into the acetonide

Scheme 8.62 Protection of critical geometry

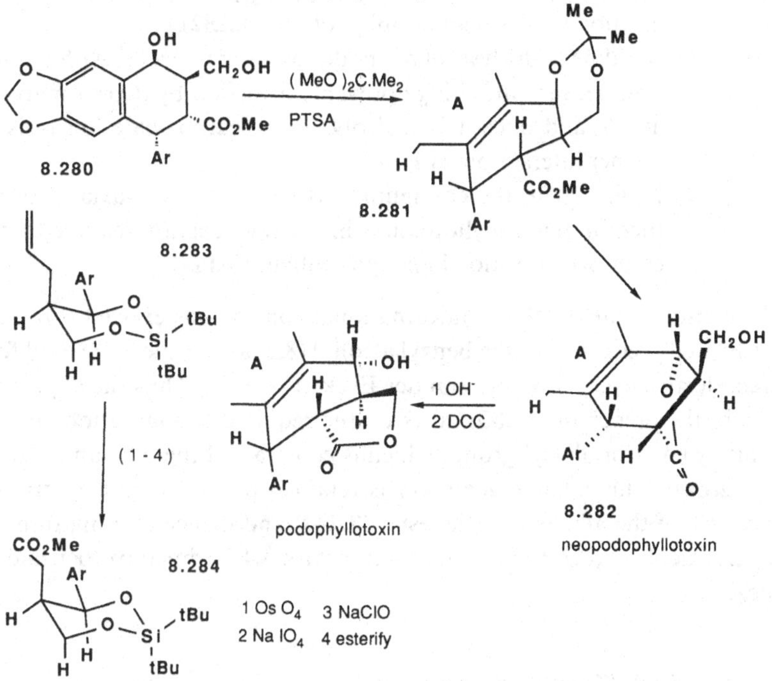

(**8.281**). Of two possible conformers, that shown is the more stable because there is no unfavourable interaction of an axial methyl group in the acetonide moiety and because the C7′-aryl substituent is in the more stable axial position (Ayres and Mundy, 1969) – hence the ester retained this configuration on hydrolysis. In the NMR spectra of these acetonides the *trans*-8,8′-configuration was confirmed by a large diaxial coupling constant. It has also been observed (Ayres and Lim, 1976) that a bulky 2′-substituent on the pendent ring leads to stabilisation of this configuration. Removal of the acetonide group under mild conditions gave epiisopodophyllic acid (cf. **8.280**), which on prolonged treatment yielded the C-8′; C-7-diaxially bridged lactone neopodophyllotoxin (**8.282**, 95%). The method of Renz *et al.* (1965) now gave podophyllotoxin in 64% yield by alkaline hydrolysis to podophyllic acid and its lactonisation with dicyclohexylcarbodiimide. These changes in configuration can be explained as follows:

(a) An equilibrium between the C7-epi-alcohols is established under the acidic conditions required for cleavage of the acetonide (**8.281**).

(b) Only the 'podo' form can be trapped by intramolecular lactonisation to give neopodophyllotoxin (**8.282**).

(c) Alkaline hydrolysis of neopodophyllotoxin occurs with retention because the bridge prevents inversion by deprotonation at C8′ and once it is hydrolysed the carboxylate ion resists further interaction with base.

(d) Note that in the configurations shown the C7′-axial substituent in podophyllotoxin is in a stable location relative to its equatorial position in neopodophyllotoxin.

It is also of interest that conformational control was exerted through the silylated precursors of the benzyl alcohol (**8.258**, Scheme 8.58) used for stereoselective cyclisation by Van der Eycken (1986b). Thus the allyl side chain in the precursor (**8.283**) takes up the required configuration *cis*- to the bulky equatorial aryl group, which is constrained there by an axial *t*-butyl group. This relative geometry is retained in the subsequent transformation of the alkene into the ester (**8.284**) and thence to a mixture of epimeric alcohols (**8.258**) by condensation with 3,4,5-trimethoxybenzaldehyde.

(a) *Bristol–Myers group syntheses*

Two high-yielding procedures have been built from precursors already described. In the first of these (Vyas *et al.*, 1986) the chalcone route (Murphy and Wattanasin, 1982) was improved by the incorporation of acetic anhydride during the rearrangement of the cyclopropyl precursor (**8.278**, Scheme 8.61, 90% yield). The derived tetralone ester (**8.285**, Scheme 8.63) was hydrolysed and converted into the *trans*-dihydronaphthalene benzhydryl ester (**8.286**). Enol quenching (cf. **8.267**, Scheme 8.59) of this compound gave the *cis*-7′,8-ester (**8.287**, 70% yield) whose bulky ring B substituents then directed cyclo-addition of bromonitrile oxide (Vyas *et al.*, 1984) to the β-face to give the adduct (**8.288**). Hydrolysis of the benzhydryl group followed by hydrogenation gave the nitrile (**8.289**) in which the cyano group was preferentially reduced to the aminoacid. The diazo-derivative then lactonised to afford epipodophyllotoxin in 25% overall yield.

The second sequence was developed (Kaneko and Wong, 1987) by separation of the *cis*-epimer of the tetralone ester (**8.285**), which formed 16% of the base-equilibrated mixture, and converting it into the enol ether (**8.290**) with trimethylsilyl iodide/HMDS. The enol ether was alkylated with chloromethyl benzyl ether to give the precursor (**8.291**) with the protected primary alcohol in place. A significant feature of this reaction is the

Scheme 8.63 First Bristol - Myers route

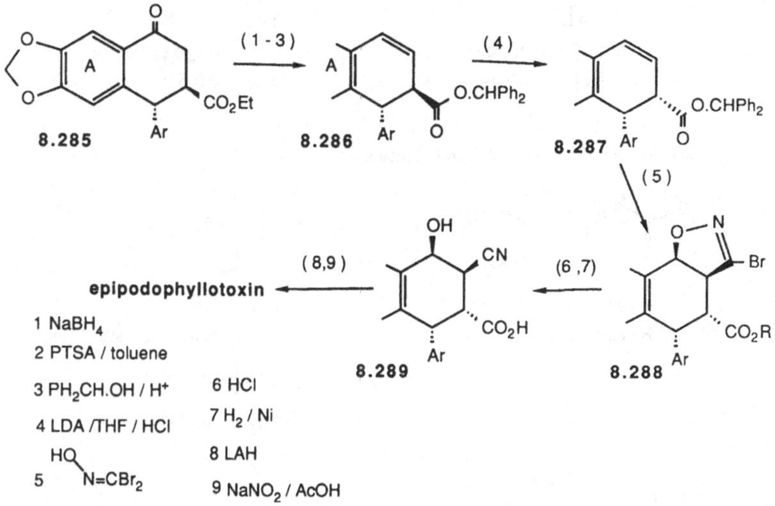
8.285
8.286
8.287
8.288
8.289
(1-3)
(4)
(5)
(6,7)
(8,9)
epipodophyllotoxin
1 NaBH4
2 PTSA / toluene
3 PH2CH.OH / H+
4 LDA /THF / HCl
5 HO N=CBr2
6 HCl
7 H2 / Ni
8 LAH
9 NaNO2 / AcOH

survival of the methylenedioxy ether group in the presence of the Lewis acid. Hydride donation to the carbonyl group of **8.291** was directed by the pendent ring to the β-face to give the 7α-ol (25% yield) together with the product (**8.292**, 43%) from its transannular lactonisation. Removal of the protecting benzyl group by hydrogenolysis now gave neo-podophyllotoxin (**8.282**), which was transformed into podophyllotoxin using the Renz *et al.* (1965) procedure.

The basis of an enantioselective synthesis of podophyllotoxin has been established (Brown and Daugan, 1985) by a route which should be compared with that of Tomioka *et al.* (1982, Scheme 8.14). In this later approach the half-ester (**8.293**) was resolved as its salt with (−)-ephedrine and also as the salt with (+)-ephedrine. The product characteristics were as shown in Scheme 8.65. Racemisation of these resolved half-esters was

Scheme 8.64 Second Bristol - Myers route

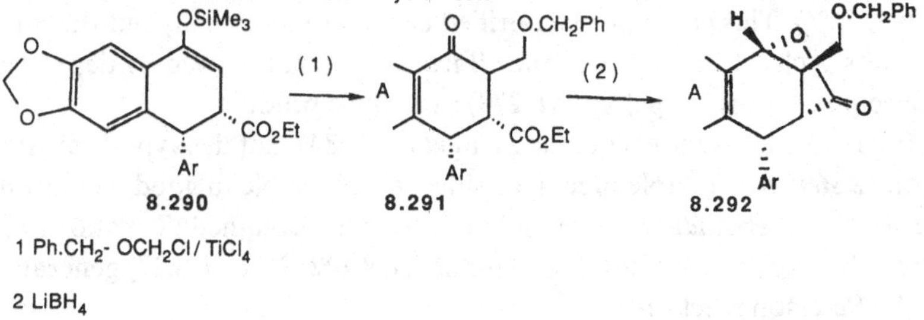
8.290
8.291
8.292
(1)
(2)
1 Ph.CH2- OCH2Cl/ TiCl4
2 LiBH4

(−)-ephedrine salt
$[\alpha]_D + 2.4°$

(R)-(+)-(**8.293**)
$[\alpha]_D + 30.4°$

(−)-ephedrine salt
$[\alpha]_D$ − −

(S)-(−)-(**8.293**)
$[\alpha]_D - 30.5°$

Scheme 8.65 Synthesis of chiral tetralins

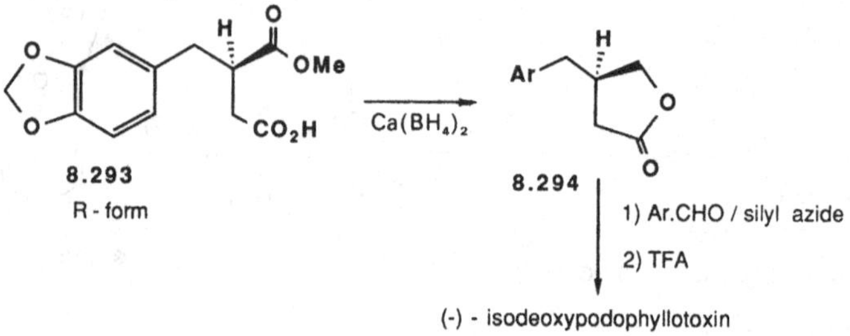

OMe
CO_2H
$Ca(BH_4)_2$
Ar
8.293
R - form
8.294
1) Ar.CHO / silyl azide
2) TFA
(-) - isodeoxypodophyllotoxin

slow in that 2-h contact with hot sodium methoxide was required; hence calcium borohydride could be used to obtain the enantiomeric lactones (**8.294**). The condensation of the (*R*)-lactone with 3,4,5-trimethoxybenaldehyde was catalysed by the weak base hexamethyldisylazide, whence cyclisation gave the *trans*-cyclised (−)-IDPT. The (*S*)-lactone yielded the antipode (+)-IDPT by a parallel sequence.

(iv) *Derivatives of podophyllotoxins*

The α- and β-peltatins are treated in this context. They are important as they are readily available in good yield from *P. peltatum* yet have found only limited application in the search for anti-cancer agents.

In a recent study (Loriot *et al.*, 1984) 2-benzyloxypiperonal (**8.295**, Scheme 8.66) gave the half-ester (**8.296**) in the Stobbe reaction; this was reduced selectively with borohydride and hydrogenated with the unavoidable loss of the protecting group in the product (**8.297**). It was reinserted using benzyl chloride/sodium iodide and the ether was condensed (cf. Scheme 8.65) with trimethoxybenzaldehyde to afford a mixture of epimers (**8.298**, 71%). This could not be purified chromatographically but the mixture was cyclised with TFA to the all *trans*-tetralin, followed by deprotection to give (±)-iso-β-peltatin (**8.299**) as a single product.

It has also been shown (Yamaguchi *et al.*, 1984) that deoxypodophyllotoxin (**2.360**) is a suitable precursor since it is available in good yield from the seeds of *Hernandia ovigera* and has also been obtained (Takano *et al.*, 1985b) by trapping of a quinodimethane (cf. **8.102**, Scheme 8.27) generated by the Peterson reaction.

Scheme 8.66 Synthesis of chiral tetralins

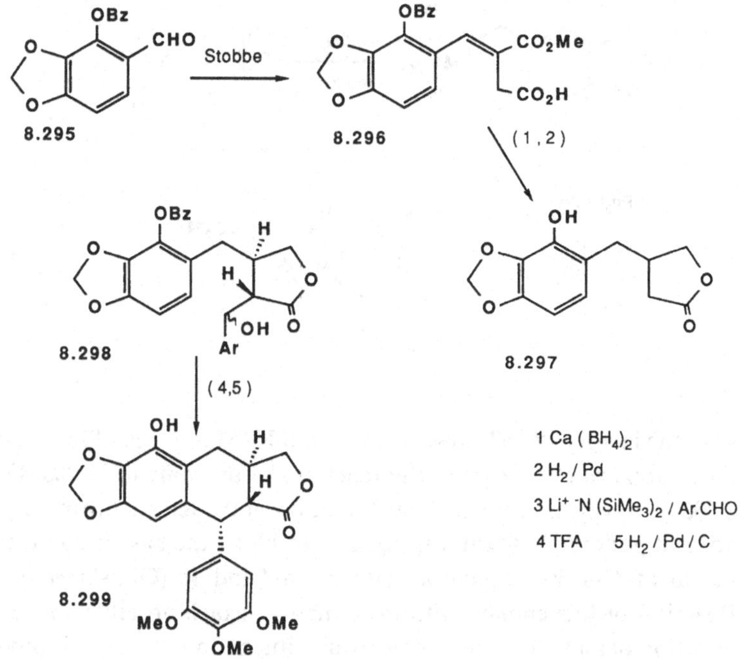

The difficulties which attend the α-bromination of tetralin lignans have been mentioned (p. 365); however if ring A is activated by demethylenation a controlled reaction of the catechol gives 6-bromodeoxypodophyllotoxin (**8.300**). Reinstatement of this ether with methylene iodide and an alkali fluoride proceeded in 70% yield, but substitution of the bromine by hydroxyl or methoxyl proved to be difficult and was attended by isomerisation to picrocompounds.

In a search (Mann *et al.*, 1984) for analogues which retain their physiological activity in contact with bases the C8′-blocked derivative (**8.301**) was obtained by the cyclo-addition of methylmaleic anhydride to a quinodimethane. Reduction with borohydride or, preferably, potassium selectride

Scheme 8.67 Controlled bromination

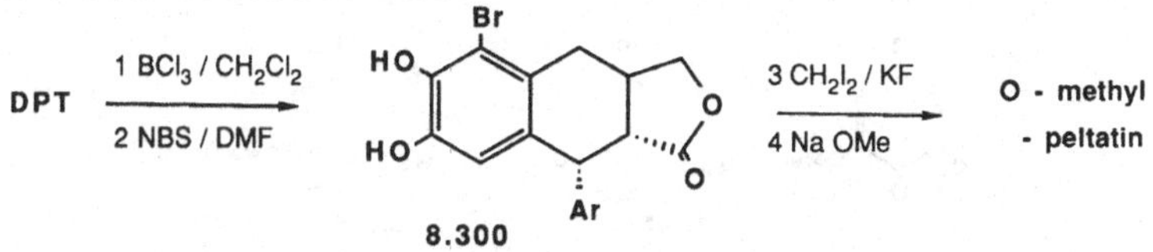

Scheme 8.68 Analogues blocked at C-8'

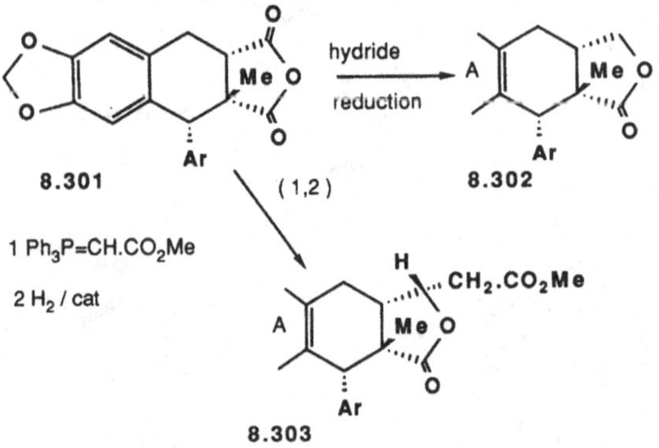

gave the lactone (**8.302**). It was also possible (Mann *et al.*, 1985) to develop the lactone ring through Wittig reactions of the anhydride (**8.301**) to give **8.303** as a typical product. After lithiation of the tetrahydropyranyl derivative of podophyllotoxin it is possible to block the base-induced epimerisation at C-8′ by alkylation with methyl iodide (Glinski *et al.*, 1987). Reaction of this enolate with bromine or hexachloroethane leads to the insertion of an 8′-bromo- or chlorosubstituent and the latter is physiologically active and less toxic than podophyllotoxin itself.

A range of alkali-stable 2′-azido-, bromo- and chloro-podophyllotoxins has been prepared (Ayres and Ritchie, 1988) by nucleophilic addition to pendent ring *ortho*-quinones of type **8.304**. These products (**8.305A, B**) all possess the 4′-dimethyl position required for specific DNA attack (Long *et al.*, 1984) but the orientation of substituents varies in a manner which may reflect a difference in the mechanism of addition. Thus azide and chloride afford derivatives substituted as for **8.305A** but a bromo-substituent takes up the position in **8.305B**.

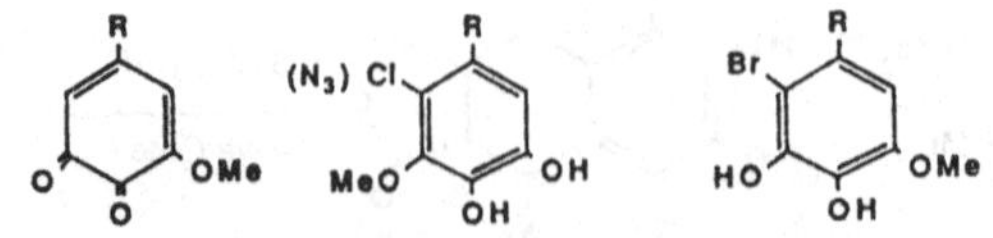

Glycosyl units may be efficiently incorporated at C7 in podophyllotoxins by the reaction with tetra-acetyl-β-D-bromo-glycosides (Kuhn and Von Wartenburg, 1968; Scheme 4.1) followed by deacetylation with zinc chloride. The scope of this reaction has been discussed by Doyle (1984) and modern methods of O- and C-glycosylation are reviewed annually (Theilheimer, 1988). Procedures for computer-aided searches using programmes such as LHASA and SYNCHEM have been reviewed by Wolman (1988).

There has been an impressive refinement in procedures for synthesis of the podophyllotoxin group. A good example is the work of the Bristol chemists whose better first synthesis led to epipodophyllotoxin in 25% overall yield. This is however reduced to about half when one considers the early-stage losses (Murphy and Wattanasin, 1980). On this account one still prefers to extract these lignans, especially the peltatins, from their plant source because the *in vivo* yields are so high.

References

Achiwa, K., Ohga, Y., Iitakay, Y. and Saito, H. (1978). The mechanism of asymmetric hydrogenations catalysed by chiral pyrrolidinophosphine–rhodium complexes. *Tetrahedron Lett.* 4683–6.

Achiwa, K. (1979). Effective catalytic asymmetric synthesis of (*S*)-(−)-3-methoxycarbonyl-4-(3,4-methylenedioxyphenyl)-butanoic acid. A simple & effective route to chiral lignans. *Heterocycles* **12**, 515–17.

Agnihotri, P.W., Pasarkar, V.R. and Bagavant, G. (1982). Synthesis of pericarbonyl lignans & their analogues. *J. Indian Chem. Soc.* **59**, 869–76.

Ahmed, R., Lehrer, M. and Stevenson, R. (1973a). Synthesis of thomasidioic acid. *Tetrahedron Lett.* 747–50.

Ahmed, R., Lehrer, M. and Stevenson, R. (1973b). Synthesis of thomasic acid. *Tetrahedron* **29**, 3753–9.

Ahmed, R., Schreiber, F. G., Stevenson, R., Williams, J.R. and Yeo, H.M. (1976). Oxidative coupling of bromo- and iodoferulic acid derivatives: synthesis of (±)-veraguensin. *Tetrahedron* **32**, 1339–44.

Aiyar, V.N. and Chang, F.C. (1977). Podophyllotoxin derivatives. Part 3. The remaining diastereoisomeric C-7 alcohols & ketones of the L series. *J. Org. Chem.* **42**, 246–9.

Anjanamurthy, C. and Rai, K.M.L. (1985). Synthesis of podophyllotoxin & related analogues. Part 2. Synthesis of β-apopicropodophyllin analogues with modified hydroaromatic ring B. *Indian J. Chem.* **24B**, 505–8.

Anjaneyulu, A.S.R., Raghu, P. and Rao, K.V.R. (1979a). Synthetic experiments in lignans. Part 6. Photocyclisation of sterically hindered 1,2-diarylidenesuccinic anhydrides. *Indian J. Chem.* **18B**, 535–7.

Anjaneyulu, A.S.R., Raghu, P., Rao, K.V.R., Sastry, C.V.M., Umasundari, P. and Satyanarayana, P. (1984). On the stereoselectivity of Stobbe condensation with ortho-substituted aromatic aldehydes: the (E, Z) configuration of monobenzylidene succinates & dibenzylidenesuccininc anhydrides. *Curr. Sci.* **53**, 239–43.

Anjaneyulu, A.S.R., Ragu, P., Rao, K.V.R. and Row, L.R. (1978). Synthetic experiments in lignans. Part 4. Observation of unusual loss of bromo- or methoxyl substituents during thermal cyclisation of 1,2-diarylidene succinic anhydrides. *Tetrahedron Lett.* 3467–8.

Anjaneyulu, A.S.R., Raghu, P., Rao, K.V.R. and Row, L.R. (1979b). Synthetic experiments in lignans. Part 5. Thermal cyclisation of sterically hindered 1,2-diarylidenesuccinic anhydrides. *Indian J. Chem.* **18B**, 391–4.

Anjaneyulu, A.S.R., Raghu, P. and Rao, K.V.R. (1980). Synthetic experiments in lignans. Part 7. *Indian J. Chem.* **19B**, 511–12.

Arnold, B.J., Mellows, S.M. and Sammes, P.G. (1973). Photochemical reactions.

Part 1. A new route to tetradehydropodophyllotoxin, tawanin E & related compounds. *J. Chem. Soc. Perkin Trans. 1*, 1266–70.

Arnold, B.J., Mellows, S.M., Sammes, P.G. and Wallace, T.W. (1974). Photochemical reactions. Part 2. Cycloaddition reactions with photoenols from 2-methoxybenzaldehyde & related systems. *J. Chem. Soc. Perkin Trans. 1*, 401-9.

Ayres, D.C. and Denney, R.C. (1961). Lignans. Part 1. Acylation in polyphosphoric acid as a route to intermediates, *J. Chem. Soc.* 4506–9.

Ayres, D.C., Carpenter, B.G. and Denney, R.C. (1965). Lignans. Part 3. Synthesis of 1-arylnaphthalenes related to podophyllotoxin. *J. Chem. Soc.* 3578–82.

Ayres, D.C. and Mhasalkar, S.E. (1965). Lignans & related phenols. Part 5. (−)-Olivil & (+)-cycloolivil. *J. Chem. Soc.* 3586–9.

Ayres, D.C. and Mundy, J.W. (1969). Lignans & related phenols. Part 8. A proof of the structures of some intermediates in arylnaphthalene synthesis. *J. Chem. Soc.* (*C*) 637–46.

Ayres, D.C. and Lim, C.K. (1972). Lignans & related phenols. Part 13. Halogenated derivatives of podophyllotoxin. *J. Chem. Soc. Perkin Trans.*, 1350–5.

Ayres, D.C. and Lim, C.K. (1976). Lignans & related phenols. Part 15. Remote substituent effects on the rates & products of some reactions of aryltetrahydronaphthalenes. *J. Chem. Soc. Perkin Trans. 1*, 832–6.

Ayres, D.C. (1978). In *Chemistry of Lignans*, ed. Rao, C.B.S. Andhra Univ. Press, pp. 123–73.

Ayres, D.C. & Lim, C.K. (1982). Modification of the pendent ring of podophyllotoxin. *Cancer Chemother. & Pharmacol.* **7**, 99–101.

Ayres, D.C. and Ritchie, T.J. (1988). Lignans & related phenols. Part 18. The synthesis of quinones from podophyllotoxin & its analogues. *J. Chem. Soc. Perkin Trans. 1*, 2573–8.

Bagavant, G. and Ganeshpure, P.A. (1978). Cyclolignans through Stobbe condensation. *Current Sci.* **47**, 338–9.

Banerjee, J., Das, B., Chatterjee, A. and Shoolery, J.N. (1984). Gadain, a lignan from *Jatropha gossypifolia*. *Phytochemistry* **23**, 2323–7.

Banerjee, J. and Das, B. (1985). Synthesis of (±)-gadain, a new lignan from *Jatropha gossypifolia* Linn. (*Euphorbiaceae*). *Heterocycles* **23**, 661–5.

Bashall, A.P. and Collins, J.F. (1975). A convenient high-yielding method for the methenylation of catechols. *Tetrahedron Lett.* 3489–90.

Batterbee, J.E., Burden, R.S., Crombie, L. and Whiting, D.A. (1969). Chemistry & synthesis of the lignan (−)-cubebin. *J. Chem. Soc.* 2470–7.

Becker, D., Hughes, L.R. and Raphael, R.A. (1977). The total synthesis of the antileukemic lignan (±)-steganacin. *J. Chem. Soc. Perkin Trans. 1*, 1674–81.

Belletire, J.L. and Fry, D.F. (1987). Oxidative coupling of carboxylic acid dianions: the total synthesis of (±)-hinokinin & fomentaric acid. *J. Org. Chem.* **52**, 2549–55.

Biftu, T., Hazrfa, B.G., Stevenson, R. and Williams, J.R. (1978a). Synthesis of the lignan (±)-deoxyschizandrin. *J. Chem. Soc. Chem. Commun.* 491–2.

Biftu, T., Hazra, B.G., Stevenson, R. and Williams, J.R. (1978b). Synthesis of lignans from 2,3-diaryoylbutanes. *J. Chem. Soc. Perkin Trans, 1*, 1147–50.

Biftu, T., Hazra, B.G. and Stevenson, R. (1979). Synthesis of deoxyschizandrin. *J. Chem. Soc. Perkin Trans. 1*, 2276–81.

Birch, A.J., Milligan, B., Smith, Mrs E. and Speake, R.N. (1958). Some stereochemical studies of lignans. *J. Chem. Soc.* 4471–6.

Blears, J.G. and Haworth, R.D. (1958). The constituents of natural phenolic resins. Part 24. A synthesis of galgravin. *J. Chem. Soc.* 1985–7.

Briggs, L.H., Cambie, R.C. and Hoare, J.L. (1959). Chemistry of the *Podocarpaceae*. Part 3. A new lignan, seco-isolariciresinol & further constituents of the heartwood of *Podocarpus spicatus*. *Tetrahedron* **7**, 261–76.

Brown, D. and Stevenson, R. (1965). Synthesis of dehydrootobain. *J. Org. Chem.* **30**, 1759–63.

Brown, E.H. and Robin, J-P. (1978). A new route to the bis-benzocyclooctadiene lignan skeleton: total synthesis of (*RS*)-picrostegane, (*RS*)-isopicrostegane & (*RS*)-isostegane. *Tetrahedron Lett.* 3613–16.

Brown, E.H., Robin, J-P. and Dhal, R. (1982). Total synthesis &⁻ studies of biologically active lignans. Application of the Ullmann reaction to the synthesis of precursors. *Tetrahedron* **38**, 2569–79.

Brown, E.H. and Daugan, A. (1985). A convenient preparation of (−) & (+)-β-piperonyl-γ-butyrolactones, key intermediates for the synthesis of optically active lignans. *Tetrahedron Lett.* **26**, 3997–8.

Brownbridge, P. and Chang, T-H. (1980a). Chemistry of 2,5-*bis*-(trimethylsilyloxy) furans. Part 2. Reactions with carbonyl compounds & synthesis of 2,6-diaryl-3,7-dioxabicyclo-[3:3:0]-octane-4,8-diones. *Tetrahedron Lett.* 3427–30.

Brownbridge, P. and Chang, T-H. (1980b). Chemistry of 2,5-*bis*-(trimethylsilyloxy) furans. Part 1. Preparation & Diels-Alder reactions. *Tetrahedron Lett.* 3423–6.

Camps, P., Font, J. and Ponsati, O. (1981). A short synthesis of (*S*)-5-hydroxymethyl-(5H)-furan-2 one and derivatives from D-ribonolactone. *Tetrahedron Lett.* **22**, 1471–2.

Cardellach, J., Font, J. and Ortuno, R.M. (1984). A facile & general entry to optically active pheromones. *J. Hetero. Chem.* **21**, 327–31.

Cartright, N.J. and Haworth, R.D. (1944). The constituents of natural phenolic resins. Part 19. The oxidation of ferulic acid. *J. Chem. Soc.* 535–9.

Chakraborty, D.P., Roy, S., Sinha Roy, S.P. and Majumber, S. (1979). Podotoxin, a germination inhibiting lignan from *Zanthoxylum accananthapodium* D.C. *Chem. & Ind.* 667–8.

Charlton, J.L. and Durst, T. (1984). *ortho*-Quinodimethanes from 3,6-dihydrobenzo[b]-1,2-oxathiin-2-oxides. *Tetrahedron Lett.* **25**, 5827–90.

Charlton, J.L. and Alauddin, M.M. (1986). Asymmetric lignan synthesis; isolariciresinol dimethyl ether. *J. Org. Chem.* **51**, 3490–3.

Cole, J.R., Bianchi, E. and Trumbull, E.R. (1969). Anti-tumour agents from *Bursera microphylla* (Burseraceae) Part 2. Isolation of a new lignan – burseran. *J. Pharm. Sci.* **58**, 175–6.

Cooley, G., Farrant, R.D., Kirk, D.N. and Wynn, S. (1981). Chemical synthesis of the first lignans to be found in humans & animals. *Tetrahedron Lett.* 349–50.

Cooper, R., Gottlieb, H.E., Lavie, D. and Levy, E.C. (1979). Lignans from *Aegilops ovata* L. Synthesis of a 2,4- and a 2,6-diaryl-monoepoxylignanolide. *Tetrahedron* **35**, 861–8.

Damon, R.E., Schlessinger, R.H. and Blount, J.F. (1976). A short synthesis of isostegane. *J. Org. Chem.* **41**, 3772–3.

Das, K.G., Afzal, J., Hazra, B.G. and Bhawal, B.M. (1983). A novel synthetic route to 1-aryltetralin lignan lactones via intermediate cycloaddition with *ortho*-quinodimethanes. *Synth. Commun.* **13**, 787–96.

De Carvalho, M.G., Yoshida, M., Gottlieb, O.R. and Gottlieb, H.G. (1987). Lignans from *Nectandra turbacensis*. *Phytochemistry* **26**, 265–7.

Dhal, R., Brown, E. and Robin, J-P. (1983). Total synthesis of (±)-steganone and its congeners and synthesis of (±)-stegane. *J. Org. Chem.* **41**, 3772–3.

Dhal, R., Nabi, Y. and Brown, E. (1986). Study of lignans. Part 7. Total synthesis of (±)-α- & β-conidendrins and methyl (±)-α - & β-conidendrals. *Tetrahedron* **42**, 2005–16.

Doyle, T.W. (1984). In *Etoposide*, ed. Issell, B.F., Muggia, F.M. and Carter, S.K. Academic Press, pp. 15–32.

Drake, N.L. and Price, E.H. (1951). Podophyllotoxin & picropodophyllin. Part 1. Their reduction by lithium aluminium hydride. *J. Amer. Chem. Soc.* **73**, 201–5.

Elson, I.H. and Kochi, J.K. (1973). Thallium(III) in one-electron oxidation of arenes by electron spin resonance. *J. Amer. Chem. Soc.* **95**, 5060–2.

Erdtman, H. (1935). Phenol oxidation. Part 6. Oxidative coupling of guaiacol derivatives. *Svensk. Kem. Tidskr.* **47**, 223–30.

Fetizon, M. and Jurion, M. (1972). Aldehydes & ketones from thioacetals. *J. Chem. Soc. Chem. Commun.* 382–3.

Fonseca, S., Nielsen, L.T. and Ruveda, E.A. (1979). Lignans of *Araucaria angustifolia* & CMR analysis of some phenyltetralins. *Phytochemistry* **18**, 1703–8.

Freudenberg, K. and Dietrich, H. (1953). On syringaresinol a dehydrogenation product of sinapyl alcohol. *Chem. Ber.* **86**, 4–10.

Freudenberg, K. and Knof, L. (1957). The lignans of spruce wood. *Chem. Ber.* **90**, 2857–69.

Freudenberg, K., Harkin, J.M., Reichert, M. and Fukuzuki, T. (1958). The dehydrogenation of sinapyl alcohols. *Chem. Ber.* **91**, 581–90.

Freudenberg, K. (1959). Biosynthesis & constitution of lignin. *Nature* **183**, 1152–5.

Freudenberg, K. (1965). Lignin, its constitution & formation from *p*-hydroxycinnamyl alcohols. *Science* **148**, 595–600.

Ganeshpure, P.A. (1979). Cyclolignans through Stobbe condensation. Part 3. Stereochemistry & mechanism of diarylidenesuccinic acid formation in the Stobbe condensation: a synthesis of Justicidin E. *Indian J. Chem.* **17B**, 202–6.

Ganeshpure, P.A. and Stevenson, R. (1981a). Synthesis of HPMF: the first lignan isolated from animal species. *Chem. & Ind.* 778.

Ganeshpure, P.A. and Stevenson, R. (1981b). Synthesis of aryltetralin & dibenzylbutyrolactone lignans: (±)-lintetralin, (±)-phyltetralin & (±)-kusunokinin. *J. Chem. Soc. Perkin Trans. 1*, 1681–4.

Gardner, J.A.F., Swan, E.P., Sutherland, S.A. and MacLean, O. (1966). Polyoxyphenols of western red cedar (*Thuja plicata* Donn.), Part 3. Structure of plicatic acid. *Can. J. Chem.* **44**, 52–8.

Gensler, W.J. and Johnson, F. (1955). Podophyllotoxone, picropodophyllone & dehydropodophyllotoxin. *J. Amer. Chem. Soc.* **77**, 3674–5.

Gensler, W.J., Samour, C.M., Wang, S.Y. and Johnson, F. (1960). Compounds related to podophyllotoxin. Part 10. Synthesis of picropodophyllin. *J. Amer. Chem. Soc.* **82**, 1714–27.

Gensler, W.J. and Gatsonis, C.D. (1962). Synthesis of podophyllotoxin. *J. Amer. Chem. Soc.* **84**, 1748–9.

Gensler, W.J. and Gatsonis, C.D. (1966). The podophyllotoxin–picropodophyllin equilibrium. *J. Org. Chem.* **31**, 3224–7.

Ghera, E., Ben-David, Y. and Becker, D. (1977). Deoxyschizandrin, stereochemistry & total synthesis. *Tetrahedron Lett.* 463–6.

Ghera, E. & Ben-David, Y. (1978). Total synthesis of the lignan (±)-schizandrin. *J. Chem. Soc. Chem. Commun.* 480–1.

Ghosal, S. and Banerjee, S. (1978). Synthesis of monoarylidene & arylnaphthalide lignans of *Polygala chinensis*. *J. Indian Chem. Soc.* **55**, 1201–3.

Ghosal, S. and Banerjee, S. (1979). Synthesis of retrochinensin; a new naturally occurring 4-aryl-2,3-naphthalide lignan. *J. Chem. Soc.. Chem. Commun.* 165–6.

Glinski, M.B. and Durst, T. (1983). Synthesis of epiisopodophyllotoxin. *Canad. J. Chem.* **61**, 573–5.

Glinski, M.B., Freed, J.C. and Durst, T. (1987). Preparation of 2-substituted podophyllotoxin derivatives. *J. Org. Chem.* **52**, 2749–53.

Gonzalez, A.G., Perez, J.P. and Trujillo, J.M. (1978). Synthesis of two arylnaphthalene lignans. *Tetrahedron* **34**, 1011–13.

Gonzalez, A.G., Rafael, D. la R. and Trujillo, J.M. (1986). Synthesis of new aryltetralin lignans. *Tetrahedron* **42**, 3899–904.

Harkin, J.M. (1967). Lignin – a natural polymeric product of phenol oxidation. In *Oxidative Coupling of Phenols*, ed. Taylor, W.I. and Battersby, A.R. Arnold, London, pp. 243–321.

Hart, R.J. and Heller, H.G. (1972). Overcrowded molecules. Part 7. Thermal and photochemical reactions of photochromic (*E*) & (*Z*)-benzylidene (diphenylmethylene) succinic anhydrides and imides. *J. Chem. Soc. Perkin Trans. 1*, 1321–4, & subsequent papers.

Hartwell, J.L. and Schrecker, A.W. (1958). The chemistry of podophyllum. In *Progr. in Chem. Org. Nat. Prod.* **15**, 83–166.

Hasek, R.H. (1964). In Kirk-Othmer, *Encyclopedia of Chem. Tech.* **13**, 874–93.

Haworth, R.D. and Kelly, W. (1936). The constitution of natural phenolic resins. Part 7. Arctigenin. *J. Chem. Soc.* 998–1003.

Haworth, R.D. and Kelly, W. (1937). The constituents of natural phenolic resins. Part 8. Lariciresinol, cubebin & some stereochemical relationships. *J. Chem. Soc.* 384–91.

Haworth, R.D. (1942). The chemistry of the lignan group of natural products. *J. Chem. Soc.* 448–56.

Hearon, W.M. and MacGregor, W.S. (1955). The naturally occurring lignans. *Chem. Rev.* **55**, 957–1068.

Hughes, L.R. and Raphael, R.A. (1976). Synthesis of the antileukemic lignan precursor (±)-steganone. *Tetrahedron Lett.* 1543–6.

Ikeya, Y., Taguchi, H., Yosioka, I. and Kobayashi, H. (1978). The constituents of *Schizandra chinensis* Baill. The structures of two new lignans, Gomisin N & tigloylgomisin P. *Chem. Pharm. Bull.* **26**, 3257–60.

Ikeya, Y., Taguchi, H. and Yoshioka, I. (1979a). The constituents of *Schizandra chinensis* Baill. The cleavage of the methylenedioxy moiety with lead tetraacetate in benzene & the structure of angeloylgomisin Q. *Chem. Pharm. Bull.* **27**, 2536–8.

Ikeya, Y., Taguchi, H., Yosioka, I. and Kobayashi, H. (1979b). The constituents of *Schizandra chinensis* Baill. Part 5. The structures of four new lignans; gomisin N, gomisin O, epigomisin O & gomisin E & transformation of gomisin N to deangeloylgomisin B. *Chem. Pharm Bull.* **27**, 2695–709.

Ikeya, Y., Taguchi, H. and Yosioka, I. (1982). The constituents of *Schizandra chinensis* Baill. Part 12. Isolation & structure of a new lignan gomisin R, the absolute structure of wuweizisu C & isolation of schisantherin D. *Chem. Pharm. Bull.* **30**, 3207–11.

Ishibashi, F. and Taniguchi, E. (1986). Synthesis & absolute configuration of (+)-phrymarolin, a lignan. *Chem. Lett.* 1771–4.

Ishiguro, T., Mizuguchi, H., Tomioka, K. and Koga, K. (1985). Stereoselective reactions. Part 8. Stereochemical requirement for the benzylic oxidation of a lignan lactone. A highly selective synthesis of the anti-tumour lactone steganacin by the oxidation of stegane. *Chem. Pharm. Bull.* **33**, 609–17.

Johnson, W.S. and Daub, G.H. (1951). The Stobbe condensation. *Org. React.* **6**, 1–73.

Jung, M.E., Lam, P-Y., Mansuri, M.M. and Speltz, L.M. (1985). Stereoselective synthesis of an analogue of podophyllotoxin by an intramolecular Diels–Alder reaction. *J. Org. Chem.* **50**, 1087–105.

Kaneko, T. and Wong, H. (1987). Total synthesis of (±)-podophyllotoxin, *Tetrahedron Lett.* **28**, 517–520.

Kende, A.S. and Liebeskind, L.S. (1976). Total synthesis of (±)-steganacin. *J. Amer. Chem. Soc.* **98**, 267–8.

Kende, A.S., Liebeskind, L.S., Mills, J.E., Rutledge, P.S. and Curran, D.P. (1977). Oxidative aryl–benzyl coupling. A biomimetic entry to podophyllin lignan lactones. *J. Amer. Chem. Soc.* **99**, 7082–3.

Kende, A.S., King, A.L. and Curran, D.P. (1981). Total synthesis of (±)-4′-demethyl-4-epipodophyllotoxin by insertion cyclisation. *J. Org. Chem.* **46**, 2826–8.

Khan, Z. and Durst, T. (1987). Preparation of 1-thio & 1-amino-substituted 1,3-dihydrobenzo-[c]-thiophene-2,2-dioxides. *Canad. J. Chem.* **65**, 482–6.

King, F.E. and Wilson, J.G. (1964). The chemistry of extractives from hardwoods. Part 36. The lignans of *Guaiacum officinale* L. *J. Chem. Soc.* 4011–24.

Klemm, L.H., Gopinath, K.W., Lee, D.H., Kelly, F.W., Trod, E. and McGuire, T.M. (1966). The intramolecular Diels–Alder reaction as a route to synthetic lignan lactones. *Tetrahedron* **22**, 1797–808.

Klemm, L.H., Tran, V.T. and Olson, D.R. (1976). Intramolecular Diels–Alder reactions. Part 11. Modal selectivity in the synthesis of some parent lignan lactones. *J. Hetero. Chem.* **13**, 741–4.

Klemm, L.H. (1978). In *Chemistry of Lignans*, ed. Rao, C.B.S. Andhra Univ., pp. 196–202.

Kofod, H. and Jorgensen, C. (1954). Dehydropodophyllotoxin, a new compound isolated from *Podophyllum peltatum* L. *Acta Chem. Scand.* **8**, 1296–7.

Krow, G.M., Damodaran, K.M., Michener, E., Wolf, R. and Guare, J. (1978). Dibenzocyclooctadiene antileukemic lignan synthesis. (±)-steganone. *J. Org. Chem.* **43**, 3950–3.

Kuhn, M. and von Wartburg, A. (1967). Podophyllum lignans. Structure & absolute configuration of podorhizol β-D-glucoside. *Helv. Chim. Acta* **50**, 1546–65.

Kuhn, M. and von Wartburg, A. (1968). Synthesis of podophyllotoxin-β-D-glucoside. *Helv. Chim. Acta* **51**, 163–8.

Kupchan, M., Britton, R.W., Ziegler, M.F., Gilmore, C.J., Restivo, R.J. and Bryan, R.F. (1973). Steganacin & Steganangin novel antileukemic lignan lactones from *Steganotaenia araliacea*, *J. Amer. Chem. Soc.* **95**, 1335–6.

Landais, Y. and Robin, J-P. (1986). The tetrakis (trifluoroacetate) of ruthenium(IV). A new catalyst for ambient temperature non-phenolic biaryl coupling. First biomimetic total synthesis of neoisostegane. *Tetrahedron Lett.* **27**, 1785–8.

Larson, E.R. and Raphael, R.A. (1982). Synthesis of (−)-steganone. *J. Chem. Soc. Perkin Trans. I*, 521–5.

Linstead, R.P., Doering, W.E., Davis, S.B., Levine, P. and Whetstone, R.R. (1942). The stereochemistry of catalytic hydrogenation. Part 1. The stereochemistry of the hydrogenation of aromatic rings. *J. Amer. Chem. Soc.* **64**, 1985–91.

Long, B.H., Musial, S.T. and Brattain, M.G. (1984). Comparison of cytotoxicity & DNA breakage activity of congeners of podophyllotoxin including

VP16–213 & VM26: a quantitative structure activity relationship. *Biochemistry* **23**, 1183–8.

Loriot, M., Brown, E. and Robin, J-P. (1983). Total synthesis & studies of biologically active lignans. Total synthesis of (*RS*)-attenuol. *Tetrahedron* **39**, 2795–8.

Loriot, M., Robin, J-P. and Brown, E. (1984). Total synthesis & studies of biologically active lignans. Total synthesis of (*RS*)-iso-β-peltatin and analogues. *Tetrahedron* **40**, 2529–35.

Loriot, M., Thomas, D. and Brown, E. (1985). Total synthesis & study of biologically active lignans. Part 8. Total synthesis of 5′-methoxyisolariciresinol. *Bull. Soc. Chim. Fr.* 871–5.

Macdonald, D.I. and Durst, T. (1986a). A synthesis of *trans*-2-aryl-benzocyclobuten-1-ols. *Tetrahedron Lett.* **27**, 2235–6.

Macdonald, D.I. and Durst, T. (1986b). A highly stereoselective Diels–Alder based synthesis of (±)-podophyllotoxin. *J. Org. Chem.* **51**, 4749–50.

McKillop, A., Turrell, A.G. and Taylor, E.C. (1977). Thallium in organic synthesis. Part 46. Coupling of aromatic compounds using thallium(III) trifluoroacetate. Synthesis of biaryls. *J. Org. Chem.* **42**, 764–5.

Mahalanabis, K.K., Mumtaz, M. and Snieckus, V. (1982a). Dimetalated tertiary succinamides. Alkylation & annelation reactions. *Tetrahedron Lett.* **23**, 3971–4.

Mahalanabis, K.K., Mumtaz, M. and Snieckus, V. (1982b). Dimetalated tertiary succinamides. Synthesis of several classes of lignans including the mammalian urinary lactones enterolactone & enterodiol. *Tetrahedron Lett.* **23**, 3975–8.

Mann, J. and Piper, S.E. (1982). The preparation of lignans via intermolecular cycloaddition with quinodimethanes. *J. Chem. Soc. Chem. Commun.* 430–2.

Mann, J., Piper, S.E. and Yeung, L.K.P. (1984). The synthesis of lignans & related structures using quinodimethanes & isobenzofurans: approaches to the podophyllotoxins. *J. Chem. Soc. Perkin Trans. 1*, 2081–8.

Mann, J., Wong, L.T.F. and Beard, A.R. (1985). The synthesis of structural analogues of podophyllotoxin lignans: selective Wittig reactions of 1-aryl-2-methyl-tetrahydronaphthoic acid anhydride. *Tetrahedron Lett.* **26**, 1667–70.

Matsuura, S. and Iinuma, M. (1985). Lignan from calyces of *Diospyros kaki*. *Phytochemistry* **24**, 626–8.

Mervic, M., Ben-David, Y. and Ghera, E. (1981). A new total synthesis of (±)-steganone. *Tetrahedron Lett.* **22**, 5091–4.

Meyers, A.I. and Avila, W.B. (1981). Chemistry of aryloxazolines. Applications to the synthesis of lignan lactone derivatives. *J. Org. Chem.* **46**, 3881–6.

Meyers, A.I., Flisak, J.R. and Aitken, D.A. (1987). An asymmetric synthesis of (−)-steganone. Further application of chiral biaryl synthesis. *J. Amer. Chem. Soc.* **109**, 5446–52.

Michael, A. and Butcher, J.E. (1898). On the action of acetic anhydride on phenylpropiolic acid. *Amer. Chem. J.* **20**, 89–120.

Minato, A., Tamao, K., Suzuki, K. and Kumadi, M. (1980a). Synthesis of a lignan skeleton via nickel & palladium–phosphine complex catalysed Grignard coupling reaction of halothiophenes. *Tetrahedron Lett.* **21**, 4017–20.

Minato, A., Tamao, K., Suzuki, K. and Kumadi, M. (1980b). Selective monoalkylation and arylation of aromatic dihalides by palladium catalysed cross coupling with Grignard & organozinc reagents. *Tetrahedron Lett.* **21**, 845–8.

Momose, T., Nakamura, T. and Kanai, K-I. (1977). Synthesis of β-apoplicatitoxin trimethyl ether. *Heterocycles* **6**, 277–80.

Momose, T., Kanai, K-I. and Hayashi, K. (1978). Synthetic studies on lignans & related compounds. Part 8. Synthesis of justicidin B, diphyllin & of tawanin C & E from 2,3-dibenzylidene butyrolactones via β-apolignans. *Chem. Pharm. Bull.* **26**, 3195–8.

Mpango, G.B. and Snieckus, V. (1980). Tandem conjugate addition – α-alkylation of unsaturated amides. Synthesis of aryltetralin lignans. *Tetrahedron Lett.* **21**, 4827–30.

Munakata, K., Marumo, S. and Ohta, K. (1967). The synthesis of justicidin B & related compounds. *Tetrahedron Lett.* 3821–5.

Murphy, W.S. and Wattanasin, S. (1980). An improved procedure for the preparation of chalcones & related enones. *Synthesis* 657–50.

Murphy, W.S. and Wattanasin, S. (1982). Total synthesis of picropodophyllone. *J. Chem. Soc. Perkin Trans. 1*, 271–6.

Nishibe, S., Chiba, M., Sakushima, A., Hisada, S., Yamanouchi, S., Takido, M., Sankawa, U. and Sakakibara, A. (1980). Introduction of an alcoholic hydroxyl group into 2,3-dibenzylbutyrolactone lignans with oxidising agents and CMR spectra of the oxidation products. *Chem. Pharm. Bull.* **28**, 850–60.

Nonhebel, D.C. and Walton, J.C. (1974). *Free-radical Chemistry*. Cambridge, Cambridge Univ. Press, pp. 190–1.

Ogawa, M. and Ogihara, Y. (1976). Studies on the constituents of *Ekianthus nudipes*. Part 5. A new xyloside from the stems. *Chem. Pharm. Bull.* **24**, 2102–5.

Pelter, A., Ward, R.S., Watson, D.J. and Collins, P. (1982). Synthesis of 2,6-diaryl-4,8-dihydroxy-3,7-dioxabicyclo-[3:3:*O*]-octanes. *J. Chem. Soc. Perkin Trans. 1*, 175–81.

Pelter, A., Ward, R.S., Satyanarayana, P. and Collins, P. (1983a). Synthesis of lignan lactones by conjugate addition of thioacetal carbanions to butenolide. *J. Chem. Soc. Perkin Trans., 1*, 643–7.

Pelter, A., Satyanarayana, P. and Ward, R.S. (1981). A short efficient synthesis of *trans*-dibenzylbutyrolactones exemplified by the synthesis of di-*O*-methyl compound X(HMPF) & an anti-tumour extractive. *Tetrahedron Lett.* **22**, 1549–50.

Pelter, A., Ward, R.S., Collins, P., Venkateswarlu, R. and Kay, I.T. (1983b). A general synthesis of 2,6-diaryl-3,7-dioxabicyclo-[3:3:*O*]-octane lignans applicable to the synthesis of unsymmetrical lignans. *Tetrahedron Lett.* **24**, 523–6.

Pelter, A., Ward, R.S., Collins, P. and Venkateswarlu, R. (1985). A general synthesis of 2,6-diaryl-3,7-dioxabicyclo-[3:3:*O*]-octane lignans applicable to unsymmetrically substituted compounds. *J. Chem. Soc. Perkin Trans. 1*, 587–94.

Pelter, A., Ward, R.S., Pritchard, M.C. and Kay, I.T. (1985b). Synthesis of deoxyisopodophyllotoxin and epiisopodophyllotoxin. *Tetrahedron Lett.* **26**, 6377–80.

Perry, C.W., Kalnins, M.U. and Deitcher, K.M. (1972). Synthesis of lignans. Part 1. Nor-dihydroguaiaretic acid. *J. Org. Chem.* **37**, 4371–6.

Rajapaksa, D. and Rodrigo, R. (1981). A stereocontrolled synthesis of anti-neoplastic podophyllum lignans. *J. Amer. Chem. Soc.* **103**, 6208–9.

Rao, B.V.G., Row, L.R. and Satyanarayana, P. (1978). Synthetic experiments in lignans. Part 2. Synthesis of chinensin. *Indian J. Chem.* **16b**, 68–70.

Rao, K.V. and Alvarez, F.M. (1983). Manassantins A, B & saucerneol: novel biologically active lignoids from *Saururus cernuus*. *Tetrahedron Lett.* **24**, 4947–50.

Rao, K.V. and Alvarez, F.M. (1985). Chemistry of *Saururus cernuus*. Part 13. Some reactions of the diarylbutane type neolignans. *J. Nat. Prod.* **48**, 592–7.

Ravid, U., Silverstein, R.M. and Smith, L.R. (1978). Synthesis of the enantiomers of 4-substituted γ-lactones with known absolute configuration. *Tetrahedron* **34**, 1449–52.

Renz, J., Kuhn, M. and von Wartburg, A. (1965). Synthesis of neopodophyllotoxin from podophyllic acid. *Liebigs Ann. Chem.* **681**, 207–24.

Robin, J-P., Gringore, O. and Brown, E. (1980). Asymmetric total synthesis of the antileukemic lignan precursor (−)-steganone & revision of its absolute configuration. *Tetrahedron Lett.* 2709–12.

Robin, J-P., Dhal, R. and Brown, E. (1982). Application of α-hydroxy-alkylation to β-benzyl – butyrolactones & creation of the skeleton of phenyltetralins & bisbenzyl cyclooctadienes. Total synthesis of (±)-podorhizol, (±)-podorhizone & (±)-isodeoxypodophyllotoxin. *Tetrahedron* **38**, 3667–71.

Robin, J-P., Dhal, R. and Brown, E. (1984). Total synthesis & study of biologically active lignans. Part 3. α-hydroxyalkylation of β-benzyl-butyrolactones & the creation of the skeleton of phenyltetralins & bisbenzylcyclooctadienes. First synthesis of picrosteganes, formal synthesis of (±)-steganacin. *Tetrahedron* **40**, 3509–20.

Rodrigo, R. (1980). A stereo- & regiocontrolled synthesis of *Podophyllum* lignans. *J. Org. Chem.* **45**, 4538–50.

Satyanarayana, P. and Rao, P.K. (1985). DDQ-cyclodehydration: syntheses of 1-phenylnaphthalene lignans. *Indian J. Chem.* **24B**, 151–3.

Satyanarayana, P., Subramanyam, P., Viswanatham, K.N. and Ward, R.S. (1988). New seco- & hydroxylignans from *Phyllanthus niruri*. *J. Nat. Prod.* **51**, 44–9.

Schneiders, G.E. and Stevenson, R. (1980). Synthesis of the *bis*-benzocyclo-octadiene lignan (±)-wuweizisu C. *Chem. & Ind.* 538–9.

Schrecker, A.W. and Hartwell, J.L. (1953). Components of podophyllin. Part 12. The configuration of podophyllotoxin. *J. Amer. Chem. Soc.* **75**, 5916–24.

Schrecker, A.W. and Hartwell, J.L. (1955). Application of tosylate reductions and molecular rotations to the stereochemistry of lignans. *J. Amer. Chem. Soc.* **77**, 432–7.

Seki, H., Koga, K., Matsuo, H., Ohki, S., Matsui, I. and Yamada, S-I. (1965). Studies on optically active aminoacids. Part 5. Synthesis of optically active α-aminoalcohols. *Chem. Pharm. Bull.* **13**, 995–1000.

Semmelhack, M.F. and Ryono, L.S. (1975). Nickel promoted synthesis of cyclic biphenyls. Total synthesis of alnusone dimethyl ether. *J. Amer. Chem. Soc.* **97**, 3873–5.

Stobbe, H. (1893). A new synthesis of teraconic acid. *Chem. Ber.* **26**, 2312–19.

Stevenson, R. and Williams, J.R. (1977). Synthesis of tetrahydrofuran lignans, (±)-galbelgin & (±)-grandisin. *Tetrahedron* **33**, 285–8.

Taafrout, M., Rouessac, F. and Robin, J-P. (1983). Prestegane A from *Steganotaenia araliacea* Hochst. The first natural dibenzylbutanolide lignan with a *meta*-phenol. A short synthesis of *cis*- and *trans*-(±)-arctigenin. *Tetrahedron Lett.* **24**, 3237–40.

Taafrout, M., Landais, Y. and Robin, J-P. (1986). Isolation, stereochemical study & biomimetic synthesis of steganolide A. A new bisbenzocyclo-octadienyl lignan of *Stegataenea araliacea*. *Tetrahedron Lett.* **27**, 1781–4.

Takano, S., Tamura, N. and Ogasawara, K. (1981a). Synthesis of a potential synthon for the chiral synthesis of the coryanthe-type indole alkaloids: enantioselective synthesis of (−)-antirhine. *J. Chem. Soc. Chem. Comm.* 1155–6.

Takano, S., Goto, E., Hirama, M. and Ogasawara, K. (1981b). Efficient synthesis of (*R*)-(−)-4-benzyloxymethyl-4-butyrolactone from (D)-(+)-mannitol. *Heterocycles* **16**, 381–5.

Takano, S., Otaki, S. and Ogasawara, K. (1985a). A new route to 1-phenylnaphthalenes by cycloaddition: a simple & selective synthesis of some naphthalene lignan lactones. *Tetrahedron Lett.* **26**, 1659–60.

Takano, S., Otaki, S. and Ogasawara, K. (1985b). Stereocontrolled synthesis of (±)-deoxypodophyllotoxin via the benzyl equivalent of the Peterson reaction. *J. Chem. Soc. Chem. Commun.* 485–7.

Takano, S., Sato, N., Otaki, S. and Ogasawara, K. (1987). Total synthesis of (±)-sikkimotoxin via the benzo-Peterson reaction. *Heterocycles* **25**, 69–73.

Takaoka, D., Takamatsu, N., Saheki, Y., Kono, K., Nakaoka, C. and Hiroi, M. (1975). *Nippon Kagaku Kaishi* (**12**), 2192–6.

Takaoka, D., Watanabe, K. and Hiro, M. (1976). Studies on lignoids in *Lauraceae*. Part 2. Studies of lignans in the leaves of *Machilus japonica* Sieb. et Zucc. *Bull. Chem. Soc. Japan* **49**, 3564–6.

Takei, Y., Mori, K. and Matsui, M. (1973). Synthesis of *dl*-matairesinol dimethyl ether, dehydrodimethyl conidendrin & dehydrodimethyl retrodendrin from ferulic acid. *Agric. Biol. Chem.* (*Japan*) **37**, 637–41.

Takei, T., Kotani, E. and Tobinaga, S. (1983). New reagent systems containing chromium trioxide & synthesis of lignans. *J. Chem. Soc. Chem. Commun.* 98–9.

Takeya, T., Matsumoto, H., Kotani, E. and Tobinaga, S. (1983). New reagent systems containing chromium trioxide provide precursors for neolignans. *Chem. Pharm. Bull.* **31**, 4364–7.

Tanoguchi, M., Arimoti, M., Saika, H. and Yamaguchi, H. (1987). Studies on the seeds of *Hernandia oivigera* L. Part 6. Isolation & determination of three lignans. *Chem. Pharm. Bull.* **35**, 4162–8.

Tayyeb Hussain, S.A.M., Ollis, W.D., Smith, C. and Stoddart, J.F. (1975). The stereochemistry of 2,4- & 2,3-disubstituted-butyrolactones. *J. Chem. Soc. Perkin Trans. 1*, 1480–92.

Tedder, J.M. and Walton, J.C. (1980). The halogenation of cycloalkanes & their derivatives. In *Adv. in Free-radical Chem.* **6**, 155–84.

Theilheimer, O. (1988). *Synthetic Methods of Organic Chemistry*, ed Finch, A.F., **42** & preceding vols.

Till, C.P. and Whiting, D.A. (1984). A short stereospecific synthesis of a 2,6-diarylmonoepoxylignanolide. *J. Chem. Soc. Chem. Commun.* 590–1.

Tomioka, K., Mizuguchi, H. and Koga, K. (1978). Studies directed towards the asymmetric total synthesis of anti-leukemic lignan lactones. Synthesis of (−)-podorhizone. *Tetrahedron Lett.* (**47**), 4687–90.

Tomioka, K., Mizuguchi, H. and Koga, K. (1979). Novel isomerisation of dibenzocyclooctadiene lignan lactones – first synthesis of (±)-stegane. *Tetrahedron Lett.* 1409–10.

Tomioka, K., Ishiguro, T. and Koga, K. (1980). First asymmetric total synthesis of (+)-steganacin, determination of absolute stereochemistry. *Tetrahedron Lett.* **21**, 2973–6.

Tomioka, M., Mizuguchi, H. and Koga, K. (1982). Stereoselective reactions. Part 5. Design of the asymmetric synthesis of lignan lactones. Synthesis of optically pure podorhizone & deoxypodorhizone by 1,3-asymmetric induction. *Chem. Pharm. Bull.* **30**, 4304–13.

Tomioka, K., Ishiguro, T. and Koga, K. (1985). Stereoselective reactions. Part 10.

Total synthesis of the optically pure anti-tumour lignan, burseran. *Chem. Pharm. Bull.* **33**, 4333–7.

Trost, B.M. and Melvin, L.S. (1975). *Sulphur Ylides*, Academic Press, New York.

Van der Eycken, J., De-Clercq, P. and Vandewalle, M. (1985). Total synthesis of (±)-podophyllotoxin & (±)-epipodophyllotoxin. *Tetrahedron Lett.* **26**, 3871–4.

Van der Eycken, J., De-Clercq, P. and Vandewalle, M. (1986a). Total synthesis of podophyllum lignans: an exploratory study. *Tetrahedron* **42**, 4285–95.

Van der Eycken, J., De-Clercq, P. and Vandewalle, M. (1986b). Total synthesis of (±)-podophyllotoxin & (±)-epipodophyllotoxin. *Tetrahedron* **42**, 4297–308.

Vyas, D.M., Chiang, Y. and Doyle, T.W. (1984). A short efficient total synthesis of (±)-acivicin & bromo-acivicin. *Tetrahedron Lett.* **25**, 487–90.

Vyas, D.M., Skonezny, P.M., Jenks, T.A. and Doyle, T.W. (1986). Total synthesis of (±)-epipodophyllotoxin via a (3 + 2)-cyclo-addition strategy. *Tetrahedron Lett.* **27**, 3099–102.

Wallis, A.F.A. (1973). Oxidation of (*E*) & (*Z*)2,6-dimethoxy-4-propenylphenol with ferric chloride – a facile route to the 2-aryl ethers of 1-arylpropan-1,2-diols. *Aust. J. Chem.* **26**, 585–94.

Wang, C-L. J. and Ripka, W.C. (1983). Total synthesis of (*RS*)-justicidin P. A new lignan lactol from *Justicia extensa. J. Org. Chem.* **48**, 2555–7.

Ward, R.S., Satyanarayana, P. and Rao, B.V.G. (1981). Reactions of aryltetralin lignans with DDQ – an example of DDQ oxidation of an allylic ether group. *Tetrahedron Lett.* **22**, 3021–4.

Ward, R.S. (1982). The synthesis of lignans & neolignans. *Chem. Soc. Rev.* **11**, 75–125.

Weinges, K. and Spanig, R. (1967). Lignans & cyclolignans. In *Oxidative Coupling of Phenols*, ed Taylor, W. I. & Battersby, A.R. Arnold, London, pp. 323–35.

Wolman, Y. (1988). *Chemical Information – a Practical Guide to Utilisation.* Wiley, Chichester, 2nd edit., pp. 260–5.

Wood, C.S. and Mallory, F.B. (1964). Photochemistry of stilbenes. Part 4. The preparation of substituted phenanthrenes. *J. Org. Chem.* **29**, 3373–7.

Yamaguchi, H., Arimoti, M., Tanoguchi, M., Ishida, T. and Inoue, M. (1982). Studies on the seeds of *Hernandia ovigera* L. Part 3. Structures of two new lignans. *Chem. Pharm. Bull.* **30**, 3212–18.

Yamaguchi, H., Nakajima, S., Arimoti, M., Tanoguchi, M., Ishida, T. and Inoue, M. (1984). Studies on the constituents of the seeds of *Hernandia ovigera* L. Part 4. Synthesis of *β*-peltatin A & B methyl ethers from desoxypodophyllotoxin. *Chem. Pharm. Bull.* **32**, 1754–60.

Ziegler, F.E. and Fowler, K.W. (1976). Substitution reactions of specifically *ortho*-metallated piperonal cyclohexylimine. *J. Org. Chem.* **41**, 1564–6.

Ziegler, F.E. and Schwartz, J.A. (1978a). Synthetic studies on lignan lactones: aryldithiane route to (±)isopodophyllotoxone and approaches to the stegane skeleton. *J. Org. Chem.* **43**, 985–91.

Ziegler, F.E., Fowler, K.W. and Sinha, N.D. (1978b). A total synthesis of (±)-steganone via the modified Ullman reaction. *Tetrahedron Lett.* 2767–70.

Botanical index

General index